Progress in Theoretical Computer Science

Dario Bini
Victor Y. Pan

Polynomial and Matrix Computations

Fundamental Algorithms
Vol. 1

Birkhäuser 1994
Boston • Basel • Berlin

Dario Bini
Dipartimento di Matematica
Università di Pisa
56127 Pisa
Italy

Victor Y. Pan
Department of Mathematics
and Computer Science
Lehman College
Bronx, NY 10468
USA

Library of Congress Cataloging-in-Publication Data

Bini, Dario.
Polynomial and matrix computations / Dario Bini, Victor Pan.
p. cm. -- (Progress in theoretical computer science)
Includes bibliographical references and index.
Contents: v. 1. Fundamental algorithms
ISBN 0-8176-3786-9 (v. 1 : acid-free)
1. Polynomials--Data processing. 2. Matrices--Data processing.
3. Numerical calculations--Data processing. I. Pan, Victor.
II. Title. III. Series.
QA241.B56 1994 94-27577
512.9'42--dc20 CIP

Printed on acid-free paper

Birkhäuser ®

ISBN 0-8176-3786-9
ISBN 3-7643-3786-9
Typeset by the authors.
Printed and bound by Quinn-Woodbine, Woodbine, NJ.
Printed in the U.S.A.

9 8 7 6 5 4 3 2 1

Contents

Preface

Our Subjects and Objectives.

This book is about algebraic and symbolic computation and numerical computing (with matrices and polynomials). It greatly extends the study of these topics presented in the celebrated books of the seventies, [AHU] and [BM] (these topics have been under-represented in [CLR], which is a highly successful extension and updating of [AHU] otherwise). Compared to [AHU] and [BM] our volume adds extensive material on parallel computations with general matrices and polynomials, on the bit-complexity of arithmetic computations (including some recent techniques of data compression and the study of numerical approximation properties of polynomial and matrix algorithms), and on computations with Toeplitz matrices and other dense structured matrices. The latter subject should attract people working in numerous areas of application (in particular, coding, signal processing, control, algebraic computing and partial differential equations). The authors' teaching experience at the Graduate Center of the City University of New York and at the University of Pisa suggests that the book may serve as a text for advanced graduate students in mathematics and computer science who have some knowledge of algorithm design and wish to enter the exciting area of algebraic and numerical computing. The potential readership may also include algorithm and software designers and researchers specializing in the design and analysis of algorithms, computational complexity, algebraic and symbolic computing, and numerical computation. The book may be used as a reference volume on algorithms and complexity in the area of matrix and polynomial computations. Due to the applications of algebra to combinatorial computing, the book may be of some interest to researchers in the latter area too.

The main subjects of this volume are the design and analysis of algorithms for *sequential and parallel computations with univariate polynomials, general matrices and dense structured matrices* (together with extensions and applications to computations with power series, rational functions, integers and multivariate polynomials), with the focus on the major issue of the asymptotic complexity of the computations and on demonstration of fundamental techniques of the design and analysis of algebraic and numerical algorithms.

Our main goal and the main challenge for us was a systematic treatment of algorithms and complexity in the areas of matrix and polynomial computations, which are fundamental for both theory and practice of computing for sciences, engineering and statistics. In spite of its importance, most of the material of this book (with the exception of a part of chapter 1) has not been collected in books yet and so far remained scattered in journal articles and technical reports or even remained unpublished.

Our second major goal was to contribute to bridging or narrowing the two gaps: a) between the design of numerical and algebraic algorithms, and b) between matrix and polynomial computations. Historically, numerical algorithms for matrix computations and algebraic algorithms for polynomial computations have been developed and implemented by two distinct groups of people, having too little interaction with each other. (Among a few notable exceptions, Schur's celebrated algorithm exploits the reduction to polynomial computations in order to yield fast and numerically stable solution of Hankel and Toeplitz linear systems; an excellent and most recent treatment of this algorithm, together with extensive bibliography, can be found in [GG].) The book exposes both topics and shows various correlations between them, in particular, via the study of computations with dense structured matrices and of their applications to computations with both polynomials and general matrices, and also via the study of Newton's iteration and the applications of algebraic and data compression techniques to bounding and decreasing the precision of approximate numerical computation (compare Remarks 3.3, 3.4, 4.2 and 9.2 of chapter 3).

We wrote this book as an organic whole; in particular, some algorithms of chapters 1 and 2 are cited and modified in chapter 4, but selective reading is still possible, with some additional efforts. In particular, some advanced readers from various areas of computer science, algebraic computing and numerical computational mathematics may wish to focus on our parallel algorithms for computations with general and structured matrices (including the solution of linear systems of equations, computing the characteristic and minimum polynomials of a matrix and matrix rank). Computing a maximal linearly independent subset of a set of vectors could turn out to be a topic singled out by those readers who work on combinatorial computations. Multivariate polynomial multiplication, computing polynomial gcd, and the extended Euclidean scheme may become the topics selected by some algebraic computing readership, whereas the new analysis of the errors and numerical stability of FFT, the compact multigrid techniques of data compression for PDE's, and the reduction of a matrix to block tridiagonal form could be of special interest to numerical analysts.

We start each chapter with some introductory material and/or formal definitions, but some readers may advance faster if they initially skip this material (particularly, in chapter 4) and consult with it selectively, as soon as their reading requires this.

Chapters Review.

In the first two chapters we avoid excessive detail by assuming sequential computations (with infinite precision) in the complex field or its subfields, but we also survey the extensions (sometimes straightforward but sometimes nontrivial) to the case of computations over arbitrary fields and rings of constants. Furthermore, we revisit our study assuming finite precision in chapter 3 and parallel models of computing in chapter 4.

Our high level presentation does not cover, however, the implementation of the algorithms on specific computers and computer systems; moreover, by focusing our study on sequential (arithmetic and Boolean) time-complexity and parallel time-processors complexity, we left out the important but technically different subjects of data structures, space-complexity and processor communication and synchronization. To limit the size of the volume, we also omitted some straightforward proofs and replaced some other proofs by references to the bibliography, and we postponed (until volume 2) our in-depth study of some major selected topics. On the other hand, this volume includes many exercises, from the trivial to ones involving some advanced research results and even open problems. A substantial part of material of this book has been surveyed in [P92a].

In chapter 1 we study the most fundamental operations with univariate polynomials (such as their evaluation, multiplication and division as well as computing the greatest common divisor of two or several polynomials) and their immediate extensions to computations with integers and power series and to interpolation to a function by a polynomial and by a rational function. All these computational problems are classical, but their solution constitutes a major part of scientific and engineering computations on modern computers; it also constitutes an area of active research. We revisit the known fundamental results and techniques (such as the *evaluation-*

interpolation method, Newton's iteration, p-adic lifting, extended Euclidean algorithm, and Chinese remainder computations).

The last decades have witnessed rapid progress in the design of effective algorithms for polynomial computations. Several classical algorithms in this area have been improved, in particular, by means of exploiting the fast Fourier transform (FFT). Untill recently, this progress has had little influence on the packages and computer libraries of subroutines available for polynomial computations. In our presentation, we focus on the polynomial computations that can be ultimately reduced to polynomial multiplications and divisions and discrete Fourier transforms (DFT's), which can be effectively performed by means of FFT's or by means of the techniques of binary segmentation, covered in chapter 3. We hope that our presentation will support the interest to the implementation of the fast polynomial arithmetic, which seems to be on rise now (compare [Sh,a]).

The advantages of using FFT seem to be easiest to understand in the case of polynomial multiplication whose reduction to DFT's is simple. In this case we arrive at the algorithm that (like FFT) is numerically stable and allows effective parallel implementation. Note, however, that even FFT itself is numerically stable only in the sense that its output error vector has an upper bound on its norm proportional to the product of the norm of the output vector and of the unit round-off error (see Proposition 4.1 of chapter 3), but such a good error bound does not generally hold for a fixed entry of the output vector. If the reduction to DFT's is more involved (as in the cases of polynomial division and computing the greatest common divisor of polynomials), the effective parallelization and/or the numerical stability of polynomial computations become problems whose study is included in chapters 3 and 4.

In chapter 2 we first survey the complexity of computations with general matrices, by ultimately reducing most of these computations to matrix-by-vector and matrix-by-matrix multiplication and matrix inversion, whose role for matrix computations is thus similar to the fundamental role of DFT and polynomial multiplication and division for polynomial computations. This way of presentation seems particularly appropriate because of a dramatic increase in the popularity of block matrix computations among users, due to recent development in computer hardware. Then we focus on computations with dense structured matrices having a wide range of major applications to problems in sciences and engineering. The class of dense structured matrices studied in this volume includes Toeplitz, Hankel, Hilbert, Sylvester,

Bezout, Vandermonde and Frobenius matrices as well as matrices with similar structure defined by means of the associated operators. We present a *unified treatment* of the computations with matrices of these classes, show their correlations to computations with polynomials, and exploit these correlations to employ FFT and to arrive at a dramatic acceleration of the algorithms versus the case of general matrices. We hope that our exposition in chapter 2 fills a disturbing gap in the computer science literature and will be of interest for a large group of scientists dealing with Toeplitz and other structured matrices. Furthermore, a substantial improvement of general matrix computations can be obtained based on their reduction (shown in chapters 1 and 2) to computations with dense structured matrices and polynomials.

In chapter 3, motivated by practical computations with finite precision, we study the Boolean computational complexity of arithmetic computations and survey the main techniques available for estimating it. This study leads us to revisiting some recent algebraic and data compression techniques introduced for bounding the precision of computations. We also deal with such topics as numerical conditioning of a computational problem and numerical stability of an algorithm; we propose a formal (but simple) quantitative definition of these concepts, which is consistent with the customary qualitative concepts, widely used by numerical analysts. We present some new techniques for improving the analysis of the numerical stability of FFT. We then consider integer and rational computations (with the infinite precision), approximate computations (with roundoff), and several tools for their analysis, such as residue (modular) arithmetic, Chinese remaindering, Newton-Hensel's lifting, randomization, summation reordering and some data compression techniques (of compact multigrid and binary segmentation). We also show how some basic algebraic computations with polynomials and rational functions can be substantially accelerated by means of combining the techniques of their exact (algebraic) and approximate (numerical) solution.

In chapter 4 we deal with parallel computations and their complexity analysis. For convenience, we use the PRAM models, but our algorithms are described at a high level, which allows their simple reinterpretation in more realistic data processing environments. We review the parallel arithmetic complexity of the algorithms from chapters 1 and 2 for computations with polynomials and general and structured matrices and modify some of these algorithms, to reach both parallel acceleration and processor efficiency. We also consider parallel Boolean complexity and parallel computations over finite fields; describe recent progress in parallel computation of a maximal linearly independent subset of a vector set, which has further applications to combinatorial and graph algorithms; and cover some fundamental recent techniques for upgrading the performance of parallel algorithms. Some recent unpublished results are included in this chapter.

Further Reading.

The omitted proofs, details and further results can be found in the cited bibliography. Here are some references to related material left outside this book (the references contain further references, of course, and our book contains further bibliography too): [Gies], [KKS], [KKS90], [Par], [P87b], [BCLR81], [P84], [P84a], [P84b], [P92a], [LRT], [Var], [V], [GL] and [W65], on matrix computations; [Alt], [Bre76], [BB] and [LO], on approximate evaluation of elementary trascendental functions; [AHU], [BM], [B84a], [B-O], [MST88], [MST89], [Yao], [PSh] (section 1.4), [SSmo], [Mul94] and [Bshouty], on the lower bounds on computational complexity; [Ka83], [Kn], [Ka90], [Ka92], [LLL], [GS], [Mil] and [Z93], on polynomial factorization over the field of rationals and finite fields; [BKR], [R88], [R89], [R89a] and [R92a], on algebraic theories; [ACM], [B-OT], [Buch], [CE], [EC], [Gies], [GKS], [Ho], [Ka87a], [KKS], [SY], [Gies], [Schw], [Vi], [Z93] and [Z], on various other topics of sparse and/or multivariate polynomial computations, including algebraic areas of computational geometry; [BCL], [LN], [Shp], [BP86], [Be], [E], [Kam89], [Rab,a], [Rab] and [Wied], on computations in arbitrary fields of constants and probabilistic polynomial computations; [Ang], [Raz] and [Smo], on Boolean complexity of algebraic computations and on computational learning theory; and [BLR], [GLRSW], [ALRS], [GeS] and [RuSu], on checking/correcting algebraic computations and related topics. [AHU], [BM], [Kn], [GCL], [Z93], [M93], [G88] and [Ka87a] survey various areas of algebraic computations.

Statement Enumeration.

We enumerate algorithms and statements [such as the displayed formulae (inequalities, equations), propositions, theorems, facts, lemmas, corollaries, definitions and computational problems] by triples indicating chapter numbers, section numbers and algorithm or statement numbers; we drop the chapter numbers (and thus stay with pairs, rather than triples of numbers) in our references confined to the current chapter. The statements of each of the above classes (lemmas, facts, definitions and so on) are enumerated separately from other classes, and we parenthesize the triples (or pairs) of numbers representing formulae.

Fields of Constants.

The fundamental algorithms of this volume are usually presented assuming computations in the complex number field or in its subfields, where, in particular, the complex conjugate of a number and the Hermitian transpose of a matrix are immediately defined. Such fields of constants (including the fields of real and rational numbers) are actually assumed by default in the literature on numerical computing.

Actually in this volume, several results are stated over a slightly more general class of the fields of characteristic 0, and in these cases, we implicitly assume that the Hermitian transpose of a matrix is readily available. (If the fields are subfields of the real number fields, then the Hermitian transpose turns into the usual transpose).

Our study in chapter 3 and the well-known success of the finite field computations motivated us to include the extension of the algorithms of this volume to computations in finite fields, and more generally, in any field. We have not, however, included into this volume (except for Section 3 of Chapter 3) the algorithms that have been specifically devised for computations over finite fields (for instance, for polynomial factorization); for this, we refer the reader to our section on *Further Reading.*

Acknowledgments

We are grateful to many people and institutions who encouraged and supported us and helped us in the completion of this work. In particular, from the start Allan Borodin and Ronald Book encouraged our efforts. James Davenport, Joachim von zur Gathen, Erich Kaltofen, Michael Monagan and John Reif kindly spent their time in discussions with one of us on the models of computations and on some aspects of implementation of algorithms for polynomial computations. The referee made several important and helpful comments. We thank the copyeditor Seth Maislin for an excellent job and the staff of Birkhäuser Boston for their cooperation.

The National Science Foundation and Professional Staff Congress of CUNY generously supported this work under the grants DCR 85-07573, CCR 88-05782, CCR 90-20690; under the PSC–CUNY Awards 668541, 669290, 661340, 662478 and 664334; and under a George N. Shuster Fellowship. Sally Goodall, Joan Bentley and Connie Engle provided superb assistance with typing this monograph, and Isdor Sobze at CUNY Graduate Center also assisted us with typing at the final stage of the preparation of volume 1. The Computer Science Department of SUNYA, Albany, NY, and particularly the Chairman, Dr. Dean Arden, as well as Pat Keller and Quei-Len Lee, helped to ensure secretarial services and computer and network support for our work. Dipartimento di Matematica of the University of Pisa, Lehman College of CUNY, Bronx, NY, and the International Computer Science Institute, Berkeley, CA, also provided computer and network support. Dario Bini is also grateful to MURST for the additional financial support of 40% and 60% funds, and to the ESPRIT BRA project 6846 POSSO (POlynomial System SOlving) under whose support part of the work has been performed. Victor Pan is pleased to recall the Craftsbury Center, Vermont, as an ideal place for assembling the material of this book.

1

Fundamental Computations with Polynomials

Summary.

In this chapter we recall the known effective algorithms for the major problems of polynomial computations and demonstrate some fundamental techniques of the algorithm design, according to the following line: we first recall fast Fourier transform (FFT), leading to a dramatic improvement of the classical algorithms for the discrete Fourier transform (DFT) and polynomial multiplication (also called convolution). Then we systematically extend such an improvement to other polynomial computations.

1. Introduction, Models of Computation, Definitions.

Subjects and their presentation.

The reader hardly needs to be reminded of the importance of the univariate polynomial computations but may ask if these computations are too simple to deserve a book. Actually, the basic question of how to represent a univariate polynomial $p(x)$ already leads to nontrivial algorithms. We may, in particular, use the coefficients of $p(x)$ [or only nonzero coefficients if $p(x)$ is sparse], or its values, or more generally, the remainders in its modular reductions, or its various other representations, say, Taylor's, Newton's or Lagrange's. The transition from one representation to another leads to such classical operations as multipoint evaluation of $p(x)$, interpolation, division, Chinese remainder computations and extended Euclidean algorithm. For all these operations, dramatic asymptotic acceleration of the classical methods can be obtained simply by reducing the computations to DFT and convolution by means of few fundamental techniques, which can be compared with few powerful keys opening numerous locks. We open more such locks by demonstrating further applications to computations with integers, with power series and (in the next chapter) with matrices.

Starting with FFT (which we present as a special case of multipoint evaluation and interpolation) and fast polynomial multiplication, we solve all the major computational problems cited above, as well as their close neighbors, whose list includes some fundamental computations with integers, power series and rational functions.

We call the resulting algorithms fast, because they use $O(K \log K)$ to $O(K \log^2 K)$ arithmetic operations versus the order of K^2 in the classical algorithms, where K is the number of input parameters, such as the coefficients or the values of the input polynomials. The crossover point K where the new algorithms become superior is reasonably low ($K = 64$ for convolution in the complex field), so that the improvement promises some practical value or already has it, as is the case for FFT (see more comments on the practical aspects in Appendix D). Our exposition substantially updates and extends the previous presentations in [AHU], [BM] and [Kn].

By no means is this study claimed to be complete even for univariate polynomials. In particular, we omit the study of the lower bounds, of the space complexity (see Remarks 1.3 and 1.4), and of the two large subjects of polynomial factorization over the rational and finite fields and of algebraic theories (see our preface and the end of section 7 on the bibliography). We, however, present several *general techniques of the design of algebraic algorithms* and demonstrate their power, and this may serve as both the basis and a motivation for the study of algebraic computing. The latter comments apply to our next chapters as well.

In this chapter, we mostly deal with major classical problems of algebraic and symbolic computing, so we usually skip extensive introductions and comments to these problems and sometimes skip their solutions, referring the reader to the bibliography, in particular, to [AHU], [BM], [Kn].

Contents of chapter 1.

The chapter is organized as follows: in the remainder of this section we introduce and discuss the models of computation and some basic definitions. Then, in section 2, we consider polynomial evaluation and interpolation on a set of points, with particular attention to the case of Fourier points (roots of unity). This case brings us to the discrete Fourier transform and to FFT algorithms. Then the evaluation-interpolation technique is introduced and is applied to the computation of a polynomial product (convolution of their coefficient vectors).

Polynomial division with a remainder is the problem dealt with in section 3, solved there by using the associated power series and Newton iteration, and more extensively studied in chapter 6.

In section 4 we apply fast polynomial multiplication and division in order to accelerate the classical solution of polynomial evaluation and interpolation on any set of points. We also consider some related computations, such as Hermite interpolation, partial fraction decomposition, modular representation of a polynomial, Chinese remainder computation, composition of polynomials and formal power series. These computations are related to various ways of representing a polynomial.

In section 5 the extended Euclidean scheme and the computation of the gcd and the lcm of two or several polynomials are considered, together with some related problems, such as computing the reciprocal modulo a polynomial, rational interpolation, Padé approximation and computing the minimum span of a linear recurrence sequence. Sections 5 and 7 also include some simple randomization techniques.

In section 6, the techniques of Newton's iteration for manipulating with power series is applied to solve several computational problems, such as the reversion of power series, the computation of the m-th root of a polynomial, polynomial decomposition and the evaluation of algebraic functions.

In section 7, we discuss the extension of the algorithms to any field of constants and, in sections 8 and 9, multivariate polynomial multiplication and interpolation, respectively. (These two sections contain some more recent results.)

Appendix A contains a brief discussion on the alternative customary models of computation, particularly on arithmetic and Boolean circuits and straight-line programs and on the extension of the models to the cases of approximate and/or randomized computations. Appendix B lists the computational problems of this chapter, together with some of their complexity bounds and reductions to each other that we study and use. Appendix C shows the fast integer multiplication algorithm of [ScSt]. In Appendix D we comment on practical impact of the fast convolution algorithms.

Model of computation and some related issues.

We will assume the customary sequential and parallel *random access machine* (RAM) *models* (see [AHU], [EG88], [Fich] and [KR]). For the computational problems that we study, however, the RAM asymptotic complexity estimates can be easily translated to other customary models of computation, such as the straight line programs and the (uniform) families of circuits (specified in Appendix A to this chapter); our (high level) presentation of the algorithms is model independent.

The RAM models allow us to read one memory location, to execute an operation from a fixed set, or to write into one memory location. To each operation, its time-cost is assigned. In particular, assigning the unit cost to each operation is said to define the *uniform cost criterion.* Generally, the cost may vary depending on the precision of the operands and on the kind of operations involved (say, multiplication may cost more than addition/subtraction and less than division). The *overall sequential time-cost* is defined as the sum of the time-cost values assigned to all the operations involved.

If the basis operations are Boolean, the RAM is called *Boolean.* If they are arithmetic, the RAM is called *arithmetic.* If they are arithmetic but also complemented with the operation of computing or numerically approximating the zeros of any polynomial of degree less than a fixed constant, then we will call the RAM *algebraic.* We will only use the algebraic RAM in one of the chapters of volume 2.

For the arithmetic and algebraic RAM's, the input consists of one or more than one sets of variables, the set of constants and the set of operations allowed in the computations, which will be the set of arithmetic operations, unless we specify otherwise. The *input size* is given by the sizes of all the sets of variables. The size of a set S of variables is usually given by its cardinality, but sometimes by a pair or a k-tuple of numbers (if, say, S is a matrix, a pair of matrices or a k-dimensional tensor). Under the Boolean model, the input includes all the binary digits (bits) of each variable. Then the input size typically equals the number of bits required in order to represent the input. We similarly define and use the *output size* of a computational problem.

In our first two chapters, we will (as a rule) measure the computational cost by the number of arithmetic operations (ops) required in order to compute or to approximate (within a prescribed error bound) the desired output values. We will assume infinite precision computations with no errors over the real or complex numbers (compare Remark 1.1 below). In addition, in section 7 of this chapter, in Remarks 2.3.2 and 4.2.1 and in section 6 of chapter 4, we will comment on the extensions of our algorithms to computation over other fields of constants. In chapters 3 and 4 we will introduce finite precision computations and parallelism, respectively (compare Remark 1.5). For Problems 2.7 ($INT \cdot MULT$) and 3.6 ($INT \cdot DIVIDE$) of integer multiplication and division, we will use the Boolean model of computations. In some solutions of Problems 5.1b ($EXT \cdot EUCLID$) and 6.3 ($POL \cdot DECOMP$) of this chapter and more in chapters 2–4, we will use random parameters and will arrive at some probabilistic (randomized) algorithms, which with a low probability may fail to output the correct solution (see Appendix A on the distinction between the randomized algorithms of *Las Vegas* and *Monte Carlo* types). In particular, we will extensively use randomization as an effective general means of avoiding singularities in algebraic computations (see Procedure 7.1).

Remark 1.1. For many problems, such as the search for polynomial zeros and matrix eigenvalues, the exact solution cannot be computed in finite time by rational algorithms, which only involve ops and comparisons (see Definition 1.8). In such cases one may shift to *approximation algorithms* (see chapter 3) that compute the desired output values within a fixed bound $\epsilon > 0$ on the output errors. Even for the computational problems whose solutions can be computed by rational algorithms, as in the case of linear systems of equations, we sometimes may decrease the computational cost by approximating within ϵ, rather than exactly computing the output values. The behavior of the approximation algorithms (in particular, their output error bounds and the speed of the convergence of their iteration loops) depends on the domain where the input variables are defined.

Remark 1.2. We rely on the widely accepted models of computing, which capture the characteristic features of numerous existent computers (see also Remark 3.1.1). The readers should realize, however, that no formal model of computation can be more than an idealization of the reality, particularly if they take into account the variety of the existent computers and the technological progress. For instance, many pipelined computers multiply and add numbers equally fast, as we assume in our models in chapters 1 and 2, but, say, Connection Machines CM1 and MASPAR are capable of adding substantially faster than they multiply. Moreover, many computers compute the expressions $\pm a \pm bc$ for three real input vales a, b and c as fast as they perform a single addition or multiplication, which again is not reflected in our model. Furthermore, our model of parallel computations has a substantial omission, which is traditional for the analysis of algebraic algorithms: it does not account for the complexity of data flow. This complexity may compete with or even dominate the computational complexity, particularly in parallel implementation of the algorithms, but again, such a correlation greatly varies among the existent computers and is likely to continue changing in the future. Finally, as a striking example of the influence of the progress in the computer hardware on the evaluation of the performance of various algorithms for matrix and polynomial computations, we recall the very fast performance of block matrix algorithms on multiprocessors, which was closely related to the problem of data flow complexity and substantially influenced the practice of matrix computations but has not been fully captured by the existent general models of computing. All these and other limitations on the practical value of our complexity analysis do not extend to the algorithms presented in this book and to the techniques of their design, whose high-level presentation in this book is architecture independent; many of these algorithms have already become or promise to become effective tools for practical computations.

Remark 1.3. We have the straightforward informational lower bound max{(*input size*)/2, (*output size*)} on the complexity of a computational problem. For many problems this bound can be improved, although usually by constant or logarithmic factors at best, by using more advanced techniques (see [ASt], [AHU], [BM], [B-O], [Kn], [MST88], [MST89], [P66]). The lower bounds enable the algorithm designers to rule out any attempt to devise "overly good" algorithms.

Remark 1.4. An important measure for the complexity of computational problems is the amount of memory space needed in order to solve them, and in practice, in many cases, this is even a more crucial parameter than the running time (see [J83] on time-space trade-off for arithmetic computations). There are two customary approaches to measuring space-complexity depending on whether this measure should or should not include the space for the allocation of the input and the output. In some applications, the input can be overwritten by the output, and this can be exploited in order to decrease the overall requirements to the memory space for the entire computation. We leave estimating the space-complexity of the computational problems of this book as a challenging exercise for the readers (exercise 31).

Remark 1.5. The important issue of numerical stability is postponed until chapter 3, where we will show, in particular, some methods for bounding and/or decreasing the precision of numerical computations.

In chapter 4, we will reexamine the algorithms of chapters 1 and 2 and will present some new algorithms under the PRAM models of parallel computing. The parallel random access machine models (called *Parallel RAM* or *PRAM* models) consist of a fixed number p of processors, working concurrently, each capable of performing a unit cost operation, each having its own private (local) memory and each communicating with other processors through a shared global memory. The processors are allowed to read from and to write into both local and global memory. There is a further subdivision of this model according to the rules of the processor access to the memory (see Remark 4.1.1). The parallel complexity of computations under the PRAM model is represented by the *parallel time* and the *number of processors*, generally defined as two functions in the input size; their product is called the *potential work* of a parallel algorithm, defining an upper bound on the time of its sequential implementation.

In the beginning of chapter 4, we will further comment on the correlation among the parallel complexity estimates obtained on the various PRAM models and circuit models, defined in the appendix. For further descriptions of these and other customary models of computations, we refer the reader to [AHU], [Kn], [Sa], [J], [Le92], [KR], [EG88], [LLMPW], [Cook85], [Ruz], [Val], [E], [G86], [Ka88] and [BSS].

Some Definitions

We will conclude this section with some definitions.

Definition 1.1. Given a real function $f(n_1, \ldots, n_k)$ in k variables, $n_1, \ldots, n_k$, let $O(f(n_1, \ldots, n_k))$ denote any function in $n_1, \ldots, n_k$ whose absolute value is at most $c|f(n_1, \ldots, n_k)|$ for a positive *overhead constant* c. In particular, $O(1)$ is any function whose absolute value is bounded from above by a constant, and $n^{O(1)}$ is any function in n whose absolute value is bounded from above by a polynomial in n. We will write $g(n) = o(f(n))$ if $\lim_{n \to \infty} g(n)/f(n) = 0$.

Definition 1.2. $\lfloor v \rfloor$ and $\lceil v \rceil$ are the two integers such that $\lceil v \rceil = \lfloor v \rfloor = v$ for an integer v; $v - 1 < \lfloor v \rfloor < v < \lceil v \rceil < v + 1$ for a real v which is not an integer.

Definition 1.3. *"ops"* will be our shorthand for "arithmetic operations" (which usually will be applied to real or complex operands).

Definition 1.4. $W = [w_{ij}]$ will denote the matrix of a fixed size, with $w_{ij} = w_{i,j} = (W)_{ij}$ denoting the common entry of row i and column j. $\mathbf{v} = [v_i]$ will denote the column vector of a fixed dimension with the i-th component (entry) $v_i = (\mathbf{v})_i$.

Definition 1.5. $W^T = [w_{ji}]$ will denote the transpose of a matrix or a vector $W = [w_{ij}]$. $W^H = [\bar{w}_{ji}]$ will denote the conjugate (Hermitian) transpose of W, where $\bar{z}$ denotes the complex conjugate of a scalar z.

Definition 1.6. log will denote logarithms to the base 2; ln will denote natural logarithms to the exponential base; deg $p(x)$ will stand for the degree of a polynomial $p(x)$.

Definition 1.7. The customary character "■" will indicate the end of a proof.

Definition 1.8. An algorithm is called *rational* if it only involves ops and comparisons with 0. Such an algorithm can be performed in any fixed field. Rational algorithms performed in the fields of rational, real and complex numbers, may involve comparisons of the magnitudes of pairs of numbers.

In the following, "bounds" and "estimates" will mean "upper bounds" and "upper estimates" unless we specify otherwise.

2. Evaluation, Interpolation and Multiplication.

We will start our list of the major problems of practical and theoretical importance with polynomial evaluation, historically the first topic of the algebraic computational complexity theory (see [AHU], [BM] and/or [Kn]).

Problem 2.1 ($POL \cdot EVAL \cdot S$), *evaluation of a polynomial at a single point* ([Kn], [P66]). *Given the coefficients of a polynomial*

$$p(x) = p_0 + p_1 x + \cdots + p_n x^n, \tag{2.1}$$

compute its value at a point x.

Solution *(Horner's rule).* We have

$$p(x) = (\cdots((p_n x + p_{n-1})x + p_{n-2})x + \cdots + p_1)x + p_0,$$

or equivalently,

$$\begin{aligned} q_0 &= p_n, \\ q_i &= q_{i-1}x + p_{n-i}, \ i = 1, \ldots, n, \\ p(x) &= q_n. \end{aligned}$$

■

Horner's rule involves n multiplications and n additions, and this can be proven to be optimum unless we evaluate the same polynomial at many points. If we write a subroutine for a fixed polynomial $p(x)$, say, for one approximating $\ln(x+1)$ or $\sin x$, we may assume cost-free preconditioning of the input values $p_0, \ldots, p_n$, so that x and any functions in $p_0, \ldots, p_n$ (but not in x) may serve as the input. Then about 50% of multiplications can be saved ([Kn], [P66], [P78]).

Very frequently, one needs to evaluate a polynomial on a fixed set of points.

Problem 2.2 ($POL \cdot EVAL$), *evaluation of a polynomial on a set of points* ([AHU] or [BM]). *Given the coefficients of a polynomial $p(x)$ of (2.1) and a set of K points, $x_0, \ldots, x_{K-1}$, $K > n$, compute $p(x_h)$ for $h = 0, 1, \ldots, K-1$.*

Solutions. The problem can be reduced to K applications of Horner's rule; then the solution would require Kn multiplications and Kn additions. Actually, $O(K \log^2 K)$ ops suffice (see section 4), and $\lceil K \log n \rceil$ ops are necessary for computing the exact values of $p(x)$ on any set $\{x_0, \ldots, x_{K-1}\}$ of K points (see section 4 or [BM]). ∎

This striking improvement of the classical algorithms is due to fast Fourier transform (FFT), which solves Problems 2.2 ($POL \cdot EVAL$) extremely fast in a special case [see Problems 2.2a (DFT) and 2.5 ($GEN \cdot DFT$)].

Let ω be a *primitive K-th root of 1*, that is,

$$\omega^K = 1, \quad \omega^h \neq 1, \quad h = 1, \ldots, K-1,$$

defining the set $\{1, \omega, \omega^2, \ldots, \omega^{K-1}\}$ of K *Fourier points.*

Problem 2.2a (DFT), *discrete Fourier transform. Let $K = 2^k$, for a natural k. Given the coefficients of a polynomial $p(x)$ of (2.1) (of degree at most n) and the set $\{1, \omega, \omega^2, \ldots, \omega^{K-1}\}$ of K Fourier points, compute $p(\omega^h)$, $h = 0, \ldots, K-1$. [If $K = n+1$, then the vector $[p(\omega^0), \ldots, p(\omega^{K-1})]^T$ is said to be the discrete Fourier transform (DFT) of the coefficient vector $[p_0, \ldots, p_n]^T$ of $p(x)$.]*

Solution ([AHU], or [BM]). The problem is solved by applying the *fast Fourier transform (FFT)* algorithm, which is an important example of *recursive algorithms* based on the *divide-and-conquer (divide-et-impera) method.* Let $p_0, p_1, \ldots, p_n, 1, \omega, \omega^2, \ldots, \omega^{K-1}$ be given and $t_A(K, n)$ denote the minimum number of ops sufficient for the DFT. Denote $y = x^2$ and observe that

$$p(x) = (p_0 + p_2 x^2 + \cdots + p_{n-1} x^{n-1}) + x(p_1 + p_3 x^2 + \cdots + p_n x^{n-1}) =$$
$$q(x^2) + x s(x^2) = q(y) + x s(y).$$

The polynomials $q(y) = p_0 + p_2 y + \cdots + p_{n-1} y^{(n-1)/2}$ and $s(y) = (p_1 + p_3 y + \cdots + p_n y^{(n-1)/2})$ have degrees at most $(n-1)/2$. Therefore, to evaluate $p(x)$ at $x = \omega^h$ for $h = 0, 1, \ldots, n$, it suffices to compute first $q(y)$ and $s(y)$ at $y = (\omega^2)^h$ and then $q(\omega^{2h}) + xs(\omega^{2h})$ at $x = \omega^h$. Observe that ω^2 is a $(K/2)$-nd root of 1, so that there exist only $K/2$ distinct values among all the integer powers of ω^2. We have reduced the problem to performing: at first, DFT (twice at $K/2$ points) and then K multiplications (by $x_h = \omega^h$ for $h = 0, 1, \ldots, K-1$) and K additions. 50% of these multiplications can be saved because $x_i = -x_{i+K/2}$ for all i. Therefore, $t_A(K, n) \leq 2t_A(\frac{K}{2}, \frac{n-1}{2}) + 1.5K$. Let us now consider the simpler case where $n + 1 = K = 2^k$. Recursively apply this estimate for $K, K/2, K/4, \ldots$ and arrive at the bound $t_A(K, n) \leq 1.5K \log K$ because, surely, $t_A(1, 0) = 0$. The $1.5K \log K$ ops are split into $K \log K$ additions and $0.5K \log K$ multiplications.

If the DFT problem is stated for $K = 2^k \leq n$, then $n+1-K$ subtractions suffice to reduce the problem to the case of $n = K - 1$ (see Remark 3.2). The algorithm can also be applied where $K = 2^k > n+1$; then the recursive process ends in $\lceil \log(n+1) \rceil$ steps, that is, in at most $1.5K \lceil \log(n+1) \rceil$ ops.

∎

In practice of computations, DFT is frequently performed on $K \geq 10000$ or even $K \geq 100000$ points, so the practical impact of FFT was immense.

Remark 2.1. We assume that all the roots of 1 that we need are readily available. In actual computations, we may use approximations to the complex numbers $\omega^i = \exp(2\pi i \sqrt{-1}/K)$, $i = 0, \ldots, K-1$ (see [Kn]). In applications to computations where all the input values are integers, we sometimes may avoid the round-off problems by means of a special technique (see Remark 7.1) or by performing the computations over a ring of integers $\mathbf{Z}_m$ modulo an appropriate natural m, where a desired primitive root of 1 is readily available (see the end of section 3 of our chapter 3; [AHU], pp. 265-269; or [BM], pp. 86-87). In practice, the choice of an appropriate natural m requires special care. In the applications where the output is integer, as in the case of multiplication of polynomials with integer coefficients, a floating point computation can be used with no round-off error in the output provided that the precision of computation is sufficiently high and that the computed output values are rounded to the integer parts (see sections 4 and 7 of chapter 3). Performing FFT over finite fields or rings is also required in several applications, for instance, to integer multiplication [see Problem 2.7 $(INT \cdot MULT)$]. Actually, for performing FFT over the finite fields or rings

$\mathbf{Z}_m$, it suffices to use any *principal* n-th root ω of 1 (for a fixed n), that is, a root satisfying the equations

$$\omega^n = 1, \quad \sum_{i=0}^{n-1} \omega^{ij} = 0, \quad i = 1, 2, \ldots, n-1$$

(every primitive n-th root of 1 is a principal n-th root of 1 over the fields, but generally not so over the rings), provided that this root ω has its reciprocal in $\mathbf{Z}_m$. To perform the inverse DFT (defined below), we also need to have the reciprocal of K in the ring $\mathbf{Z}_m$.

Next, we consider the converse transition from the values of $p(x)$ to its coefficients.

Problem 2.3 ($POL \cdot INTERP$), *interpolation to a function by a polynomial* ([AHU] or [BM]). *Given two sets of values,*

$$i) \quad \{x_i \ : \quad i = 0, \ldots, n; \quad x_i \neq x_j \text{ for } i \neq j\},$$
$$ii) \quad \{r_i \ : i = 0, \ldots, n\},$$

evaluate the coefficients $p_0, p_1, \ldots, p_n$ *of the polynomial* $p(x)$ *of (2.1) such that*

$$p(x_h) = r_h, \quad h = 0, 1, \ldots, n.$$

Solutions. The problem always has a unique solution that can be computed by means of the classical interpolation algorithms in $O(n^2)$ ops (see [CdB] or [GL], pp. 180-182), but $O(n \log^2 n)$ ops actually suffice (see section 4). Moreover, at least $\lceil (n+1) \log n \rceil$ ops are needed (V. Strassen, see [BM] and compare [B-O]). ∎

We may think of the r_h as of the values of a function $r(x)$ at the nodes of interpolation, $x_0, \ldots, x_n$.

Such a factor $n / \log^2 n$ asymptotic improvement of the classical methods is ultimately due to FFT. Problem 2.3 ($POL \cdot INTERP$) can be solved particularly fast in the special case where the nodes of interpolation are the Fourier points.

Problem 2.3a ($IDFT$), *Inverse Discrete Fourier Transform. Let* $K = n + 1 = 2^k$, *for a natural k. Given the Fourier points* ω^h, $h = 0, \ldots, n$, *and the values* $r_0, \ldots, r_n$, *compute the coefficients* $p_0, \ldots, p_n$ *of the polynomial* $p(x)$ *of (2.1) such that* $p(\omega^h) = r_h$, $h = 0, \ldots, n$. *The vector* $[p_0, \ldots, p_n]^T$ *is said to be the inverse discrete Fourier transform (IDFT) of the vector* $[r_0, \ldots, r_n]^T$.

Solution. The problem is solved in at most $K + 1.5K \log K$ ops by applying the *inverse FFT* algorithm. For the direct derivation of this algorithm, again apply the divide-and-conquer method, that is, represent $p(x) = q(y) + xs(y)$ and observe that, for each h, the values $q(y)$ and $s(y)$ at $y = \omega^{2h} = (-\omega^h)^2$ can be recovered from the values of $p(x)$ at $x = x_h = \omega^h$ and $x = x_{\frac{K}{2}+h} = -\omega^h$. The resulting algorithm is very similar to forward FFT. To directly reduce DFT and IDFT to each other, *we will represent them as matrix computations.* Associate these two problems with the following vector equation: $\Omega\mathbf{p} = \mathbf{r}/\sqrt{K}$, where the matrix $\Omega = [\omega^{ij}/\sqrt{K}]$ will be referred to as the *Fourier matrix*, $\mathbf{p} = [p_j]$, $\mathbf{r} = [r_i]$, $i, j = 0, 1, \dots, n$. Then $\Omega^{-1} = [\omega^{-ij}/\sqrt{K}]$, since $\sum_{h=0}^{n} \omega^{hs} = 0$ if $1 \leq s \leq n$. Therefore, $\mathbf{p} = \Omega^{-1}\mathbf{r}/\sqrt{K} = [\omega^{-ij}]\mathbf{r}/K = [\tilde{\omega}^{ij}]\mathbf{r}/K$ where $\tilde{\omega} = 1/\omega$ is a primitive K-th root of 1, since ω is. Thus, the product $[\tilde{\omega}^{ij}]\mathbf{r}$ can be computed by using FFT, and then K divisions of the resulting vector by K will complete performing the inverse DFT on K points. ■

By means of FFT's, we will next accelerate several basic operations with polynomials.

Problem 2.4 (*SEV · POL · MULT*), *multiplication of several polynomials,* also called *computing the iterated product of polynomials. Given positive integers $s > 1$ and d_j, $j = 1, \dots, s$, and the coefficients of s polynomials, $u_j(x) = \sum_{i=0}^{d_j} u_{i,j}x^i$, $j = 1, 2, \dots, s$, compute the coefficients of their product, $P(x) = \prod_{j=1}^{s} u_j(x) = \sum_{i=0}^{D} P_i x^i$, where $D = \sum_{j=1}^{s} d_j$.*

Solutions. We may recursively compute the coefficients of the polynomials $P_{j+1}(x) = P_j(x)u_{j+1}(x)$, $j = 1, \dots, s-1$, where $P_1(x) = u_1(x)$, $P_s(x) = P(x)$, by multiplying $s - 1$ pairs of polynomials. By applying the straightforward algorithm for these pairwise multiplications, we may solve Problem 2.4 in $\sum_{j=2}^{s}(d_j + 1)(\sum_{i=1}^{j-1} d_i + 1)$ multiplications and about as many additions. Let us greatly accelerate the solution by reducing the original problem to FFT's. Let ω denote a primitive K-th root of 1 where $D < K \leq 2D$, $K = 2^k$, k is an integer and then:

a) evaluate the s polynomials $u_j(x)$, $j = 1, \dots, s$, on the K Fourier points ω^h, $h = 0, 1, \dots, K-1$ (by means of s applications of FFT on K points),
b) compute $P(\omega^h) = \prod_{j=1}^{s} u_j(\omega^h)$ for $h = 0, 1, \dots, K-1$,
c) recover all the coefficients of $P(x)$ by applying the inverse FFT on the K points ω^h, $h = 0, 1, \dots, K-1$.

The computations at all these stages a), b) and c) require less than $1.5(s+1)K \log K + sK$ ops. ■

This dramatic asymptotic improvement of the classical algorithm is due to using FFT and the *evaluation-interpolation method* (of [To]), which applies to several other polynomial computations as well. The actual practical impact of this acceleration, however, is more limited than in the case of FFT, since frequently the input size of practical polynomial computations is smaller and since the overhead constant grows from 1.5 to $1.5(s+1)$ in the ops asymptotic count (also compare Appendix D).

Let us display a special case of Problem 2.4 ($SEV \cdot POL \cdot MULT$), for $s = 2$.

Problem 2.4a ($POL{\cdot}MULT$), *multiplication of two polynomials (convolution of two vectors). Given the input values $u_0, \ldots, u_m$, $v_0, \ldots, v_n$, compute*

$$w_i = \sum_{j=0}^{i} u_j v_{i-j}, \quad i = 0, 1, \ldots, m+n, \tag{2.2}$$

where we assume that $u_j = 0$, for $j > m$, and $v_k = 0$, for $k > n$, that is, compute the coefficients of the polynomial $w(x) = \sum_{i=0}^{m+n} w_i x^i = u(x)v(x)$, $u(x) = \sum_{j=0}^{m} u_j x^j$, $v(x) = \sum_{k=0}^{n} v_k x^k$ or, equivalently, compute the convolution $[w_0, \ldots, w_{m+n}]$ of two vectors $[u_0, \ldots, u_m]$ and $[v_0, \ldots, v_n]$.

Solution. Substitute $s = 2$ and $D = m + n$ into the above cost bound and obtain that all the values w_i in (2.2) can be computed by using at most

$$M(K) = 4.5K \log K + 2K \tag{2.3}$$

ops where $m + n < K \leq 2(m+n)$, $K = 2^k$, k is an integer. ∎

For comparison, the straightforward algorithm for this problem requires $(m+1)(n+1)$ multiplications and $(m+1)(n+1) - m - n - 1$ additions.

An extension of the above problem is given by

Problem 2.4b ($\pm CONVOL$). *Compute the positive and/or negative wrapped convolutions, that is, given the values $u_0, v_0, u_1, v_1, \ldots, u_n, v_n$, compute the coefficients w_i^+ and/or w_i^- of the polynomials*

$$w^+(x) = \sum_{i=0}^{n} w_i^+ x^i, \quad w_i^-(x) = \sum_{i=0}^{n} w_i^- x^i$$

such that

$$w_i^+ = w_i + \hat{w}_i, \quad w_i^- = w_i - \hat{w}_i,$$

$$w_i = \sum_{j=0}^{i} u_j v_{i-j}, \quad \hat{w}_i = \sum_{j=i+1}^{n} u_j v_{n+1+i-j}, \quad i = 0, 1, \ldots, n.$$

Here we assume that $\hat{w}_n = 0$. *In fact,* $w^+(x) = u(x)v(x) \bmod (x^{n+1} - 1)$, $w^-(x) = u(x)v(x) \bmod (x^{n+1} + 1)$, $u(x) = \sum_{j=0}^{n} u_j x^j$, $v(x) = \sum_{j=0}^{n} v_j x^j$ (compare Remark 3.2 or [AHU]).

Note that the solution of Problem 2.4a ($POL \cdot MULT$) for $m = n$ is immediately reduced to the evaluation of the positive and negative wrapped convolutions.

Solution. To compute w_i^+, again apply Toom's evaluation-interpolation method. Specifically, first compute the values $u(\omega^j)$, $v(\omega^j)$ and then $w(\omega^j)$, for $j = 0, 1, \ldots, n$, where ω is a primitive $(n+1)$-th root of 1. Then observe that $w(x) = \sum_{i=0}^{n} w_i x^i + x^{n+1} \sum_{i=0}^{n-1} \hat{w}_i x^i$ and thus $w^+(\omega^j) = w(\omega^j)$, $j = 0, 1, \ldots, n$, and interpolate to the polynomial $w^+(x)$ by means of the inverse DFT. All the three DFT's are on $n+1$ points, which saves about 50% of ops and some memory space versus the usual convolution algorithm.

To compute the negative wrapped convolution (also called *cyclic convolution*), one may, of course, compute the convolution w_i and the positive wrapped convolution $w_i + \hat{w}_i$ and then recover $w_i - \hat{w}_i$ for all i. There is a simpler way, however: first compute the values $\sum_{i=0}^{n} u_i \psi^i \omega^{ik}$ and $\sum_{i=0}^{n} v_i \psi^i \omega^{ik}$ by applying DFT, and then compute their pairwise products, where ψ is a primitive $(2+2n)$-th root of 1. These products give us the values of the polynomial $\sum_{i=0}^{n} (w_i^-) \psi^i x^i$ at $x = \omega^k$, for $k = 0, 1, \ldots, n$. Finally, recover the coefficients $(w_i^-)\psi^i$ by means of the inverse DFT. Again, all of the three DFT's involved are on $n+1$ points. ■

Next, we will show two surprising applications of polynomial multiplication, where two computational problems of high independent interest are first reduced to the format (2.2) by means of appropriate substitutions of variables and then solved by the algorithm for Problem 2.4a ($POL \cdot MULT$). Then again, this improves the classical algorithms by decreasing their cost from the order of n^2 to $O(n \log n)$ ops. Since one of these two problems includes DFT, we may, in principle, *ultimately reduce all our polynomial computations to polynomial multiplication.*

Problem 2.5 ($GEN \cdot DFT$), *generalized DFT at any number of points* (L.I. Bluestein; see [Kn], pp. 300 and 588). *Given the values* $w, p_0, p_1, \ldots, p_{K-1}$, w *having the reciprocal* w^{-1} *(such that* $ww^{-1} = 1$*), compute* $r_h = \sum_{j=0}^{K-1} p_j w^{hj}$ *for* $h = 0, 1, \ldots, K-1$.

Solutions.

a) In particular, if w is a primitive K-th root of 1, the problem turns into DFT. The FFT algorithm gives us a solution in $O(K \log K)$ ops for

$K = 2^k$, k integer, and can be easily extended to the case of $K = c^k$, for any fixed integer $c > 1$; furthermore, nonasymptotic (but practically useful) improvement of FFT has been proposed in these cases in [W78], with its further extension to $K = \prod_{i=1}^{s} K_i$ where $K_i = c_i^{k(i)}$, $c_i > 1$, $k(i) \geq 1$ are integers, $i = 1, \ldots, s$.

b) To reach the bound of $O(K \log K)$ ops for any K, rewrite

$$r_h = w^{-h^2/2} \sum_{j=0}^{K-1} w^{(j+h)^2/2} w^{-j^2/2} p_j, \quad h = 0, 1, \ldots, K-1.$$

Substitute $m = K - 1$, $n = 2K - 2$, $w_{n-h} = r_h w^{h^2/2}$, $u_j = w^{-j^2/2} p_j$ for $j \leq m$, $u_j = 0$ for $j > m$, $v_s = w^{(n-s)^2/2}$, and rewrite the expressions for r_h as follows: $w_i = \sum_{j=0}^{i} u_j v_{i-j}$, $i = m, m+1, \ldots, n$. This reduces generalized DFT on K points to the evaluation of w_i for $i = m, m+1, \ldots, n$, which is a part of the convolution problem, defined by the equation (2.2). Therefore, generalized DFT on K points involves at most $(3K-3)(9\log(3K-3)+13)$ ops, not counting the $2K$ multiplications $u_j = w^{-j^2/2} p_j$ and $r_h = w_{n-h} w^{-h^2/2}$. ■

The best available lower bounds are of the order of K, even for DFT on K Fourier points (K-th roots of 1).

Problem 2.6 ($VAR \cdot SHIFT$), *shift of the variable* ([ASU]). *Given a scalar* Δ *and the coefficients* $p_0, p_1, \ldots, p_n$ *of a polynomial* $p(x)$ *of (2.1), compute the coefficients* $q_0(\Delta)$, $q_1(\Delta), \ldots, q_n(\Delta)$ *of the polynomial*

$$q(y) = p(y + \Delta) = \sum_{h=0}^{n} p_h (y + \Delta)^h = \sum_{g=0}^{n} q_g(\Delta) y^g,$$

or equivalently, compute the values at $x = \Delta$ *of the polynomial* $p(x)$ *and of all its derivatives* $p^{(g)}(\Delta) = q_g(\Delta) g!$ *for* $g = 1, 2, \ldots, n$.

Solution. Expand the powers $(y + \Delta)^h$ and obtain that

$$q_g(\Delta) = \sum_{h=g}^{n} p_h \Delta^{h-g} \frac{h!}{g!(h-g)!}, \quad g = 0, 1, .., n.$$

Substitute $w_{n-g} = g!\, q_g(\Delta)$, $u_{n-h} = h! p_h$, $v_s = \Delta^s / s!$, $j = n - h$ and $i = n - g$, and arrive at the following expressions [which are a part of the convolution problem defined by (2.2) with $m = n$]:

$$w_i = \sum_{j=0}^{i} u_j v_{i-j}, \quad i = 0, 1, \ldots, n.$$

Thus, the solution involves at most $4.5K\log K + 2K$ ops for $2n < K \leq 4n$, not counting $4n-4$ multiplications and divisions by Δ and $s!$ for $s = 2, 3, \ldots, n$. ■

Remark 2.2. The shift of x is naturally complemented by the simple operation of *scaling the variable* x by a constant factor a, that is, of computing the coefficients of the polynomial $s(z) = p(az)$; the scaling costs $2n-1$ ops. Shift and scaling together enable us to make a linear transformation of the variable, that is, for given coefficients $p_0, \ldots, p_n$ of a polynomial $p(x)$ of (2.1), to compute, at the cost of $O(n \log n)$ ops, the coefficients of the polynomial $\sum_{i=0}^{n} q_i x^i = p(ax+b)$, for any pair of constants a and b. Similarly, at the cost of $O(n \log n)$ ops, we may compute the coefficients of the polynomial $\sum_{i=0}^{2n} r_i x^i = p(ax^2+bx+c)$ for any triple of constants a, b and c [note that $ax^2+bx+c = a(x+d)^2+e$, for $a \neq 0$, $d = b/(2a)$, $e = c - ad^2$]. Consequently, the solution of Problem 2.5 ($GEN \cdot DFT$) can be extended to the evaluation [in $O(n \log n)$ ops] of $p(x)$ on the set of points $\{ah^{2i} + fh^i + g,\ i = 0, 1, \ldots, n-1\}$ for any fixed 4-tuple of constants (a, f, g, h) ([ASU]).

We will now consider some applications to the extensions of FFT. Among many applications of the shift and scaling of the variable, note the extension of FFT from the Fourier set $\{x_h = \omega^h\}$ to the *adjusted (expanded) Fourier set* $\{x_h = a\omega^h + b\}$, for any pair of complex a and b and for $h = 0, 1, \ldots, K-1$. Less trivial is the extension to the solution of Problems 2.2 ($POL \cdot EVAL$) and 2.3 ($POL \cdot INTERP$) for the *adjusted (expanded) Chebyshev sets* of real points (nodes) $x_h = \frac{a+b}{2} + \frac{b-a}{2}\cos(\frac{2h+1}{K}\pi)$, $h = 0, 1, \ldots, K-1$, which are the real parts of the complex points of the adjusted Fourier sets, with real a and b. Problems 2.2 ($POL \cdot EVAL$) and 2.3 ($POL \cdot INTERP$) for the adjusted Chebyshev sets of nodes are encountered in the approximation theory (see [CdB] and [DB]). Since we may shift and scale the variable at a low computational cost, it suffices to consider the Chebyshev unadjusted sets, where $b = -a = 1$. Let $K = 2H$, $y = 1 - 2x^2$ and represent $p(x) = p_1(x^2) + xp_2(x^2) = p_1(y) + xp_2(y)$. Then the $2H$ point Chebyshev x-set is transformed into the H point Chebyshev y-set, for $2\cos^2 x - 1 = \cos(2x)$, so that the original problem is reduced to two problems of half-size. The reduction involves two shifts of the variable by 1 [at the cost of $O(n \log n)$ ops], as well as $O(K)$ ops for multiplications by x_h and additions, so the overall cost is $O((K + n\log n)\log K)$ ops if $K < n$ and $O((K + n\log n)\log n)$ ops otherwise. This asymptotic cost bound only by a constant factor improves the bound for the general input set of $2n$ points (see section 4). The above algorithm, however, has a little better numerical stability (compare chapter 3). In [Ger], a numerically stable $O(n \log n)$ algorithm (for $K = n$) is presented for this problem.

Coming last in this section is an important application of polynomial multiplication to integer multiplication.

Problem 2.7 ($INT \cdot MULT$), *integer multiplication* ([AHU], [BM] or [Kn]). *Given a pair of N-bit integers u and v,*

$$u = \sum_{i=0}^{N-1} u_i 2^i, \quad v = \sum_{i=0}^{N-1} v_i 2^i, \quad u_i, v_i \in \{0, 1\},$$

compute the integer uv.

Solution. For such computations with integers, we will assume the Boolean circuit model (see appendix A to this chapter). The classical solution requires Boolean circuits of sizes of the order of N^2, but A. Schönhage and V. Strassen [ScSt] decreased this bound to

$$\mu(N) = O(N \log N \log \log N). \tag{2.4}$$

As well as the preceding algorithm of A. Toom [To], which reached the bound $O(N^{1+\epsilon})$ for any positive ϵ, their algorithm relies on multiplication of u and v as polynomials $u(x)$ and $v(x)$ at $x = 2^L$, for an appropriate natural L, that is, the algorithm computes the coefficients of $w(x) = u(x)v(x)$ and then outputs the value $w(2^L) = u(2^L)v(2^L) = uv$. This effectively reduces the original problem to multiplication of shorter binary integers, at a lower cost. The natural L roughly equals the length of such auxiliary integers. The best choice for L turned out to be a power of 2 close to $\sqrt{N}$. To decrease the Boolean cost, the algorithm relies on *FFT's over finite rings of constants* [specifically, over integers modulo $L(2^{2L} + 1)$] and exploits the negative wrapped convolution (see more in Appendix C). ■

Remark 2.3. Schönhage and Strassen's algorithm supports the asymptotic bound (2.4) where the overhead constant is considerable. At present, the algorithms used in practice support the estimates

$$\mu(N) = O(N^2) \quad \text{or} \quad \mu(N) = O(N^{\log 3}), \quad \log 3 = 1.5849\ldots, \tag{2.5}$$

with small overhead constants (see exercise 7), except that recently some faster algorithms (DFT based and of Schönhage-Strassen's type) have been implemented on modern computers, in particular, by D. H. Bailey [Bai,a; Bai,b] and by A.K. Lenstra.

Figure 1 shows our current progress in expanding the domain of polynomial computations, for which we have obtained effective algorithms by reducing the computations to DFT and polynomial multiplication. By adding polynomial division, extended Euclidean algorithm, fan-in method, repeated squaring, and manipulation with power series and with partial fractions to the set of our basic tools for such a reduction, we will make such a domain grow like a snowball.

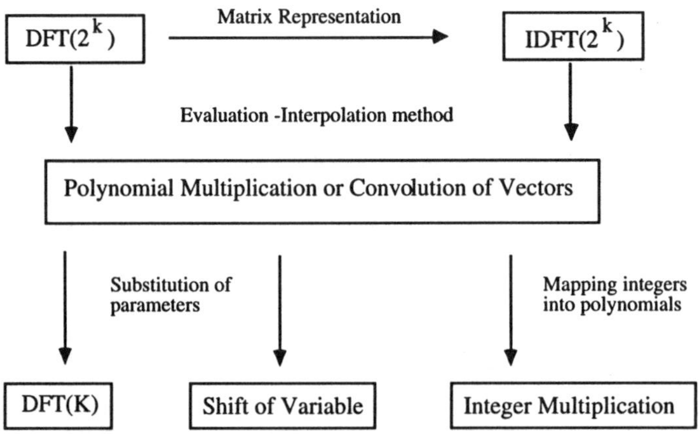

Figure 1. Effective polynomial computations and some techniques used

3. Polynomial Division and Modular Reduction. An Application of Manipulation with Formal Power Series and Newton's Iteration.

Next, manipulation with formal power series and Newton's iteration will enable us to reduce polynomial division to polynomial multiplication, which again means acceleration from the classical n^2 to $O(n \log n)$ ops, and we will immediately show some further applications (with more of them coming in section 4).

Problem 3.1 ($POL{\cdot}DIVIDE$), *polynomial division with a remainder* ([AHU] or [BM]). *Given the coefficients of two polynomials*

$$s(x) = \sum_{i=0}^{m} s_i x^i, \quad t(x) = \sum_{i=0}^{n} t_j x^j,$$

$s_m t_n \neq 0$, $m \geq n$, compute the coefficients of the two polynomials $q(x) = \sum_{g=0}^{m-n} q_g x^g$ and $r(x) = \sum_{h=0}^{n-1} r_h x^h$ such that

$$s(x) = q(x)t(x) + r(x), \quad \deg r(x) < n. \tag{3.1}$$

$q(x)$ is called the quotient, and $r(x)$ is called the remainder of the division of $s(x)$ by $t(x)$.

Fact 3.1. *There always exists unique solution $q(x)$ and $r(x)$ to Problem 3.1 ($POL \cdot DIVIDE$).*

The remainder $r(x)$ is called the *residue* of $s(x)$ modulo $t(x)$ and is denoted $s(x) \bmod t(x)$. Hereafter, *computing* $s(x) \bmod t(x)$ will mean computing the coefficients of the polynomial $s(x) \bmod t(x)$. We will assume hereafter that $a \circ b \bmod t$ denotes $(a \circ b) \bmod t$ (that is, the polynomial or the power series $a \circ b$ reduced $\bmod t$), for polynomials a, b and t and for any operation $\circ$. In particular, we will use the notation $w(x) = u(x)/v(x) \bmod t(x)$ as long as $\deg w(x) < \deg t(x)$ and $v(x)w(x) = u(x) \bmod t(x)$. Note the different meanings of $a \circ b \bmod t = (a \circ b) \bmod t$, of $a \circ (b \bmod t)$, and of $(a \bmod t) \circ b$. Where this causes no ambiguity, we will also write $u(x) = v(x) \bmod t(x)$ whenever $t(x)$ divides $u(x) - v(x)$, even if $\deg u(x) \geq \deg t(x)$.

Remark 3.1. Problem 3.1 ($POL \cdot DIVIDE$) and all the above comments and definitions can be immediately extended to the case where the dividend $s(x) = \sum_{i=0}^{+\infty} s_i x^i$ is a formal power series.

Remark 3.2. Reduction modulo a polynomial $t(x)$ is a general customary tool of algorithm design. We have already used this tool. The positive and negative wrapped convolutions are nothing else but the products of a pair of n-th degree polynomials modulo $x^{n+1} - 1$ and $x^{n+1} + 1$, respectively. Furthermore, reduction modulo x^k does not change any polynomial of degree less than k. Thus, we may interpret Problem 2.4a ($POL \cdot MULT$) as the problem of computing the product $u(x)\,v(x) \bmod (x^{m+n+2} - 1)$. Letting $m = n$ and decomposing $x^{m+n+2} - 1 = x^{2n+2} - 1 = (x^{n+1} - 1)(x^{n+1} + 1)$, we reduce the problem to computing $u(x)\,v(x) \bmod (x^{n+1} - 1)$ (positive wrapped convolution) and $u(x)\,v(x) \bmod (x^{n+1} + 1)$ (negative wrapped convolution). Indeed, $uv \bmod (x^{2n+2} - 1) = (1/2)((uv \bmod (x^{n+1} - 1))(x^{n+1} + 1) - (uv \bmod (x^{n+1} + 1))(x^{n+1} - 1))$. [Moreover, $((a + b)p) \bmod (ab) = a(p \bmod b) + b(p \bmod a)$, for any triple of polynomials a, b and p.] Also note that $p(\omega^h) = r(\omega^h)$ for any integer h provided that $r(x) = p(x) \bmod (x^K - 1)$ and $\omega^K = 1$, so that FFT on K points for $p(x)$ is immediately reduced to

FFT on K points for $r(x)$, and $\deg r(x) < K$. If $n+1 = 2^k$, k is an integer, we may recursively reduce the evaluation of $u(x)\,v(x) \bmod (x^{n+1} - 1)$ to the evaluation of $(u(x) \bmod (x^{2^h}+1))\,(v(x) \bmod (x^{2^h}+1)) \bmod (x^{2^h}+1)$, $h = k-1, k-2, \ldots, 0$, and of $u(x)\,v(x) \bmod (x-1) = u(1)\,v(1)$, since $x^{n+1} - 1 = (x-1)\prod_{h=0}^{k-1}(x^{2^h}+1)$.

Classical Solution of Problem 3.1 (*POL·DIVIDE*). The classical algorithm for "synthetic division" ([Kn], p. 402) relies on the recursive reduction of the degree of $s(x)$. Specifically, let $s(x)$ and $t(x)$ be monic, $d = \deg s(x) - \deg t(x)$, $\hat{s}(x) = s(x) - x^d t(x)$, $\hat{s}(x) = \hat{q}(x)t(x) + \hat{r}(x)$, $\deg \hat{r}(x) < \deg t(x)$. Then $\deg \hat{s}(x) < \deg s(x)$, and it suffices to compute $\hat{q}(x)$ and $\hat{r}(x)$ and to set $r(x) = \hat{r}(x)$, $q(x) = \hat{q}(x) + x^d$. In order to compute $\hat{q}(x)$ and $\hat{r}(x)$, scale $\hat{s}(x)$ and recursively repeat this stage.

Algorithm 3.1.

Input: two natural numbers m, n, $m \geq n$, and the coefficients $s_0, \ldots, s_m$, $t_0, \ldots, t_n$ of two polynomials $s(x) = \sum_{i=0}^{m} s_i x^i$, $t(x) = \sum_{i=0}^{n} t_j x^j$ of degrees m and n, respectively.

Output: the coefficients of the quotient $q(x) = \sum_{g=0}^{m-n} q_g x^g$ and of the remainder $r(x) = \sum_{h=0}^{n-1} r_h x^h$ of the division of $s(x)$ by $t(x)$, defined in (3.1).

Computation:

1. (Initialization.) Set $k = m - n$, $s_{i,m-n} = s_i$ for all i.
2. For $k = m-n, m-n-1, \ldots, 0$, set $s_{j,k-1} = s_{j,k}$ for $j < k$, compute $q_k = s_{m,k}/t_n$, then $s_{j,k-1} = s_{j,k} - q_k t_{j-k}$ for $j = n+k-1, n+k-2, \ldots, k$, that is, compute the coefficients $s_{j,k-1}$ of the polynomial $s_{k-1}(x) = s_k(x) - q_k x^k t(x)$ where $s_g(x) = \sum_j s_{j,g} x^j$, for $g = k-1, k$.
3. Output the coefficients q_k, $k = 0, \ldots, m-n$, of the quotient and the coefficients $r_h = s_{h,-1}$, $h = 0, \ldots, n-1$, of the remainder. ■

This algorithm involves no divisions if $t(x)$ is monic, and we may avoid divisions for any $t(x)$ if we replace $s(x)$ by $t_n^n s(x)$. The algorithm, however, may involve up to $(2n+1)(m-n+1)$ ops, that is, up to $(2n+1)(n+1)$ if $m = 2n$. If $r(x)$ is identically zero, then $q(x)$ can be found by using $O(m + (m-n+1)\log(m-n+1))$ ops, by means of the evaluation of $s(x)$ and $t(x)$ on the $m-n+1$ Fourier points, division of the resulting values $s(x)$ by $t(x)$, and interpolation. If $r(x) \neq 0$, then [PLS] proposed applying these techniques to the polynomials $s_H(x) = s(Hx)$ and $t_H(x) = t(Hx)$ for large H and immediately deduce for the output polynomial $q_H(x) = \sum_{i=0}^{m-n} q_{H,i} x^i$ that $|q_{H,i} - q_i| = O(1/H^{i+1})$, $i = 0, \ldots, m-n$, as $H \to \infty$. Furthermore, [PLS] proposed replacing $s(x)$ by $x^K s(x)$ and thus $q(x)$ by

$x^K q(x)$, $K = m - n + 1$. This decreases the above bounds to $O(1/H^{K+i+1})$ and gives us a quite effective scheme *for approximate polynomial division*, which may ensure any fixed upper bound $\epsilon > 0$ on the output errors [compare (A.3), (A.4)]. The resulting complexity bound of $O((m + K) \log K)$ ops for approximating $q(x)$ decreases to $O(K \log K)$ ops if we replace $s(x)$ by $\sum_{i=0}^{m-n} s_{m-i} x^{2(m-n)-i}$ and $t(x)$ by $\sum_{i=0}^{m-n} t_{n-i} x^{m-n-i}$, which leaves $q(x)$ unchanged.

Next, we will obtain the same complexity bound, $O(K \log K)$, for the exact evaluation of $q(x)$, by exploiting the *correlation between the polynomial and formal power series computations* (see some alternate ways in chapter 6). We will first demonstrate such a correlation by extending Problem 2.4 ($SEV \cdot POL \cdot MULT$) of polynomial multiplication to multiplication of formal power series.

Problem 3.2 ($SER \cdot MULT$). *Given two formal power series* $u(x) = \sum_{i=0}^{+\infty} u_i x^i$ *and* $v(x) = \sum_{j=0}^{+\infty} v_j x^j$, *compute the first* K *coefficients of the formal power series* $w(x) = \sum_{h=0}^{+\infty} w_h x^h = u(x)v(x)$, *that is, compute the values* w_i *of (2.2) where* $K = m + n + 1$ *and where* u_j *and* v_k *are generally nonzero for* $j > m$ *and* $k > n$.

Solution. Reduce the problem to polynomial multiplication,

$$w(x) = (u(x) \bmod x^K)(v(x) \bmod x^K) \bmod x^K.$$

■

Problem 3.3 ($\pm SER \cdot MULT$), [compare (2.2)]. *Given a polynomial* $u(x) = \sum_{i=0}^{m} u_i x^i$ *and a formal power series* $v(x) = \sum_{j=-\infty}^{+\infty} v_j x^j$, *compute the coefficients* $w_0, \ldots, w_K$ *of* $w(x) = \sum_{h=-\infty}^{+\infty} w_h x^h = u(x)v(x)$. *This amounts to computing the values*

$$w_i = \sum_{j=0}^{m} u_j v_{i-j}, \quad i = 0, 1, \ldots, K.$$

Solution. Let $\sum_{h=0}^{2m+K} s_h x^h = (\sum_{i=0}^{m} u_i x^i)(\sum_{j=0}^{m+K} v_{j-m} x^j)$, then $w_h = s_{h+m}$, $h = 0, 1, \ldots, K$. ■

As an extension of Problem 3.3 ($\pm SER \cdot MULT$), suppose that, under the assumptions of Problem 3.3 ($\pm SER \cdot MULT$), we seek the coefficients $w_G, w_{G+1}, \ldots, w_K$ of $w(x)$ for some $G \leq K$. Then $w_h = s_{h+m-G}$ where

$$\sum_{h=0}^{2m+K-G} s_h x^h = (\sum_{i=0}^{m} u_i x^i)(\sum_{j=0}^{m+K-G} v_{j-m+G} x^j).$$

Now, we will again turn to Problem 3.1 ($POL \cdot DIVIDE$) and will reduce it to the problem of the division of two formal power series. Let $x = 1/z$ and equivalently rewrite (3.1) as follows:

$$S(z) = Q(z)T(z) + z^{m-n+1}R(z) = Q(z)T(z) \bmod z^{m-n+1}, \tag{3.2}$$

where $S(z) = z^m s(1/z)$, $Q(z) = z^{m-n}q(1/z)$, $T(z) = z^n t(1/z)$, $R(z) = z^{n-1}r(1/z)$ denote the polynomials obtained by reversing the order of the coefficients of the polynomials $s(x)$, $q(x)$, $t(x)$ and $r(x)$, respectively. Given $s(x)$ and $t(x)$, it suffices to compute the $K = m - n + 1$ coefficients of the polynomial $Q(z)$ satisfying (3.2) in order to find $q(x)$ satisfying (3.1). Given $q(x)$, we immediately obtain $r(x)$ from (3.1). To evaluate $Q(z)$ satisfying (3.2), we will first solve the following auxiliary problem for $K = m - n + 1$:

Problem 3.4 ($POL \cdot RECIPR$), *computing the reciprocal of a polynomial. Given a natural K and a polynomial $T(z)$, $T(0) \neq 0$, compute the first K coefficients of the formal power series $w(z)$ such that $w(z)T(z) = 1$. We will write $w(z) = 1/T(z)$.*

Solution [Sieveking-Kung, but compare Cook's earlier solution to Problem 3.6 ($INT \cdot DIVIDE$), see [BM], pp. 94-98]. Denote $T = T(z)$, consider the equation $f(w) = T - 1/w = 0$ over the ring of formal power series in z, $w = w(z)$, and apply Newton's iteration to this equation, $w_{j+1} = w_j - f(w_j)/f'(w_j)$, where $f'(w)$ is the formal derivative of $f(w)$. The iteration takes the form $w_0 = 1/T(0)$, $w_{j+1} = w_j(2 - Tw_j)$, $j = 0, 1, \ldots$, so that

$$(1/T) - w_{j+1} = T((1/T) - w_j)^2. \tag{3.3}$$

Therefore, $w_j = w_j(z) = w(z) \bmod z^J$, $J = 2^j$ for $j = 0, 1, \ldots, \lceil \log K \rceil$ (see section 6 for more details on Newton's iteration). Without changing the solution, the polynomials $w_j(z)$, $T = T(z)$ and $2 - T(z)w_j(z)$ can be reduced modulo z^{2J}, before performing iteration step j, so this iteration step essentially amounts to two convolutions of two pairs of vectors of dimensions at most $2J = 2^{j+1}$. Thus, the overall cost of computing $w(z) \bmod z^K$ is at most $18K_0 \log K_0 + 8K_0$ ops where $K \leq K_0 < 2K$, $K_0 = 2^{k_0}$, k_0 is an integer. ∎

Fast solution of Problem 3.1 ($POL \cdot DIVIDE$). Problem 3.1 can be easily reduced to Problem 3.4 ($POL \cdot RECIPR$). In fact, under (3.1) and (3.2), multiplication modulo z^K of $w(z)$ by $S(z)$ gives us $Q(z)$ and $q(x)$ at the cost of at most $9K_0 \log K_0 + 4K_0$ ops. Given the coefficients of $q(x)$, the coefficients of the remainder $r(x)$ are immediately computed from (3.1) at the cost of at most $4.5n_0 \log n_0 + 3n_0$ ops. Here, $m < n_0 \leq 2m$, $n_0 = 2^s$ for an integer s, so that Problem 3.1 ($POL \cdot DIVIDE$) can be solved at the cost of $O(m \log m)$ ops, with a small overhead constant. ∎

Remark 3.3. For simplicity assume that $T(0) = 1$. Then $w_j(z) = \sum_{i=0}^{+\infty}(1 - T(z))^i = \sum_{i=0}^{J}(1 - T(z))^i \mod z^J$, $J = 2^j$. In chapter 6 we will devise some alternative polynomial division algorithms based on these relations and their extensions.

Remark 3.4. There is the converse reduction of Problem 2.4a ($POL \cdot MULT$) of polynomial multiplication, first to polynomial squaring (exercise 6) and then to Problem 3.4 ($POL \cdot RECIPR$), based on the identity

$$p^2 = (\frac{1}{p} - \frac{1}{p+1})^{-1} - p, \tag{3.4}$$

where we may assume that $p = p(x)$ is a polynomial, $p(0) \neq 0$, $p(0) \neq -1$ [if $p(0) = 0$ or $p(0) = -1$, replace p by $p \pm 1$ in (3.4) and then recover $p^2 = (p \pm 1)^2 \mp 2p - 1$].

Polynomial division is used in many polynomial computations, in particular, in the algorithms for fast polynomial evaluation and interpolation (see the next section) and for the following extension of Problem 2.6 ($VAR \cdot SHIFT$) of the shift of the variable:

Problem 3.5 ($TAYLOR \cdot EXP$), *generalized Taylor expansion* ([GKL]). *Given the coefficients of two polynomials: $u(x)$ of degree N and $t(x)$ of degree n, let $m = \lceil (N+1)/n \rceil - 1$ and compute the coefficients of the polynomials $s_i(x)$, $i = 0, 1, \ldots, m$, of degrees less than n, defining the Taylor expansion of $u(x)$ as a polynomial in $t(x)$,*

$$u(x) = \sum_{i=0}^{m} s_i(x)(t(x))^i.$$

Solution. Again apply the divide-and-conquer method, assuming (without loss of generality) that $m + 1 = (N+1)/n = 2^h$ is an integer power of 2. Compute $k = \lceil m/2 \rceil$, divide $u(x)$ by $(t(x))^k$, and obtain that $u(x) = q(x)(t(x))^k + r(x)$. This reduces the original problem to computing generalized Taylor expansions of two polynomials $q(x)$ and $r(x)$ of degrees less than $nk = (N+1)/2$. Recursively apply this algorithm and note that only the powers $(t(x))^J$ for $J = 2^j$, $j = 1, 2, \ldots, h-1$, need to be computed (we may compute them by means of *repeated squaring* or by means of the evaluation and interpolation on the set of Fourier points). Then the desired generalized Taylor expansion of $u(x)$ is computed in $m+1$ polynomial divisions. The cost of these divisions is at most $\sum_{j=1}^{h-1} 2^j D((N+1)/2^j) = O(N \log N \log m)$ ops (with a small overhead constant). This bound dominates the cost $O(N \log N)$ of repeated squaring. Here, $D(s) = O(s \log s)$ denotes the cost of the division with a remainder of two polynomials of degrees at most s. ■

Observe that if $t(x) = x - a$, we arrive at Problem 2.6 ($VAR{\cdot}SHIFT$).

The next problem is similar to polynomial division in both its statement and solutions:

Problem 3.6 ($INT{\cdot}DIVIDE$), *integer division* ([AHU], [BM], pp. 97-98, or [Kn]). *Given four integers m, n, S and T, $2^{n-1} \le T < 2^n \le S < 2^m$, compute the unique pair of integers Q and R such that*

$$S = QT + R, \quad 0 \le R < T. \tag{3.5}$$

Solutions. Both classical "synthetic division" and Sieveking-Kung algorithm have their counterparts for integer division. The sizes of the Boolean circuits for these counterparts-algorithms are of the order of mn and $O(\mu(m))$, respectively, where $\mu(m)$ denotes the Boolean complexity of multiplication of m-bit integers [see (2.4)]. The latter bound for integer division is due to S. Cook, whose algorithm preceded Sieveking-Kung's. Cook's algorithm first computes a rational value w_k that approximates $s^{-1} = 2^n T^{-1}$ within 2^{n-m-1}, so that $|Q_k - S/T| < 2^{-m-1}S < 1/2$ for $Q_k = 2^{-n} w_k S$, and, therefore, $|Q_k T - S| < 0.5T$. Then Q is obtained by rounding off Q_k to the nearest integer, and R is recovered by using (3.5). The value w_k is computed by means of *numerical Newton's iteration*, $w_0 = 2$, $w_{j+1} = \lceil 2w_j - w_j^2 s \rceil_j$ where $\lceil x \rceil_j$ denotes the value x rounded up, to $2^{j+1} + 1$ bits. It can be immediately verified that $0 \le w_j - s^{-1} < 2^{1-2^j}$, for $j = 0$, and then we may extend these bounds to all j by using a simple extension of (3.3) (see [Kn]). Therefore, we may set $k = \lceil \log(m - n + 2) \rceil$ and thus arrive at w_k by using Boolean circuits of the size $O(\mu(m - n))$, which implies the desired complexity bound $O(\mu(m))$ for integer division. ■

4. Manipulation with Partial Fractions and Fast Polynomial Evaluation and Interpolation via Polynomial Multiplication and Division. Further Extensions.

In this section we will apply fast polynomial multiplication and division in order to accelerate the classical solution of Problems 2.2 ($POL \cdot EVAL$)

and 2.3 ($POL \cdot INTERP$) of polynomial evaluation and interpolation and some related computations, which frequently amount to the transition from one to another representation of the same polynomial (compare Remark 4.4).

We will start with *modular reduction and evaluation of a polynomial by means of polynomial division,* based on the observation that $s(x) \bmod (x - a) = s(a)$.

Fast solution of Problem 2.2 ($POL{\cdot}EVAL$), *polynomial evaluation on a set of points* ([AHU], [BM] or [Fi72]). For simplicity, let $K \leq 2^k \leq n+1$ be an integer power of 2. Denote that $m_i = x - x_i$, $i = 0, \ldots, K-1$. Our objective is to compute $p(x_i) = p(x) \bmod (x - x_i)$ for $i = 0, 1, \ldots, K-1$. Instead of the division of $p(x)$ by the K "moduli" $m_i = x - x_i$, we will recursively divide $p(x)$ by fewer "supermoduli" of higher degrees, that is, by appropriate products of m_i. Specifically, we will first compute the coefficients of the "supermoduli" $m_i^{(j)} = m_{2i}^{(j-1)} m_{2i+1}^{(j-1)}$, for $i = 0, \ldots, \frac{K}{2^j} - 1$ and $j = 1, \ldots, k-1$, where $m_i^{(0)} = m_i$ for all i. Essential is this order of the computation, and such a computational scheme can be called the *fan-in method.* With fast polynomial multiplication [see (2.3)], the overall cost of this preprocessing stage is $O(M(K) \log K) = O(K \log^2 K)$ ops. Then we will compute $r_1(x) = p(x) \bmod m_0^{(k-1)}$, $r_2(x) = p(x) \bmod m_1^{(k-1)}$. Now, since $m_0^{(k-1)} = \prod_{i=0}^{K/2-1} m_i$ and $m_1^{(k-1)} = \prod_{i=K/2}^{K-1} m_i$, we have $p(x) \bmod m_i = r_1(x) \bmod m_i$ for $i = 0, \ldots, K/2 - 1$, $p(x) \bmod m_i = r_2(x) \bmod m_i$ for $i = K/2, \ldots, K-1$. This is another demonstration of the divide-and-conquer method: the original problem has been reduced to two problems of half size by means of two polynomial divisions. Continuing recursively, we solve Problem 2.2 ($POL \cdot EVAL$) in k steps by using $O(D(K) \log K + D(n)) = O(K \log^2 K + n \log n)$ ops [compare (2.3)]. The latter divide-and-conquer scheme can be called the *fan-out method.* This algorithm can be applied where $n + 1 < K$ too. In this case it involves $O(K \log^2 n)$ ops, since we may reduce the original problem to $\lceil K/(n+1) \rceil$ subproblems of the evaluation of $p(x)$ on a set of $n+1$ points. ∎

Fast solution of Problem 2.3 ($POL \cdot INTERP$), *polynomial interpolation* ([Fi87]). Recall the Lagrange interpolation formula,

$$p(x) = L(x) \sum_{i=0}^{n} r_i w_i / (x - x_i) \tag{4.1}$$

where r_i and x_i are the input values,

$$L(x) = \prod_{i=0}^{n}(x - x_i),\ w_i = 1/\prod_{k=1,\ k\neq i}^{n}(x_i - x_k) = 1/L'(x_i),\ i = 0, 1, \dots, n\ , \tag{4.2}$$

and $L'(x)$ is the formal derivative of $L(x)$ (see [Hen82], p. 238). Then proceed as follows:

1) compute the coefficients of the polynomials $L(x)$ and $L'(x)$,
2) compute the values $L'(x)$ and $1/L'(x)$ at the points x_i for all i,
3) recursively sum all the partial fractions $r_i w_i/(x - x_i)$ together.

$L(x)$ is the denominator of the output sum, whereas $p(x)$ is its numerator. Thus we employed *manipulation with partial fractions* as a means of polynomial computations. To keep the cost of the computations lower, order the summation of the partial fractions according to the fan-in method, that is, first pairwise sum the original partial fractions and then recursively sum the pairs of the resulting sums, which are partial fractions themselves. At Stage 1 we multiply the "moduli" $m_i = x - x_i$ together by using the fan-in method again [as in the above solution of Problem 2.2 $(POL \cdot EVAL)$], at the cost of $O(M(n)\log n) = O(n\log^2 n)$ ops. Stage 2 is reduced to Problem 2.2 $(POL \cdot EVAL)$ of polynomial evaluation for $L'(x)$, so its cost is $O(M(n)\log n) = O(n\log^2 n)$ ops. The computation at Stage 1 by the fan-in method also gives us all the denominators required at Stage 3 (at no additional cost). Computing the numerator of the sum of each pair of partial fractions summed in Stage 3 is reduced to two polynomial multiplications, and the overall cost of Stage 3 is $O(M(n)\log n) = O(n\log^2 n)$. ■

If the input set is the Fourier set, $\{x_j = \omega^j,\ j = 0, 1, \dots, K-1\}$, ω being a primitive K-th root of 1, the above algorithms for Problems 2.2 $(POL \cdot EVAL)$ and 2.3 $(POL \cdot INTERP)$ of polynomial evaluation and interpolation can be greatly simplified and reduced to FFT. This is an alternative derivation of FFT, which solves Problems 2.2a (DFT), 2.3a $(IDFT)$. Specifically, forming the "supermoduli" in this case, we should group the nodes $x_j = \omega^j$, so that all the "supermoduli" remain binomials of the form $x^H - \omega^{jH}$, $H = 2^h$. Then we have the coefficients of the "supermoduli" cost-free. Furthermore, the division by such binomials is much simpler than by the general polynomials, so that the above solution of Problem 2.2 $(POL \cdot EVAL)$ amounts to forward FFT. Similarly, the algorithm for Problem 2.3 $(POL \cdot INTERP)$ of interpolation is simplified in the case of the Fourier set of nodes $\{x_j\}$. Then the summation of partial fractions takes the form

$$\frac{g(x)}{x^H - \omega^{jH}} + \frac{g^*(x)}{x^H + \omega^{jH}} = \frac{(g(x) + g^*(x))x^H + (g(x) - g^*(x))\omega^{jH}}{x^{2H} - \omega^{2jH}},$$

where $g(x)$ and $g^*(x)$ are polynomials of degree less than H, and computing the numerators amounts to the inverse FFT.

Problem 2.2 ($POL \cdot EVAL$) and its fast solution are immediately extended to the following problem and to its solution using $O(K \log K \log k)$ ops:

Problem 4.1 ($POL{\cdot}MODULI$), *modular representation of a polynomial. Given the coefficients of polynomials* $p(x), m_0(x), \ldots, m_k(x)$, *compute the polynomials* $p(x) \bmod m_i(x)$ *for all* i.

It is instructive to extend to integers both this problem and its solution (see Remark 4.2), as well as the following problem, originated in China centuries ago (see [Kn]) and being both an important extension of Problem 2.3 ($POL{\cdot}INTERP$) of polynomial evaluation and the converse to Problem 4.1 ($POL{\cdot}MODULI$) under the natural assumption that the "supermoduli" are pairwise relatively prime:

Problem 4.2 ($CH \cdot REMNDR$), *Chinese remainder computation* ([AHU], [BM], [Kn]). *Given the coefficients of polynomials* $m_i(x), r_i(x)$, $i = 0, \ldots, k$, *where* $\deg m_i(x) = N_i > \deg r_i(x)$ *and where the polynomials* $m_i(x)$ *are pairwise relatively prime (that is, pairwise have only constant common divisors), compute the unique polynomial* $p(x) \bmod L(x)$, *such that* $r_i(x) = p(x) \bmod m_i(x)$, $i = 0, 1, \ldots, k$, $L(x) = \prod_{i=0}^{k} m_i(x)$.

Problem 2.3 ($POL{\cdot}INTERP$) is the special case of Problem 4.2 ($CH \cdot REMNDR$) where $m_i(x) = x - x_i$.

Fact 4.1. *There always exists a unique solution to Problem 4.2 ($CH \cdot REMNDR$).*

A fast algorithm for Problem 4.2 ($CH \cdot REMNDR$) is shown below. It involves the following auxiliary problem (which we must not solve in the case of interpolation):

Problem 4.3 ($MOD{\cdot}POL{\cdot}RECIPR$). *Compute the reciprocal* $w(x)$ $\bmod\ m(x)$ *of a polynomial* $v(x)$ *such that* $w(x)v(x) = 1 \bmod m(x)$, *provided that the polynomials* $v(x)$ *and* $m(x)$ *are relatively prime.*

Solution. The solution for any polynomial $v(x)$ and $m(x)$ is shown in Remark 5.1 in the next section. It involves the extended Euclidean algorithm (see Algorithm 5.1b) and costs $O(M(k)+M(n)\log n) = O(k\log k+n\log^2 n)$ ops, $k = \deg v(x)$, $n = \deg m(x)$. ■

Observe that Problem 3.4 $(POL \cdot RECIPR)$ is the simple special case where $m(x) = x^K$. For the interpolation, we only need the case where $m(x) = x - a$; in this case the solution is trivial: $w(x) = 1/v(a)$.

Solution of Problem 4.2 $(CH \cdot REMNDR)$. Extend the fast solution of Problem 2.3 $(POL \cdot INTERP)$, with $m_i(x)$ replacing the "moduli" $x-x_i$. The Lagrange interpolation formula (4.1), (4.2) is extended as follows:

$$p(x) = \sum_{i=0}^{k} (r_i(x)w_i(x) \bmod m_i(x))L(x)/m_i(x), \tag{4.3}$$

where the polynomials $w_i(x)$ satisfy the equations

$$v_i(x)w_i(x) = 1 \bmod m_i(x), \tag{4.4}$$

$$v_i(x) = \prod_{h\neq i} m_h(x) \bmod m_i(x) = (L(x)/m_i(x)) \bmod m_i(x). \tag{4.5}$$

The latter equations have solutions because the polynomials $m_i(x)$ are pairwise relatively prime.

Now proceed as follows:

1) compute the coefficients of the polynomial $L(x)$,
2) compute the coefficients of the polynomials $w_i(x)$, $i = 0, \ldots, k$,
3) recursively sum together all the partial fractions

$$\frac{r_i(x)w_i(x) \bmod m_i(x)}{m_i(x)}$$

and output the numerator of the resulting partial fraction [note the similarity with the manipulation with partial fractions in the fast solution of Problem 2.3 $(POL \cdot INTERP)$].

Compared with interpolation, computation of the coefficients of the polynomials $w_i(x)$ of (4.4) at Stage 2 is a little harder, but it can be performed in two substages. At the first substage, we evaluate the polynomials $v_i(x)$ of (4.5) according to the following scheme: First, the polynomials $L(x) \bmod m_i^2(x) = v_i(x)m_i(x)$ of degrees less than $2N_i$ are computed for all i; here we proceed similarly to the solution of Problem 2.2 $(POL \cdot EVAL)$ of polynomial evaluation; in particular, we use the fan-in method. Then we divide the polynomials $v_i(x)\ m_i(x)$ by $m_i(x)$ in order to recover the polynomials $v_i(x)$.

We may use any algorithm for Problem 3.1 ($POL \cdot DIVIDE$), for which the remainders vanish. In particular, these polynomial divisions may be effectively performed by means of the evaluation of the polynomials $v_i(x)m_i(x)$ and $m_i(x)$ on the K_i Fourier points, division of the results by each other for all i, and the subsequent interpolation, provided that $K_i > \deg v_i(x)$. At the second substage, at the cost $O(\sum_{i=0}^{k} N_i \log^2 N_i)$, we arrive at the polynomials $w_i(x)$ by solving Problem 4.3 ($MOD \cdot POL \cdot RECIPR$), where $w(x) = w_i(x)$, $v(x) = v_i(x)$, $m(x) = m_i(x)$, $i = 0, \ldots, k$.

The recursive summation of all the partial fractions is performed as in the solution of Problem 2.3 ($POL \cdot INTERP$), which is the special case of Problem 4.2 ($CH \cdot REMNDR$) where $m_i(x) = x - x_i$, for all i, that is, we recursively sum the consecutive pairs of the partial fractions: we first sum $\frac{(r_i(x)w_i(x) \bmod m_i(x))}{m_i(x)}$, for $i = 0$ and $i = 1$, for $i = 2$ and $i = 3$, and so on, and then we recursively sum the consecutive pairs of the resulting sums, which are partial fractions themselves. If $K = \sum_{i=0}^{k} N_i$ denotes the degree of $L(x)$, then the overall cost of the solution of Problem 4.2 ($CH \cdot REMNDR$) is $O(M(K)\log k) = O(K \log K \log k)$. ∎

Remark 4.1. In the special case of Problem 4.2 ($CH \cdot REMNDR$) where $m_i(x) = x - x_i$, $r_i(x) = r_i$, we have $w_i(x) = \dfrac{1}{\prod_{j \neq i}(x_i - x_j)} = w_i$, and we arrive at the Lagrange original formula (4.1), (4.2).

Remark 4.2. Fact 4.1, Problems 4.1 ($POL \cdot MODULI$) and 4.2 ($CH \cdot REMNDR$), their solution algorithms and the complexity estimates can be extended from the case of polynomials to the case of binary integers, so that the precision $d = d(i)$ of the input integers replaces the degrees N_i of the input polynomials and, in the transition from the number of ops for polynomials to the Boolean circuit size estimates for integers, the value $\mu(\sum_i d(i))$ for μ of (2.4) replaces $M(\sum_i N_i)$ ops for M of (2.3).

We will now consider some special cases and extensions of Problem 4.2 ($CH \cdot REMNDR$).

Problem 4.2a ($CH \cdot REMNDR1$), *Chinese remainder computation modulo powers. Solve Problem 4.2 ($CH \cdot REMNDR$) in the special case where $m_i(x) = (g_i(x))^{n_i}$ and natural n_i and the coefficients of the polynomials $g_i(x)$ are given for $i = 0, 1, \ldots, k$, so that $N_i = \deg m_i(x) = n_i \deg g_i(x)$.*

Solution. The above algorithm for Problem 4.2 ($CH \cdot REMNDR$) applies, because we may immediately compute the coefficients of $m_i(x)$, for instance, by means of the evaluation-interpolation algorithm. However, its Stage 2, that is, the stage of the evaluation of the polynomials $w_i(x) = 1/v_i(x) \bmod m_i(x)$ of (4.4), can now be reduced to evaluating the polynomials $1/v_i(x) \bmod g_i(x)$. This will simplify the auxiliary Problem 4.3 ($MOD \cdot POL \cdot RECIPR$) and thus will decrease the overall computational cost of the solution, because $\deg g_i(x) = (\deg m_i(x))/n_i$. Specifically, we are given $v_i(x) = (L(x)/m_i(x)) \bmod m_i(x)$, and we evaluate the coefficients of $w_i(x)$ and $g_i(x)$ and the powers n_i as follows:

1) Divide $v_i(x)$ by $g_i(x)$ to represent $v_i(x) = v_{i0}(x) + g_i(x)v_{i1}(x)$, where $\deg v_{i0}(x) < \deg g_i(x)$; then solve Problem 4.3 ($MOD \cdot POL \cdot RECIPR$), for $v(x) = v_{i0}(x)$, $m(x) = g_i(x)$ and $w(x) = w_{i0}(x)$, that is, compute the polynomial $w_{i0}(x) = 1/v_{i0}(x) \bmod g_i(x)$ [recall that $v_i(x)$ and $m_i(x)$ are mutually prime and, therefore, $v_{i0}(x)$ and $g_i(x)$ are mutually prime].
2) Compute the coefficients of $s_i(x) = 1 - v_i(x)w_{i0}(x)$ and of the polynomial $z_i(x) = 1/(v_i(x)w_{i0}(x)) \bmod m_i(x) = 1/(1 - s_i(x)) \bmod m_i(x)$ by means of the formula:

$$z_i(x) = \sum_{j=0}^{n_i-1} s_i(x)^j \bmod m_i(x) = \prod_{h=1}^{\lceil \log n_i \rceil} (1 + s_i(x)^{2^h}) \bmod m_i(x)$$

[we use the identity $1/(1-s) = 1+s+s^2+\cdots = (1+s)(1+s^2)(1+s^4)\cdots$ and the relations $1 - v_i(x)w_{i0}(x) = 0 \bmod g_i(x)$, $m_i(x) = g_i(x)^{n_i}$].
3) Compute $w_i(x) = w_{i0}(x)z_i(x) \bmod m_i(x)$. ■

Problem 4.2b ($H \cdot INTERP$), *Hermite interpolation* ([G86a] and [GY]). *Solve Problem 4.2a ($CH \cdot REMNDR1$) in the special case where* $m_i(x) = (x - x_i)^{n_i}$, $g_i(x) = x - x_i$.

Solution. The above solution of Problem 4.2a ($CH \cdot REMNDR1$) applies. In this special case, the polynomials $r_i(x)$ can be expressed either as $r_i(x) = \sum_{j=0}^{n_i-1} r_{ij}x^j$ or as $r_i(x) = \sum_{j=0}^{n_i-1} r^*_{ij}(x - x_i)^j$. The transition between these two representations of $r_i(x)$ amounts to Problem 2.6 ($VAR \cdot SHIFT$) of the shift of the variable. The auxiliary Problem 4.3 ($MOD \cdot POL \cdot RECIPR$) degenerates to the evaluation of $1/v_i(x)$ at $x = x_i$. ■

Problem 2.3 ($POL \cdot INTERP$) is the special case of Problem 4.2b ($H \cdot INTERP$) where $n_i = 1$ for all i.

Problem 4.2c ($PART \cdot FRACT$), *partial fraction decomposition* ([G86a]). *Given polynomials $m_0(x), \ldots, m_k(x)$ (pairwise relatively prime) and $p(x)$ such that $L(x) = \prod_{i=0}^k m_i(x)$, $\deg p(x) < \deg L(x)$ (the latter degree bound can be relaxed by means of the division of $p(x)$ by $L(x)$ with* *a remainder), compute the coefficients of all the polynomials of the unique set* $\{s_1(x), \ldots, s_k(x)\}$ *such that*

$$\frac{p(x)}{L(x)} = \sum_{i=1}^{k} \frac{s_i(x)}{m_i(x)},$$

$\deg s_i(x) < \deg m_i(x)$ *for all* i.

Solution. The solution follows the pattern of the solution of Problem 4.2 ($CH \cdot REMNDR$) and has the same asymptotic cost bound. Specifically, first compute $r_i(x) = p(x) \bmod m_i(x)$, for $i = 0, \ldots, k$, so that

$$r_i(x) = L(x)s_i(x)/m_i(x) \bmod m_i(x).$$

Then successively compute the coefficients of the polynomials

$$L(x), \quad v_i(x) = L(x)/m_i(x) \bmod m_i(x), \quad w_i(x) \bmod m_i(x)$$

such that $w_i(x)v_i(x) = 1 \bmod m_i(x)$ [Problem 4.3 ($MOD \cdot POL \cdot RECIPR$)], and finally compute $s_i(x) = w_i(x)r_i(x) \bmod m_i(x)$. ■

In particular, if $m_i(x) = (g_i(x))^{n_i}$, then continuing the above computation, we may solve Problem 3.5 ($TAYLOR \cdot EXP$) of generalized Taylor expansion for $u(x) = s_i(x)$, $t(x) = g_i(x)$ and finally compute the coefficients of the polynomials $s_{ij}(x)$ such that $s_i(x)/(g_i(x))^{n_i} = \sum_{j=0}^{n_i} s_{ij}(x)/(g_i(x))^j$, $\deg s_{ij}(x) < \deg g_i(x)$ for all i. This would define a more refined partial fraction decomposition of $p(x)/L(x)$. Note also the special case where $g_i(x) = x - x_i$ for all i.

By solving Problems 2.6 ($VAR \cdot SHIFT$), 3.5 ($TAYLOR \cdot EXP$) and 4.2 ($CH \cdot REMNDR$), we may arrive at modular representation of the same polynomial $p(x)$ with different basis sets of the moduli $m_i(x)$, in particular, at the generalized Taylor expansion for $m_i(x) = (g_i(x))^{n_i}$, $i = 0, 1, \ldots, k$, which turns into the representation (2.1) if $m_0(x) = x$, $n_0 = n + 1$, $k = 0$.

Our next problem is related to Hermite interpolation [Problem 4.2b ($H \cdot INTERP$)] where the polynomials $r_i(x)$ are identically 0 for all i; actually, we only require to compute the product $L(x) = \prod_i m_i(x)$ in this case; furthermore, we assume that $m_i(x) = x - x_i$, $i = 1, \ldots, n$, for a given set $\{x_1, \ldots, x_n\}$.

Problem 4.2d ($SYMM \cdot FUNCT$), *computing the elementary symmetric functions. Given the set* $\{x_1, \ldots, x_n\}$ *of cardinality* n, *compute the elementary symmetric functions,*

$$s_i = \sigma_i(x_1, \ldots, x_n) = \sum_{j_1 < \ldots < j_i} \prod_{h=1}^{i} x_{j_h},$$

such that

$$x^n - s_1 x^{n-1} + \cdots + (-1)^n s_n = \prod_{j=1}^{n} (x - x_j).$$

Solution. The problem can be immediately solved by means of the fan-in method applied to the evaluation of the product $\prod_j (x - x_j)$. ∎

Here is the converse problem:

Problem 4.4 ($POL \cdot ZEROS$), *computing polynomial zeros. Given the set $\{p_0, p_1, \ldots, p_n\}$, $p_n \neq 0$, compute the zeros $x_1, \ldots, x_n$ of the polynomial*

$$p(x) = \sum_{i=0}^{n} p_i x^i = p_n \prod_{j=1}^{n} (x - x_j). \tag{4.6}$$

The problem always has a unique solution $x_1, \ldots, x_n$ over the field of complex numbers. In this solution not all of the zeros must be distinct, so there are n zeros of $p(x)$, counted with their *multiplicities.* There is, however, no rational solution algorithm, that is, no algorithm involving only a finite number of arithmetic operations, for $n > 1$. We will study approximation algorithms for this problem in chapter 7.

We will continue this section with the problems of composition of polynomials and formal power series, for which we will use FFT, the solutions to Problems 2.2 ($POL \cdot EVAL$) and 3.1 ($POL \cdot DIVIDE$) of polynomial evaluation and division, and the evaluation-interpolation method.

Problem 4.5 ($POL \cdot COMP$), *composition of polynomials. Given the coefficients of two polynomials $s(x)$ (of degree m) and $t(x)$ (of degree n), compute the coefficients of the polynomial $w(x) = s(t(x))$.*

Solution. First apply FFT in order to compute the values $t(x)$ on the $K = 2^k$ Fourier points $x = \omega^h$ for $h = 0, 1, \ldots, K-1$, where ω is a primitive K-th root of 1, $K = 2^{\lceil \log(mn+1) \rceil}$. Then compute $w(\omega^h) = s(t(\omega^h))$ for $h = 0, 1, \ldots, K-1$, that is, solve Problem 2.2 ($POL \cdot EVAL$) of the evaluation of the polynomial $s(t)$ on the set of points $t = t(\omega^h)$. Finally, interpolate to $w(x)$ by applying the inverse FFT. The overall cost of the solution is $O(mn(\log(mn) + \log^2 m))$ ops [compare the fast solution of Problem 2.2 ($POL \cdot EVAL$)]. ∎

The above solution can be immediately extended to the recursive composition of several polynomials.

Problem 4.6 ($SER \cdot COMP$), *composition of power series* ([BKu] or [Kn], p. 657). *Given a natural n and two formal power series $s(t)$ and $t(x)$*

such that $t(0) = 0$, *compute the polynomial* $w(x) = s(t(x)) \bmod x^{n+1}$. *Note that we may replace the power series by polynomials, that is, replace* $t(x)$ *by* $t(x)$ *modulo* x^{n+1} *and* $s(t)$ *by* $s(t)$ *modulo* t^{n+1} *due to the assumption that* $t(0) = 0$.

Solutions. Problem 4.6 ($SER \cdot COMP$) is a special case of Problem 4.5 ($POL \cdot COMP$), whose solution involves $O(n^2 \log^2 n)$ ops. Since we now compute modulo x^{n+1}, we decrease the latter bound to $O((n \log n)^{3/2})$ as follows: Fix an integer h of the order of $\sqrt{n/\log n}$. Represent $t(x)$ as the sum $t_0(x) + x^h t_1(x)$ where $t_0(x) = t(x) \bmod x^h$. Write the generalized Taylor expansion

$$s(t(x)) = s(t_0(x)) + s'(t_0(x))x^h t_1(x) + s''(t_0(x))(x^h t_1(x))^2/2! + \cdots + s^{(g)}(t_0(x))(x^h t_1(x))^g/g!$$

where $g = \lceil n/h \rceil$ and $s^{(g)}(t)$ denotes the derivative of $s(t)$ of the order g. Evaluate the coefficients of the polynomial $s(t_0(x))$ by solving Problem 4.5 ($POL \cdot COMP$) at the cost $O(nh \log^2 n) = O((n \log n)^{3/2})$. Then compute the polynomials $(1/j!)s^{(j)}(t_0(x)) \bmod x^{n-jh+1}$ for $j = 1, 2, \ldots, g$, by means of the recursive differentiation of the polynomial $s(t_0(x))$ in x, with the j-th differentiation step followed by the division of the resulting derivative by $jt_0'(x)$ for $j = 1, 2, \ldots, g$. Then multiply modulo x^{n+1} the results by the precomputed powers of $x^h t_1(x)$ and sum all the products. This amounts to $O(n/h) = O(\sqrt{n \log n})$ multiplications of polynomials modulo x^{n+1}, so the overall cost is $O((n \log n)^{3/2})$. ∎

Some special cases are listed below:

1) $s(t) = \ln(1+t) = t - \frac{t^2}{2} + \frac{t^3}{3} - \frac{t^4}{4} + \cdots$, ln denoting the natural logarithm to the exponential base, so that $s(t(x)) = \ln(1 + \sum_{i=1}^{+\infty} t_i x^i)$ for $t(x) = \sum_{i=1}^{+\infty} t_i x^i$.
2) $s(t) = \exp(t) = 1 + t + \frac{t^2}{2!} + \frac{t^3}{3!} + \cdots$, so that $s(t(x)) = \exp(t(x))$.

In both cases $O(n \log n)$ ops (with a small overhead constant) suffice in order to evaluate the coefficients of the polynomial $s(t(x)) \bmod x^n$ ([Bre76], [BB], [Kn], p. 514, exercise 4).

Another important representation of a monic polynomial can be obtained in terms of the *power sums* of its zeros. Let $p(x)$ be the polynomial defined in (4.6) where $p_n = 1$ and define the power sums

$$s_k = \sum_{j=1}^{n} x_j^k, \quad k = 0, \pm 1, \pm 2, \ldots . \tag{4.7}$$

Problem 4.7 (*POWER · SUMS*). *Given the coefficients $p_0, p_1, \dots, p_{n-1}$ of the monic polynomial $p(x)$ of (4.6) and an integer m, compute the power sums s_k, $k = 1, 2, \dots, m$, of (4.7).*

Solution. The coefficients of $p(x)$ and the power sums $s_1, s_2, \dots, s_m$ satisfy the following linear system of *Newton's identities* ([H70]):

$$s_k + \sum_{i=1}^{k-1} p_{n-i} s_{k-i} + k p_{n-k} = 0, \quad k = 1, 2, \dots, n, \tag{4.8}$$

$$s_{n+k} + \sum_{i=1}^{n} p_{n-i} s_{n+k-i} = 0, \quad k = 1, 2, \dots, m-n, \tag{4.9}$$

where the sum in (4.8) vanishes if $k = 1$. [Note that the equations (4.9) for $p(x)$ are among the equations (4.8) for $x^{m-n} p(x)$.] Therefore, the power sums can be computed by solving an $m \times m$ linear system of equations. Due to its special structure, this system can be solved in $O(m \log m)$ ops (see Proposition 2.6.1). ∎

The converse problem has the same complexity bound.

Problem 4.8 (*I·POWER·SUMS*). *Given the power sums $s_1, s_2, \dots, s_n$, compute the coefficients $p_0, p_1, \dots, p_{n-1}$ of the monic polynomial $p(x)$ of (4.6) such that (4.7) holds.*

Solution (compare [Sc82], [P90], [BP93]). Denote the *reverse polynomial* $x^n p(x^{-1}) = \sum_{i=0}^{n} p_{n-i} x^i$ of $p(x)$ as follows:

$$x^n p(x^{-1}) = 1 + g(x) = 1 + \sum_{i=1}^{n} p_{n-i} x^i = \prod_{j=1}^{n} (1 - x x_j),$$

and deduce that

$$\begin{aligned} (\ln(1 + g(x)))' &= g'(x)/(1 + g(x)) = \\ -\sum_{j=1}^{n} x_j/(1 - x x_j) &= -\sum_{j=1}^{k} s_j x^{j-1} \bmod x^k \end{aligned} \tag{4.10}$$

for all $k \le n + 1$. Suppose that the polynomial $g_r(x) = g(x) \bmod x^{r+1}$ is available and that we want to compute $g_{2r}(x) = g(x) \bmod x^{2r+1}$. Observe that the polynomials $g_1(x) = p_{n-1} x$ and $g_2(x) = p_{n-1} x + p_{n-2} x^2$ are readily available, for $p_{n-1} = -s_1$, $2p_{n-2} = -s_2 + s_1^2$. We will express $g_{2r}(x)$ as follows:

$$1 + g_{2r}(x) = (1 + g_r(x))(1 + h_r(x)) \bmod x^{2r+1}, \tag{4.11}$$

where

$$h_r(x) = h_{r+1}x^{r+1} + \cdots + h_{2r}x^{2r} \tag{4.12}$$

is an unknown auxiliary polynomial. We observe the two following identities [the first of them is immediately implied by the equation (4.12) and the second by the two equations (4.11) and (4.12)]:

$$h_r'(x)/(1+h_r(x)) = h_r'(x) \bmod x^{2r+1},$$

$$(\ln(1+g_{2r}(x)))' = \frac{g_{2r}'(x)}{1+g_{2r}(x)} = \frac{g_r'(x)}{1+g_r(x)} + \frac{h_r'(x)}{1+h_r(x)} \bmod x^{2r}.$$

Combining these identities with the equations (4.10) for $k = 2r+1$, we obtain that

$$\frac{g_r'(x)}{1+g_r(x)} + h_r'(x) = -\sum_{j=1}^{2r} s_j x^{j-1} \bmod x^{2r}. \tag{4.13}$$

We assumed that we know the coefficients of $g_r(x)$, as well as the values s_j for all $j \le 2r$. Now, we compute the polynomial $1/(1+g_r(x)) \bmod x^{2r}$, multiply it by $g_r'(x) \bmod x^{2r}$, subtract the result from the right side of the equation (4.13) to obtain $h_r'(x)$, and then compute $h_r(x)$ by using the equation (4.12) and $g_{2r}(x)$ by using the equation (4.11). We start this recursive evaluation of $g_{2r}(x)$ with $r = 1$ or $r = 2$, for the polynomials $g_1(x)$ and $g_2(x)$ are readily available. Then we recursively compute the polynomials $g_4(x)$, $g_8(x)$, $g_{16}(x)$, and so on, until this procedure yields all the coefficients of the polynomial $p(x)$. We will recursively use each reciprocal polynomial $1/(1+g(x)) \bmod x^r = 1/(1+g_r(x)) \bmod x^r$ when we compute modulo x^{2r} the reciprocal of $1+g(x)$ in the next iteration, so the overall cost of the computation is $O(n \log n)$. ∎

Remark 4.3. The reciprocals $1/x_j$ are the zeros of the reverse polynomial $q(z) = p_0 z^n + p_1 z^{n-1} + \cdots + p_n$, so that Newton's identities for $q(z)$ relate $s_0, s_{-1}, \ldots, s_{-m}$ to $p_0, p_1, \ldots, p_n$, provided that $p_0 \neq 0$.

Remark 4.4. We have observed several ways of representing a polynomial: by its coefficients, by its zeros, by their power sums, by its values on a fixed set of points, or more generally, by a set of the remainders of its division by polynomials of a fixed set. Several algorithms that we studied can be unified as algorithms for the transition from one such representation to another (see [G86a]). There are further important extensions of this general problem (see [Ka87], [Ka88], [KaT]), including, in particular, the interpolation problems for sparse multivariate polynomials (see section 9, as well as [B-OT], [GKS], [KLW]).

5. Polynomial GCD and LCM, Extended Euclidean Algorithm, Rational Interpolation and Padé Approximation.

In this section we will solve the problem of computing the *greatest common divisor* (the *gcd*) of two polynomials, having numerous applications. The solution relies on the *Euclidean algorithm*, which is a major general tool for many algebraic and numerical computations and, historically, is the oldest nontrivial algorithm still in use. We will first present its slow classical version, then its fast version, and then its extended version, which, in addition to the gcd, computes some other useful polynomials such as the reciprocal of a polynomial modulo a polynomial [see Problem 4.3 ($MOD \cdot POL \cdot RECIPR$)]. Then we will show further applications to computing the *least common multiple (lcm)* of two or several polynomials, to *rational interpolation*, which extends Problem 2.3 ($POL \cdot INTERP$) of polynomial interpolation, and to other computations. We ultimately reduce all the problems to polynomial multiplications and thus arrive at the fast solutions. The gcd problem and its solution algorithms can be extended to integers, and the reader may enjoy in working out the details (see Remark 5.2 and [HW79]). In addition to our exposition, see [AHU], [BGY], [BT], [HW79], [Kn], [McES], [Schw] and [G84] on computing the gcd and the other entries of the extended Euclidean scheme.

Problem 5.1 ($POL \cdot GCD$), *gcd computation. Given the coefficients of two polynomials*

$$u(x) = \sum_{i=0}^{m} u_i x^i, \quad v(x) = \sum_{j=0}^{n} v_j x^j, \quad m \geq n \geq 0, \quad u_m v_n \neq 0, \tag{5.1}$$

compute $\gcd(u(x), v(x))$, *that is, a common divisor of* $u(x)$ *and* $v(x)$ *having the highest degree in* x. *The gcd of* $u(x)$ *and* $v(x)$ *is unique up to within constant factors, or it can be assumed monic and unique.*

Solutions. A $\gcd(u(x), v(x))$ can be immediately obtained if a complete factorization of an input polynomial, say, of $u(x)$, is available over a field containing all the coefficients of $u(x)$ and $v(x)$: $\gcd(u(x), v(x))$ is just the product of all the monic factors in this factorization of $u(x)$ that divide $v(x)$. On the other hand, the polynomial $p(x)/\gcd(p(x), p'(x))$ has exactly the same zeros as $p(x)$, but all of them are simple, so that the general problem of polynomial factorization can be narrowed respectively. The $\gcd(u(x), v(x))$ can be computed *without factorization* of $u(x)$ and $v(x)$, by using Algorithm 5.1.

Algorithm 5.1, *Euclid's Algorithm.*

Input: natural numbers m and n and the coefficients $u_0, \dots, u_m$, $v_0, \dots, v_n$ of two polynomials $u(x) = \sum_{j=0}^{m} u_j x^j$, $v(x) = \sum_{k=0}^{n} v_k x^k$, of degrees m and n, respectively.

Output: the coefficients of the polynomial $\gcd(u(x), v(x))$.

Computation:

1. Set $u_0(x) = u(x)$, $v_0(x) = v(x)$.

2. Compute

$$u_{i+1}(x) = v_i(x),$$
$$v_{i+1}(x) = u_i(x) \bmod v_i(x) = u_i(x) - q_{i+1}(x)v_i(x), \; i = 0, 1, \dots, \ell - 1, \tag{5.2}$$

where $q_{i+1}(x)$ is the quotient polynomial, and ℓ is such that $v_\ell(x) \equiv 0$.

3. Output $u_\ell(x) = \gcd(u(x), v(x))$. ∎

The correctness of the algorithm can be deduced from the following equation:

$$\gcd(u_i(x), v_i(x)) = \gcd(u_{i+1}(x), v_{i+1}(x)),$$

which holds for all i. Observe that each stage i is reduced to the division of $u_i(x)$ by $v_i(x)$ (with a remainder) and that $\ell \le n + 1$, that is, there can be at most $n + 1$ stages of this algorithm, since $\deg v_{i+1}(x) < \deg v_i(x)$.

Hereafter assume that $m = n$ [otherwise replace $u(x)$ or $v(x)$ by $u(x) + v(x)$]. The algorithm involves $O(n^2)$ ops (see exercise 16), but we will present its faster version (compare [AHU, pp. 303–308], [Mo], [BGY] and [Schw]), which only requires $O(n \log^2 n)$ ops. This fast version exploits matrix representation of the recurrence (5.2), and the desired acceleration of the original algorithm is achieved because, for the smaller values of i, the quotients $q_{i+1}(x)$ of (5.2) only depend on relatively few leading terms of the dividend $u_i(x)$ and the divisor $v_i(x)$. Specifically, we will first rewrite the recursive equations (5.2) in the form of vector and matrix equations:

$$\begin{pmatrix} u_i(x) \\ v_i(x) \end{pmatrix} = \tilde{Q}_i(x) \begin{pmatrix} u_{i-1}(x) \\ v_{i-1}(x) \end{pmatrix} = \hat{Q}_i(x) \begin{pmatrix} u(x) \\ v(x) \end{pmatrix}, \tilde{Q}_i(x) = \begin{pmatrix} 0 & 1 \\ 1 & -q_i(x) \end{pmatrix}, \tag{5.3}$$

$$\hat{Q}_i(x) = \tilde{Q}_i(x)\tilde{Q}_{i-1}(x) \cdots \tilde{Q}_1(x), \; i = 1, \dots, \ell, \quad \ell \le n + 1. \tag{5.4}$$

Due to this *matrix representation of Euclid's algorithm*, the computation of $\gcd(u(x), v(x))$ has been reduced to computing the first row of the matrix $\hat{Q}_\ell(x)$ and to multiplication of its first row by the vector $[u(x), v(x)]^T$, that is, to two polynomial multiplications. Given the matrices $\tilde{Q}_i(x)$, $i =$

$1, \dots, \ell$, the matrix $\hat{Q}_\ell(x)$ can be recursively evaluated in $\lceil \log \ell \rceil \le \lceil \log(n+1) \rceil$ recursive steps of 2×2 matrix multiplications, by using (5.4) and the fan-in method, where we shall balance the degrees of the auxiliary products of the matrices $\tilde{Q}_j(x)$. Here, the degree of a matrix over polynomials is the maximum degree of its entries. Thus, we will extend the sequence of triples $\{[u_i(x), v_i(x), \hat{Q}_i(x)],\ i = 1, \dots, \ell\}$ by replacing the i-th triple by its $s(i)$ copies, that is, by repeating the i-th triple $s(i)$ times in the sequence, where $s(i) = \deg v_{i-1}(x) - \deg v_i(x)$ for $i = 1, \dots, \ell$, and we will let $\{[w_k(x), z_k(x), Q_k(x)],\ k = 1, \dots, n\}$ denote the resulting sequence of $n+1$ triples. Then

$$\begin{aligned} &[w_k(x), z_k(x)]^T = Q_k(x)[u(x), v(x)]^T, \\ &\deg Q_k(x) \le n - \deg w_k(x) \le k, \quad k = 1, \dots, n. \end{aligned} \tag{5.5}$$

Next, by inductive application of (5.2) for $i = 0, 1, \dots, \ell - 1$, we verify the *key observation: the matrix* $Q_k(x)$ *only depends on the* $2k$ *leading terms of each of the polynomials* $u(x)$ *and* $v(x)$. We assume that $n+1$ is a power of 2 [otherwise, replace $u(x)$ by $x^s u(x)$ and $v(x)$ by $x^s v(x)$ where $s + n + 1$ is a power of 2] and specify a fast version of Euclid's algorithm as follows:

Algorithm 5.1a.

Input: a natural number k and the coefficients of two polynomials $u(x)$ and $v(x)$ of degree $n = 2^k$.

Output: the coefficients of $\gcd(u(x), v(x))$ defined within a scalar factor.

Computation:

1. (Initialization.) Evaluate the matrix $Q_1(x) = \tilde{Q}_1(x)$. This involves just two leading coefficients u_m and v_n if $m = n$ or at most four leading coefficients, u_m, u_{m-1}, v_n and v_{n-1} if $m > n$.

2. Assume that a procedure for the evaluation of the matrix $Q_H(x)$ is available for $H = 2^h$ [for $h = 0$ such a procedure amounts to the division of $u(x)$ by $v(x)$] and recursively define a procedure for evaluating the matrix $Q_{2H}(x)$, for $h = 0, 1, \dots, \log(n+1)$, as follows:

a) Note that the entries of $Q_H(x)$ only depend on the $2H$ leading terms of each of the two polynomials $u(x)$ and $v(x)$. Keep these terms and also the H next terms of each polynomial $u(x)$ and $v(x)$ and drop all other terms when you apply the matrix equation (5.5) for $k = H$. Denote the resulting polynomials $u^*(x)$ and $v^*(x)$. Compute the matrix $Q_H(x)$ and then the pair of polynomials $[u_H^*(x), v_H^*(x)]^T = Q_H(x)[u^*(x), v^*(x)]^T$.

b) Apply step a) starting with the pair $[u^*_H(x), v^*_H(x)]$, rather than with the pair $[u(x), v(x)]$, in order to compute the matrix $Q^*_H(x)$. Note that the resulting matrix is the same as in the case where step a) is applied starting with the pair $[w_H(x), z_H(x)]$.

c) Compute the matrix $Q_{2H}(x) = Q^*_H(x)Q_H(x)$.

3. Compute $[w_n(x), z_n(x)]^T = Q_n(x)[u(x), v(x)]^T$ and output the polynomial $w_n(x)$, a scalar multiple of $\gcd(u(x), v(x))$.

The h-th recursive step of Stage 2 amounts to few multiplications and divisions of polynomials having at most $3 * 2^h$ nonzero terms (which are the leading terms, since we drop the other terms) and thus only requires $O(h2^h)$ ops, so that the overall cost of the computations is $O(n \log^2 n)$ ops.

Euclid's algorithm only involves arithmetic operations, so that the output coefficients of the gcd are rational functions in the input coefficients.

Problem 5.1a ($POL \cdot LCM$), *lcm computation. Given the coefficients of the two polynomials of (5.1), compute the coefficients of* $\operatorname{lcm}(u(x), v(x))$, *that is, of a common multiple of* $u(x)$ *and* $v(x)$ *having the minimum degree.*

Solution. Given $\gcd(u(x), v(x))$, we may immediately compute the least common multiple (lcm) of two polynomials $u(x)$ and $v(x)$ relying on the following equation:

$$\operatorname{lcm}(u(x), v(x)) = u(x)v(x)/\gcd(u(x), v(x)),$$

which also expresses the gcd through the lcm. ∎

We refer the reader to exercise 24, to Algorithms 5.3 and 5.4, and to sections 8 and 9 of chapter 2 on some other approaches to computing the gcd and lcm of two polynomials and to the extensions to the case of several polynomials.

Euclid's algorithm enables us to compute some other useful polynomials, in addition to the gcd and to the lcm.

Algorithm 5.1b *(Extended Euclidean Scheme).*

Input: natural numbers m and n, $m \geq n$, and the coefficients of the polynomials $u(x)$, $v(x)$ of (5.1).

Output: the coefficients of the polynomials $q_{i-1}(x)$, $r_i(x)$, $s_i(x)$ and $t_i(x)$, $i = 2, 3, \ldots, \ell$, where the integer ℓ and the polynomials are defined below.

Computation:

1. Set $s_0(x) = t_1(x) = 1$, $s_1(x) = t_0(x) = 0$, $r_0(x) = u(x)$, $r_1(x) = v(x)$.

2. For $i = 2, 3, \ldots, \ell + 1$, compute the coefficients of the polynomials $q_{i-1}(x)$, $r_i(x)$, $s_i(x)$, $t_i(x)$ by using the following relations:

$$r_i(x) = r_{i-2}(x) - q_{i-1}(x)r_{i-1}(x), \tag{5.6}$$

$$s_i(x) = s_{i-2}(x) - q_{i-1}(x)s_{i-1}(x), \tag{5.7}$$

$$t_i(x) = t_{i-2}(x) - q_{i-1}(x)t_{i-1}(x). \tag{5.8}$$

Here $\deg r_i(x) < \deg r_{i-1}(x)$, $i = 2, 3, \ldots, \ell$, that is, $r_i(x) = r_{i-2}(x) \bmod r_{i-1}(x)$, and ℓ is such that $r_{\ell+1} \equiv 0$, that is, $r_{\ell-1}(x) = q_\ell(x)r_\ell(x)$, and therefore, $r_\ell(x) = \gcd(u(x), v(x))$.

Note that the sequence $\{r_0(x), r_1(x), r_2(x), \ldots, r_\ell(x)\}$ consists of the polynomials $u_i(x)$, $v_i(x)$, $i = 1, 2, \ldots, \ell$, that appear in the above solution to Problem 5.1 ($POL \cdot GCD$). There are several important computations involving the polynomials $s_i(x)$ and $t_i(x)$ (see Remark 5.1 below and section 8 of chapter 2). Now we define **Problem 5.1b ($EXT \cdot EUCLID$)** as the problem of computing the polynomials r_i, s_i, t_i, q_{i-1}, $i = 2, \ldots, \ell+1$ of (5.6)-(5.8), generated in the extended Euclidean scheme for two given polynomials $r_0(x)$ and $r_1(x)$.

Fact 5.1. *The following relations hold:*

$$s_i(x)r_0(x) + t_i(x)r_1(x) = r_i(x), \quad i = 1, \ldots, \ell, \tag{5.9}$$

$$\deg r_{i+1}(x) < \deg r_i(x), \quad i = 1, \ldots, \ell - 1,$$

$$\deg s_{i+1}(x) > \deg s_i(x) = \deg (q_{i-1}(x)s_{i-1}(x)) =$$
$$\deg v(x) - \deg r_{i-1}(x), \quad i = 2, \ldots, \ell,$$

$$\deg t_{i+1}(x) > \deg t_i(x) = \deg (q_{i-1}(x)t_{i-1}(x)) =$$
$$\deg u(x) - \deg r_{i-1}(x), \quad i = 2, \ldots, \ell,$$

where we let $\deg s_{\ell+1}(x) = \deg t_{\ell+1}(x) = +\infty$.

This simple fact is deduced from (5.6)-(5.8) by means of induction on i.

In chapter 2 we will also use the following fact:

Fact 5.2 ([G84], [IKo]). *For $i = 1, \ldots, \ell$, the polynomials $s(x) = s_i(x)$ and $t(x) = t_i(x)$ represent a unique solution to the relation* $\deg (s(x)u(x) + t(x)v(x) - r_i(x)) < \deg r_i(x)$, *provided that* $\deg s(x) < \deg v(x) - \deg r_i(x)$, $\deg t(x) < \deg u(x) - \deg r_i(x)$.

The relations (5.6) imply that $q_i(x)$ are the quotients and $r_i(x)$ are the remainders of the polynomial divisions performed in the Euclidean algorithm, so that $\gcd(u(x), v(x))$ is a scalar multiple of $r_\ell(x)$. In some applications, (5.6) is replaced by its scaled version, where the scaled sequence $r_0(x), \ldots, r_i(x)$ is called *the pseudo remainder sequence* of polynomials. All these polynomials have integer coefficients if $u(x)$ and $v(x)$ have integer coefficients (see section 8 of chapter 2).

Remark 5.1, *on computing reciprocals modulo a polynomial [Problem 4.3 (MOD · POL · RECIPR)].* Given three polynomials $r_0(x)$, $r_1(x)$ and $r_i(x)$, such that $d_0 \geq d_1 > 0$, $d_h = \deg r_h(x) - \deg r_i(x)$, for $h = 0, 1$, the equation (5.9) is satisfied by the unique pair of polynomials $s_i(x)$, $t_i(x)$ such that $\deg s_i(x) < d_1$, $\deg t_i(x) < d_0$. In particular, three polynomials $r_0(x), r_1(x)$ and $r_\ell(x) = \gcd(r_0(x), r_1(x))$ define the unique pair $s_\ell(x), t_\ell(x)$ satisfying (5.9) for $i = \ell$ and such that $\deg s_\ell(x) < \deg\,(r_1(x)/r_\ell(x))$, $\deg t_\ell(x) < \deg\,(r_0(x)/r_\ell(x))$. If $\gcd(r_0(x), r_1(x)) = 1$, then (5.9) defines the unique polynomial $t_\ell(x) = 1/r_1(x) \mod r_0(x)$, which gives us the solution to Problem 4.3 ($MOD \cdot POL \cdot RECIPR$).

The extended Euclidean algorithm uses $O(m^2)$ ops to compute the coefficients of $q_i(x), r_i(x), s_i(x)$ and $t_i(x)$ for all i. This estimate is obvious if $\deg q_i(x) = 1$ for all i, but it is not hard to obtain it in the general case too, since $\sum_i \deg q_i(x) \leq n$ (see exercise 16). On the other hand, $(n+1)(n+2)/2$ is a lower bound on the worst-case cost of computing the coefficients of the polynomials $r_0(x), \ldots, r_\ell(x)$. These polynomials form the *polynomial remainder sequence* for $u(x)$ and $v(x)$, and together they may have up to $(n+1)(n+2)/2$ distinct coefficients. On the other hand, Algorithm 5.1a can be extended to computing the coefficients of the three polynomials $r_i(x), s_i(x)$ and $t_i(x)$ for *any fixed* i, as well as the coefficients of all the quotients $q_1(x), \ldots, q_\ell(x)$, at the overall cost $O(m \log^2 m)$ ops [BGY]. The latter sequence is given by the matrices $\tilde{Q}_i(x)$ of (5.3). The matrices $\tilde{Q}_i(x)$ are recovered from their products $Q_H(x)$ and $Q_H^*(x)$ by means of the *fan-out method.* The cost bound is within logarithmic factor from the lower bound of [St81] on the cost of computing the sequence $q_1(x), \ldots, q_\ell(x)$, called the *continued fraction* because of the identity $u(x)/v(x) = q_1(z) + 1/(q_2(x) + 1/(\cdots + 1/q_i(x))\cdots)$, which holds for $i = \ell$. For $i < \ell$, the expressions on the right side, defined by the extended Euclidean scheme, give us approximations to $u(x)/v(x)$.

Remark 5.2. All the development of the Euclidean and the extended

Euclidean algorithms for computing the gcd and lcm, presented above for polynomials, can be extended to integers. In fact, it is customary to refer to the *continued fraction approximation* algorithm (which is a well-known tool of computations for the number theory) when the extended Euclidean scheme is applied to a pair of positive integers u and v, replacing the polynomials $u(x)$ and $v(x)$. We will omit the detailed comparison, for which we refer the reader to [AHU], [Sch86], [Kn] and [HW79], but will just point out that *the gcd of a pair of n-bit binary integers can be computed on the Boolean circuits of size* $O(\mu(n)\log n) = O(n\log^2 n\log\log n)$, [compare (2.4)].

Problem 5.1c ($SEV \cdot POL \cdot GCD$), *gcd of several polynomials. Given natural m and n and the coefficients of m polynomials $u_1(x),\dots,u_m(x)$ of degrees at most n, compute the coefficients of the polynomial $r(x) = gcd(u_1(x),\dots,u_m(x))$.*

Solution. Set $v_{i,0}(x) = u_i(x)$ for $i = 1,\dots,m$, $v_{i,0}(x) = 0$ for $i = m+1,\dots,2^g$, $g = \lceil\log m\rceil$. Then successively, for $k = 0,1,\dots,g-1$, compute the coefficients of $v_{j,k+1}(x) = \gcd(v_{2j-1,k}(x), v_{2j,k}(x))$ for $j = 1,2,\dots,m/2^{k+1}$, and finally, output $r(x) = v_{1,g}(x)$. The cost of this solution is $O(mn\log^2 n)$ ops. ∎

The problem, its solution and the complexity estimate can be immediately extended to

Problem 5.1d ($SEV \cdot POL \cdot LCM$), *lcm of several polynomials. Under the assumptions of Problem 5.1c ($SEV \cdot POL \cdot GCD$), compute the coefficients of* $\operatorname{lcm}(u_1(x),\dots,u_m(x))$.

We may improve the above complexity estimates for Problem 5.1c ($SEV \cdot POL \cdot GCD$) by using randomization of Las Vegas type ([IKo] and [Ka88]) if, at a low computational cost, we may choose a random element h under the uniform probability distribution on a finite set of numbers S. Indeed, for such a choice of h, we will show below, by following [IKo], that, with a probability at least

$$P = 1 - \frac{(m-2)\deg\,(u_m(x)/r(x))}{|S|}\,, \tag{5.10}$$

we have $r(x)$, the gcd of the input polynomials $u_1(x),\dots,u_m(x)$, equal to the monic polynomial $g(x) = \gcd(u_m(x), \sum_{i=1}^{m-1} u_i(x)h^{i-1})$. Here and hereafter $|S|$ denotes the cardinality of a set S. Thus, we have a randomized reduction of our computation to the case of two input polynomials, $u_m(x)$

and $\sum_{i=1}^{m-1} u_i(x)h^{i-1}$, and we may maximize its effectiveness if we enumerate the input polynomials in such a way that

$$\deg u_m(x) \leq \deg u_i(x), \quad i < m.$$

Surely, $r(x)$ always divides $g(x)$, and to verify if $g(x)$ is the desired gcd, it suffices to check if $g(x)$ divides all the input polynomials $u_i(x), i = 1, \ldots, m$. If this test shows that $g(x)$ is not the gcd, then the computation is repeated for a new value h. The expected number of trials of these computations is $1/P$ (see [IKo]), so the expected overall computational cost of this randomized algorithm is $O((\frac{1}{P})(m + \log n)n \log n)$ ops versus the record $O(mn \log^2 n)$ ops for the deterministic solution.

The proof of the above bounds on P relies on a result from [Schw] and [Z], which we are going to reproduce. This result is widely used for the design of randomized algebraic and combinatorial algorithms, for instance, of randomized algorithms for Problems 2.2.10 ($M \cdot RANK$), 2.2.10a ($SUBMATRIX$), and 2.2.10b ($M \cdot PRECNDTN$) (also compare Theorem 4.7.2). First we need to introduce the following definition.

Definition 5.1. Let $p = p(x_1, \ldots, x_m) = \prod_{i=1}^{m} x_i^{k_i}$, for natural k_i, $i = 1, \ldots, m$, be a multivariate monomial. The *total degree* of p is $\sum_{i=1}^{m} k_i$. The *total degree of a multivariate polynomial* $q(x_1, \ldots, x_m)$ is the greatest total degree of its monomials.

Lemma 5.1 ([DL], [Schw], [Z]). *Let $p(x_1, \ldots, x_m)$ be a nonzero m-variate polynomial over any field $\mathbf{F}$ or ring $\mathcal{R}$ having total degree d. Let S be a finite subset of $\mathbf{F}$ or $\mathcal{R}$, and let $x_1^*, \ldots, x_m^*$ be a set of points from S, chosen at random, under the uniform and independent distribution of all the x_i^* on S, for $i = 1, \ldots, m$. Then the probability that $p(x_1^*, \ldots, x_m^*) = 0$ is at most $d/|S|$.*

A major application of Lemma 5.1 is to estimating the probability that a nonsingular $n \times n$ matrix A filled with indeterminates (variables) or with multivariate polynomials of degrees at most d remains nonsingular after assigning random values from a large fixed finite set S to all the indeterminates (variables). Indeed, the singularity of A would amount to vanishing the determinant of A, which is a multivariate polynomial of degree n in the entries of A, and the probability of its vanishing is at most $dn/|S|$, according to Lemma 5.1.

We will deduce Lemma 5.1 from the following fact and corollary, due to [DL], [Schw], [Z]:

Fact 5.3. *Let $p_1 = p_1(x_1, \ldots, x_s)$ be a nonzero s-variate polynomial over any field* $\mathbf{F}$ *or ring* $\mathcal{R}$, *in the variables $x_1, \ldots, x_s$. Let p_1 have degree d_1 in x_1, and let $p_2 = p_2(x_2, \ldots, x_s)$ be the coefficient of $x_1^{d_1}$ in p_1. Inductively, let d_j be the degree of p_j in x_j and p_{j+1} be the coefficient of $x_j^{d_j}$ in p_j, for $j = 1, \ldots, s$. Let I_j be a finite subset of* $\mathbf{F}$ *or* $\mathcal{R}$ *for $j = 1, \ldots, s$. Then p_g has at most*

$$M_g = |I_g \times I_{g+1} \times \cdots \times I_s| \sum_{j=g}^{s} d_j/|I_j|$$

zeros in the set $I_g \times I_{g+1} \times \cdots \times I_s$, for $g = s-1, s-2, \ldots, 1$, where $U \times V = \{(u, v) : u \in U, v \in V\}$ denotes the Cartesian product of two sets U and V.

Proof. We proceed by induction on i, proving Fact 5.3 for $p_{s-i}(x_{s-i}, \ldots, x_s)$ for $i = 0, 1, \ldots, s-1$. The result is obvious for $i = 0$. Let it be true for $i = 0, 1, \ldots, j-1$. Denote

$$T_k = I_{s-k} \times \cdots \times I_s\ , \quad M_k = |T_k| \sum_{h=s-k}^{s} d_h/|I_h|\ ,$$

for $k = j-1, j$. Observe that for a fixed element $(z_{s-j+1}, \ldots, z_s)$ of the set T_{j-1}, either the polynomial $p_{s-j} = p_{s-j}(x_{s-j}, z_{s-j+1}, \ldots, z_s) = 0$ identically in x_{s-j}, and then $p_{s-j+1}(z_{s-j+1}, \ldots, z_s) = 0$, or at most d_{s-j} values of x_{s-j} in I_{s-j} turn p_{s-j} into zero. Apply the induction hypothesis (for $i = j-1$) to p_{s-j+1} and deduce that the first case may occur for at most M_{j-1} elements of the set T_{j-1}, which correspond to at most $|I_{s-j}|M_{j-1}$ zeros of p_{s-j} in the set T_j. The remaining, second case cannot contribute more than $d_{s-j}|T_{j-1}|$ zeros of p_{s-j} in T_j. Therefore, p_{s-j} has a total of at most

$$\begin{aligned} |I_{s-j}|M_{j-1} + d_{s-j}|T_{j-1}| &= |I_{s-j}|(|T_{j-1}|) \sum_{k=s-j+1}^{s} d_k/|I_k| + d_{s-j}|T_{j-1}| \\ &= |T_j| \sum_{k=s-j}^{s} d_k/|I_k| = M_j \end{aligned}$$

zeros in the set T_j, which proves the induction hypothesis for $i = j$. ∎

Lemma 5.1 can be restated in the form of the following

Corollary 5.1. *Let $I = I_1 = \cdots = I_s$, and $|I| \geq c \deg p_1$. Then p_1 has at most $|I|^s/c$ zeros in the set $I_1 \times \cdots \times I_s$ (assuming that p_1 is not identically zero).*

Proof. Apply Fact 5.3 for $g = 1$; note that $\sum_{j=1}^{s} d_j/|I| \leq 1/c$ in this case. ∎

To see an application of Lemma 5.1 to the above randomized solution of Problem 5.1b ($EXT \cdot EUCLID$), observe that deg $g(x) >$ deg $r(x)$ if $g(x) \neq r(x)$, and this implies (and actually is equivalent to) vanishing a certain subresultant of the two polynomials $u_m(x)$ and $\sum_{i=1}^{m-1} u_i(x)h^{i-1}$. This subresultant is a polynomial in h of degree $(m-2)$deg $(u_m(x)/r(x))$ (see section 8 of chapter 2 on the subresultants and their properties). Lemma 5.1 now immediately implies the above estimate (5.10).

On the next three problems (that is, on Problems 5.2, 5.2a, 5.2b), we refer the reader to [GY] and [BGY].

Problem 5.2 ($RAT \cdot INTERP$), *rational interpolation; (m, n) Hermite interpolation problem. Given a set of $N+1 = m+n+1$ points $x_0, x_1, \ldots, x_N$ (not necessarily distinct) and the values of an analytic function $V(x)$ at these points, find two polynomials, $R(x)$ of degree at most m and $T(x)$ of degree at most n, such that $V(x_i) = R(x_i)/T(x_i)$ for $i = 0, \ldots, N$, that is,*

$$V(x) - R(x)/T(x) = G(x) \prod_{i=0}^{N} (x - x_i),$$

for some analytic function $G(x)$.

Remark 5.3. Here and hereafter, we require to interpolate by $R(x)/T(x)$ on the set $\{x_0, \ldots, x_N\}$ to the function $V(x)$ and, furthermore, to all its derivatives of orders less than m_i by the derivatives of $R(x)/T(x)$ where the factor $x - x_i$ has multiplicity m_i.

The (m, n) Hermite interpolation problem is called the (m, n) *Cauchy interpolation problem* if all the interpolation points $x_0, x_1, \ldots, x_N$ are distinct.

Problem 5.2a ($H \cdot INTERP1$), *the modified (m, n) Hermite interpolation problem. Given a set of $N+1 = m+n+1$ points $x_0, \ldots, x_N$ (not necessarily distinct) and the values of an analytic function $V(x)$ at these points, find two polynomials, $R(x)$ of degree m or less and $T(x)$ of degree n or less, such that*

$$R(x) = S(x)U(x) + T(x)V(x) = T(x)V(x) \bmod U(x),$$
$$U(x) = \prod_{i=0}^{N} (x - x_i) \qquad (5.11)$$

for some analytic function $S(x)$ [compare (5.9)].

If a pair $[R(x), T(x)]$ is a solution to the (m, n) Hermite interpolation problem, then this pair satisfies (5.11); the converse is also true, unless the (m, n) Hermite interpolation problem has no solution. The modified (m, n) Hermite interpolation problem (5.11) always has solutions $[R(x), T(x)]$ such that the ratio $R(x)/T(x)$ is the unique rational function.

Problem 5.2b ($PADE$), *Padé approximation. Solve Problem 5.2a* ($H \cdot INTERP1$) *in the case where* $x_0 = x_1 = \cdots = x_N = 0$, *that is, where the interpolation is at a single node* 0 *of multiplicity* $N+1$, *and then (5.11) takes the following form:*

$$R(x) - T(x)V(x) = 0 \bmod x^{N+1},\ N = m+n,\ \deg T(x) \le n,\ \deg R(x) \le m.$$

In other words, in the (m, n) *Padé approximation problem, we are given two natural numbers* m *and* n *and the first* $N+1 = m+n+1$ *Taylor coefficients of an analytic function* $V(x)$ *decomposed at* $x = 0$, *and we are seeking two polynomials* $R(x)$ *and* $T(x)$ *satisfying the above relations.*

Hereafter, let an analytic function $V(x)$ and the points $x_0, \ldots, x_N$ (not necessarily distinct) be fixed. Then the solutions to the modified interpolation problem (5.11) for all natural pairs (m, n) form the *rational interpolation table,* associated with $x_0, \ldots, x_N$ and $V(x)$. Comparison of (5.9) with (5.11) shows that the entries of each antidiagonal of this table are related to each other via the equations (5.6)-(5.9) and, consequently, via Euclid's extended algorithm. Therefore, for a fixed pair $[m, n]$, the entry $[R(x), T(x)]$ of this table, that is, a solution to (5.11), can be computed in $O(N \log^2 N)$ ops $(N = m + n)$, as follows:

Algorithm 5.2, *rational interpolation table.*

Input: a natural number N, $N+1$ points $x_0, \ldots, x_N$ (not necessarily distinct) and the values $V(x_0), \ldots, V(x_N)$ of an analytic function $V(x)$ and its derivatives, defined according to Remark 5.3.

Output: the $[N-k, k]$ entries for $k = 0, 1, \ldots, N$ of the rational interpolation table associated with $x_0, .., x_N$ and $V(x)$.

1. Compute the coefficients of the polynomial $U(x) = \prod_{j=0}^{N}(x - x_j)$ by using the fan-in method; then solve the $(N, 0)$ Hermite interpolation problem with the input sets $\{x_0, \ldots, x_N\}$ and $\{V(x_0), \ldots, V(x_N)\}$ [this is Problem 4.2b ($H \cdot INTERP$)]. The solution is given by two polynomials, $R(x)$ and $T(x) = 1$. [$R(x)$ is called the *Hermite interpolation* *polynomial* of degree at most N; if all the points $x_0, \ldots, x_N$ are distinct, $R(x)$ is the *Lagrange interpolation polynomial* of degree at most N.]

2. Let $r_0(x) = U(x)$, $r_1(x) = R(x)$, $t_0(x) = 0$, $t_1(x) = 1$ and define the polynomials $r_i(x)$, $t_i(x)$ for $i = 2, 3, \ldots \ell+1$ by applying (5.6) and (5.8), that is, apply Algorithm 5.1b (extended Euclidean scheme) to the pair $u(x) = r_0(x) = U(x)$, $v(x) = r_1(x) = R(x)$.

3. Output the pairs $[r_i(x), t_i(x)]$ of the successive entries of the extended Euclidean scheme. They successively fill up the entries $[N - i, i], i = 0, 1, \ldots, N$, of the antidiagonal of the rational interpolation table, associated with $x_0, x_1, \ldots, x_N$ and $V(x)$ and having the pair $[r_1(x), t_1(x)] = [R(x), 1]$ at the entry [N,0].

Observe that the pair $[r_i(x), t_i(x)]$ fills exactly deg $q_{i-1}(x)$ successive entries of this antidiagonal where $q_{i-1}(x)$ is the polynomial of (5.6)-(5.8).

Rational interpolation, and, in particular, Padé approximation, can be used for computing the lcm and the gcd of two polynomials:

Algorithm 5.3, *lcm of two polynomials* ([P92b]).

Input: the coefficients of two polynomials $u(x)$ and $v(x)$, such that (with no loss of generality) $u(x) + v(x) \not\equiv 0$ and $u(0) = v(0) = 1$.

Output: $\text{lcm}(u(x), v(x))$.

Computation:

1. Compute $m = \text{deg}\ (u(x)v(x))$, $n = \text{deg}\ (u(x) + v(x))$, $N = m + n + 1$.

2. Compute the N first coefficients $a_0, a_1, \ldots, a_{N-1}$ of Taylor's power series $a(x) = (\frac{1}{u(x)} + \frac{1}{v(x)})^{-1} = \sum_{i=0}^{+\infty} a_i x^i$.

3. Compute the $(m - r, n - r)$ Padé approximation $R(x), T(x)$ to $a(x)$ for the maximum integer r, $r \le \min\{m, n\}$, such that $R(x)/T(x) = a(x)$. Output $R(x) = \text{lcm}(u(x), v(x))$.

Remark 5.4. By using a randomization of Las Vegas type, we may extend Algorithm 5.3 to the case of several input polynomials (see exercise 24).

Algorithm 5.4, *gcd of two polynomials.*

Input: natural m and n and the coefficients of two polynomials $u(x)$ and $v(x)$, $v(0) \neq 0$, of degrees m and n, respectively.

Output: $\text{gcd}(u(x), v(x))$.

Computation:

1. Compute the first $N = n + m$ Taylor's coefficients $a_0, \ldots, a_{N-1}$ of the power series

$$a(x) = \frac{u(x)}{v(x)} = \sum_{i=0}^{+\infty} a_i x^i.$$

2. Compute the $(m - r, n - r)$ Padé approximation $R(x), T(x)$ to $a(x)$ for the maximum integer r, $r \leq \min\{m, n\}$, such that $R(x)/T(x) = u(x)/v(x)$.
3. Compute and output $\gcd(u(x), v(x)) = u(x)/R(x) = v(x)/T(x)$.

The latter algorithm, of [P92b], has been independently derived based on the properties of Hankel matrices in [BG90], [BG92]. The i-th coefficient h_i of the power series at Stage 1 may grow exponentially in i. In section 9 of chapter 2, we overcome this problem by applying a distinct algorithm (Algorithm 2.9.1) having the same input and output and based on specific properties of Hankel and Bezout matrices (see the definitions of such matrices in sections 5 and 9 of chapter 2).

Instead of the ratio $a(x) = u(x)/v(x)$, we may use the ratio $a(x) = r_i(x)/r_j(x)$ for any pair of distinct polynomials in the polynomial remainder sequence $\{r_i(x)\}$ of (5.6).

Our first solution to Problem 5.1a ($POL{\cdot}LCM$) enables us to substitute Algorithms 5.3 and 5.4 for each other.

We will conclude this section with another important application of Padé approximation.

Problem 5.3 ($RECUR \cdot SPAN$), *computing the minimum span for a linear recurrence* (Berlekamp-Massey). *Given a natural s and $2s$ numbers $v_0, \ldots, v_{2s-1}$, compute the minimum natural $n \leq s$ and n numbers $t_0, \ldots, t_{n-1}$ such that $v_i = t_{n-1}v_{i-1} + \cdots + t_0 v_{i-n}$, for $i = n, n+1, \ldots, 2s-1$.*

Solution ([Be] or [BGY]). The solution n, $t_0, \ldots, t_{n-1}$ is unique and is given by the degree n and the coefficients $t_0, \ldots, t_{n-1}$ of the minimum span polynomial $T(x) = x^n - \sum_{i=0}^{n-1} t_i x^i$, such that the pair of polynomials $[R(x), T(x)]$ is an (m, n) Padé approximation to $V(x) = \sum_{i=0}^{2s-1} v_i x^i$, where $\deg R(x) < s$, $N = 2s - 1$, $m = s - 1$, $n = s$ [see alternative solutions of both of these Problems 5.2b ($PADE$) and 5.3 ($RECUR{\cdot}SPAN$) in section 5 of chapter 2]. ∎

This problem and its solution applies to coding theory ([Be]), to sparse polynomial interpolation ([B-OT], [GKS], [KLW] and our section 9) and to the parallel matrix computations (chapter 4).

6. Power Series Manipulation and Algebraic Newton's Iteration.

In this section, in addition to the first example given in the solution of Problem 3.4 ($POL \cdot RECIPR$) of computing the reciprocal of a polynomial modulo a power, we will present more applications of manipulation with power series and of algebraic Newton's iteration being powerful solution methods (compare numerical Newton's iteration of section 3). These applications will not be used in this book, but they are of interest in their own right.

In these examples and in numerous other cases, polynomial computations are reduced to the following problem: given a rational function $f(x)$, with coefficients in the ring of power series in the variable z, and a natural K, compute the first K coefficients of a formal power series

$$w = w(z) = \sum_{i=0}^{+\infty} w_i z^i \tag{6.1}$$

satisfying the equation

$$f(w) = 0, \tag{6.2}$$

provided that there exists such a formal power series. Consider the solution of the problem in two stages.

a) First, for a fixed nonnegative integer j_0, the polynomial

$$w_{j_0}(z) = w(z) \bmod z^{j_0+1}$$

is computed. [$j_0 = 0$ in the solution of Problem 3.4 ($POL \cdot RECIPR$).]

b) Then, *algebraic Newton's iteration*

$$w_{j+1}(z) = w_j(z) - f(w_j(z))/f'(w_j(z)) \bmod z^K, \quad j = j_0, j_0+1, \ldots, \tag{6.3}$$

is applied. Here and hereafter, f' denotes the derivative of f.

Step j of the iteration (6.3) essentially amounts to computing, at first, the polynomials $f(w_j(z)) \bmod z^K$ and $f'(w_j(z)) \bmod z^K$, and then their ratio modulo z^K. Since f and f' are rational functions, the computations are reduced to polynomial multiplication modulo z^K [see Problem 2.4 ($SEV \cdot POL \cdot MULT$)] and to computing the reciprocal of a polynomial modulo z^K [see Problem 3.4 ($POL \cdot RECIPR$)].

Newton's iteration converges very fast. Indeed, define the auxiliary rational function $g(x) = f(x)/(x-w)$, where w is the power series satisfying (6.1), (6.2). Then, apply (6.3), and obtain the equation

$$w_{j+1} - w = (w_j - w)^2 \frac{g'(w_j)}{f'(w_j)} \bmod z^K.$$

Therefore, if the power series $f'(w)$ has its inverse, that is, if

$$f'(0) \neq 0, \tag{6.4}$$

then each iteration step (6.3) doubles the number of the computed coefficients of the formal power series w. In other words,

$$w_j(z) = w(z) \bmod z^{(j_0+1)J}, \quad J = 2^{j-j_0},\ j = j_0, j_0+1, \ldots, \tag{6.5}$$

and thus $\lceil \log(K/(j_0+1)) \rceil$ iteration steps (6.3) suffice in order to compute $w(z) \bmod z^K$. This convergence is called *quadratic* or *exponential.* Note that the iteration (6.3) can be simplified under (6.4), for we may reduce the right side of (6.3) $\bmod z^{(j_0+1)J}$, rather than $\bmod z^K$, as we have done in the case of Problem 3.4 ($POL \cdot RECIPR$) of the evaluation of the reciprocal of a polynomial. Other simplifications of the steps (6.3) and/or the Stage a) are also possible in many special cases.

Newton's iteration (6.3) is a customary powerful tool for numerous polynomial computations that can be reduced to solving equations of the form (6.2). To motivate our ad hoc derivation of Newton's iteration, first expand the function $f(x)$ as the Taylor power series in x,

$$f(x) = f(y) + \sum_{h=1}^{+\infty} \frac{(x-y)^h}{h!} f^{(h)}(y).$$

Now, suppose that we approximate $f(x)$ by the partial sum formed by the first two terms of this expansion,

$$f_1(x) = f(y) + (x-y)f'(y) = f(x) \bmod (x-y)^2.$$

By setting $f_1(x) = 0$, we obtain an approximation $x = y - f(y)/f'(y)$ to the zero of $f(x) = 0$. The latter expression for w turns into (6.3) if we substitute $x = w_{j+1}(z)$, $y = w_j(z)$ and reduce the resulting equation modulo z^K.

Newton's iteration (6.3) can be replaced by several other iterative processes having similar properties ([BKu] and [KT]); in particular, we may rely on the approximation to $f(x)$ by higher order partial sums of Taylor's series.

Next, we will demonstrate the power of Newton's iteration (6.3) by solving five problems of polynomial computations.

Problem 6.1 ($SER \cdot REVERSE$), *generalized reversion of power series* ([BKu] and [Kn], p. 510). *Given a natural K and two formal power series,*

$$s(z) = \sum_{i=1}^{+\infty} s_i z^i, \quad t(w) = w + \sum_{i=2}^{+\infty} t_i w^i,$$

find the polynomial $w(z)$ of degree less than K satisfying the equation $s(z) = t(w(z)) \bmod z^K$.

Problem 6.1 ($SER \cdot REVERSE$) is a generalization of the customary reversion problem where $s(z) = z$ and where for a given $t(w)$, we seek $w(z)$ such that $z = t(w(z))$.

Solution. Since $s(0) = t(0) = 0$, the terms $t_i w^i$ of $t(w)$ for $i \geq K$ do not influence the solution and can be ignored, as well as the terms $s_i z^i$ of $s(z)$ for $i \geq K$, so that the problem can be stated for polynomials rather than for the power series $s(z)$ and $t(w)$. Apply Newton's method for $j_0 = 1$, $w_1(z) = s_1 z = w(z) \bmod z^2$ to the equation $f(w) = s(z) - t(w) = 0$ and arrive at the following iteration:

$$w_{j+1}(z) = w_j(z) + \frac{s(z) - t(w_j(z))}{t'(w_j(z))} \bmod z^K, \quad j = 1, 2, \ldots . \tag{6.6}$$

Here, $t'(w) = 1 + \sum_{i=1}^{K-1} i t_i w^{i-1}$. Then $w_{j+1}(z) = w(z) \bmod z^J$, $J = 2^j$, $j = 1, \ldots, \lceil \log K \rceil$. In this case, $t'(0) = 1$, so that we may perform the iteration step (6.6) modulo z^{2J} rather than modulo z^K. Given $w_j(z) \bmod z^J$, we will compute the coefficients of the polynomial $w_{j+1}(z) \bmod z^{2J}$ as follows: First, we will compute $r_j(z) = s(z) - t(w_j(z)) \bmod z^{2J}$, by applying the algorithm given earlier for Problem 4.6 ($SER \cdot COMP$) of computing the composition of power series. Then we will compute the polynomial $q_j(z) = t'(w_j(z)) \bmod z^{2J}$ and its reciprocal modulo z^{2J} [see Problem 3.4 ($POL \cdot RECIPR$)]. Finally, we will compute the polynomial $w_{j+1}(z) = w(z) \bmod z^{2J} = w_j(z) + r_j(z)/q_j(z) \bmod z^{2J}$. The overall cost bound is $O((K \log K)^{3/2})$, with a substantial overhead constant hidden in O. The cost bound is dominated by the cost of the stages of the composition of power series, and the composition and the reversion can actually be reduced to each other, so that they have the same asymptotic complexity (see p. 509 of [Kn] for the solution of the reversion problem by using about $K^3/6$ ops, which is a better bound for smaller K). ■

Problem 6.2 ($POL \cdot ROOT$), *evaluation of a root of a polynomial* ([Y] and [KT]). *Given natural d and n and a polynomial $p(z)$ of degree at most n such that $p_0 = p(0) = 1$, compute the unique polynomial $w(z)$ of degree at most n such that $w_0 = w(0) = 1$ and $(w(z))^d = p(z) \bmod z^{n+1}$.*

Solution. Apply Newton's iteration (6.3) with $j_0 = 0$, $w_0(z) = 1$ to the equation $f(w) = w^d - p = 0$. The cost of the solution is $O(n \log n)$ with a small overhead. This result will be used in the solution of our next problem. ■

Problem 6.3 ($POL \cdot DECOMP$), *polynomial decomposition* ([GKL]). *Given two natural numbers m and n and the coefficients of a monic polynomial $u(z)$ of degree $N = mn$, find two monic polynomials, $s(t)$ of degree m and $t(z)$ of degree n, such that $u(z) = s(t(z))$, or prove that there exist no such pairs of polynomials.*

Solutions. First assume that $u(z) = s(t(z))$ for some monic polynomials, $s(t)$ of degree m and $t(z)$ of degree n. Then the polynomials $u(z)$ and $(t(z))^m$ must agree on their n highest degree terms, that is, the polynomial $u(z) - (t(z))^m$ has degree at most $N - n$. Replace $u(z)$ and $t(z)$ by the reverse polynomials, $U(z) = z^N u(1/z)$ and $T(z) = z^n t(1/z)$. Then $T(0) = 1$, $U(z) - (T(z))^m = 0 \bmod z^n$. Due to the latter relations, we may evaluate the coefficients of $T(z) \bmod z^n$ at the cost of $O(n \log n)$ ops [see Problem 6.2 ($POL \cdot ROOT$)] and then immediately recover the coefficients of the candidate polynomial $t(z)$. Then we may solve Problem 3.5 ($TAYLOR \cdot EXP$) of computing the generalized Taylor expansion at the cost of $O(N \log N \log m)$ ops with a small overhead; the cost of this stage dominates the asymptotic cost of computing $t(z)$. This will define the unique set of polynomials $s_i(z)$, $i = 0, 1, \ldots, m$, of degrees less than m satisfying the polynomial identity of Problem 3.5 ($TAYLOR \cdot EXP$). If the polynomials $s_i(z) = s_i$ are constants for all i, then we arrive at the desired polynomial $s(t) = \sum_{i=0}^{m} s_i t^i$, such that $u(z) = s(t(z))$; otherwise, we conclude that there exists no desired decomposition $u(z) = s(t(z))$ for given $u(z)$, m and n.

In an alternative method for the evaluation of the coefficients of $s(t)$, [this method is numerically less stable but asymptotically faster if $\log m = o(\log N)$], first compute $t(z)$ and $u(z)$ at K points z_i, $i = 0, 1, \ldots, K-1$, such that $t(z)$ takes at least $m+1$ distinct values at these points. To ensure this property of $t(z)$, it suffices to choose the expanded Fourier points $z_i = a\omega^i$, $i = 0, 1, \ldots, K-1$, ω being a primitive K-th root of 1, a being positive and sufficiently large, and K exceeding m and being mutually prime with n. (If we extend the model of RAM by allowing randomization of Las Vegas type, then we may alternatively choose $z_i = a\omega^i + b$ either for $b = 0$ and for a random a or for random a and b assuming, as usual in randomized algorithms, that *the cost of choosing random parameters is dominated by the cost of other computations involved.*) Then recover the coefficients of $s(t)$, by using interpolation on the set $\{t_i = t(z_i),\ i = 0, 1, \ldots, K-1\}$, and, finally, evaluate the coefficients of the polynomial $s(t(z))$, as in the solution of Problem 4.5 ($POL \cdot COMP$) of composition of polynomials, to test if $u(z) = s(t(z))$. The cost $O(N(\log^2 m + \log N))$ of the latter test dominates the overall asymptotic cost of this solution. [Let us again allow randomization, of Monte Carlo type this time. Then the following alternative randomized test for $u(z) = s(t(z))$ can be suggested within the second solution: evaluate $u(z)$ and $t(z)$ at a random point $z = z_0$, then compute $s(t(z_0))$ and compare with $u(z_0)$; if these two values are distinct, there is no desired decomposition of $u(z)$; otherwise, $u(z) = s(t(z))$ with a high probability; the test can be repeated for a few random values of z.] The overall cost of such a randomized algorithm is $O(N \log N + m \log^2 m)$ ops (compare exercise 25)]. ■

The two solution methods above show that the coefficients of the polynomials $s(t)$ and $t(z)$ are rational functions in the coefficients of $u(z)$ as long as there exists decomposition $u(z) = s(t(z))$.

Problem 6.4 ($COMPL \cdot POL \cdot DECOMP$), *complete decomposition of a polynomial* ([GKL], [G90]). *Given the coefficients of a polynomial $u(z)$ of degree N and a prime factorization of N, compute a complete decomposition of $u(z)$ into a composition of indecomposable polynomials.*

Solution. Solve Problem 6.3 ($POL \cdot DECOMP$) for the input polynomial $u(z)$ and for all natural m dividing N, until a decomposition $s(t(z))$ is found [unless for all the divisors m of N, there is no decomposition of $u(z)$, in which case we arrive at a trivial complete decomposition $u(z) = u(z)$]. Then recursively apply the same process to $s(t)$ and $t(z)$ until the complete decomposition of $u(z)$ is obtained. It can be estimated that the overall cost of this solution is $O(N^{1+\epsilon})$ ops, for any fixed positive ϵ. ■

Observe that Problem 6.4 ($COMPL \cdot POL \cdot DECOMP$) is an extension of Problem 6.3 ($POL \cdot DECOMP$). It can be proved that the complete decomposition of a polynomial is essentially unique.

Theorem 6.1 ([Eng], [Ritt]). *Complete decomposition of any polynomial is unique up to within the trivial ambiguities of the three following groups:*

i) *$z = s(t(z))$ for linear polynomials $s(t) = t - a$, $t(z) = z + a$ and for any constant a;*
ii) *$u(z) = s(t(z)) = s_1(t_1(z))$ provided that $s(t) = t^m$, $t(z) = z^h q(z^m)$, $s_1(t) = t^h q^m(t)$, $t_1(z) = z^m$, h is a natural number, and $q(z)$ is a polynomial;*
iii) *$s(t(z)) = t(s(z))$ for $s(t) = T_m(t)$, $t(z) = T_n(z)$ where $T_h(x)$ denotes the h-th degree Chebyshev polynomial, $T_{h+1}(x) = 2xT_h(x) - T_{h-1}(x)$, $h = 1, 2, \ldots$, $T_0(x) = 1$, $T_1(x) = x$.*

Problem 6.5 ($ALG \cdot FUNCTS$), *evaluation of algebraic functions ([KT]). Given the coefficients of $n+1$ polynomials $f_i(z)$, $i = 0, \dots, n$, compute the first K coefficients of a fractional power series $w = w(z) = \sum_{i=0}^{+\infty} w_i z^{i/d}$, $d \leq n$, satisfying the equation*

$$f(w) = f_n(z)w^n + \cdots + f_0(z) = 0. \tag{6.7}$$

[$w(z)$ may have several expansions into the fractional power series; one of them is to be computed.]

Solution. The equation (6.7) is a special case of (6.2), so the general scheme of the solution is applied [see (6.3), (6.5)].

Stage a). Without loss of generality, assume that $f_n(0) \neq 0$ (see [KT]). Further, if $w_0 = w(0)$, then by reducing (6.7) modulo z, we obtain that $f_n(0)w_0^n + f_{n-1}(0)w_0^{n-1} + \cdots + f_0(0) = 0$, and the value w_0 can be computed as a zero of the above n-th degree polynomial. In the singular case, where $f'(w)$ is not invertible, i.e. $f'(w) = 0 \bmod z$, we apply *Newton's polygon process* ([KT]), to reduce the original problem to a regular one, where $f'(w) \neq 0 \bmod z$; the subsequent application of the algebraic function theory yields that $d = 1$ and $w = w(z)$ is a power series (6.1). Newton's polygon process defines one of the expansions of $w(z)$ by computing the j_0 coefficients of its lowest degree terms for a fixed j_0. The cost of computing each coefficient is dominated by the cost of computing a zero of a polynomial of degree at most n, and the integer $j_0 - 1$ does not exceed the degree in z of the term of the lowest degree in z of the resultant of $f(w)$ and $f'(w)$ (see definition of the resultant in section 8 of chapter 2); this degree does not exceed $m(2n-1)$ where m is the maximum degree of $f_i(z)$ for all i [see (6.7)].

Stage b). Given $w(z) \bmod z^{j_0+1}$, the polynomial $w(z) \bmod z^K$ is computed in at most $\log(K/(j_0+1)) \leq \log K$ iterations (6.3). Iteration j is reduced to $O(n)$ multiplications of pairs of polynomials modulo $z^{2(j_0+1)J}$, $J = 2^j$, due to (6.5) and (6.7), so the overall cost of the Stage b) is $O(nK \log K)$ ops, which dominates the cost of the Stage a) for large K. A trade-off is possible: we may avoid solving a polynomial equation at the Stage a) (to decrease its cost) and carry out the polynomial zeros symbolically, by operating in an algebraic extension $\mathbf{E}$ of the original field $\mathbf{F}$. Here, $\mathbf{E}$ is the field of polynomials modulo the (irreducible) minimum polynomial of these zeros, that is, the minimum degree d polynomial with coefficients in $\mathbf{F}$, irreducible in $\mathbf{F}$ and having these zeros (compare the next section and note that $d \leq n$). Then the cost of the Stage b) increases by the factor of $n \log n$, representing the cost of arithmetic operations with polynomials modulo $m(z)$, rather than with their zeros. (We will revisit the algebraic extensions of fields and rings in the next section.) ■

Here are some sample problems that can be reduced to the solution of the equation (6.7) [compare Problems 6.1 ($SER \cdot REVERSE$) and 6.2 ($POL \cdot ROOT$)]:

1. *Computation of an n-th root $w(x)$ of a polynomial $p(x)$ satisfying* (6.7), such that $f_0(z) = p(x)$, $f_n(z) = 1$, $f_i(z) = 0$ for $i = 1, 2, \ldots, n-1$ in (6.7).
2. *Reversion of a polynomial modulo z^K* : $f_0(z) = -s(z)$, $f_i(z) = t_i$, where t_i are constants for $i = 1, 2, \ldots, n$; $t_1 = 1$.
3. *Computing the Legendre polynomials:* $n = 2$, $f_2(z) = 1 - 2tz + z^2$, $f_1(z) = 0$, $f_0(z) = -1$. The i-th coefficient of $w(z)$ is the i-th degree Legendre polynomial in t, for $(1-2tz+z^2)^{-\frac{1}{2}}$ is the generating function of Legendre polynomials.

7. Extension to the Computations Over Any Field or Ring of Constants. Regularization via Randomization.

For practical implementation of many algorithms for polynomial and matrix computations with rational numbers (representing real values on computers), it is crucial to extend the computations so as to perform them over finite fields or rings (see, for instance, [AHU] or any general algebra text, such as [Ja], [vdW], on rings and fields). In particular, according to the customary recipe for overcoming the problem of numerical stability in computer algebra (see section 3 of chapter 3), the users first temporarily reduce the computations with rational or integer numbers to computations modulo a prime p, a prime power p^k, several primes, or several pairwise prime natural numbers $p_1, \ldots, p_h$, and then, at the end, recover the desired output from its values modulo p, p^k, or $p_1, \ldots, p_h$. This leads us to computations in the ring $\mathbf{Z}_M$ of integers modulo a fixed integer $M > 1$ (such a ring is a field if and only if M is a prime), and further on, in the rings of polynomials, formal power series and rational functions in s variables over $\mathbf{Z}_M$ (for fixed integers $M > 1$ and $s > 0$). The latter rings and fields are infinite. Unlike the infinite number fields, however, they contain exactly M distinct integers, that is, $0, 1, \ldots, M-1$, and such rings and fields are said to *have characteristic* M, whereas the rings and fields containing all the integers $0, \pm 1, \pm 2, \ldots$ are said to *have characteristic* 0.

Can we extend our study to computations over abstract fields and rings? Some of our computational problems, such as Problems 2.3 ($POL \cdot INTERP$) and 2.5 ($GEN \cdot DFT$), must be performed in the rings (or fields) of a sufficiently high cardinality; if the original ring $\mathcal{R}$ has a lower cardinality, we may just shift to an algebraic extension of $\mathcal{R}$ (see more comments on this later on in this section). Now we recall that the algorithms of this chapter involve only arithmetic operations and (rarely) also comparisons and, apart from the above restriction, can usually be easily extended to computations over any field. Moreover, they can be further extended to any (commutative) ring $\mathcal{R}$ of constants in the cases where either divisions are not involved or all the divisors have reciprocals in or over the ring $\mathcal{R}$. Ignoring the latter requirement may lead to pitfalls. For instance, recall the algorithm of section 4 that computes the coefficients of a polynomial of degree n, where the power sums of its zeros are given as the input [see Problem 4.8 ($I \cdot POWER \cdot SUMS$)]. This algorithm involves divisions by $2, 3, \dots, n$ and can be performed if and only if the field or ring of its constants allows such divisions. In particular, the algorithm fails over any ring of integers modulo M, for $M \leq n$. Further examination shows that not only this algorithm cannot be extended, but generally, the power sums do not well define the coefficients of a polynomial over such a ring or field (exercise 27). On the other hand, a computational problem can be well defined over any field, but certain algorithms may work over some fields and fail over other ones (exercise 26).

The extension of the algorithms of this chapter to any field, and of many of them to any ring (with only a small increase of their asymptotic computational cost), relies, in particular, on the following fundamental result ([CKa], [Nus], [Sc77]):

Theorem 7.1. *Given two polynomials of degree n with coefficients in any ring $\mathcal{R}$ of constants, the coefficients of their product can be computed by using $O(n \log n \log\log n)$ additions and subtractions in $\mathcal{R}$ and $O(n \log n)$ multiplications in $\mathcal{R}$, and these operations can be implemented on arithmetic PRAM by using $O(\log n)$ parallel arithmetic steps and $O(n \log\log n)$ processors.*

Proof. We will first give a proof, by following [Nus], in the case of rings $\mathcal{R}$ allowing divisions by 2. Furthermore, we will recall Remark 3.2 and reduce the original problem to computing $w(x) = u(x)\,v(x) \bmod (x^J + 1)$ where $J = 2^j$, $j = 0, \dots, k$; $K = n + 1 = 2^k$. It is sufficient to show how to compute

$$w(x) = w_K(x) = u(x)\,v(x) \bmod (x^K + 1) \; .$$

We will further assume that $r = k/2$ is an integer, denote that $R = 2^r$, $y = x^R$, and $i = gR + h$, for $i = 0, 1, \ldots, n$ and for $g = g(i)$, $h = h(i)$ ranging from 0 to $R - 1$. Note that $K = R^2$, consider the bivariate polynomials

$$u(x, y) = \sum_{g,h=0}^{R-1} u_{gR+h} x^h y^g \; ,$$

$$v(x, y) = \sum_{g,h=0}^{R-1} v_{gR+h} x^h y^g \; ,$$

$$w(x, y) = u(x, y)\,v(x, y) \; ,$$

and observe that $u(x) = u(x, y)$, $v(x) = v(x, y)$, for $y = x^R$. We will hereafter use the equations $u(x) = u(x, y) \bmod (x^R - y)$, $v(x) = v(x, y) \bmod (x^R - y)$.

The polynomial $w(x, y)$ has degree at most $2R - 2$ in x, so that

$$u(x)\,v(x) = w(x, y) \bmod (x^{2R} - 1) \bmod (x^R - y) \; .$$

Thus, we represent $w(x) = u(x)\,v(x) \bmod (x^K + 1)$ as follows:

$$w(x) = \tilde{w}(x, y) \bmod (x^R - y) \; ,$$

$$\tilde{w}(x, y) = u(x, y)\,v(x, y) \bmod (x^{2R} - 1) \bmod (y^R + 1) \; .$$

$\tilde{w}(x, y)$ has degrees at most $2R - 2$ in x and at most $R - 1$ in y. To make transition from $\tilde{w}(x, y)$ to $w(x)$, first substitute y for x^R in the expansion of $\tilde{w}(x, y)$ as a polynomial in x and y, so that each monomial $x^{R+g} y^h$ turns into $x^g y^{h+1}$, whereas the monomials $x^g y^h$ remain invariant for $g, h = 0, 1, \ldots, R - 1$. In this way, we represent the resulting polynomial as $w_0(x, y) + y\,w_1(x, y)$ where $w_i(x, y) = \sum_{g,h=0}^{R-1} w_{i,g,h} x^g y^h$, $i = 0, 1$, so that

$$w_0(x, y) = w_0(x, y) \bmod x^R \; , \quad y\,w_1(x, y) = \sum_{g,h=0}^{R-1} w_{1,g,h} x^g y^{h+1} \; .$$

Substituting $y^R = -1$, obtain that

$$y\,w_1(x, y) = \sum_{g=0}^{R-1} x^g \left(\sum_{h=0}^{R-2} w_{1,g,h} y^{h+1} - w_{1,g,R-1} \right) .$$

Thus, $w(x)$ can be obtained from $\tilde{w}(x,y)$ by means of R^2 additions and subtractions.

To compute the coefficients of the polynomial $\tilde{w}(x,y)$, first represent it as a polynomial in x over the ring $\mathcal{R}[y]/(y^R+1)$ of polynomials in y modulo y^R+1, then observe that y is a primitive $2R$-th root of 1 in this ring and apply the FFT-based evaluation-interpolation techniques using the Fourier points y^i, $i=0,1,\ldots,2R-1$. Employ the reduction modulo y^R+1 to simplify this computation.

Let $M(S)$ and $A(S)$ denote the numbers of multiplications and additions/subtractions, respectively, required for the evaluation modulo x^S+1 of the polynomial product $u(x)\,v(x)$. Then simple inspection of the above approach gives us the bounds

$$M(R^2) \le 2R\,M(R)\ , \tag{7.1}$$

$$A(R^2) \le 4R^2 \log R + 3R^2 + R + (2R)A(R)\ , \tag{7.2}$$

and actually, we may improve (7.1) to $(2R-1)\,M(R) \ge M(R^2)$ (see [Nus]). Recursive application of these bounds for $R = 2, 4, 16, 256, \ldots$ gives us Theorem 7.1, in the case where divisions by 2 are allowed in the ring of constants.

In the above proof, we need divisions by 2 in order to perform the inverse DFT and the transition from $w(x)$ mod $(x^{n+1}-1)$ and $w(x)$ mod $(x^{n+1}+1)$ to $w(x)$ mod $(x^{2n+2}-1)$. On the other hand, we may extend the above proof of Theorem 7.1 to any ring that allows divisions by any fixed natural $c \ge 2$, simply by replacing $K=2^k$ by $K=c^k$, $k = \lceil \log(n+1)/\log c \rceil$ [compare the solution a) to Problem 2.5 $(GEN \cdot DFT)$]. The paper [CKa] gives a unified proof for all rings. The latter proof exploits DFT's in the rings of polynomials modulo $(x^R - y^{t_i})$ for natural t_i, $i=1,2$, $t_1 \ne t_2$, and recovers the desired polynomial product by means of Chinese remainder computations over abstract rings. This can be regarded as a generalization of the idea of recovering convolution from the positive and negative wrapped convolutions. ■

Remark 7.1. Theorem 7.1 enables us to solve (at a low computational cost) Problem 2.5 $(GEN \cdot DFT)$ in any ring of constants, in which this problem can be defined. Thus, for this problem and for many DFT based computations, the evaluation of primitive roots of 1 can be avoided, at the cost of adding the factor of $\log\log n$ to the estimated numbers of ops involved (or to the number of processors in parallel implementation).

Remark 7.2. Performing only one or few recursive steps of the algorithm presented in the above proof of Theorem 7.1, we may reduce the evaluation of $u(x)\,v(x) \bmod (x^{n+1}+1)$ to several convolution problems, which have smaller sizes. Recursive application of (7.1) and (7.2) gives us specific estimates for the sizes and the number of these convolution problems.

Remark 7.3. The algorithms supporting the proof of Theorem 7.1 can be immediately extended to the solution of Problem 2.4 ($SEV \cdot POL \cdot MULT$).

Theorem 7.1 implies that, by using by $O(\log\log n)$ times more ops, we may extend the fast polynomial multiplication algorithm of section 2 to the most general case of computation over an arbitrary ring $\mathcal{R}$, and consequently, we may extend numerous other computations reducible to polynomial multiplication, including polynomial division and the generalized DFT of Problem 2.5 ($GEN \cdot DFT$).

Let us next examine some further aspects of and recipes for the extension to abstract fields and rings. In particular, recall that, to apply the generalized DFT on K points in a (commutative) ring $\mathcal{R}$, we need to have a unity 1 in $\mathcal{R}$ and to select an invertible element w having at least K distinct powers, $1, w, \ldots, w^{K-1}$, in $\mathcal{R}$.

The first issue is immediately resolved since any ring $\mathcal{R}$ can be embedded into the ring $\mathbf{Z} \times \mathcal{R}$, where $\mathbf{Z}$ is the ring of integers and where in the new ring, $(1,0)$ is a unit element,

$$(m,a)+(n,b)=(m+n,a+b)\ ,$$

$$(m,a)\times(n,b)=(mn,ab+mb+na)$$

(see [Ja]). Thus, hereafter we will assume that the given ring $\mathcal{R}$ has a unit element.

On the other hand, in order to be able to select an invertible element w in $\mathcal{R}$, having at least K distinct powers in $\mathcal{R}$, we need to have $|\mathcal{R}|$, the cardinality of $\mathcal{R}$, exceeding $K-1$. The same problem also arises when we wish to apply interpolation on K points in $\mathcal{R}$, in particular, when we wish to use the evaluation-interpolation techniques of Toom [To]. If the cardinality $|\mathcal{R}|$ of $\mathcal{R}$ is less than K, we need to extend $\mathcal{R}$, and similarly, where we apply Lemma 5.1 to a subset S of a larger cardinality. In such cases the routine is the techniques of *algebraic extension*: we simply embed $\mathcal{R}$ into the ring (or field) $\mathbf{E}$ of univariate polynomials over $\mathcal{R}$ modulo a polynomial $m(x)$ of degree d over $\mathcal{R}$, such that $d \geq \log K/\log|\mathcal{R}|$; $m(x)$ has to be an irreducible polynomial if we need to allow divisions in **E**. (We refer to [Be] and [Sh] on how to find an irreducible polynomial in $\mathbf{Z}_m$.) Then each arithmetic operation "$\circ$" is performed in the extension ring (or field) **E** and amounts to $c(\circ, d)$ operations in $\mathcal{R}$ performed at the cost $O_A(\log d, c(\circ, d)/\log d)$ in $\mathcal{R}$. If $\circ$ is an addition or a subtraction, then $c(\circ, d) = O(d)$; if $\circ$ is a multiplication or a division, then $c(\circ, d) = O(d \log d \log\log d)$, due to Theorem 7.1.

Thus, the growth of the computational complexity (due to the algebraic extension) is rather small: for instance, to increase the cardinality of $\mathcal{R}$ by $n^{O(1)}$ times, we only increase the parallel time bound by $O(\log\log n)$ times, the processor bound by $O(\log n \log\log\log n)$ times, and the potential work and the sequential time bounds by $O(\log n \log\log n \log\log\log n)$ times.

The techniques of the algebraic extension are used, in particular, in the following routine for the application of Lemma 5.1 (with the goal of avoiding singularities, such as vanishing denominators and singular matrices, which may appear for certain input instances of the original algorithm):

Procedure 7.1, *regularization via randomization.*

1. Restate the original algorithm in its *generic version*, involving one or several indeterminates $x_0, \ldots, x_{k-1}$, so as to parametrize the potentially vanishing values by turning them into nonvanishing polynomials in $x_0, \ldots, x_{k-1}$.

2. Choose a sufficiently large set S in the given field or ring $\mathcal{R}$ of constants or in its algebraic extension **E** and replace these indeterminates by random values independently and uniformly chosen from S, thus arriving at a *randomized version* of the original algorithm.

3. Apply Lemma 5.1 to estimate the probability P of encountering a singularity and repeat Stage 2 of the procedure for a larger set S if P is too large.

4. Perform the computations to solve the problem in its randomized version and estimate their complexity in the original ring or field $\mathcal{R}$, either directly or by extending the estimates first obtained in **E**.

Hereafter, in our calls for Procedure 7.1, we will usually specify only its Stage 1, thus working with the generic version of the problem and assuming that the given field or ring and the set S are large enough to ensure the desired regularity with a high probability. As a rule, the specific routine estimates for the probability and for the extension of the complexity estimates due to the algebraic extension of the fields or rings will be left to the reader as an exercise.

For many computational problems, we may, in principle, complement

Procedure 7.1 by its certification stage, which would verify the correctness of the output, so as to define the Las Vegas type randomization. Such a certification stage can be omitted, however, if the computational cost of this stage exceeds the cost of the randomized solution itself.

Let us now assume that $|\mathcal{R}| > K$ and revisit the problem of the application of generalized DFT on K points in $\mathcal{R}$.

For an element y of a ring $\mathcal{R}$, let $c(y)$ denote the minimum positive integer c such that $y^c = 1$ provided that there exists such an integer. Let an invertible element x in $\mathcal{R}$ be available such that $c(x) \geq K$. Then generalized DFT on K points in $\mathcal{R}$ can be defined by the powers $1, x, \ldots, x^{K-1}$ of such a *generator* x and can be applied, in particular, to reduce the evaluation and interpolation on K points to polynomial multiplication.

Let us show that a desired generator x can be computed in any ring $\mathcal{R}$ with unity where K distinct invertible elements $x_0 = 1, x_1, \ldots, x_{K-1}$ are available. Indeed, if $c(x_i) \geq K$ for some i, then simply set $x = x_i$. Otherwise, for the multiplicative group G generated by $x_0 = 1$, $x_1, x_2, \ldots, x_{K-1}$, compute a single generator g, such that $g \in G$ and $x_1, \ldots, x_{K-1}$ are powers of g, and set $x = g$. [The reader may verify that $c(g) = m$ if $g = \prod_{i=1}^{K-1} x_i^{a_i}$ where $a_1, \ldots, a_{K-1}$ are integers such that $m \sum_{i=1}^{K-1} a_i/c(x_i) = 1$, $m = \text{lcm}(c(x_1), c(x_2), \ldots, c(x_{K-1})) \geq K$.]

Now, the reader may extend the asymptotic estimates of this chapter to the case of computations over any field and, in many cases, any ring of constants. Computing over the rings, one frequently may avoid divisions by means of scaling, as in the case of divisions by nonmonic polynomials (see section 3), and/or by applying the Chinese remainder theorem for integers, as in [CKa], for polynomial multiplication over any ring with unity.

In few cases, the existence of the solution requires special consideration. In particular, a polynomial $p(x)$ of degree n over $\mathcal{R}$ may have less than n zeros in $\mathcal{R}$ (counting them with their multiplicity) and may even have no zeros in $\mathcal{R}$, so that over an arbitrary field $\mathbf{F}$ or ring $\mathcal{R}$, Problem 4.4 ($POL \cdot ZEROS$) should be replaced by the problem of factorization of $p(x)$ into the product of irreducible factors (see [Kn], [LLL], [Ka83], [GSh92], [Ka92], [Mil92], and also Algorithm 3.3.4 on the solution of the latter problem).

In Remarks 2.3.2, 4.2.1 and in section 6 of chapter 4 we will further comment on the computations over arbitrary fields.

8. Multiplication of Multivariate Polynomials.

We may immediately extend Theorem 7.1 to the case of multiplication of two multivariate polynomials, $u = u(x_1, \dots, x_m)$ and $v = v(x_1, \dots, x_m)$, such that

$$p = uv = p(x_1, \dots, x_m) = \sum_{i_1, \dots, i_m} p_{i_1 i_2 \dots i_m} x_1^{i_1} x_2^{i_2} \dots x_m^{i_m}, \tag{8.1}$$

by applying the Kronecker substitution,

$$x_1 = y \ , \quad x_{k+1} = y^{D_1 \dots D_k} \ , \quad k = 1, \dots, m-1 \ , \tag{8.2}$$

where the product of a pair of m-variate polynomials has degree less than D_k in x_k. If the two input polynomials have degrees at most d_j in x_j, then we may set $D_j = 2d_j + 1$, $j = 1, 2, \dots, m$, and thus map these polynomials and their product into univariate polynomials of degrees at most $N = \prod_{i=1}^{m}(2d_i + 1)$, from which we may go back to the original m-variate polynomials and to their product by means of the same Kronecker map (8.2), in the reverse direction. Thus, we arrive at the complexity bounds of

$$M = M(d_1, \dots, d_m) = O(N \log N \log \log N), \quad N = \prod_{i=1}^{m}(2d_i + 1) \ , \tag{8.3}$$

operations in a fixed ring or field $\mathcal{R}$ of constants or $O(\log N)$ parallel steps and $O(N \log \log N)$ processors for the multiplication over $\mathcal{R}$ of two such polynomials.

In this section, we will recall the results of [P89d] which improve the bound (8.3), over an arbitrary field $\mathbf{F}$ [see (8.7)]. (The omitted details can be found in [P89d].)

In comparison with the estimate (8.3), obtained via the reduction to the univariate case and based on the mapping (8.2), our approach is a strict improvement. Indeed, all the known algorithms that support Theorem 7.1 recursively reduce the computation to the multivariate case, that is, to polynomials of smaller degrees and with more variables. Thus, if we first go to the univariate case and then back to the multivariate case, we will lose the operations involved in this transition, as this is also clear from the comparison of (8.3) with (8.7), for a smaller c.

For the polynomial u of (8.1), and similarly for the polynomials v and p of (8.1), we have:

$$u = u(x_1, \dots, x_m) = u_{x_1}(x_2, \dots, x_m), \tag{8.4}$$
$$u_{x_1}(x_2, \dots, x_m) = \sum_{i_2, \dots, i_m} u_{i_2 \cdots i_m}(x_1) x_2^{i_2} \cdots x_m^{i_m},$$
$$u_{i_2 \cdots i_m}(x_1) = \sum_{i_1} u_{i_1 \cdots i_m} x_1^{m}.$$

Let d_j denote the degree of $p(x_1, \ldots, x_m)$ in x_j, and let $G = G(c_1, \ldots, c_m)$ denote an m-dimensional grid (or lattice) where the variable x_j takes on c_j distinct values in $\mathbf{F}$ or in its algebraic extension.

To simplify the subsequent estimates, let $d_j + 1 \le d$ and $c_j = 2d_j + 1 \le c = 2d - 1$, for an integer $d > 1$ and for all j, and let $E(c, m)$ and $I(c, m)$ denote the numbers of ops required in order both to evaluate, on the grid G, a polynomial $u(x_1, \ldots, x_m)$ in m variables of degrees at most $d - 1$ in each of them, and, respectively, to solve the converse interpolation problem, so that

$$\max\{E(c,1)\,,\, I(c,1)\} \le \gamma M(c,1) \log c\,, \tag{8.5}$$

for a fixed constant γ. By using the identity (8.4), we deduce that

$$\begin{aligned} &E(c,m) \le d^{m-1}E(c,1) + cE(c,m-1) \le \\ &(c+d)d^{m-2}E(c,1) + c^2E(c,m-2) \le \cdots \le \\ &E(c,1)\sum_{i=0}^{m-1} c^i d^{m-1-i} \le h(d)c^{m-1}E(c,1)\,, \end{aligned}$$

where $c = 2d - 1$, $h(d) < \frac{2d-1}{d-1} \le 3$ for $d \ge 2$. The complexity of the evaluation of $v = v(x_1, \ldots, x_m)$ on G is bounded in the same way.

Similarly, we deduce the bound $I(c, m) \le mc^{m-1}I(c, 1)$.

Given two multivariate polynomials, we will apply the evaluation-interpolation technique of [To] to compute the coefficients of their product. In this way, we arrive at the upper bound

$$M(c,m) = O(N \log N \log c \log\log c),\ N = c^m, \tag{8.6}$$

over any field $\mathbf{F}$, such that $|\mathbf{F}| \ge c$ (see the details in [P89d]). This bound improves (8.3) if, say, $c = O(1)$ as $m \to \infty$.

Moreover, we may remove the factor of $\log c$ from the right sides of (8.5) and, consequently, of (8.6), by applying the techniques of the proof of Theorem 7.1. Namely, if $\mathbf{F}$ allows divisions by a natural $h > 1$, we may create a principal c-th root of 1, as in the proof of Theorem 7.1, and apply DFT on $h^{\lceil k/2 \rceil}$ points, $k = \lceil \log c / \log h \rceil$, in order to reduce the original evaluation and interpolation for univariate polynomials of degrees at most $c - 1$ to similar computations with polynomials of degrees of the order of $2\sqrt{c}$. Recursively repeating such a process, as in the proof of Theorem 7.1, we arrive at the following bound over any field $\mathbf{F}$:

$$M(c,m) = O(N \log N \log\log c),\ N = c^m\,. \tag{8.7}$$

By using the techniques of [CKa], which we cited in the proof of Theorem 7.1, we may extend this bound to the case of computations over any (commutative) ring of constants.

Remark 8.1. Over the fields of characteristic 0, an alternative algorithm of [CKL89] evaluates any polynomial in m variables having at most T monomials (each of total degree at most T) by using $O(M(T,1)\log T)$ ops. (Recall that the total degree of a monomial $x_1^{i_1}\cdots x_m^{i_m}$ is $i_1+\cdots+i_m$.) This bound is superior to (8.7) on the class of input polynomials in m variables with a fixed upper bound d on the total degree of all its monomials, provided that d and m are large. The same bound is inferior to (8.7) on the class of m-variate input polynomials with the bound $d = O(1)$ on their degrees in all variables where m is large. Technically, the algorithm of [CKL89] relies on the polynomial evaluation and interpolation on the special set of points $\{(a_1^i, a_2^i, \ldots, a_m^i),\ i = 0, 1, \ldots, T-1\}$, and on fast multiplication of a Vandermonde matrix and its inverse and transpose by a vector (see section 6 of chapter 2).

9. Sparse Multivariate Polynomial Interpolation.

In practical computations with sparse multivariate polynomials, it is efficient to define them by a (black box) subroutine available for their evaluation at the cost bounded by a fixed constant B. At some point, we may need to recover the coefficients of the polynomial. Typically, we may assume the representation

$$h(\mathbf{x}) = \sum_{i=1}^{n} c_i m_i(\mathbf{x})$$

where c_i are unknown nonzero coefficients; $m_i(\mathbf{x}) = \prod_{j=1}^{k} x_j^{e_{i,j}}$ are monomials in the components of the vector $\mathbf{x}$ defined by the unknown exponents $e_{i,j}$, $i = 1, \ldots, n$; $\mathbf{x} = [x_1, \ldots, x_k]^T$ is the vector of variables; and n is the unknown number of terms of $h(\mathbf{x})$ bounded by a fixed natural $N \geq n$. Practically important is the case where $\max_i \sum_{j=1}^{k} e_{i,j}$ is much greater than n, which means that $h(\mathbf{x})$ is a sparse polynomial. In this case, a surprisingly simple algorithm for the recovery of the exponents $e_{i,j}$, at first, and then of the coefficients c_i for all i and j [by using the black box subroutine for the evaluation of $h(\mathbf{x})$] was given in [B-OT] (compare also a later exposition in [KLW]).

Denote that

$$v_i = \prod_{j=1}^{k} p_j^{e_{i,j}}, \quad h_s = h(p_1^s, \ldots, p_k^s), \quad s = 0, 1, \ldots, 2N-1, \tag{9.1}$$

for k fixed distinct primes $p_1, \dots, p_k$, so that

$$v_i \neq v_j \text{ for } i \neq j, \; h_s = \sum_{i=1}^{n} c_i v_i^s. \tag{9.2}$$

The values v_i are immediately available at the cost of

$$C = O(\sum_{i,j} \log(e_{i,j} + 1)) \tag{9.3}$$

ops, as soon as the exponents $e(i,j)$ are known. On the other hand, the values h_s can be computed, at the cost $2NB$, by the (black box) subroutine for the evaluation of $h(\mathbf{x})$. Consequently, given a natural n and the exponents $e(i,j)$, we may compute the coefficient vector $\mathbf{c} = [c_i]$ at the cost of $2NB + C + O(n \log^2 n)$ ops by solving the special linear system

$$V^T \mathbf{c} = \mathbf{h}, \; V^T = [v_i^j], \; \mathbf{h} = [h_s],$$

where $V^T = [v_i^j]$ is an $n \times n$ transposed Vandermonde matrix [see our solution of Problems 2.6.1 ($V^T \cdot SOLVE$) and 2.6.2 ($V^T \cdot VECTOR$)].

To compute the exponents $e_{i,j}$, consider the auxiliary polynomial

$$p(y) = \prod_{i=1}^{n} (y - v_i) = \sum_{s=0}^{n} u_s y^s, \; u_n = 1, \tag{9.4}$$

where the values v_i are available at the cost bounded by (9.4), but n and the values u_s are not known. Then $p(v_i) = 0$ for all i, $\sum_{i=1}^{n} c_i v_i^j p(v_i) = 0$ for all j. Substitute (9.4) and deduce that

$$\sum_{s=0}^{n} u_s \sum_{i=1}^{n} c_i v_i^{s+j} = 0, \; j = 0, \dots, n-1.$$

Now substitute (9.2) and obtain that

$$\sum_{s=0}^{n} h_{s+j} u_s = 0, \; j = 0, \dots, n-1.$$

We first compute the values $h_0, \dots, h_{2N-1}$ at the cost $2NB$, recall that $u_n = 1$, and then compute n and u_s for $s = 0, 1, \dots, n-1$. This amounts to solving Problem 5.3 ($RECUR \cdot SPAN$) of computing the minimum span for a linear recurrence sequence. A solution algorithm for this problem was shown in section 5 (also see section 5 of chapter 2).

When the coefficient vector $\mathbf{u}$ of $p(y)$ is available, we obtain the integer zeros $v_1, \dots, v_n$ of the polynomial $p(y)$ by using any zerofinding algorithm for polynomials having only integer zeros (see [Z90] and/or Algorithm 3.3.4 in chapter 3). Having v_i, we recover the exponents $e_{i,j}$ for all i and j (see exercise 33).

Exercises to Chapter 1.

1. Specify the constants hidden in the "O" notation in the asymptotic estimates of this chapter (and of all other chapters too).
2. Apply all the algorithms of this chapter to some specific polynomials, such as $u(x) = x^3 + 2x^2 - 3x + 4$, $v(x) = x^2 + 2x - 3$, $s(x) = 2x^2 - x + 3$, $t(x) = 3x^3 + 2x^2 + x - 1$. (Extend this exercise to other chapters as well.)
3. Extend the reducibility digraph for polynomial computations of Figure 1 as much as you can. (Devise similar digraphs for chapters 2 and 4.)
4. Given the polynomial $x^3 + 2x^2 - x + 5$, rewrite it as $(x-2)^3 + a_2(x-2)^2 + a_1(x-2) + a_0$ specifying the integer constants a_0, a_1 and a_2. Reduce this problem to multiplication of two polynomials.
5. Show that the computation of the *sine transform*,

$$x_i = \sum_{j=1}^{n} y_j \sin \frac{\pi i j}{n+1}, \quad i = 1, \ldots, n,$$

can be performed in $O(n \log n)$ ops by means of FFT.

6. Reduce polynomial (integer) multiplication to squaring a polynomial (an integer).
7. To multiply two integers p and q modulo 2^{2d} by using $O(d^{\log 3})$ Boolean operations, recursively apply the following relation: $pq = (p_0 + 2^d p_1)(q_0 + 2^d q_1) = p_0 q_0 (1 - 2^d) + (p_1 + p_0)(q_1 + q_0)2^d + p_1 q_1 (2^{2d} - 2^d)$ ([KO]). Extend this algorithm to multiplication of a pair of polynomials of degrees at most n using $O(n^{\log 3})$ ops in an algorithm involving only 0's and 1's as the constants.
8. For n ranging from 1 to 100, compare the number of ops for multiplication of a pair of n-th degree polynomials with real coefficients in the classical algorithm and in the fast algorithm of this chapter. Find a threshold value n for which the classical algorithm becomes inferior. Redo this problem in the case of polynomials with complex coefficients assuming that each complex addition amounts to 2 real ops and that one complex multiplication amounts to 6 real ops. Solve the similar problem for the division of a polynomial of degree $2n$ by a polynomial of degree n.
9. Complete the omitted details of the description of Schönhage-Strassen's algorithm for integer multiplication (see Appendix C and [AHU]).
10. Compute $w(z)$ such that $w(z)(1 + 2z - 3z^2) = 1 \bmod z^3$.

11. Prove that the coefficients of the generalized Taylor expansion of Problem 3.5 ($TAYLOR \cdot EXP$) are unique.
12. Extend the Chinese remainder computation to integers.
13. Prove the existence and the uniqueness of the Chinese remainder polynomial that solves Problem 4.2 ($CH \cdot REMNDR$) for polynomials. Extend the solution to integers.
14. Let $m_i(x)$, for $i = 0, 1, \dots, k$, be polynomials of degree 3, $L(x) = \prod_{i=0}^{k} m_i(x)$. To compute $v_i(x) = (L(x)/m_i(x)) \bmod m_i(x)$, we may first perform $k+1$ divisions of $L(x)$ by $m_i(x)$ for all i and then reduce the results modulo $m_i(x)$. Alternatively, we may proceed as in the solution of Problem 4.2 ($CH \cdot REMNDR$). Estimate the computational cost as a function in k (as $k \to \infty$) in both cases.
15. [This exercise extends Problems 2.2 ($POL \cdot EVAL$) and 2.3 ($POL \cdot INTERP$) of polynomial evaluation and interpolation.] Let $p(x) = \sum_{i=0}^{n} p_i x^i$ be a polynomial of degree n. Given the numbers $c_0, \dots, c_{n-1}$, consider Newton's representation of $p(x)$:

$$\begin{aligned} p(x) =& q_0 + q_1(x - c_0) + q_2(x - c_0)(x - c_1) + \cdots + \\ & q_n(x - c_0)(x - c_1) \cdots (x - c_{n-1}). \end{aligned}$$

   a) Describe an algorithm based on Horner's scheme for the evaluation of $p(x)$ at a point x, for given $q_0, \dots, q_n$ and $c_0, \dots, c_{n-1}$.

   Then show that the following computational problems can be solved in $O(n \log^2 n)$ ops:

   b) Given $c_0, \dots, c_{n-1}$ and the coefficients $p_0, \dots, p_n$, compute the *divided differences* $q_0, \dots, q_n$ [compare Problem 3.5 ($TAYLOR \cdot EXP$)];

   c) Given $c_0, \dots, c_{n-1}$, $q_0, \dots, q_n$, compute the coefficients $p_0, \dots, p_n$ (precompute some "supermoduli" and then use a divide-and-conquer technique);

   d) Given $c_0, \dots, c_{n-1}$, $q_0, \dots, q_n$, $x_0, \dots, x_n$, compute $p(x_i)$, $i = 0, \dots, n$.
16. Show that both Euclid's algorithm and the extended Euclidean scheme (Algorithms 5.1 and 5.1b, respectively), in which polynomial divisions are performed by means of the classical algorithm for "synthetic division," involve $O(n^2)$ ops, for two input polynomials of degrees n and m, $m = O(n)$.
17. Recursively define the Chebyshev polynomials as follows: $T_0(x) = 1$,

$T_1(x) = x$,

$$\begin{pmatrix} T_s(x) \\ T_{s-1}(x) \end{pmatrix} = \begin{pmatrix} 2x & -1 \\ 1 & 0 \end{pmatrix} \begin{pmatrix} T_{s-1}(x) \\ T_{s-2}(x) \end{pmatrix}, \quad s = 2, 3, \ldots .$$

Compute the coefficients of $T_{33}(x)$ without computing $T_s(x)$ for $1 < s < 33$. Devise an algorithm that computes the coefficients of T_s, for a given s, in $O(s \log s)$ ops. *Hint:* compute $\begin{pmatrix} 2x & -1 \\ 1 & 0 \end{pmatrix}^s$ for $x = \omega^i$, $i = 0, \ldots, s$, by means of repeated squaring, and then perform interpolation. Improve this bound by using the explicit expressions for such coefficients (see Remark 3.4.2).

18. Extend the fast gcd computation algorithm from the case of polynomials to the case of integers ([AHU]).
19. Given $m_0(x) = x^2 + 1$ and $m_1(x) = 2x + 5$, compute $w_0(x) \bmod m_0(x)$ and $w_1(x) \bmod m_1(x)$ such that $w_0(x)m_1(x) = 1 \bmod m_0(x)$, $w_1(x)m_0(x) = 1 \bmod m_1(x)$.
20. Compute the partial fraction decomposition for the ratio $r(x) = \frac{7x^2+4x-12}{(x^2+1)(2x+5)}$, that is, find the rational a, b and c such that $r(x) = \frac{ax+b}{x^2+1} + \frac{c}{2x+5}$.
21. Find the polynomial $p(x)$ of degree 2 such that $p(x) \bmod (x^2 + 1) = 5x - 16$, $p(x) \bmod (2x + 5) = 20$.
22. Prove Facts 5.1 and 5.2.
23. Compute the (2,2) Padé approximation to $1 + 2x - 3x^2 + x^3 - 2x^4 + x^5$.
24. Devise a randomized algorithm for computing the lcm of k polynomials $u_1(x), \ldots, u_k(x)$ based on the observation that the $(m - r, m - r)$ Padé approximation $(R(x), T(x))$ to $a(x) = (\sum_{i=1}^{k} c_i/u_i(x))^{-1}$ (with a probability p close to 1) defines $R(x) = \mathrm{lcm}(u_1(x), \ldots, u_k(x))$ provided that $m = \sum_{i=1}^{k} \deg u_i(x)$, $n = \max_{i=1,\ldots,k} \deg (U(x)/u_i(x))$, $U(x) = \prod_{i=1}^{k} u_i(x)$, and that $c_1, \ldots, c_k$ have been chosen at random from a sufficiently large finite set S of constants. Use Lemma 5.1 to estimate p as a function in $|S|$.
25. Extend the alternative randomized test for $u(z) = s(t(z))$ in the solution of Problem 6.3 ($POL \cdot DECOMP$) to verification of other polynomial identities, such as $u(z) = v(z)q(z)$, $r(z) = u(z)s(z) + v(z)t(z)$. Estimate and compare the complexity of randomized and deterministic verification of such identities (also see [Kam89]).
26. Consider the two following questions on Newton's iteration:
    a) In order to compute modulo x^m the square root $w(x)$ of a polynomial $a(x)$ of degree n in x, apply Newton's iteration to the equations $f(w) = w^2 - a = 0$ and $g(w) = a/w^2 - 1 = 0$, obtaining the

recurrences $w_{i+1} = \frac{1}{2}w_i + \frac{a}{2w_i}$, $w_{i+1} = \frac{1}{2}(3w_i - \frac{w_i^3}{a})$, respectively. Estimate the computational costs of the two algorithms.

b) Show that the polynomial $x+1$ is the square root of x^2+1 over $\mathbf{Z}_2$ (the field of integers modulo 2) and that Newton's iteration cannot compute this square root over $\mathbf{Z}_2$.

27. Show that power sums may nonuniquely define the coefficients of a polynomial over a finite field. Consider a) the polynomials of degree 3 over $\mathbf{Z}_2$ and b) the polynomials x^m and $(x+1)^m$ over $\mathbf{Z}_k$ where k divides m.
28. Prove that 3 is a primitive 16-th root of 1 in $\mathbf{Z}_{17}$. Use this property in order to compute the coefficients of the product of the two polynomials $p(x) = \sum_{i=0}^{7}(i+1)x^i$, $q(x) = \sum_{i=0}^{7}(i+2)x^i$ in $\mathbf{Z}_{17}$.
29. Extend the asymptotic complexity estimates of this chapter (and of chapters 2 and 3) to the case of computations over any field of constants.
30. Given any ring $\mathcal{R}$ and a natural $n \geq 2$, extend $\mathcal{R}$ to the ring $\bar{\mathcal{R}} = \mathcal{R}/(x^n - 1)$ of polynomials modulo $x^n - 1$ over $\mathcal{R}$. Observe that x is a primitive n-th root of 1 in the ring $\bar{\mathcal{R}}$. Perform DFT(n) in $\bar{\mathcal{R}}$ on the Fourier set $\{1, x, x^2, \ldots, x^{n-1}\}$. Estimate that $O(n^2 \log n)$ ops in $\mathcal{R}$ suffices. Hint: $O(n \log n)$ ops in $\bar{\mathcal{R}}$ suffice, where ops are additions/subtractions and multiplications by powers of x.
31. Estimate the space-complexity of the computational problems of this book under both customary assumptions, with and without accounting for the space for the input and output allocation (compare Remark 1.4). You may first apply the algorithms of this book and then try to improve the space complexity estimates associated with them, by allowing some limited increase of the running time of the algorithms. Consider, in particular, Problems 2.2a (DFT) and 2.4a ($POL \cdot MULT$), with integer inputs. Allow the outputs to overwrite the inputs.
32. Extend the estimates of section 8 by expressing them in terms of the degrees $d_1, \ldots, d_m$ of the input polynomials in all the m variables.
33. Estimate the complexity of the evaluation of the exponents e_{ij} of (9.1) for $i = 1, \ldots, n$; $j = 1, \ldots, k$, given the set $\{n, v_1, \ldots, v_n, p_1, \ldots, p_k\}$.

Appendix A. Some Customary Models of Computation.

Some other customary models of computing imply the same asymptotic complexity estimates as RAM and PRAM models for most of our computations. In particular, this is the case for the computations of the first two chapters if we rely on the model of *straight line programs*, defined over an arbitrary field of constants $\mathbf{F}$ as follows:

Straight Line Program:

Input: $\mathbf{F}$, a field of constants; $\mathbf{a} = [a_h]$, a vector (or a set) of n variables, whose dimension (respectively, cardinality) n is said to be the *input size* (as under the RAM model); and Σ, a set of operations, each allowed at the unit cost. The constants from $\mathbf{F}$ and all the variables a_h are supposed to be available cost-free (compare Remark 3.1.1).

Output: a vector $\mathbf{w} = [w_k(\mathbf{a})]$ of m functions in $\mathbf{a}$.

Computation: a sequence $f_1, f_2, \ldots, f_L$ of functions in $\mathbf{a}$ such that

$$f_j = q_j \circ r_j, \quad \circ \in \Sigma, \tag{A.1}$$

for every j. Here each q_j and r_j is a constant from $\mathbf{F}$, a variable a_h, or f_i for $i < j$, and for every k, there exists j_k such that, identically in $\mathbf{a}$,

$$w_k(\mathbf{a}) = f_{j_k}(\mathbf{a}), \text{ for } 1 \le k \le m. \tag{A.2}$$

The *computational cost* of such a straight line program is given by its length L.

We refer to [Ka88] for an extension to integral domains.

The set Σ defines the set of the functions that can be computed by the straight line program. For instance, if $\Sigma = \{+, -, \times, \div\}$, then the functions f_j, $j = 1, \ldots, L$, generated in this model are *rational functions* in the set of variables $\{a_h\}$ and with coefficients in $\mathbf{F}$. We call the computations and the algorithms *rational* in this case (compare Definition 1.8). If $\circ = \div$, in (A.1), then r_j shall not be the null function. If $\Sigma = \{\wedge, \vee\}$, $\mathbf{F} = \{0, 1\}$, and besides (A.1) we allow also functions f_j such that $f_j = \neg r_j$ (where $\neg 0 = 1$, $\neg 1 = 0$), then the functions f_j, $j = 1, \ldots, L$, generated in this model, are *Boolean functions*, and we have a straight line *Boolean model.*

The straight line program is too restrictive to be used in the later chapters, in which we employ PRAM models and their customary extensions, including, in particular, approximation, randomization, nonrational algorithms, comparisons of the field elements with zeros, comparisons of the magnitudes of pairs of real numbers and branchings of the program. These extensions enable us to make the computations more effective, that is, of a lower complexity. On the other hand, the resulting new models simulate the actual computations better than the original RAM or straight line program; in particular, we may simulate finite precision numerical computations as soon as we extend our original model by allowing approximations. Furthermore, Problem 4.4 ($POL \cdot ZEROS$) of this chapter, Problems 2.3.1 ($EIGENVALUES$), 2.3.3 (SVD) and many other problems of polynomial and matrix computations do not generally have rational solutions but are practically solved by using approximation algorithms. We will next describe some of these extensions, which can be similarly applied to the RAM, PRAM, and circuit models.

(A.1) and (A.2) still define the straight line program if we include nonrational operations into the set Σ, but we arrive at the *algebraic computation tree* model if we allow comparisons and use their results to branch the program.

Over any subfield $\mathbf{F}$ of the complex field, we may replace computing the exact solution by the simpler problem of computing its approximation within a fixed tolerance $\epsilon > 0$ to the output errors; this is performed by *approximation algorithms.* (Note that here we keep assuming that all the field operations are performed with no errors.) To modify the straight line program to capture such a realistic assumptions, we just introduce the domain Ω where all the output functions $w_k(\mathbf{a})$, $k = 1, \ldots, m$, and $f_j(\mathbf{a})$, $j = 1, \ldots, L$, are defined, include ϵ into the input set, and replace (A.2) by the relations

$$|w_k(\mathbf{a}) - f_{j_k}(\mathbf{a})| \le \epsilon, \ 1 \le k \le m, \ \mathbf{a} \in \Omega, \tag{A.3}$$

or

$$|w_k(\mathbf{a}) - f_{j_k}(\mathbf{a})| \le \epsilon |w_k(\mathbf{a})|, \ 1 \le j_k \le L, \ \mathbf{a} \in \Omega, \tag{A.4}$$

which amount to (A.2), if $\epsilon = 0$.

Note that (A.4) implies that

$$f_{j_k}(\mathbf{a}) = w_k(\mathbf{a}) = 0 \quad \text{if} \quad w_k(\mathbf{a}) = 0 \ ,$$

that is, (A.4) is a stronger requirement than (A.3) in this case.

Further, we may allow randomization, that is, we may allow to include into the set of variables some auxiliary random parameters, assuming that the cost of choosing their values from a fixed set is negligible (compare the end of section 2 of chapter 3). Then we shall require that the relations

(A.3) or (A.4) (for $\epsilon = 0$ or $\epsilon > 0$) hold with a probability at least p_0 for a random choice of the parameter values, where p_0 is a fixed input value, $0 < p_0 \leq 1$. Some randomized algorithms verify the correctness of the solution and report $FAILURE$ unless the solution is proven to be correct; they are customarily called *Las Vegas algorithms.* Other algorithms may leave the output errors undetected (although with low probability $1 - p_0$); they are called *Monte Carlo algorithms.*

To estimate the Boolean cost (bit-cost) of rational or even nonrational algorithms, we may replace each $\circ$ in (A.1) by a set of (unit cost) *Boolean operations* (also called *bit-operations*) whose application supports such an operation $\circ$ performed with rounding off its operands and its output to a certain number of binary digits (bits) in their fixed or floating point binary representation. Then the *input size* is defined (as under the RAM model) as the number of bits required in order to represent all the input values, and the cost of an operation $\circ$ is not unit anymore but amounts to the number of Boolean (bit-)operations supporting it. In particular, this makes the cost of arithmetic operations depend on the precision of the operands and relates the cost of the algorithms to their numerical stability (see chapter 3). Recall that the numerical stability is the crucial requirement to practical numerical algorithms.

Next, let us briefly review the *circuit models*, widely used in the literature on algorithms. For our subjects, the reader may use these models as a substitution for the RAM models. To define the circuit models, represent the computations by an acyclic digraph (identified with a network or a circuit) whose vertices have indegrees 0, 1 or 2. The vertices of indegrees 1 and 2 are identified with the gates of the circuit and represent the unary or binary operations from a fixed set of operations. They also represent the values computed in these operations. The set of all these values must contain the desired output or its approximation. The vertices having indegree zero (the leaves) are associated with the input variables and constants. Assigning a set of values to such variables, that is, to the leaves, induces certain values to all the vertices of the digraph (via the operations associated with its vertices of positive indegrees). The acyclic digraph is called an *arithmetic* or *Boolean circuit* if its every vertex represents an arithmetic or Boolean operation, respectively. The circuit represents an algorithm for a fixed input size. The circuits for all the sizes form a family where some uniformity conditions on computing the input constants are customarily required (see [Bor77], [G86]).

The time required to perform an operation is assigned to the associated vertex as its time-cost. The overall sequential time of the computations is the sum of the costs of all the vertices; this sum is said to be the *size of the circuit.* The size equals the cardinality of the vertex set under the quite customary assumption that all the vertices have unit cost. For our computations the circuit size will usually remain within a constant factor from the sequential time estimate for the same algorithm implemented on the RAM model. Under the circuit model, the *parallel time of the computations* is represented by the circuit *depth*, defined as the length of the longest directed path in the digraph, that is, the depth equals the number of the unit cost parallel steps (or the parallel time) of the algorithm assuming that sufficiently many processors are available. For the algorithms that we will study, the size of the circuit will usually coincide (within a constant factor) with the potential work of the same algorithm implemented on a PRAM model, so that the ratio (size/depth) corresponds to the number of processors.

Finally, we recall that some arithmetic computations only require few nonarithmetic operations, such as branching, and then we may count both arithmetic operations and Boolean operations (supporting these few nonarithmetic operations) as unit cost operations, thus arriving at the *arithmetic-Boolean circuit model* of computing ([G86]).

Appendix B. The Problems and Their Reductions to Each Other.

In this appendix we will list the computational problems of this chapter, together with some of their reductions to other problems. Let $a_{\mathcal{P}}$ denote the minimum number of ops that are sufficient to use in order to solve Problem $\mathcal{P}$; let us write $(\mathcal{A}, \mathcal{B}) \succeq \mathcal{C}$ if Problem $\mathcal{C}$ can be reduced to Problems $\mathcal{A}$ and $\mathcal{B}$ so that $a_{\mathcal{C}} = O(a_{\mathcal{A}} + a_{\mathcal{B}})$. Some of the reductions in the list below use forward referencing, in particular, to Problems 4.3 ($MOD \cdot POL \cdot RECIPR$) and 5.1b ($EXT \cdot EUCLID$). To keep the list more compact, we have not specified the numbers of calls to some auxiliary subroutines needed for the solution via the reductions shown below. Such numbers of calls are implicit (or sometimes explicit) in our exposition in the main body of this chapter.

– Problem 2.1 ($POL \cdot EVAL \cdot S$), polynomial evaluation at a single point.
– Problem 2.2 ($POL \cdot EVAL$), polynomial evaluation on a set of points.

$$POL \cdot EVAL \preceq (POL \cdot MULT\text{'s}, POL \cdot DIVIDE\text{'s}).$$

– Problem 2.2a (DFT), discrete Fourier transform.

$$DFT \preceq IDFT.$$

– Problem 2.3 ($POL \cdot INTERP$), polynomial interpolation.

$$POL \cdot INTERP \preceq (POL \cdot EVAL, POL \cdot MULT\text{'s}).$$

– Problem 2.3a ($IDFT$), inverse discrete Fourier transform.

$$IDFT \preceq DFT.$$

– Problem 2.4 ($SEV \cdot POL \cdot MULT$), multiplication of several polynomials.

$$SEV \cdot POL \cdot MULT \preceq (DFT\text{'s}, IDFT).$$

– Problem 2.4a ($POL \cdot MULT$), multiplication of two polynomials.

$$POL \cdot MULT \preceq (DFT\text{'s}, IDFT).$$

– Problem 2.4b ($\pm CONVOL$), positive or negative wrapped convolutions.

$$\pm CONVOL \preceq (DFT\text{'s}, IDFT).$$

– Problem 2.5 ($GEN \cdot DFT$), DFT at any number of points.

$$GEN \cdot DFT \preceq POL \cdot MULT.$$

– Problem 2.6 ($VAR \cdot SHIFT$), shift of the variable.

$$VAR \cdot SHIFT \preceq POL \cdot MULT.$$

– Problem 2.7 ($INT \cdot MULT$), integer multiplication.
– Problem 3.1 ($POL \cdot DIVIDE$), polynomial division.

$$POL \cdot DIVIDE \preceq (POL \cdot RECIPR, POL \cdot MULT).$$

– Problem 3.2 ($SER \cdot MULT$), multiplication of power series.

$$SER \cdot MULT \preceq POL \cdot MULT.$$

– Problem 3.3 ($\pm SER \cdot MULT$), multiplication of a bi-infinite power series and a polynomial.

$$\pm SER \cdot MULT \preceq POL \cdot MULT.$$

– Problem 3.4 ($POL \cdot RECIPR$), reciprocal of a polynomial.

$$POL \cdot RECIPR \preceq POL \cdot MULT\text{'s}.$$

– Problem 3.5 ($TAYLOR \cdot EXP$), generalized Taylor expansion of a polynomial.

$$TAYLOR \cdot EXP \preceq (POL \cdot DIVIDE\text{'s}, POL \cdot MULT\text{'s}).$$

– Problem 3.6 ($INT \cdot DIVIDE$), integer division.
– Problem 4.1 ($POL \cdot MODULI$), modular representation of a polynomial.

$$POL \cdot MODULI \preceq (POL \cdot DIVIDE\text{'s}, POL \cdot MULT\text{'s}).$$

– Problem 4.2 ($CH \cdot REMNDR$), Chinese remainder computation.

$$CH \cdot REMNDR \preceq$$
$$(POL \cdot DIVIDE\text{'s}, POL \cdot MULT\text{'s}, MOD \cdot POL \cdot RECIPR).$$

– Problem 4.2a ($CH\cdot REMNDR1$), Chinese remainder computation modulo a power.

$$CH\cdot REMNDR1 \preceq (POL\cdot DIVIDE\text{'s}, POL\cdot MULT\text{'s}, MOD\cdot POL\cdot RECIPR).$$

– Problem 4.2b ($H\cdot INTERP$), Hermite interpolation.
– Problem 4.2c ($PART\cdot FRACT$), partial fraction decomposition.
– Problem 4.2d ($SYMM\cdot FUNCT$), evaluation of the symmetric functions.
– Problem 4.3 ($MOD\cdot POL\cdot RECIPR$) reciprocal of a polynomial modulo a polynomial.

$$MOD\cdot POL\cdot RECIPR \preceq EXT\cdot EUCLID.$$

– Problem 4.4 ($POL\cdot ZEROS$), approximation of polynomial zeros.
– Problem 4.5 ($POL\cdot COMP$), composition of polynomials.
– Problem 4.6 ($SER\cdot COMP$), composition of power series.
– Problem 4.7 ($POWER\cdot SUMS$), evaluation of the power sums of the zeros of a polynomial.
– Problem 4.8 ($I\cdot POWER\cdot SUMS$), evaluation of the power sums of the inverses of the zeros of a polynomial.
– Problem 5.1 ($POL\cdot GCD$), greatest common divisor of two polynomials.
– Problem 5.1a ($POL\cdot LCM$), least common multiple of two polynomials.
– Problem 5.1b ($EXT\cdot EUCLID$) extended Euclidean scheme.
– Problem 5.1c ($SEV\cdot POL\cdot GCD$), gcd of several polynomials.
– Problem 5.1d ($SEV\cdot POL\cdot LCM$), lcm of several polynomials.
– Problem 5.2 ($RAT\cdot INTERP$), rational interpolation.
– Problem 5.2a ($H\cdot INTERP1$), Hermite interpolation.
– Problem 5.2b ($PADE$), Padé approximation.
– Problem 5.3 ($RECUR\cdot SPAN$), minimum span of a linear recurrence.

$$RECUR\cdot SPAN \preceq PADE.$$

– Problem 6.1 ($SER\cdot REVERSE$), reversion of power series.
– Problem 6.2 ($POL\cdot ROOT$), root of a polynomial.
– Problem 6.3 ($POL\cdot DECOMP$), decomposition of a polynomial.
– Problem 6.4 ($COMPL\cdot POL\cdot DECOMP$), complete decomposition of a polynomial.
– Problem 6.5 ($ALG\cdot FUNCTS$), evaluation of algebraic functions.

Finally, here are the computational complexity estimates for the solution of some of the listed problems, whose input and/or output polynomials have degrees $O(n)$, that is: $O(n \log n)$ ops, for Problems 2.2a, 2.3a, 2.4a, 2.4b, 2.5, 2.6, 3.1, 3.4, 4.7, 4.8, 6.2; $O(n \log^2 n)$ ops, for Problems 2.2, 2.3, 4.2d, 4.3, 5.1, 5.1a, 5.2, 5.2a, 5.2b, 5.3.

Appendix C. Fast Integer Multiplication.

By following [AHU], pp. 270–274, we will now outline the algorithm of Schönhage–Strassen [ScSt] for multiplication modulo (2^s+1) of two positive integers u and v in

$$\mu^*(s) = O(s \log s \log\log s) \qquad (C.1)$$

Boolean operations. For simplicity, assume that

$$s = h^2 = 2^{2^d}, \quad \text{for a natural } d \qquad (C.2)$$

[later on we will show how to relax (C.2)]. Since $uv \bmod (2^s+1) = 2^s+1-v$ if $u = 2^s$, we assume that $u < 2^s$ and, similarly, $v < 2^s$ and denote that

$$u = \sum_{i=0}^{h-1} u_i 2^{ih}, \quad v = \sum_{i=0}^{h-1} v_i 2^{ih}, \quad 0 \le u_j, v_j < 2^h, \quad \text{for all } j; \qquad (C.3)$$

$$w = uv = \sum_{i=0}^{2h-2} w_i 2^{ih}, \quad w_i = \sum_{j=0}^{i} u_j v_{i-j}, \quad u_j = v_j = 0, \quad \text{for } j \ge h; \qquad (C.4)$$

$$y = w \bmod (2^s+1) = \sum_{i=0}^{h-1} y_i 2^{ih}, \qquad y_i = w_i - w_{h+i}, \quad \text{for all } i. \qquad (C.5)$$

Note that

$$-(h-1-i)2^{2h} < y_i < (i+1)2^{2h}, \quad i = 0, \ldots, h-1,$$

so it suffices to compute $y_i \bmod (h2^{2h})$, $i = 0, 1, \ldots, h-1$. Moreover, it suffices to compute $y_i^* = y_i \bmod h$, $x_i = y_i \bmod (2^{2h}+1)$, and then we immediately obtain

$$y_i = (2^{2h}+1)((y_i^* - x_i) \bmod h) + x_i, \quad \text{for all } i.$$

a) *Computation of* $y_0^*, \ldots, y_{h-1}^*$.

Denote: $u_i^* = u_i \bmod h$, $v_i^* = v_i \bmod h$, $i = 0, \ldots, h-1$,

$$\tilde{u} = \sum_{i=0}^{h-1} u_i^* h^{3i}, \quad \tilde{v} = \sum_{i=0}^{h-1} v_i^* h^{3i}, \quad 0 \le u_i^*, v_i^* < h \text{ for all } i.$$

Compute

$$\tilde{w} = \tilde{u}\tilde{v} = \sum_{i=0}^{2h-2} \tilde{w}_i h^{3i}, \quad 0 \le \tilde{w}_i \le \sum_{j=0}^{i} \tilde{u}_j \tilde{v}_{i-j} < h^3, \quad i = 0, \ldots, 2k-2,$$

$$\tilde{u}_j = \tilde{v}_j = 0 \quad \text{for all } j \geq h ,$$

and output $y_i^* = \tilde{w}_i - \tilde{w}_{h+i} \bmod h$, $i = 0, \ldots, h-1$ [compare (C.5)].

Observe that all the above computations require $O(s)$ Boolean operations [recursively use (C.1) or apply the algorithm of exercise 7 to multiply the positive integers $\tilde{u}$ and $\tilde{v}$, $\max\{\tilde{u}, \tilde{v}\} < h^{3h}$; the straightforward integer multiplication would require the order of $h^2 \log^2 h$ Boolean operations at this stage, which is the order of $s \log^2 s$ since $h^2 = s$].

b) *Computation of* $x_0, \ldots, x_{h-1}$.

Due to (C.5), the computation is immediately reduced to the evaluation of the values $w_i \bmod (2^{2h}+1)$, $i = 0, \ldots, 2h-2$ [see (C.4)], that is, to computing the negative wrapped convolution of the $(2h)$-dimensional vectors $[u_0, \ldots, u_{h-1}, 0, \ldots, 0]$ and $[v_0, \ldots, v_{h-1}, 0, \ldots, 0]$. Observe that $\omega = 16$ is a principal h-th root of 1 modulo $(2^{2h}+1)$ and that ω and h have reciprocals modulo $(2^{2h}+1)$ (compare Remark 2.1). Compute the negative wrapped convolution by applying the solution algorithm of section 2 for Problem 2.4b ($\pm$CONVOL) and using $\omega = 16$ as a principal h-th root of 1 and $\psi = 4$ as a principal $(2h)$-th root of 1. Due to the bounds of (C.3) on u_j and v_j, the three FFT's, involved here, require $O(h^2 \log h)$ Boolean operations. The multiplication stage of the solution to Problem 2.4b ($\pm$CONVOL) amounts to h multiplications of integers modulo $(2^{2h}+1)$.

Summarizing the above algorithm, we obtain that $\mu^*(s) \leq cs \log s + h\mu^*(2h)$. Recursively applying this relation to $\mu^*(2h)$, we arrive at (C.1).

Instead of (C.2), we may assume that $s = 2^d$ for a natural d [otherwise, we may replace s by $2^{\lceil \log s \rceil}$ and observe that this does not change the right side of (C.1)]. Then we may replace (C.3) as follows:

$$u = \sum_{i=0}^{g-1} u_i 2^{ih} , \quad v = \sum_{i=0}^{g-1} v_i 2^{ih} , \quad 0 \leq u_j, v_j < 2^h , \text{ for all } j ,$$

where $g = 2^{\lfloor d/2 \rfloor}$, $h = s/g$, and extend the above algorithm (setting, in particular, $\psi = 4^{h/g}$, $\omega = 16^{h/g}$) to deduce (C.1) for all natural s.

Appendix D. On Practical Impact of the Fast Convolution Algorithm.

The practical impact of using FFT and the fast convolution algorithm in computer algebra is more limited than one may suspect, for several reasons. Most of the computer time in this area is actually spent on computations with sparse multivariate polynomials, where the classical algorithms remain quite competitive with the fast convolution algorithms. Even in the univariate case, a large portion of computer time is spent in practice on such operations as divisions by polynomials of smaller degrees and computing the greatest common divisor of pairs of polynomials, where the advantages of using the fast convolution algorithm either disappear or become more limited. In many cases, the application of the fast convolution algorithm requires special care in order to contract the resulting growth of the precision needed in order to represent the auxiliary parameters. Finally, besides classical algorithms, strong competition to FFT and fast convolution comes from the algorithms based on the techniques of binary segmentation (see Remark 3.9.2), which is an effective and frequently superior approach for computation in finite fields and in other cases where the input parameters are represented by short binary integers.

2

Fundamental Computations with General and Dense Structured Matrices

Summary.

We survey computations with general (dense and unstructured) and with dense and structured matrices, including their numerous reductions and correlations to each other and to polynomial computations and their complexity.

The treatment of all these subjects is extensive and complete; the exposition of the algorithms for the computations with dense structured matrices follows the unified and systematic approach based on the study of the associated linear operators.

Furthermore, some recent and unpublished results are included, in particular, new and practically promising algorithms for computation of polynomial gcd and extended Euclidean scheme and for parallel computations with Toeplitz-like and Hankel-like matrices.

The subject of this chapter has substantial impact on other areas of computer science and computational mathematics.

1. Introduction, Contents, Definitions and Auxiliary Facts.

Matrix computations, particularly solving linear systems of equations and approximating matrix eigenvalues and singular values, dominate the practical computations in sciences and engineering. Most frequently, the matrices are either very large but special (sparse or dense but structured) or have smaller sizes, but large general matrices are also encountered, say, in the applications to solving integral equations.

In this chapter, we will first study the computations with *general* (that is, dense and unstructured) *matrices*; the complexity of such computations is a traditional subject of the computational complexity theory, which has further applications to some major problems of algebraic and combinatorial computing (see [P84b], Sect. 18; [GP85;88], [KUW], [GM], [Lo], [AGM], [NSV], [Sei], and Appendix A to this chapter). We will present a complete and systematic treatment of the main fundamental results and techniques in this area.

Then we will study computations with *dense structured matrices*, which are widely applied in the areas of control, signal processing, solving PDE's and algebraic computing. This study has various correlations to computations with polynomials and general matrices but has never been systematically presented in the computer science texts thus far.

We will omit the major topic of solving large sparse linear systems, on which we refer the reader to [GLi], [LRT], [P93b] (on direct methods) and to [FF], [GL], [FGN], [McC87], [SaS], [Var] (compare also chapter 3, sections 5 and 7) (on iterative methods). For various topics of matrix computations not covered in this book, we also refer the reader to [W65], [GL], [P84b].

Further material related to this chapter appears in chapter 4 (parallel matrix computations), in chapter 5 (matrix multiplication) and in chapters 8 and 9 (the matrix eigenvalue computations).

The algorithms supporting the best asymptotic complexity estimates for the computations with general matrices are not practical due to the immense overhead constants hidden in these asymptotic estimates, but the practical value of a more minor acceleration of the classical algorithms for the same computations has been recently recognized ([GL]). Furthermore, the design of the asymptotically fast algorithms for computations with general matrices involves some techniques of independent interest.

The complexity of computations with $n \times n$ dense structured matrices dramatically decreases in comparison with the case of $n \times n$ general matrices, that is, from the order of n^2 words of storage space and of n^ω ops with $2.37 < \omega \leq 3$ in the best algorithms, to $O(n)$ words of storage space and to $O(n \log^2 n)$ ops [and sometimes to $O(n \log n)$ ops], with small overhead constants. In practice of computations with dense structured matrices, n is frequently large, and we refer the readers to [Bun], [K87], [G84], [Tur], [Rok85], [ODR89] and [Ger] on some of the numerous practical and theoretical applications of computations with structured matrices.

In this chapter we included some novel techniques and results on dense structured matrix computations, in particular, on computation of the Krylov sequence for structured matrices (section 7); on the extensions to polynomial and rational computations, such as computing Padé approximants (section 5), the gcd, the lcm and the Euclidean scheme for polynomials (sections 9 and 10); and on the operator representations of structured matrices (sections 11 and 12). Our study has some novel applications to computations with general matrices: to their reduction to the tridiagonal or block tridiagonal form, which we present in sections 3 and 10; to the parallel evaluation of their determinants, ranks, inverses, and null spaces; and to solving linear systems, which we present in sections 4 and 6 of chapter 4 and in its Appendix C. The algorithm of section 9 dramatically improves the previous record upper estimates for the Boolean complexity of computing the polynomial gcd (see Remark 9.4). Likewise, in section 10, we present new fast algorithms for computing all the entries of the extended Euclidean scheme for two polynomials and for block tridiagonalization of a matrix, which are also characterized by their good numerical stability and lower Boolean complexity. The algorithms of sections 9 and 10 promise to become practically effective, since their arithmetic cost and the precision of computations (required to bound the input errors) are relatively low; in particular, this precision is substantially lower than the precision required in the previously known algorithms for the polynomial gcd and the extended Euclidean scheme.

In this chapter we will again assume rational algorithms and the infinite precision computations over the fields **F** of rational, real or complex numbers. For few computations involving orthonormalization, we need to compute square roots in **F**, whereas for defining the eigendecomposition, the SVD and the Schur decomposition of section 3, we need to assume that **F** is *algebraically closed*, that is, contains all the n zeros of any monic polynomial of degree n defined over **F**. Alternatively, we may approximate the output by using rational algorithms. In Remark 3.2, we will comment on the (frequently straightforward) extension of the algorithms of this chapter to computations over any field, and in chapter 3 we will revisit matrix (and polynomial) computations assuming that they are performed with a finite precision and will study how and to which extent the numerical stability problems can be avoided in these computations.

We will focus on estimating the arithmetic time-complexity [the space-complexity for computations with $m \times n$ matrices is $O(mn)$ words of memory in our algorithms]. The Boolean complexity will be considered in chapter 3, together with the related numerical stability issues.

We organize the material of this chapter as follows. After some preliminaries (that is, the definitions and simple results from matrix theory) presented in the remainder of this section, we survey several known estimates for the time-complexity of computations with general matrices in section 2. Specifically, we include the computations that have been asymptotically accelerated by means of their reduction to matrix multiplication and inversion, with a further impact on combinatorial computing.

In section 3 we consider the problems of approximating matrix eigenvalues and singular values, as well as some related computations and the auxiliary problem of the reduction of a matrix to the tridiagonal or block tridiagonal forms. We relate the solution of the latter problem (given by Algorithm 10.2) to computations with dense structured matrices, which we intensively study in sections 4–13, where, in particular, we demonstrate their correlations to polynomial computations. (In chapter 4, we show their applications to parallel computation with general matrices.)

In section 4 we consider computations with Vandermonde and Cauchy (generalized Hilbert) matrices and analyze their correlation to polynomial computations. In particular, we present asymptotically fast algorithms for computing the product of a Cauchy matrix and a vector and for the solution of a linear system with a nonsingular Cauchy coefficient matrix.

In section 5 we study computations with Toeplitz and Hankel matrices, present some fundamental techniques and results in this area (including some major results on circulant and f-circulant matrices, the Gohberg-Semencul formula for Toeplitz matrix inversion, and an extension of it) and show correlations to polynomial multiplication, polynomial division and Padé approximation.

In section 6 we analyze the problem of multiplying a vector by a Vandermonde matrix and by its inverse. We reduce this computation to Toeplitz and Hankel computations.

In section 7 we devise algorithms for computing the determinant and the characteristic polynomial of a Toeplitz matrix, based on the Gohberg-Semencul formula and Newton's iteration.

Section 8 deals with the computation of the Euclidean scheme, polynomial gcd and lcm, expressed in terms of computations with Toeplitz and some other structured matrices. We recall the concepts of the resultant (Sylvester) matrix and pseudo resultants and their correlations to the extended Euclidean scheme.

In Sections 9 and 10 we extend our study to some other structured matrices. We recall the main properties of Bezout, Frobenius and Hankel matrices, associated with polynomials, and their correlations to the Euclidean scheme and gcd computation. In particular, we characterize the polynomial remainder sequence (p.r.s.) generated by Euclid's algorithm in terms of the block LU factorization of a Bezout (Hankel) matrix. The base of this correlation is an important property of the Schur complement of a Bezout matrix that we prove in section 10 and then use for devising new algorithms for the computation of the polynomial gcd and the polynomial remainder sequence.

These algorithms have a low parallel arithmetic cost and low parallel and sequential Boolean costs. We also show how to apply block LU factorization of Hankel matrices in order to reduce a general matrix to block tridiagonal form (compare Propositions 4.5.3 and 4.5.4).

Section 11 is devoted to the computations with Toeplitz-like and Hankel-like matrices and to the study of the associated displacement operators. We prove some new inversion formulae for Toeplitz-like+Hankel-like matrices and extend our algorithms of section 7 (for the computation of the characteristic polynomial and for the solution of a linear system, based on Newton's iteration) from the case of Toeplitz input matrices to the case of Toeplitz-like+Hankel-like matrices.

Some results of section 11 are extended to Cauchy and Vandermonde matrices in section 12, where we present a unified and systematic approach based on the study of the associated linear operators.

In section 13 we first recall one of the "superfast" sequential algorithms for Toeplitz linear systems and then study some ways of regularization of matrices by means of preconditioning with randomization.

In section 14 we review some computations with banded Toeplitz matrices.

In Appendix A we show an application of matrix multiplication to some fundamental computations in graphs. This material is not used in our book and can be omitted by those readers who are not interested in graph algorithms.

In Appendix B we list the computational problems of this chapter, displaying their abbreviated names, their reductions to each other, and their computational complexity estimates.

Sections 6 and 8 include some recent, less known results, and in sections 7, 9–14, we present several new algorithms. The recent advanced results and techniques from [BG90], [BG92] on structured matrices (with applications to the block tridiagonalization of a general matrix and to the evaluation of the polynomial gcd and Euclidean scheme), covered in sections 9 and 10, are not used in sections 11–14, but are of independent interest themselves. Some algorithms of sections 7–11 have no effective sequential implementation, but support the current record parallel complexity estimates, as immediately follows from the simple inspection of these algorithms and substituting for their blocks the parallel complexity estimates from sections 1 and 2 of chapter 4. For reader's convenience, we postpone this inspection until chapter 4 but present the record parallel complexity estimates together with the algorithms.

In the remainder of this section we will recall some definitions (in addition to Definitions 1.1.4 and 1.1.5) and some simple results from the matrix theory (compare [A], [CdB], [DB], [GL], [Str80] and [W65]). The readers familiar with this subject, as well as the readers who want to have more motivation before delving into these preliminaries, may now go to section 2.

Definition 1.1. $\mathbf{F}_{m,n}$ is the set of all the $m \times n$ matrices $A = [a_{ij}]$ with the entries a_{ij}, $i = 0, \ldots, m-1$, $j = 0, \ldots, n-1$, in a fixed field $\mathbf{F}$. (The matrices turn into vectors if $m = 1$ or $n = 1$, and into scalars if $m = n = 1$.)

We will explicitly specify the matrix size only where such a specification is needed but is not clear from the context. As in chapter 1, the reader may for simplicity assume that $\mathbf{F}$ is the field of complex numbers $\mathbf{C}$, the field of real numbers $\mathbf{R}$ or the field of rationals $\mathbf{Q}$. We will let $(A)_{ij}$ denote the (i, j)-th entry a_{ij} of a matrix A, and $(\mathbf{v})_i$ the i-th entry v_i of a vector $\mathbf{v} = [v_i]$.

Definition 1.2. I and O are the identity and the null matrices, respectively; $\mathbf{e}^{(k)}$ is column k of I, $k = 0, 1, \ldots$; $\mathbf{e}=[1, \ldots, 1]^T$ is the vector filled with ones; $\mathbf{0}$ is the null vector. I_r is the $r \times r$ identity matrix; $O_{m,n}$ is the $m \times n$ null matrix.

Definition 1.3. Let $n \leq m$. For a vector $\mathbf{a} = [a_i]$, let $\text{diag}(\mathbf{a}) = \text{diag}([a_i]) = \text{diag}(a_0, \ldots, a_{n-1}) = [d_{ij}]$ denote the $m \times n$ (or the $n \times m$) *diagonal matrix* having entries $d_{ii} = a_i$, $i = 0, \ldots, n-1$, $d_{ij} = 0$ for $i \neq j$. If $A = [a_{ij}]$, $i, j = 0, \ldots, n-1$, the matrix $\text{diag}(A)$ is the $n \times n$ diagonal matrix $\text{diag}(a_{00}, \ldots, a_{n-1,n-1})$. An $m \times n$ matrix is *lower* (or *upper*) *triangular* if all its superdiagonal (or, respectively, subdiagonal) entries are zero. A triangular matrix A is a *unit triangular matrix* if $\text{diag}(A) = I$. A is a *permutation* matrix if, for every vector $\mathbf{v}$, the vectors $A\mathbf{v}$ and $\mathbf{v}^T A$ are obtained by permuting the components of $\mathbf{v}$.

Definition 1.4. Every $k \times k$ submatrix of a matrix A formed by its rows $i_1, i_2, \ldots, i_k$ and columns $i_1, i_2, \ldots, i_k$ is called its *principal submatrix*; the submatrix of A formed by the first k rows and by the first k columns of A is called its *leading principal submatrix* and is denoted $A_{(k,k)}$ (and in some specified cases A_k).

Definition 1.5. trace W denotes the sum of the diagonal entries of W. $\det W$ denotes the *determinant* of a square matrix W, and adj W denotes the adjugate (adjoint) matrix of W, so that $W\text{adj } W = (\text{adj } W)W = I \det W$. The matrix W is called *singular* if $\det W = 0$, called *nonsingular* otherwise, and called *strongly nonsingular* if it is nonsingular together with all its leading principal submatrices.

Clearly, the matrices RA and AS are strongly nonsingular for any fixed $n \times n$ nonsingular matrix A and for the $n \times n$ matrices R and S whose entries are indeterminates. Lemma 1.5.1 implies strong nonsingularity of RA and AS with a high probability if the values from a sufficiently large set have been independently assigned to each of these indeterminates.

Fact 1.1. $\det(AB) = \det A \det B$, $\det I = 1$, $|\det A^H| = |\det A|$ *and the Laplace rule applies:* $\det A = \sum_{i=0}^{n-1} (-1)^{i+j} a_{ij} \det A(i,j)$ *for any integer* $j, 0 \le j < n$, *where* $A(g,h)$ *denotes the* $(n-1) \times (n-1)$ *submatrix of* $A = [a_{ij}]$ *obtained by means of the deletion of row* g *and column* h *of* A; *moreover,* $(AB)^T = B^T A^T$, $(AB)^H = B^H A^H$. *Here* A^T *and* A^H *denote the transpose of* A *and the transposed conjugate matrix of* A *(compare Definition 1.1.5).*

Fact 1.2. *There exists the (unique) inverse* A^{-1} *of a nonsingular matrix* A *such that* $AA^{-1} = A^{-1}A = I$. *Furthermore,* $(A^{-1})^T = (A^T)^{-1}$. *If* AB *is nonsingular, then* $(AB)^{-1} = B^{-1}A^{-1}$. *If* A *is a permutation matrix, then* $A^{-1} = A^T$.

Corollary 1.1. $\det(A^{-1}) = (\det A)^{-1}$ *if* A *is a nonsingular matrix.*

Definition 1.6. The *characteristic polynomial* of an $n \times n$ matrix A is the polynomial $c_A(\lambda) = \det(\lambda I - A) = \sum_{i=0}^{n} c_i \lambda^i$, $c_n = 1$, $c_0 = (-1)^n \det A$, $c_{n-1} = -\text{trace}\, A$. An *eigenvalue* of A of *algebraic multiplicity* m is a zero of $c_A(\lambda)$ of multiplicity m, so that, in the complex field, A has exactly n eigenvalues counted with their algebraic multiplicities. The *eigendecomposition* of A is given by the two matrices $\Lambda = \text{diag}(\lambda_1, \ldots, \lambda_s)$ and $V \in \mathbf{C}_{n,s}$ such that s is the maximum integer (not exceeding n) and

$$AV = V\Lambda. \tag{1.1}$$

Then $\lambda_1, \ldots, \lambda_s$ are the *eigenvalues* of A, their set is called the *spectrum* of A, $\rho = \rho(A) = \max |\lambda_i|$ is the *spectral radius* of A, and the columns of V are called the *eigenvectors* of A.

Theorem 1.1 (Cayley-Hamilton). *Let* A *be an* $n \times n$ *matrix,* $c_A(\lambda) = \det(\lambda I - A)$. *Then* $c_A(A) = O$.

Remark 1.1. The set of all the zeros of any polynomial $p(\lambda)$ over $\mathbf{F}$ such that $p(A) = O$ includes the set of all the eigenvalues of the $n \times n$ matrix A. The monic polynomial $m_A(\lambda)$ of the minimum degree such that

$m_A(A) = O$ is called the *minimum polynomial* of A. $m_A(\lambda)$ divides $c_A(\lambda) = \det(\lambda I - A)$, the characteristic polynomial of A, so that $m_A(\lambda) = c_A(\lambda)$ if and only if deg $m_A(\lambda) = n$ for an $n \times n$ matrix A.

The following simple fact is implied by (1.1) ([GL]):

Fact 1.3. *If μ is a scalar, λ is an eigenvalue of A, S is a nonsingular matrix, then λ is an eigenvalue of $S^{-1}AS$, and $\lambda + \mu$ is an eigenvalue of $A + \mu I$; if, in addition, A is a nonsingular matrix, then λ^{-1} is an eigenvalue of A^{-1}.*

Definition 1.7. A matrix A is called *Hermitian* if $A^H = A$, real symmetric if $A^H = A^T = A$. A matrix A is called *Hermitian nonnegative definite (h.n.d.)* if it can be represented as a product $W^H W$ for some matrix W. If W is nonsingular, then so is A, and then A is a *Hermitian positive definite (h.p.d.)* matrix. A matrix Q is called *orthogonal* or *unitary* if $Q^H Q = I$. A vector $\mathbf{v}$ is a *unit vector* if $\mathbf{v}^T\mathbf{v} = 1$.

Remark 1.2. For real matrices A, we have $A^T = A^H$, and real Hermitian matrices are identified as real symmetric matrices. Whenever the readers restrict any relation over the complex field to the real field or its any subfield or define this relation over a finite field, they should replace W^H by W^T for every matrix W^H involved.

Fact 1.4. *The following statements are equivalent to each other:*

a) *A is an h.p.d. matrix;*
b) *A is a Hermitian matrix, and all the eigenvalues of A are positive;*
c) *all the principal submatrices of A are h.p.d.;*
d) *A^{-1} is an h.p.d. matrix.*

Besides, the statements a), b) and c) remain equivalent to each other if the word "nonnegative" replaces "positive" in b), and if "h.n.d." replaces "h.p.d." throughout.

Definition 1.8. A pair of matrices $G \in \mathbf{F}_{m,d}$ and $H \in \mathbf{F}_{n,d}$ is called a *generator of length d* for the matrix $A = GH^T$. The minimum length of a generator of A is called the *rank* of A and is denoted rank A. $A \in \mathbf{F}_{m,n}$ has the *full (maximum) rank* if rank $A = \min\{m,n\}$. Otherwise, rank $A < \min\{m,n\}$, A is *rank deficient.* rank A can be equivalently defined as the maximum number of linearly independent rows or columns of A or as the maximum integer r such that A has an $r \times r$ nonsingular submatrix. If $\det A_{(k,k)} \neq 0$, for $k = 1, 2, \ldots,$ rank A, then A is said to be a *tame* matrix. (This generalizes strong nonsingularity to rank deficient matrices.) The two sets of vectors are associated with a matrix $A \in \mathbf{F}_{m,n}$: the *null space* or *kernel* of A, defined as $\mathcal{N}(A) = \{\mathbf{v} : \quad A\mathbf{v} = \mathbf{0},\ \mathbf{v} \in \mathbf{F}^n\}$, and the *range* of A, defined as $\mathcal{R}(A) = \{A\mathbf{v} : \quad \mathbf{v} \in \mathbf{F}^n\}$. The dimension of $\mathcal{R}(A)$ equals rank A; the dimension of $\mathcal{N}(A)$ equals $n -$ rank A.

From Fact 1.4, we deduce

Corollary 1.2. *Any h.n.d. matrix is tame over any field of characteristic* 0.

Long generators are readily available (say, $A = IA$), but only short generators are of actual interest, of course. In particular, if $r =$ rank A is much smaller than $\min\{m, n\}$ and if G and H form a length r generator for A, then we may represent A by the $(m+n)r$ entries of G and H, rather than by the mn entries of A. Besides saving some memory space, this enables us to multiply A by a vector in $2(m+n)r-m-r$ ops, rather than in $2mn-m$ ops. By using short generators, we obtain the following convenient expression for updating the inverse of a nonsingular matrix after its modification obtained by adding a matrix GH^T of a small rank (see [GL]):

Theorem 1.2 (Sherman-Morrison-Woodbury formula). *Let $A \in \mathbf{C}_{n,n}$, G, $H \in \mathbf{C}_{n,r}$, $r \leq n$, A and $I_r + H^T A^{-1} G$ be nonsingular matrices. Then*

$$(A + GH^T)^{-1} = A^{-1} - A^{-1}G(I_r + H^T A^{-1} G)^{-1} H^T A^{-1}.$$

Definition 1.9. For a rank r matrix $A \in \mathbf{C}_{m,n}$, its *singular value decomposition (SVD)* is defined as a triple consisting of the $m \times n$ diagonal matrix,

$$\Sigma = \mathrm{diag}(\sigma_1, \ldots, \sigma_r, 0, \ldots, 0), \tag{1.2}$$

and of two orthogonal matrices, $U \in \mathbf{C}_{m,m}$ and $V \in \mathbf{C}_{n,n}$, such that

$$A = U\Sigma V^H, \quad U^H U = V^H V = I, \tag{1.3}$$

$\sigma_1 \geq \sigma_2 \geq \cdots \geq \sigma_r > 0$. $\sigma_1, \ldots, \sigma_r$ are called the *singular values* of A.

Note that the matrices U, Σ and V^H define minimum length generators for A.

Fact 1.5. *For any matrix A, there exists its SVD, the matrices U and V are defined by the eigendecompositions of AA^H and $A^H A$, respectively; all the singular values of A coincide with the square roots of all the nonzero eigenvalues of AA^H or $A^H A$. Furthermore, a matrix A is Hermitian if and* *only if the matrix V in (1.1) is orthogonal, that is, (1.1) and (1.3) represent the same decomposition.*

The next fact shows the third decomposition, due to Schur [not used in this volume but applied as an auxiliary step in many practical algorithms for computing the eigendecomposition (1.1)]:

Fact 1.6 ([GL]). *For any square matrix A there exist an orthogonal matrix Q and a triangular matrix T such that $T = QAQ^H$.*

Definition 1.10. Given a matrix A and its SVD (1.3), the matrix $A^+ = V\Sigma^+U^H$, where $\Sigma^+ = \text{diag}(1/\sigma_1, \dots, 1/\sigma_r, 0, \dots, 0) \in \mathbf{C}_{n,m}$, is called the *Moore-Penrose generalized inverse* of A (and is also called the *pseudo inverse* of A).

Fact 1.7. *The Moore-Penrose generalized inverse of A is the unique matrix A^+ satisfying the following equations:*

$$\begin{aligned} AA^+A &= A, \\ A^+AA^+ &= A^+, \\ (AA^+)^H &= AA^+, \\ (A^+A)^H &= A^+A. \end{aligned} \tag{1.4}$$

Moreover, if $A \in \mathbf{F}_{m,n}$ has the full (maximum) rank, then $A^+ = (A^HA)^{-1}A^H$ for $m \geq n$, $A^+ = A^H(AA^H)^{-1}$ for $m \leq n$, $A^+ = A^{-1}$ if A is nonsingular. If A is Hermitian, then $A^+A = AA^+$.

Definition 1.11 ([A], [CdB], [GL] and [W65]). For a vector $\mathbf{v} = [v_i]$ and a matrix $W = [w_{ij}]$, we define the *vector norms* $||\mathbf{v}||_\infty = \max_i |v_i|$ and $||\mathbf{v}||_q = (\sum_i |v_i|^q)^{1/q}$, which we only need for $q = 1, 2$, and the associated *matrix norms* (also called the *operator norms subordinate to and induced by* these vector norms),

$$||W||_h = \sup_{\mathbf{v} \neq \mathbf{0}} ||W\mathbf{v}||_h / ||\mathbf{v}||_h,$$

which we only need for $h = 1, 2, \infty$. Such a pair of vector and matrix norms is also called *consistent*.

Note that

$$\begin{aligned} &||U|| = 0 \quad \text{if and only if} \quad U = O, \\ &||aU|| = |a| \; ||U||, \\ &||U + V|| \leq ||U|| + ||V||, \quad ||VW|| \leq ||V|| \; ||W||, \end{aligned}$$

for any operator norm, any scalar a and any triple of matrices U, V and W such that $U + V$ and VW are defined.

The following proposition summarizes some well-known properties of the matrix and vector norms:

Proposition 1.1 ([A], [CdB], [GL] and [W65]). *For a matrix* $W = [w_{ij}]$, *a vector* $\mathbf{v} = [v_i]$, *scalars* $b_1, b_2, \ldots, b_k$, *and* $h = 1, 2, \infty$, *we have*

$$||W^T||_\infty = ||W||_1 = \max_j \sum_i |w_{ij}|, \quad ||W||_2 = \sigma_1,$$

σ_1 *being the largest singular value of* W; *moreover,*

$$||W\mathbf{v}||_h \leq ||W||_h ||\mathbf{v}||_h,$$
$$||\mathbf{v}||_\infty \leq ||\mathbf{v}||_2 \leq ||\mathbf{v}||_1,$$
$$||\mathrm{diag}(b_1, b_2, \ldots, b_k)||_h = \max_j |b_j|.$$

If W *is an* $n \times n$ *matrix, then*

$$n^{-1}||W||_h \leq \max |w_{ij}| \leq ||W||_h,$$
$$n^{-1/2}||W||_h \leq ||W||_2 \leq n^{1/2}||W||_h.$$

If $W^H W = I$, *that is, if* W *is a unitary matrix, then*

$$||W||_2 = 1, \quad ||WA||_2 = ||A||_2$$

for any matrix A *for which the product* WA *is defined.*

Definition 1.12. The *condition number* of a nonsingular matrix A associated with the h-norm of A is defined as $\mathrm{cond}_h(A) = ||A||_h ||A^{-1}||_h$, $h = 1, 2, \infty$.

The condition number of a matrix A gives a measure for the numerical conditioning of the problem of solving a linear system $A\mathbf{x} = \mathbf{b}$ (see section 1 of chapter 3).

Fact 1.3 and Proposition 1.1 together imply that $\mathrm{cond}_2(A) = \sigma_1/\sigma_r$, for any $r \times r$ nonsingular matrix A and for σ_1, σ_r defined by (1.2), (1.3). The above equation may also serve as a definition of $\mathrm{cond}_2(A)$ for any singular matrix A.

If A is an h.p.d. matrix and B is one of its principal submatrices, then it is possible to prove that

$$||A||_2 \geq ||B||_2, \quad ||A^{-1}||_2 \geq ||B^{-1}||_2. \tag{1.5}$$

Definition 1.13. A matrix A with nonzero diagonal entries is row- (or, respectively, column-) *diagonally dominant*, with the dominance coefficient at least $1/c > 1$ if $||I - (\text{diag}(A))^{-1}A||_\infty \leq c$ (or, respectively, if $||I - A(\text{diag}(A))^{-1}||_1 \leq c$).

The diagonal dominance implies nonsingularity (see exercise 3).

Hereafter, we will write $||A||$ and cond(A) in the relations that apply to any choice of a matrix norm $||A||_h$ for $h = 1, 2$ or ∞.

Definition 1.14. Let $m = pr$, $n = qs$ for natural p, r, q and s. Then any $m \times n$ matrix $A = [a_{gh}]$, $g = 0, 1, \ldots, m-1$; $h = 0, 1, \ldots, n-1$, can be represented as a $p \times q$ *block matrix* $[A_{ij}]$ whose entries A_{ij} (called *blocks* of A) are $r \times s$ matrices, $A_{ij} = [a_{ir+u,js+v}]$, $u = 0, 1, \ldots, r-1$; $v = 0, 1, \ldots, s-1$; $i = 0, 1, \ldots, p-1$; $j = 0, 1, \ldots, q-1$. More generally, we may represent an $m \times n$ matrix A as a $p \times q$ block matrix $A = [A_{ij}]$ where the blocks A_{ij} have sizes $r_i \times s_j$, such that $\sum_{i=0}^{p-1} r_i = m$, $\sum_{j=0}^{q-1} s_j = n$.

Here are some examples of how we represent block matrices:

$$[A_1, \ldots, A_k], \quad \begin{pmatrix} A \\ B \end{pmatrix}, \quad \begin{pmatrix} A & B \\ C & D \end{pmatrix}$$

are block matrices with blocks $A_1, \ldots, A_k$; A and B; A, B, C and D, respectively.

Definition 1.15. Given two natural numbers m and n, an $n \times n$ matrix A and an n-dimensional vector $\mathbf{v}$, the $n \times m$ matrix $K(A, \mathbf{v}, m) = [\mathbf{v}, A\mathbf{v}, A^2\mathbf{v}, \ldots, A^{m-1}\mathbf{v}]$ is called an $n \times m$ *Krylov matrix* ([GL], [Par]). We will write $K(A, \mathbf{v}) = K(A, \mathbf{v}, n)$.

Finally, we will recall the following technically simple but important result, which we will use in the next section (in the proof of Proposition 2.2) and in chapter 4:

Theorem 1.3 ([Li76], [BSt]). *Let a ops suffice for the evaluation at a point $\mathbf{x} = (x_1, \ldots, x_k)$ of a rational function $f(x_1, \ldots, x_k)$ by a straight line program over a field of constants* $\mathbf{F}$. *Then $4a$ ops in a straight line program over* $\mathbf{F}$ *suffice for the evaluation at a point* $\mathbf{x}$ *of the function f together with all its partial derivatives of the first order, that is, for the evaluation of $\partial f/\partial x_1, \partial f/\partial x_2, \ldots, \partial f/\partial x_k$.*

2. Matrix Multiplication and Some Related Computational Problems.

In this section and in section 3 we will recall several fundamental problems of matrix computations, some typical solution techniques, and the asymptotic estimates for the arithmetic time-complexity of the solution. The solutions of Problems 2.1 ($M\cdot VECTOR$), 2.2 ($M\cdot MULTIPLY$), 2.3 ($LIN\cdot SOLVE$), 2.5 ($INVERT$), 2.7 ($LU\cdot FACTORS$), 2.8 ($L\cdot SQUARES$), 2.9 ($QR\cdot FACTORS$), 3.1 ($EIGENVALUES$), 3.2 ($H\cdot REDUCE$), 3.2a ($T\cdot REDUCE$) and 3.3 (SVD) constitute a great part of all numerical computations performed every day on modern computers. The study of the complexity and the practice of matrix computations (similarly to the case of polynomial computations) extensively exploit *reduction of various computational problems to each other*, and we will show several examples of such reduction.

In this section, we will select some of the most fundamental matrix computational problems, which have rational solution algorithms and are ultimately related to matrix multiplication. Our first group of computational problems is formed by Problems 2.1–2.6, which seem to be most familiar to the computer science reader. This set of problems includes, in particular, Problem 2.1a ($KRYLOV$), 2.3b ($NULL\cdot SPACE$), 2.4 ($DETERMINANT$), 2.6 ($CHAR\cdot POL$), 2.6a ($MIN\cdot POL$). Then we consider several problems related to matrix factorizations. Besides factorizations with no pivoting, such as LU, QR, LL^H, we deal with matrix factorizations that involve permutation matrices, such as PLU, $PLU\tilde{P}$, QRP, LSP. Then, by using these factorizations, we solve some major singular problems, such as computing the rank ($M\cdot RANK$), the nonsingular submatrix of maximum size ($SUBMATRIX$), and the generalized inverse ($GEN\cdot INVERSE$) of a general matrix.

Many operations with matrices can be applied to block matrices (see Definition 1.14), and the computations are reduced (both in theory and practice) *to arithmetic operations with the blocks,* that is, essentially, to matrix-by-vector and matrix-by-matrix multiplication and matrix inversion, which makes the three latter operations as fundamental for the theory and practice of matrix computations as FFT and polynomial multiplication and division are for polynomial computations. This phenomenon is also reflected in our study of the complexity of matrix computations, which we begin now.

Matrix multiplication.

Problem 2.1 ($M \cdot VECTOR$), *matrix-by-vector multiplication. Given an $m \times n$ matrix A and an n-dimensional vector* $\mathbf{b}$, *compute their product* $A\mathbf{b}$.

If $A\mathbf{b}$ had to be evaluated exactly, then the straightforward solution is optimum, and it uses mn multiplications and $m(n-1)$ additions. However, the number of multiplications can be roughly halved under the model of approximate computation introduced in chapter 1 (see chapter 3, section 8).

Problem 2.2 ($M \cdot MULTIPLY$), *matrix multiplication. Compute the product of an $m \times n$ matrix by an $n \times p$ matrix.*

The straightforward solution uses $2n^3 - n^2$ ops and $O(n^2)$ words of storage space for $n \times n$ input matrices. It is possible to solve Problem 2.2 ($M \cdot MULTIPLY$) by using $O(n^\omega)$ ops where $2 \leq \omega < 2.376$ ([CW90], also see our chapter 5) and the order of n^2 words of storage space ([P84b]), although to support the latter time bound for $\omega < 2.376$, and even for $\omega < 2.77$, by using the known algorithms, we need to hide a large overhead constant in the above "O". In practice, Problem 2.2 ($M \cdot MULTIPLY$) until recently was usually solved by means of the straightforward algorithm. Some recent subroutines for $n \times n$ matrix multiplication (in particular, one for CRAY-2, due to [Bai], and another, created in the IBM Research Center, Yorktown Heights, New York) judiciously implement the algorithms of [St69] and of S. Winograd ([BM], pp. 45-46), which in such implementations use less than $4.54n^{2.81}$ and less than $3.92n^{2.81}$ ops, respectively (see also the last section of chapter 5). The weak (but sufficient in most of the practical computations) numerical stability of these algorithms, as well as of several other fast competitive algorithms (see [P84b], pages 154-162, and [LPSh]), has been both proved theoretically ([BL]) and confirmed in numerous experiments.

Hereafter let $M(m, n, p)$ denote the minimum number of ops sufficient for $m \times n$ by $n \times p$ matrix multiplication and let $M(n)$ denote $M(n, n, n)$.

Some fundamental computational problems related to matrix multiplication.

The asymptotic cost bound $O(M(n)) = O(n^\omega)$ also holds for the next four fundamental computational problems (Problems 2.3, 2.4, 2.5 and 2.6) in the case of the $n \times n$ input matrices ([AHU] and [K-G]), although in the present-day computational practice their solution involves about cn^3 ops with c ranging from 2/3 to 2 for Problems 2.3–2.5 [also see [FF], [GL], and exercise 5e on Problem 2.6 ($CHAR \cdot POL$)]:

Problem 2.3 ($LIN \cdot SOLVE$), *solving a linear system of equations. Given an $n \times n$ matrix A and an n-dimensional vector* $\mathbf{b}$, *compute the unique vector $\mathbf{x} = A^{-1}\mathbf{b}$ giving the solution to the linear system $A\mathbf{x} = \mathbf{b}$ if the coefficient matrix A is nonsingular; otherwise, output $SINGULAR$.*

If A is an $n \times n$ triangular matrix, the solution of Problem 2.3 ($LIN \cdot SOLVE$) by means of the simple substitution algorithm only requires about n^2 ops ([A], [CdB] and [GL]).

A linear system $A\mathbf{x} = \mathbf{b}$ with a singular matrix A is called *singular* and has either no solution or infinitely many solutions, which for any fixed solution vector $\mathbf{x}$, form the linear variety $\{\mathbf{x}+\mathbf{y} : \quad \mathbf{y} \in \mathcal{N}(A)\}$. We extend Problem 2.3 ($LIN \cdot SOLVE$) as follows:

Problem 2.3a ($LIN \cdot SOLVE1$). *Given an $m \times n$ matrix A and an n-dimensional vector* $\mathbf{b}$, *compute a solution vector* $\mathbf{x}$ *to the linear system $A\mathbf{x} = \mathbf{b}$ if the system is consistent; output $INCONSISTENT$ otherwise.*

Problem 2.3 ($LIN \cdot SOLVE$) can be reduced to matrix inversion and to multiplication of A^{-1} by $\mathbf{b}$ if A is nonsingular, and later on in this section Problem 2.3a ($LIN \cdot SOLVE1$) will be reduced to Problems 2.10b ($M \cdot PRECNDTN$), 2.1 ($M \cdot VECTOR$), 2.2 ($M \cdot MULTIPLY$), and 2.3 ($LIN \cdot SOLVE$). Over the fields of characteristic 0, we will alternatively reduce $LIN \cdot SOLVE1$ to Problems 2.8 ($L \cdot SQUARES$) and 2.1 ($M \cdot VECTOR$). In addition, we will reduce (over any field of constants) both Problems 2.3 ($LIN \cdot SOLVE$) and 2.3a ($LIN \cdot SOLVE1$) to Problem 2.7c ($LSP \cdot FACTORS$). The latter reduction implies the complexity bounds $O(M(n))$, $O(M(m)n/m)$ for Problems 2.3 and 2.3a, respectively. The users, however, prefer some alternative approaches to Problem 2.3 ($LIN \cdot SOLVE$), that is, Gaussian elimination (with or without pivoting) or iterative processes, whose every step essentially amounts to a single or to few matrix-by-vector multiplications (see [GL] and the next chapter). These alternative approaches lead to a higher asymptotic time bound, but with smaller overhead constants, and to better numerical stability.

Problem 2.3b ($NULL \cdot SPACE$). *Given an $m \times n$ matrix, compute a basis for its null space.*

Later on, we will reduce this problem to Problems 2.10b ($M \cdot PRECNDTN$), 2.5 ($INVERT$) and 2.2 ($M \cdot MULTIPLY$) or, alternatively, to Problems 2.7c ($LSP \cdot FACTORS$), 2.5 ($INVERT$) and 2.2 ($M \cdot MULTIPLY$).

The next three important computational problems are closely related to matrix multiplication:

Problem 2.4 (*DETERMINANT*). *Compute* $\det A$ *for an* $n \times n$ *matrix* A.

Problem 2.5 (*INVERT*), *matrix inversion. Compute* A^{-1}, *the inverse of* A *if* A *is a nonsingular matrix; otherwise, output* *SINGULAR*.

Solution of Problem 2.5 implies the solution of **Problem 2.4a** (*SINGULAR?*) of testing if A is singular, that is, if $\det A = 0$.

We may replace A by the h.n.d. matrix $A^H A$ and compute $A^{-1} = (A^H A)^{-1} A^H$ if A is nonsingular. In this way Problem 2.5 (*INVERT*) is reduced to Problem 2.2 ($M \cdot MULTIPLY$) of matrix multiplication and to inversion of an h.n.d. matrix. Moreover, Problem 2.2 ($M \cdot MULTIPLY$) has the same asymptotic complexity as Problem 2.5 (*INVERT*) (see exercise 5 and Proposition 2.2).

In addition to computing A^{-1}, we may require to compute the inverses $A^{-1}_{(h,h)}$ of all the $h \times h$ leading principal submatrices $A_{(h,h)}$ of A, for $h = 1, \ldots, n$, provided that A is a strongly nonsingular $n \times n$ matrix [we will call this **Problem 2.5a** ($ALL \cdot INVERT$)], or even more generally, for a given sequence of nonsingular submatrices $A_0 = A$, $A_1, \ldots, A_k$, such that A_{i+1} is a proper submatrix of A_i for $i = 0, 1, \ldots, k-1$, we may require to compute A_i^{-1}, for $i = 0, 1, \ldots, k$ [we will call this computational task **Problem 2.5b** ($ALL \cdot INVERT1$)]. Instead of assuming nonsingularity of the submatrices, we may ensure it with a high probability by means of randomization (see section 13).

Recursive application of Theorem 1.2 enables us to solve Problems 2.5a ($ALL \cdot INVERT$) and 2.5b ($ALL \cdot INVERT1$) for an $n \times n$ input matrix by using $O(n^3)$ ops, which is the optimum bound within a constant factor.

Problem 2.6 ($CHAR \cdot POL$). *Evaluate the coefficients of the characteristic polynomial* $c_A(\lambda) = \det(\lambda I - A) = \sum_{i=0}^{n} c_i \lambda^i$ *of a given* $n \times n$ *matrix* A.

We also recall a related problem that we will solve in Proposition 2.2 and will use in chapter 4 (compare Remark 1.1).

Problem 2.6a ($MIN \cdot POL$). *Evaluate the coefficients of the minimum polynomial* $m_A(\lambda)$ *of a given* $n \times n$ *matrix* A.

Some correlations among the fundamental matrix computations.

There are various correlations among Problems 2.1–2.6, which also involve the two following auxiliary problems:

Problem 2.1a (*KRYLOV*), *computing Krylov matrix. Given natural m and n, an $n \times n$ matrix A and an n-dimensional vector* $\mathbf{v}$, *compute the Krylov matrix $K(A, \mathbf{v}, m)$.*

Problem 2.2a ($M \cdot POWERS$), *matrix powers. Given a natural m and a square matrix A, compute the powers $A^2, A^3, \ldots, A^m$.*

Clearly, Problem 2.2a ($M \cdot POWERS$) is reduced to $m - 1$ matrix multiplications, and we also have the following result:

Proposition 2.1 (see [BM], proof of corollary 6.1.5 on p. 128; and [K-G]). *Let A and* $\mathbf{v}$ *be as in Definition 1.15 and let $m > 1$. Then the Krylov matrix $K(A, \mathbf{v}, m)$ can be computed by using $\lfloor \log(m-1) \rfloor$ multiplications of matrices of sizes $n \times n$ by $n \times n$ and $\lfloor \log(m-1) \rfloor + 1$ multiplications of matrices of sizes $n \times n$ by $n \times j$ where $j = 2^0, 2^1, \ldots, 2^{\lfloor \log(m-1) \rfloor}$.*

Proof. Let $H = 2^h$ be a power of 2 such that $H < m \leq 2H$, $h = \lfloor \log(m-1) \rfloor$. Compute the matrix powers A^J for $J = 2, 4, 8, \ldots, H$. Then compute the matrix products

$$
\begin{aligned}
&A\mathbf{v} \\
&A^2[\mathbf{v}, A\mathbf{v}] = [A^2\mathbf{v}, A^3\mathbf{v}], \\
&A^4[\mathbf{v}, A\mathbf{v}, A^2\mathbf{v}, A^3\mathbf{v}] = [A^4\mathbf{v}, A^5\mathbf{v}, A^6\mathbf{v}, A^7\mathbf{v}], \\
&\cdots\cdots\cdots\cdots\cdots\cdots\cdots\cdots\cdots\cdots \\
&A^H[\mathbf{v}, A\mathbf{v}, \ldots, A^{H-1}\mathbf{v}] = [A^H\mathbf{v}, A^{H+1}\mathbf{v}, \ldots, A^{2H-1}\mathbf{v}].
\end{aligned}
$$

■

For Problem 2.1a (*KRYLOV*) where $m = O(n)$, Proposition 2.1 implies the complexity bound of $O(M(n) \log n)$ ops.

Next, we will show some reductions of Problems 2.1–2.6 to each other. Let $a_{\mathcal{P}}$ denote the number of ops that are needed in order to solve Problem $\mathcal{P}$; let us write $(\mathcal{A}, \mathcal{B}) \succeq \mathcal{C}$ if Problem $\mathcal{C}$ can be reduced to Problems $\mathcal{A}$ and $\mathcal{B}$, so that $a_{\mathcal{C}} = O(a_{\mathcal{A}} + a_{\mathcal{B}})$. It is assumed here that $i.s.(\mathcal{A}) + i.s.(\mathcal{B}) = O(i.s.(\mathcal{C}))$ where $i.s.(\mathcal{P})$ stands for the input size of Problem $\mathcal{P}$ and $(g, h) = O((m, n))$ means that

$$g = O(m) \;, \quad h = O(n) \;.$$

Proposition 2.2. *We have*

$$CHAR \cdot POL \succeq DETERMINANT \succeq INVERT;$$
$$(M \cdot VECTOR, INVERT) \succeq LIN \cdot SOLVE;$$
$$(CHAR \cdot POL, KRYLOV) \succeq LIN \cdot SOLVE,$$
$$DETERMINANT \succeq SINGULAR?.$$

Moreover, a randomized Monte Carlo algorithm supports the reduction $(RECUR \cdot SPAN,\ M \cdot VECTOR,\ KRYLOV) \succeq MIN \cdot POL$.

Proof. The reductions $(M \cdot VECTOR, INVERT) \succeq LIN \cdot SOLVE$ and $CHAR \cdot POL \succeq DETERMINANT$ and $DETERMINANT \succeq SINGULAR?$ are obvious. $DETERMINANT \succeq INVERT$ follows from Theorem 1.3. The reduction $(CHAR \cdot POL,\ KRYLOV) \succeq LIN \cdot SOLVE$ follows from Cayley-Hamilton Theorem 1.1. To show the randomized reduction

$$(RECUR \cdot SPAN, M \cdot VECTOR, KRYLOV) \succeq MIN \cdot POL,$$

we will follow [Wied] and [KP] and will first show how to compute a divisor of the polynomial $m_A(\lambda)$. We first fix two vectors $\mathbf{u}$ and $\mathbf{v}$, then compute the Krylov matrix $K(A, \mathbf{v}, 2n)$, the vector

$$\mathbf{u}^T K(A, \mathbf{v}, 2n) = [\mathbf{u}^T\mathbf{v}, \mathbf{u}^T A\mathbf{v}, \ldots, \mathbf{u}^T A^{2n-1}\mathbf{v}],$$

and finally, the $(n-1, n)$ Padé approximation $(R(\lambda),\ T(\lambda))$ to the polynomial $\sum_{i=0}^{2n-1} \mathbf{u}^T A^i \mathbf{v} \lambda^i$. The polynomial $T(\lambda)$ of degree at most n is the minimum span polynomial for the linear recurrence sequence defined by the vector $\mathbf{u}^T K(A, \mathbf{v}, 2n)$ [see Problem 1.5.3 $(RECUR{\cdot}SPAN)$]. It divides $m_A(\lambda)$, and moreover, by Lemma 1.5.1, with a probability close to 1, $T(\lambda)$ coincides with $m_A(\lambda)$ (up to within their scaling) under the random and independent choice of the components of the two vectors $\mathbf{u}$ and $\mathbf{v}$ from any fixed sufficiently large subset S of a given field $\mathbf{F}$ or of its extension ([KP]). ∎

Note that the deterministic verification of the matrix equation $m_A(A) = O$ has higher computational cost bounds, but we may probabilistically verify this equation by checking if $m_A(A)\mathbf{w} = \mathbf{0}$ for a random vector $\mathbf{w}$.

The asymptotic cost bound $O(M(n)) = O(n^\omega)$, $2 \le \omega < 2.376$, for Problems 2.3, 2.4 and 2.5, follows *since these problems for* $n \times n$ *matrices can be reduced to Problem 2.2* $(M \cdot MULTIPLY)$ *of matrix multiplication, which can in turn be reduced to each of Problems 2.4, 2.5 and 2.6* (see exercise 5 and [St69], [AHU], [BM], [P84b], [P84], and [K-G]). In particular, the reduction of Problem 2.5 ($INVERT$) [and consequently of Problem 2.3 ($LIN \cdot SOLVE$)] to Problem 2.2 ($M \cdot MULTIPLY$), leading to the complexity estimates with the exponent ω for these problems, can be obtained in the following way ([AHU], [BM]):

Since $A^{-1} = (A^H A)^{-1} A^H$, Problem 2.5 ($INVERT$) is reduced to inverting the h.p.d. matrix $A^H A$ (over the fields of characteristic 0). So, for computations over such fields, without loss of generality, assume that A is an h.p.d. matrix. Then, we have the identities

$$A = \begin{pmatrix} I & O \\ EB^{-1} & I \end{pmatrix} \begin{pmatrix} B & O \\ O & S \end{pmatrix} \begin{pmatrix} I & B^{-1}C \\ O & I \end{pmatrix}, \tag{2.1}$$

$$A^{-1} = \begin{pmatrix} I & -B^{-1}C \\ O & I \end{pmatrix} \begin{pmatrix} B^{-1} & O \\ O & S^{-1} \end{pmatrix} \begin{pmatrix} I & O \\ -EB^{-1} & I \end{pmatrix}, \tag{2.2}$$

$$A = \begin{pmatrix} B & C \\ E & G \end{pmatrix}, \quad S = G - EB^{-1}C, \tag{2.3}$$

which hold for any nonsingular matrix A provided that B is a square and nonsingular matrix. Moreover, since A is an h.p.d. matrix, Fact 1.4 and Proposition 2.3 ensure that B and S are also h.p.d. (The matrix S is also called the *Schur complement* of B in A.) Therefore, the identities (2.2) and (2.3) can be recursively extended to the submatrices B and S.

Now assume for simplicity that $n = 2^h$ is a power of 2 and that the blocks B and G have size $n/2 \times n/2$ (for any n, we may specify that B has size $\lceil n/2 \rceil \times \lceil n/2 \rceil$ and then call the factorization *balanced*). Moreover, let $I(n)$, $M(n)$ and $A(n) = n^2$ denote the minimum numbers of ops required for $n \times n$ matrix inversion, multiplication and addition or subtraction, respectively. Then (2.2) implies that

$$I(n) \le 2I(n/2) + 6M(n/2) + 2A(n/2) \le 3 \sum_{i=1}^{h} 2^i (M(n/2^i) + A(n/2^i)),$$

and the relation $I(n) = O(M(n))$ follows, over the fields of characteristic 0 (we leave the details as exercise 5a, compare [AHU], [BM], and refer to exercise 5b on an extension to the computations over any field of constants).

Factorizations of a matrix with no pivoting.

Motivated by the successful application of (2.1)–(2.3), we will next study various matrix factorizations. In particular, the *recursive factorization* (2.1)–(2.3) deserves more comments as one of our major tools for designing algorithms for matrix computations and will be yet another example of recursive algorithms based on the divide-and-conquer approach. *Any strongly nonsingular matrix A has such a recursive factorization* (exercise 4). Moreover, in chapter 3 we will show that this factorization is a numerically stable and thus practical algorithm if A is an h.p.d. and/or diagonally-dominant matrix.

The proof in the h.p.d. case relies on (1.5) and on the next proposition, which can easily be verified based on the matrix equation (2.2) and on the observation that the Schur complement S of (2.3) can be obtained from the matrix A of (2.3) by means of the Gauss-Jordan transformation of A (compare [GL], page 103):

Proposition 2.3. *Let A and S be the matrices defined in (2.3). Let A_1 be a leading principal submatrix of S and let S_1 denote the Schur complement of A_1 in S. Then S^{-1} and S_1^{-1} form the respective southeastern blocks of A^{-1}.*

Remark 2.1. Note that $S = G$ if A is a triangular matrix, in which case the evaluation of the factorization (2.1) and (2.2) can be substantially simplified; in particular, if $C = O$, then

$$A^{-1} = \begin{pmatrix} B^{-1} & O \\ O & G^{-1} \end{pmatrix} \begin{pmatrix} I & O \\ -EB^{-1} & I \end{pmatrix}.$$

Recursive factorization (2.1)-(2.3) is closely related to the triangular factorization of the matrix A, whose computation amounts to the elimination stage of the popular algorithm of *Gaussian elimination* with no pivoting ([GL]), used for solving linear systems. In fact, both factorizations have the same domain of definition (see exercise 4c).

Problem 2.7 (*$LU \cdot FACTORS$*), *triangular factorization, LDM factorization, LU factorization. Given an $n \times n$ matrix A, compute, if they exist, three matrices L, D and M, such that*

$$A = LDM = LU, \tag{2.4}$$

L and M^T are unit lower triangular matrices, and D is a diagonal matrix.

If A is an h.p.d. matrix, the factorization LL^H, where L is a lower triangular matrix having positive diagonal entries, is called the *Choleski factorization* of A.

Note that $LU \cdot FACTORS \succeq DETERMINANT$. If there exists the triangular (Choleski) factorization, then it is unique. As in the case of the factorization (2.1)-(2.3), there exists the Choleski factorization for any h.p.d. matrix A, and there exists the triangular factorization (2.4) for any diagonally dominant and, more generally, for any strongly nonsingular matrix A, but not for any matrix A; for instance, even the simple symmetric matrix $A = \begin{pmatrix} 0 & 1 \\ 1 & 0 \end{pmatrix}$ does not have such a factorization.

Given the factorization (2.4), we have that $A^{-1} = M^{-1}D^{-1}L^{-1}$, and, therefore, in this case we may reduce Problem 2.3 ($LIN \cdot SOLVE$) of solving linear systems with a general $n \times n$ coefficient matrix A to two Problems 2.3 ($LIN \cdot SOLVE$) with triangular coefficient matrices L and M, which we may solve by using $O(n^2)$ ops. Computing the factorization (2.4), however, may cost more than this. $O(M(n)) = O(n^\omega)$ ops for $\omega < 2.376$ suffice in theory: first compute the balanced recursive factorization (2.1), representing it in the form

$$A = L_1 L_2 \cdots L_h D M_h M_{h-1} \cdots M_1$$

where $h = \lceil \log n \rceil$, L_i and M_i^T are lower triangular matrices, D is a diagonal matrix, and then compute

$$L = L_1 L_2 \cdots L_h, \quad M = M_h M_{h-1} \cdots M_1.$$

Here we need to assume the existence of the recursive factorization (2.1) [which is ensured if there exists the LU factorization of the matrix A (see exercise 4c), in particular, if A is an h.p.d. and/or diagonally dominant matrix]. Under the similar assumptions, practical algorithms compute the factorization (2.4) in $(2/3)n^3$ ops by means of Gaussian elimination; this cost decreases to $n^3/3$ ops if A is an h.p.d. matrix ([GL]).

Let us more closely examine the h.p.d. case. Suppose that the triangular factorization of an h.p.d. matrix is known. Then, from its uniqueness, it follows that $L = M^H$ and, from the positive definiteness of A, it follows that D has positive diagonal entries. Therefore, we may define $\sqrt{D}$, the diagonal matrix having positive diagonal entries, such that $\sqrt{D}\sqrt{D} = D$. Then $\tilde{L}\tilde{L}^H$ gives us the Choleski factorization of A where $\tilde{L} = L\sqrt{D}$. In addition to $O(M(n)) = O(n^\omega)$ ops, we need n evaluations of the square roots of positive numbers in this case; we shall assume that such an operation is included in the set of the unit cost operations allowed in the RAM model. The existence of the Choleski factorization of a matrix implies its property of being h.p.d. and vice versa:

Fact 2.1. *A (nonsingular) matrix is h.p.d. if and only if there exists its Choleski factorization.*

Corollary 2.1. *Computing the factorization (2.4) of a Hermitian matrix A, where $L = M^H$, can be used as a test if A is h.p.d.; this test uses $O(n^\omega)$ ops for $\omega < 2.376$.*

Factorizations with pivoting and some applications to computations with singular matrices.

For practical computations, the recursive factorization (2.1)-(2.3) is a reliable algorithm if A is a positive definite and/or diagonally dominant matrix. Otherwise, and also if A is h.p.d. and sparse, practical algorithms for Gaussian elimination and for both triangular factorizations (2.4) and the recursive factorization(2.1)-(2.3) include special policies of row and column interchange (called *pivoting*) introduced in order to ensure the existence of a factorization and numerical stability of its evaluation and/or to maintain sparsity of the original input matrix during the elimination/factorization process. This process, performed with pivoting, amounts to computing the factorization $A = PLDM\tilde{P}$ where L, D and M are as in (2.4) and $\tilde{P}$ and P are permutation matrices, whose application to a matrix amounts to the desired interchange of its rows and columns (compare [GL], [GLi], [P87b], [LRT]) and [P93b]).

By applying pivoting, we may ensure a more special form for triangular matrices L or M. In particular, for any $m \times n$ input matrix A, we may solve the following problem in $O(mnr)$ ops and $O(mnr)$ comparisons of computed values with 0 ([GL]):

Problem 2.7a ($PLU\tilde{P} \cdot FACTORS$), *triangular factorization with complete pivoting. Given an $m \times n$ matrix A, $m \geq n$, compute five matrices P, L, D, M, $\tilde{P}$ such that*

$$A = PLDM\tilde{P} = PLU\tilde{P}, \quad U = DM,$$

P and $\tilde{P}$ are permutation matrices, L and D are as in Problem 2.7 ($LU \cdot FACTORS$), and $M = \begin{pmatrix} R & S \\ O & O \end{pmatrix}$ where R is an $r \times r$ unit upper triangular matrix, $r = \operatorname{rank} A$.

We may restrict pivoting to row interchange and use $O(M(n)) = O(n^\omega)$ ops and comparisons, for $\omega < 2.376$ ([AHU]), in the solution of the following problem:

Problem 2.7b ($PLU \cdot FACTORS$), *triangular factorization with partial pivoting. Given an $n \times n$ matrix A, compute three matrices P, L, U such that*

$$A = PLU,$$

P is a permutation matrix, L is a unit lower triangular, U is an upper triangular matrix.

A slightly different factorization, which applies to $m \times n$ matrices and can be computed in time $O(M(m)n/m)$ over any field of constants, is the LSP factorization of [IMH].

Problem 2.7c ($LSP \cdot FACTORS$). *Given an $m \times n$ matrix A, $m \le n$, compute an $m \times m$ lower triangular matrix L with 1's on its main diagonal, an $m \times n$ matrix S that reduces to an upper triangular matrix with nonzero diagonal entries when its $m - r$ zero rows are deleted, where $r = \text{rank } A$, and an $n \times n$ permutation matrix P such that $A = LSP$.*

Solution ([IMH]). Without loss of generality assume that m and n are powers of 2. If $m = 1$ the problem is trivial. Let $A = \begin{pmatrix} A_1 \\ A_2 \end{pmatrix}$ where A_1 and A_2 are $\frac{m}{2} \times n$ matrices. Perform the following steps:

1. If $A_1 = O$, then set $C' = A_2$, and go to Step 5. Otherwise, compute the LSP factorization of A_1: $A_1 = L_1 S_1 P_1$. Let r denote the rank of A_1 and set $A_2^* = A_2 P_1^T$. Then recall that $P_1^T P_1 = I_n$ and set

$$A = \begin{pmatrix} L_1 & O \\ O & I_{m/2} \end{pmatrix} \begin{pmatrix} S_1 \\ A_2^* \end{pmatrix} P_1,$$

$$\begin{pmatrix} S_1 \\ A_2^* \end{pmatrix} = \begin{pmatrix} S_1' & B \\ F' & C \end{pmatrix},$$

   where F' and S_1' are $\frac{m}{2} \times r$ matrices.
2. Compute an $(m/2) \times (m/2)$ matrix $S_1^{(-1)}$ such that $S_1^{(-1)} S_1 = \begin{pmatrix} I_r & H \\ O & O \end{pmatrix}$, $S_1^{(-1)} S_1' = \begin{pmatrix} I_r \\ O \end{pmatrix}$. This step can be performed by means of row permutations, computing the inverse of an $r \times r$ triangular matrix, and multiplying this inverse by an $r \times (m/2 - r)$ matrix.
3. Let F be the $(m/2) \times (m/2)$ matrix obtained by adding $(m/2) - r$ zero columns to F'. Compute $G = F S_1^{(-1)}$. Then we have $G S_1' = F'$.
4. Compute $C^* = C - GB$. Then

$$\begin{pmatrix} S_1 \\ A_2^* \end{pmatrix} = \begin{pmatrix} S_1' & B \\ F' & C \end{pmatrix} = \begin{pmatrix} I_{m/2} & O \\ G & I_{m/2} \end{pmatrix} \begin{pmatrix} S_1' & B \\ O & C^* \end{pmatrix},$$

and therefore,

$$A = \begin{pmatrix} L_1 & O \\ G & I_{m/2} \end{pmatrix} \begin{pmatrix} S_1' & B \\ O & C^* \end{pmatrix} P_1 .$$

5. Compute the LSP factorization of C^*, $C^* = L_2 S_2 P_2$. If $A_1 = O$, then set

$$L = \begin{pmatrix} I_{m/2} & O \\ O & L_2 \end{pmatrix}, S = \begin{pmatrix} O \\ S_2 \end{pmatrix}, \quad P = P_2 ,$$
$$A = \begin{pmatrix} I_{m/2} & O \\ O & L_2 \end{pmatrix} \begin{pmatrix} O \\ S_2 \end{pmatrix} P_2 = LSP .$$

Otherwise observe that

$$\begin{pmatrix} S_1' & B \\ O & C^* \end{pmatrix} = \begin{pmatrix} I & O \\ O & L_2 \end{pmatrix} \begin{pmatrix} S_1' & B \\ O & S_2 P_2 \end{pmatrix}$$
$$= \begin{pmatrix} I & O \\ O & L_2 \end{pmatrix} \begin{pmatrix} S_1' & B P_2^{-1} \\ O & S_2 \end{pmatrix} \begin{pmatrix} I_r & O \\ O & P_2 \end{pmatrix}$$

and set

$$L = \begin{pmatrix} L_1 & O \\ G & I_{m/2} \end{pmatrix} \begin{pmatrix} I & O \\ O & L_2 \end{pmatrix} = \begin{pmatrix} L_1 & O \\ G & L_2 \end{pmatrix},$$
$$S = \begin{pmatrix} S_1' & B P_2^{-1} \\ O & S_2 \end{pmatrix}, \quad P = \begin{pmatrix} I_r & O \\ O & P_2 \end{pmatrix} P_1 ,$$
$$A = \begin{pmatrix} L_1 & O \\ G & L_2 \end{pmatrix} \begin{pmatrix} S_1' & B P_2^{-1} \\ O & S_2 \end{pmatrix} \begin{pmatrix} I_r & O \\ O & P_2 \end{pmatrix} P_1 = LSP.$$

We easily verify that the cost of computing this factorization is $O(M(m)\, n/m)$. ■

The fast solution algorithm for Problem 2.7c ($LSP \cdot FACTORS$) provides an $O(M(n))$ algorithm for Problem 2.3 ($LIN \cdot SOLVE$) over any field of constants.

Solution of Problem 2.3 ($LIN \cdot SOLVE$). Compute the LSP factorization of A. If there exists a zero row in S, then output SINGULAR. Otherwise compute $\mathbf{x} = P^T S^{-1} L^{-1} \mathbf{b}$ by solving two triangular linear systems by means of substitution. ■

Solution of Problem 2.3a ($LIN \cdot SOLVE1$). If $m \leq n$, compute the LSP factorization of A. Compute the vector $\mathbf{c} = (c_i) = L^{-1}\mathbf{b}$ by solving a triangular linear system. Consider the system $SP\mathbf{x} = \mathbf{c}$. If $c_i \neq 0$, for some i such that the i-th row of S is a zero row, output INCONSISTENT. Otherwise, choose any values for the last $n-r$ components of $\mathbf{y} = P\mathbf{x}$, where r is the number of nonzero rows of S, and solve the triangular linear system

$S\mathbf{y} = \mathbf{c}$. This solution algorithm supports the reduction $LIN \cdot SOLVE1 \preceq LSP \cdot FACTORS$ and the complexity bound $O(M(m)n/m)$ for $m \leq n$. If $m > n$, compute the LSP factorization of A^T and rewrite the original linear system $A\mathbf{x} = \mathbf{b}$ as $\mathbf{x}^T LSP = \mathbf{b}^T$. This reduces the original problem for $A\mathbf{x} = \mathbf{b}$ to the same problem for $S^T\mathbf{y} = \mathbf{d}$, with $\mathbf{d} = P\mathbf{b}$, and to solving the triangular linear system $L^T\mathbf{x} = \mathbf{y}$. Due to the structure of S, Problem 3.1a ($LIN \cdot SOLVE1$) for the system $S^T\mathbf{y} = \mathbf{d}$ is reduced to solving a triangular linear system of r equations, for $r = \text{rank } A$, and to multiplication of an $(n - r) \times r$ matrix by a vector. The overall cost of the solution of the original problem is dominated by $O(M(n)m/n)$, the complexity of the LSP factorization of A^T.

Since no fast parallel algorithm for the LSP factorization is available so far, we also recall another solution algorithm, which applies (and has fast parallel implementations) over any field of characteristic 0 and for all values of m and n. This solution reduces Problem 2.3a ($LIN \cdot SOLVE1$) to Problem 2.8 ($L \cdot SQUARES$): a least-squares solution $\mathbf{x} = A^+\mathbf{b}$ to a linear system $A\mathbf{x} = \mathbf{b}$ satisfies such a system if and only if this system is consistent. A different solution algorithm, based on the reduction

$$LIN \cdot SOLVE1 \preceq \quad (M \cdot PRCNDTN, M \cdot VECTOR,\\ M \cdot MULTIPLY, LIN \cdot SOLVE),$$

will be described later on. ∎

Solution of Problem 2.3b ($NULL \cdot SPACE$). If $m \leq n$, compute the LSP factorization of A. Let r denote the number of nonzero rows of S. A basis of the null space of A is given by the columns of P^TY, Y denoting the $n \times (n-r)$ matrix that solves the system $\tilde{S}Y = O$, where $\tilde{S}$ is the matrix obtained by deleting the zero rows of S. Rewrite the latter system as

$$(\tilde{S}_1 \quad \tilde{S}_2)(Y_1 \quad Y_2) = O,$$

where $\tilde{S}_1$ is an $r \times r$ upper triangular and nonsingular matrix, Y_1 and Y_2 are $r \times (n-r)$ and $r \times r$ matrices, respectively. Set $Y_2 = I_r$ and $Y_1 = -\tilde{S}_1^{-1}\tilde{S}_2$. If $m \geq n$, consider the LSP factorization of A^T. It is easy to check that the null space of A is obtained by multiplying the vectors of the null space of S^T by $(L^T)^{-1}$. On the other hand, the null space of S^T is made up by all the vectors having zero components corresponding to nonzero rows of S. In this way we arrive at the reduction

$$NULL \cdot SPACE \preceq (LSP \cdot FACTORS, M \cdot MULTIPLY, INVERT),$$

which leads to the complexity bounds $O(M(m)n/m)$ if $m \leq n$, $O(M(n)m/n)$ if $m > n$. A different solution algorithm, based on the reduction

$$NULL \cdot SPACE \preceq (M \cdot PRCNDTN, M \cdot MULTIPLY, INVERT),$$

will be described later on. ∎

Least-squares solution and orthogonal factorization.

We will next extend our study to some other computational problems reducible to each other and, ultimately, to matrix multiplication and inversion. This will enable us to demonstrate some typical techniques of the design of matrix algorithms.

Problem 2.3 ($LIN \cdot SOLVE$) of solving a nonsingular linear system has a natural extension (over the fields of characteristic 0) to the systems $B\mathbf{x} = \mathbf{b}$ where B is a rectangular or square but singular matrix.

Problem 2.8 ($L{\cdot}SQUARES$), *least-squares solution of a linear system. Given an $m \times n$ matrix B and an m-dimensional vector* $\mathbf{b}$, *compute a vector* $\mathbf{x}$ *that minimizes $||B\mathbf{x} - \mathbf{b}||_2$. If the solution is not unique, compute one that has the minimum 2-norm.*

Solutions. The solution is given by $B^+\mathbf{b}$, and this enables us to reduce the original problem to Problems 2.12 ($GEN \cdot INVERSE$) and 2.1 ($M \cdot VECTOR$). Problem 2.12 ($GEN \cdot INVERSE$) can be reduced to Problem 2.7c ($LSP \cdot FACTORS$), thus leading to the complexity bounds $O(M(m)n/m)$ if $m \leq n$ and $O(M(n)m/n)$ if $m > n$. Other solution algorithms can be obtained by reducing this problem to Problems 2.1 ($KRYLOV$) and 2.6 ($CHAR \cdot POL$). We will show this reduction at the end of this section (see also [GL] on the solution based on the so called *complete orthogonal decomposition* of B). In the important special case where $m \geq n$ and B has full rank, the solution can be expressed as

$$\mathbf{x} = (B^H B)^{-1} B^H \mathbf{b} \tag{2.5}$$

(see [GL]), so that the problem is reduced to Problems 2.1 ($M \cdot VECTOR$), 2.2 ($M \cdot MULTIPLY$) and 2.3 ($LIN \cdot SOLVE$) of matrix-by-vector and matrix-by-matrix multiplications and of solving the linear system of *normal equations*, $B^H B\mathbf{x} = B^H\mathbf{b}$. Likewise, $\mathbf{x} = B^H(BB^H)^{-1}\mathbf{b}$ if $m \leq n$ and if B has full rank. ∎

Observe that, by adding to $\mathbf{x}$ the vectors of $\mathcal{N}(B)$, we may obtain all the least-squares solutions to $B\mathbf{x} = \mathbf{b}$.

In practice, another approach to the solution of Problem 2.8 ($L \cdot SQUARES$) with an input matrix B of full rank is frequently preferred as having better numerical stability. This solution reduces Problem 2.8 ($L \cdot SQUARES$) to the following problem, having other important applications as well (see [GL], [Par]):

Problem 2.9 ($QR \cdot FACTORS$), *orthogonal factorization, QR-factorization. Given an $m \times n$ matrix B of full rank $n \leq m$, compute an upper triangular matrix R and an orthogonal $m \times n$ matrix Q ($Q^H Q = I$) such that*

$$B = QR. \tag{2.6}$$

Solution. $B^H B$ is nonsingular and thus an h.p.d. matrix, since B is a matrix of full rank. Therefore, Problem 2.9 ($QR \cdot FACTORS$) can be reduced to Problem 2.2 ($M \cdot MULTIPLY$) of matrix multiplication, to Problem 2.7 ($LU \cdot FACTORS$) of triangular factorization of the Hermitian matrix $A = B^H B$, and to the inversion of the triangular matrix R, because the equation $B = QR$ implies that

$$B^H B = R^H R, \tag{2.7}$$

$Q = BR^{-1}$. This leads to the complexity bound $O(M(n))$ ops for $QR \cdot FACTORS$.

We refer the reader to [GL] on other practical algorithms based on some well-known techniques of numerical linear algebra (such as Householder transformations, Givens rotations and the Gram-Schmidt orthogonalization process). These algorithms also compute QR factorization of B even where B is a rank deficient matrix. ∎

Substitute (2.6) and (2.7) into (2.5) and obtain that

$$\mathbf{x} = (R^H R)^{-1} R^H Q^H \mathbf{b} = R^{-1} Q^H \mathbf{b},$$

so that Problem 2.8 ($L \cdot SQUARES$) for a full rank matrix B is reduced to Problem 2.9 ($QR \cdot FACTORS$), to solving a linear system with a triangular coefficient matrix R, and to Problem 2.1 ($M \cdot VECTOR$).

Problem 2.9 ($QR \cdot FACTORS$) can be extended to the case of rank deficient input matrices:

Problem 2.9a ($QRP \cdot FACTORS$), *QR factorization with column pivoting. For an $m \times n$ matrix A of rank r, compute a permutation matrix*

P, an $m \times n$ orthogonal matrix Q, an $r \times r$ upper triangular matrix R, and an $r \times (n-r)$ matrix S such that

$$A = Q \begin{pmatrix} R & S \\ O & O \end{pmatrix} P.$$

A solution of [GL] relies on Householder transformations ([GL]) with an appropriate pivoting strategy and uses $O(mnr)$ ops and comparisons.

Further computations with singular matrices.

The statements of Problems 2.9 ($QR \cdot FACTORS$) and 2.9a ($QRP \cdot FACTORS$) and of the above specialization of Problem 2.8 ($L \cdot SQUARES$) involve the ranks of the matrices A and/or B, and, of course, we need to compute matrix rank in many other computations too.

Problem 2.10 ($M \cdot RANK$). *Given an $m \times n$ matrix A, compute its rank.*

Closely related to this is the following problem:

Problem 2.11 ($GENERATOR$). *Given an $m \times n$ matrix A, compute its generator of the minimum length $r =$* rank *A.*

Solutions of Problems 2.10 and 2.11. Without loss of generality, assume that $m \leq n$ (otherwise, shift to A^T). The LSP factorization of A provides both the rank and a generator of the minimum length. The rank equals the number of nonzero rows of the matrix S. A generator of the minimum length (G, H) can be obtained in the following way: H is obtained from SP by removing the zero rows; G is obtained from L by removing the columns corresponding to the zero rows of S. This reduction to Problem 2.7c ($LSP \cdot FACTORS$) leads to the complexity bound $O(M(m)n/m)$ over any field of constants for both $M \cdot RANK$ and $GENERATOR$, for $m \leq n$. Another solution can be obtained by reducing the problem to computing the SVD, which gives us both the rank and a generator of the minimum length, and this observation leads to a customary nonrational but numerically stable algorithm for both of these computations (see Remark 3.1). An alternative approach reduces the rank computation to solving Problem 2.9a ($QRP \cdot FACTORS$). The solution involves the computation of square roots, so again, the algorithm is not rational. Next, we will describe rational solutions based on the computation of the coefficients of the characteristic polynomial. Over the fields of characteristic 0, the smallest degree of the nonzero terms of the characteristic polynomial $\det(\lambda I - A)$ equals $n -$ rank A provided that

A is a Hermitian matrix. The latter assumption can be relaxed based on the equations

$$2 \text{ rank } A = \text{rank } \begin{pmatrix} O & A^H \\ A & O \end{pmatrix}$$

(which holds over any field) or

$$\text{rank } A = \text{rank } (A^H A)$$

(which holds over any field of characteristic 0). By using these two equations, we obtain the two reductions $CHAR \cdot POL \succeq M \cdot RANK$ and $(M \cdot MULTIPLY, CHAR \cdot POL) \succeq M \cdot RANK$, both of which imply the complexity bound $O(M(m+n))$ for $M \cdot RANK$ over any field of characteristic 0. In section 6 of chapter 4, we will describe a randomized algorithm leading to the same complexity estimates over any field. Over the fields of constants of characteristic 0, the equation rank $A =$ trace (A^+A) defines an alternative reduction of the problem $M \cdot RANK$ to $CHAR \cdot POL$ [see our comments to (2.9) at the end of this section]. ∎

From the LSP factorization, we may obtain a solution to the following important problem [over the complex and real fields, such a solution can also be obtained as a by-product of the solution of Problem 2.9a ($QRP \cdot FACTORS$)]:

Problem 2.10a ($SUBMATRIX$). *Given an $m \times n$ matrix A, compute its nonsingular submatrix of the maximum size $r \times r$.*

Solution. Without loss of generality assume that $m \leq n$. Let $A = LSP$ be the LSP factorization of A. Then it is easy to check that the submatrix of A having rows corresponding to the r nonzero rows of S and columns π_i, $i = 1, 2, \ldots, r$, is nonsingular, where π_i is the permutation associated with the matrix $P = (p_{i,j})$, i.e. $p_{i,\pi_i} = 1$, $p_{i,j} = 0$ if $j \neq \pi_i$. Therefore, we arrive at the complexity bound $O(M(m)n/m)$ for $SUBMATRIX$ over any field of constants. We also note that $SUBMATRIX$ has a subproblem $M \cdot RANK$, to which $SUBMATRIX$ itself is equivalent for a tame input matrix A (see Definition 1.8). ∎

The solution of this problem gives us two maximal linearly independent subsets of the column set and of the row set of A, respectively.

In some applications it suffices to solve the following problem, rather than Problem 2.10a ($SUBMATRIX$):

Problem 2.10b ($M \cdot PRECNDTN$). *Given an $m \times n$ matrix A, compute $r =$ rank A, an $r \times m$ matrix $R(r)$ and an $n \times r$ matrix $S(r)$ such that* $\det(R(r)AS(r)) \neq 0$.

Solutions. It suffices to compute $r = \text{rank } A$ (by solving $M \cdot RANK$), as well as an $m \times m$ matrix R and an $n \times n$ matrix S such that rank $(RAS) = r$ and RAS is a tame matrix. Then we may just set

$$R(r) = [I_r, O_{r,m-r}]R, \quad S(r) = S\begin{pmatrix} I_r \\ O_{n-r,r} \end{pmatrix}.$$

On the other hand, over the fields of characteristic 0, we may set $R = A^H$, $S = I$ or $R = I$, $S = A^H$, due to Corollary 1.2, whereas over an arbitrary field we may choose some special (triangular Toeplitz) matrices R and S defined by their $m+n$ random entries, according to Lemma 13.1 and Procedure 1.7.1. ■

Based on the solution of Problem 2.10b ($M \cdot PRCNDTN$), it is possible to give alternative solutions to Problems 2.3a ($LIN \cdot SOLVE1$) and 2.3b ($NULL \cdot SPACE$) (stated and solved earlier). Indeed, assume that we have solved Problem 2.10b ($M \cdot PRECNDTN$) [and obtained $r = \text{rank } A$ and two matrices $R(r)$ and $S(r)$ such that $R(r)AS(r)$ is a nonsingular $r \times r$ matrix]. Then, the linear system $A\mathbf{x} = \mathbf{b}$ is consistent if and only if, for $\mathbf{y} = (R(r)AS(r))^{-1}R(r)\mathbf{b}$ being the unique solution to the linear system

$$R(r)AS(r)\mathbf{y} = R(r)\mathbf{b},$$

the vector $\mathbf{x} = S(r)\mathbf{y}$ is a solution to the linear system $A\mathbf{x} = \mathbf{b}$, that is, if and only if

$$AS(r)\mathbf{y} = \mathbf{b}.$$

Thus Problem 2.3a ($LIN \cdot SOLVE1$) has been reduced to Problems 2.10b ($M \cdot PRECNDTN$) and 2.1-2.3 ($M \cdot VECTOR, M \cdot MULTIPLY, LIN \cdot SOLVE$), and here we may exclude $M \cdot VECTOR$, since it is reduced to $M \cdot MULTIPLY$.

To obtain an alternative solution of Problem 2.3b ($NULL \cdot SPACE$), we will follow [KP]. Assume that the $r \times r$ matrix B is a solution to Problem 2.10a ($SUBMATRIX$). With no loss of generality, we may assume that $A = \begin{pmatrix} B & C \\ D & E \end{pmatrix}$. Let

$$F = \begin{pmatrix} I_r & -B^{-1}C \\ O & I_{n-r} \end{pmatrix}.$$

Then $AF = \begin{pmatrix} B & O \\ D & O \end{pmatrix}$, and the null space of A is spanned by the columns of the matrix $F\begin{pmatrix} O \\ I_{n-r} \end{pmatrix}$. Similarly, we may reduce Problem 2.3b ($NULL \cdot$

$SPACE$) to the triple of Problems 2.10b ($M \cdot PRECNDTN$), 2.2 ($M \cdot MULTIPLY$) and 2.4 ($INVERT$).

Problem 2.10b ($M \cdot PRECNDTN$) can be extended as follows.

Problem 2.11a ($GENERATOR1$). *Given an $m \times n$ matrix A, an $r \times m$ matrix R and an $n \times r$ matrix S such that $r = \text{rank } A = \text{rank } (RAS)$, compute a generator (G, H) of length r for A. (R and S are said to be the compressors of A.)*

Fact 2.2 (compare [BA]). *There are infinitely many generators of length r for A, and one of them is given by $G = AS$ and by*

$$H^T = (RAS)^{-1} RA, \tag{2.8}$$

so that $(M \cdot MULT, INVERT) \succeq GENERATOR1$.

Proof. With no loss of generality, we may assume that A, S and R are the northwestern blocks of three square matrices (all three of the same size) filled with zeros outside these blocks. Let ϵ be a scalar parameter and $A(\epsilon)$, $S(\epsilon)$ and $R(\epsilon)$ be nonsingular matrices that converge, as $\epsilon \to 0$, to these square matrices, containing A, S and R, respectively. Let $H^T(\epsilon) = (R(\epsilon)A(\epsilon)S(\epsilon))^{-1}R(\epsilon)A(\epsilon) = S(\epsilon)^{-1}$, $G(\epsilon) = A(\epsilon)S(\epsilon)$, so that $A(\epsilon) = G(\epsilon)H^T(\epsilon)$. Let $\epsilon \to 0$ and arrive at Fact 2.2 by using a continuity argument. ∎

The proof demonstrates the powerful *method of homotopic transformation* of the solution to an auxiliary problem $\mathcal{P}_\epsilon$, readily available for every fixed positive value of the auxiliary parameter ϵ, into the solution to the original problem $\mathcal{P}_0$, so that the desired properties of the solution [in this case (2.8)] are easily preserved as $\epsilon \to 0$.

Fact 2.2 implies the complexity bound $O(M(m, n, r))$ for Problem 2.11a ($GENERATOR1$).

In sections 12 and 13 and in chapter 4, we will use the following modification of Problem 2.11 ($GENERATOR$):

Problem 2.11b ($G \cdot COMPRESS$), *compression of a generator of a matrix. Given a generator G, H of length $\hat{r}$ for an $n \times n$ matrix A and its rank $r < \hat{r} \leq n$, compute a generator of length r for $A = GH^T$.*

Solutions [compare [P92b] and the solution to Problem 3.6 ($M \cdot COMPRESS$) in the next section]. Apply the LSP factorization to the matrix G^T in order to compute an $n \times \tilde{r}$ submatrix $\tilde{G}$ (of G) having full rank $\tilde{r}$, $\tilde{r} \leq \hat{r}$, and express the remaining columns of G as linear combinations of the columns of $\tilde{G}$. (An alternative way for computing such a submatrix $\tilde{G}$ is via computing the SVD of G, [P93].) To simplify the notation, assume that the matrix $\tilde{G} = [\mathbf{g}_1, \ldots, \mathbf{g}_{\tilde{r}}]$ consists of the first $\tilde{r}$ columns of $G = [\mathbf{g}_1, \ldots, \mathbf{g}_{\hat{r}}]$. Then denote $H = [\mathbf{h}_1, \ldots, \mathbf{h}_{\hat{r}}]$ and write the expressions: $\mathbf{g}_j = \sum_{i=1}^{\tilde{r}} a_{ji}\mathbf{g}_i$, $j = \tilde{r}+1, \ldots, \hat{r}$. It follows that

$$A = GH^T = \sum_{i=1}^{\hat{r}} \mathbf{g}_i\mathbf{h}_i^T = \sum_{i=1}^{\tilde{r}} \mathbf{g}_i\mathbf{h}_i^T + \sum_{j=\tilde{r}+1}^{\hat{r}} \sum_{i=1}^{\tilde{r}} a_{ji}\mathbf{g}_i\mathbf{h}_j^T = \sum_{i=1}^{\tilde{r}} \mathbf{g}_i\tilde{\mathbf{h}}_i^T,$$

and we arrive at a generator of length $\tilde{r}$ for A,

$$A = \tilde{G}\tilde{H}^T, \quad \tilde{H} = [\tilde{\mathbf{h}}_1, \ldots, \tilde{\mathbf{h}}_{\tilde{r}}],$$

$$\tilde{\mathbf{h}}_i = \mathbf{h}_i + \sum_{j=\tilde{r}+1}^{\hat{r}} a_{ji}\mathbf{h}_j, \quad i = 1, \ldots, \tilde{r}.$$

If $\tilde{H}$ has full rank $\tilde{r}$, then r = rank $A = \tilde{r}$, and $G = \tilde{G}$, $H = \tilde{H}$ is a desired solution. Otherwise, we similarly express the columns of $\tilde{H}$ as linear combinations of the columns of its full rank $\bar{r}$ submatrix $\bar{H}$, then transform $\tilde{G}$ into an $n \times \bar{r}$ matrix $\bar{G}$ such that $A = \bar{G}\bar{H}^T$, observe that both $\bar{G}$ and $\bar{H}$ have full rank $\bar{r}$, and deduce that $\bar{r} = r$, so that $G = \bar{G}$, $H = \bar{H}$ is a desired generator of length r for A. The solution shows that

$$(SUBMATRIX, INVERT, M\cdot MULTIPLY) \succeq G\cdot COMPRESS,$$

which implies the deterministic complexity bound $O(M(\hat{r})n/\hat{r})$. Let us also show an alternative way, which uses randomization instead of the LSP factorization and leads to the same (although probabilistic) complexity bound: Compute $\tilde{r}$ = rank G, a random $\hat{r} \times \hat{r}$ matrix S, its inverse S^{-1}, and a new generator GS, $H(S^{-1})^T$, such that with a high probability the first $\tilde{r}$ columns of GS form a full rank matrix. Then proceed as in the above algorithm but also use randomization to transform the matrix $\tilde{H}$ into a matrix having its first r columns linearly independent with a high probability. ∎

The next problem, the last in this section, generalizes Problem 2.5 ($INVERT$) of matrix inversion; its solution also gives us solutions to Problems 2.8 ($L\cdot SQUARES$) and 2.10 ($M\cdot RANK$).

Problem 2.12 ($GEN\cdot INVERSE$), *computing the generalized inverse of a matrix. Given a matrix $A \in \mathbf{C}_{m,n}$, compute its Moore-Penrose generalized inverse A^+, uniquely defined by the relations (1.4).*

Solutions. The problem can be reduced to Problem 2.7c ($LSP \cdot FACTORS$). For simplicity assume that $m \leq n$ (otherwise replace A with A^T) and rewrite the LSP factorization of A in the following way: $A = LQUP$, where $S = QU$ and Q is the permutation matrix such that $U = Q^T S$ is the upper triangular matrix obtained by moving to the bottom all the zero rows of S. Then it is easy to check [IMH] that all the four matrix equations (1.4) hold for the matrix $A^+ = F^H(FF^H)^{-1}(G^H G)^{-1}G^H$, where F and G are the matrices made up by the first r rows of UP and by the first r columns of LQ, respectively, and r is the rank of A. This leads to the complexity bound $O(M(m)n/m)$. In the particular case where rank $A = \min\{m, n\}$, the problem is reduced to Problems 2.2 ($M \cdot MULTIPLY$) and/or 2.5 ($INVERT$) of matrix multiplication and inversion. Indeed, $A^+ = (A^H A)^{-1}A^H$ if rank $A = n$, $A^+ = A^H(AA^H)^{-1}$ if rank $A = m$, $A^+ = A^{-1}$ if rank $A = m = n$.

Yet another solution is given by means of the so-called *complete orthogonal decomposition of A,* which applies even to a rank deficient matrix A [GL].

An alternative approach is by the reduction of Problem 2.12 ($GEN \cdot INVERSE$) to Problem 2.6 ($CHAR \cdot POL$) of computing the coefficients $c_0, c_1, \ldots, c_{n-1}$ of the characteristic polynomial $\det(\lambda I - A)$ and to Problem 2.2a ($M \cdot POWERS$) [P90]. We may assume that A is a Hermitian matrix, for otherwise, we will first compute the Hermitian matrix $A^H A$, then its generalized inverse $(A^H A)^+$, and finally, $A^+ = (A^H A)^+ A^H$. An alternative way to this symmetrization is to recover A^+ from the generalized inverse

$$\begin{pmatrix} O & A \\ A^H & O \end{pmatrix}^+ = \begin{pmatrix} O & (A^H)^+ \\ A^+ & O \end{pmatrix}$$

of the Hermitian matrix $\begin{pmatrix} O & A \\ A^H & O \end{pmatrix}$; its size doubles the size of $A^H A$, but its condition number generally has a much smaller upper bound. Now, recall Theorem 1.1 and deduce that $A^{n-r}\sum_{i=n-r}^{n} c_i A^{i-n+r} = O$ where $\det(\lambda I - A) = \sum_{i=n-r}^{n} c_i \lambda^i$, $c_n = 1$, $c_{n-r} \neq 0$ for some r (and where, actually, $r = \text{rank } A$, over the fields of characteristic 0). Postmultiply this matrix equation by $(A^+)^{n-r}$, recursively apply Fact 1.7 and deduce that $c_{n-r}A^+A = -\sum_{i=n-r+1}^{n} c_i A^{i-n+r}$. Postmultiply the latter equation by A^+

and obtain that

$$c_{n-r}A^+ = -c_{n-r+1}A^+A - \sum_{i=n-r+2}^{n} c_i A^{i-n+r-1} =$$

$$(c_{n-r+1}/c_{n-r}) \sum_{i=n-r+1}^{n} c_i A^{i-n+r} - \sum_{i=n-r+2}^{n} c_i A^{i-n+r-1} = \tag{2.9}$$

$$\sum_{i=n-r+1}^{n-1} ((c_{n-r+1}/c_{n-r})c_i - c_{i+1})A^{i-n+r} + (c_{n-r+1}/c_{n-r})A^r.$$

This reduces the evaluation of A^+ to Problem 2.6 ($CHAR \cdot POL$) and to matrix powering [Problem 2.2a ($M \cdot POWERS$)], so that ($CHAR \cdot POL$, $M \cdot POWERS$)$\succeq$ $GEN \cdot INVERSE$, which implies the complexity bound $O((m+n)M(m+n))$ for $GEN \cdot INVERSE$. Matrix powering can be replaced by computing a Krylov matrix [Problem 2.1a ($KRYLOV$)] if we only need to compute $A^+\mathbf{v}$ for a fixed vector $\mathbf{v}$, that is, ($CHAR \cdot POL$, $KRYLOV$)$\succeq$ $L \cdot SQUARES$. (2.9) enables us to express trace(A^+) via the traces of A^k, $k = 0, 1, \ldots, r$, which can be easily computed from the coefficients $c_0, c_1, \ldots, c_{n-1}$ by using Newton's identities (1.4.8) and Proposition 6.1. Also observe that (2.9) implies the uniqueness of the matrix A^+ satisfying (1.4). We refer the reader to [Grev] for yet another alternative rational algorithm.

Finally, even if $r = \text{rank } A < \min\{m, n\}$, A^+ can be effectively approximated by using iterative algorithms, either by means of computing the SVD of A or by using the Newton iteration algorithm of section 5 of the next chapter and of [PS] (compare also the third alternative iterative approach based on [Chan]). ∎

Remark 2.2. Expressions similar to (2.9) can be based on any matrix equation $p(A) = O$ where $p(\lambda)$ is a polynomial over $\mathbf{F}$, in particular, on the equation $m_A(A) = O$ (see exercise 7).

Remark 2.3. Computing $A^+\mathbf{b}$ gives us a least-squares solution to the linear system $A\mathbf{x} = \mathbf{b}$, but for testing if this system is consistent [which is a part of Problem 2.3a ($LIN \cdot SOLVE1$)], it suffices to compute rank A and rank $[A, \mathbf{b}]$:

Fact 2.3. *A linear system* $A\mathbf{x} = \mathbf{b}$ *is consistent if and only if* rank $A =$ rank $[A, \mathbf{b}]$.

3. Further Matrix Computations, Tridiagonal Reduction and Matrix Computations over an Arbitrary Field of Constants.

This section will complete our review of computations with general matrices and will end with a remark on performing such computations over an arbitrary field of constants.

Generally, the solutions to Problems 3.1 ($EIGENVALUES$), 3.1a ($EXTREME \cdot EIGENVALUES$), 3.3 ($SVD$), to be studied in this section, as well as 3.4 ($M \cdot NORM$) and 3.5 ($M \cdot CONDITION$) for $h = 2$, can only be computed approximately by rational algorithms, and their approximate solutions are also a substantial part of the customary numerical computations routinely performed in practice. We will postpone the study of these approximation algorithms until chapter 8. In this section we will consider some rational algorithms that simplify these problems, in particular, by means of reducing the input matrix to a special (Hessenberg, tridiagonal or block tridiagonal) form. We will relate the solution of the latter problem to computations with dense structured matrices, which we will intensively study in sections 4–13.

The eigenvalue problem.

Problem 3.1 ($EIGENVALUES$), *the matrix eigenvalue problem. Compute the eigendecomposition (1.1), that is, compute all the eigenvalues and the associated eigenvectors of an $n \times n$ matrix A.*

It is customary to treat Problem 3.1 ($EIGENVALUES$), as two separate problems, that is, as the (numerically well-conditioned) *symmetric eigenvalue problem* in the cases where the matrix A is real symmetric or Hermitian, and as the *unsymmetric eigenvalue problem* otherwise [GL], [Par].

Given all the s eigenvalues of A, (1.1) defines s singular linear systems, each of n equations, for computing the eigenvectors. Such systems can be solved, for instance, by means of orthogonal factorization with pivoting, by means of $PLDM\tilde{P}$ factorization of their coefficient matrices [compare Problems 2.7a ($PLU\tilde{P} \cdot FACTORS$), 2.9a ($QRP \cdot FACTORS$)], or by using methods for Problem 2.8 ($L \cdot SQUARES$). In practice, some other, nonrational algorithms are preferred, since they have better numerical stability [GL].

Computation of the eigenvalues can be reduced first to Problem 2.6 ($CHAR{\cdot}POL$) of computing the coefficients of the characteristic polynomial of A and then to Problem 1.4.4 ($POL \cdot ZEROS$) of computing the zeros of the monic polynomial $p(x) = \sum_{i=0}^{n} p_i x^i$, $p_n = 1$. The complexity of the latter problem is estimated in chapter 7.

The dependence of the zeros of a polynomial on its coefficients is, however, ill-conditioned, and as a result, this way of the solution of Problem 3.1 (*EIGENVALUES*) leads to numerical stability problems (see more on this issue in section 1 of chapter 3 and in [W65]), so that distinct approaches are currently used in practice (see [GL]). In fact, the eigenvalues of a Hermitian (real symmetric) matrix are *well-conditioned* functions of its entries, whereas the zeros of its characteristic polynomial are generally *ill-conditioned* functions of the coefficients (see [W65]). Conversely, Problem 1.4.4 (*POL · ZEROS*) for a monic polynomial $p(x)$ is a special case of the problem of computing the eigenvalues of the associated *Frobenius matrix*:

$$F = \begin{pmatrix} 0 & 1 & 0 & \dots & 0 \\ \vdots & 0 & 1 & \ddots & \vdots \\ & & \ddots & \ddots & 0 \\ 0 & \dots & \dots & 0 & 1 \\ -p_0 & -p_1 & \dots & \dots & -p_{n-1} \end{pmatrix}. \tag{3.1}$$

Both F and F^T are also called the *companion matrices* of $p(x)$.

Due to the latter reduction, Problem 3.1 (*EIGENVALUES*) has no general rational solution algorithm even in the symmetric case; for its approximate solutions, see our chapter 8 and [GL], [W65], [Par].

Problem 3.1 (*EIGENVALUES*) is sometimes restricted to computing either a single eigenvalue or few eigenvalues:

Problem 3.1a (*EXTREME · EIGENVALUES*), *computing the extremal eigenvalues of a matrix. Compute the g absolutely largest eigenvalues and/or the h absolutely smallest eigenvalues and the associated eigenvectors of a given matrix A for fixed natural g and/or h.*

Most popular practical algorithms for Problems 3.1 (*EIGENVALUES*) start with reducing the input matrix A to a simpler form based on the *similarity transformation* of A into a *similar matrix* $S^{-1}AS$ where S is a readily invertible matrix, usually an orthogonal matrix (compare Facts 1.3 and 1.6).

Problem 3.2 (*H · REDUCE*), *Hessenberg reduction. Given an $n \times n$ matrix A, compute an orthogonal matrix Q and the matrix $H = [h_{ij}]$ such that $Q^H Q = I$, $Q^H AQ = H$, $h_{ij} = 0$ if $i - j > 1$. (Such a matrix H is said to be in the upper Hessenberg form. If A is Hermitian, so is H, and then*

$h_{ji} = 0$ if $i-j > 1$, so that H is tridiagonal, and then there exists a diagonal matrix D such that DHD^{-1} is a real symmetric tridiagonal matrix.)

A solution using about $(10/3)n^3$ ops for any input matrix A can be found in [GL]; an asymptotically faster solution is shown in section 5 of chapter 4. It relies on a reduction of Problem 3.2 ($H \cdot REDUCE$) to Problem 2.9 ($QR \cdot FACTORS$) (see Fact 4.5.1).

Seeking the reduction of a general matrix to the tridiagonal form we first have to relax the condition of the orthogonality of the similarity transformation.

Problem 3.2a ($T \cdot REDUCE$), *tridiagonal reduction. Given an $n \times n$ matrix A, compute four matrices P, Q, D and T, such that*

$$\begin{aligned} P^H Q &= D = \mathrm{diag}(d_0, \ldots, d_{n-1}), \\ P^H AQ &= T, \end{aligned}$$

D, P and Q are nonsingular, T is a complex tridiagonal symmetric matrix, so that A is similar to the tridiagonal matrix $D^{-1}T$.

If we restrict our task to computing the matrices T and D and do not require to compute P and Q, we arrive at the problem that we call **Problem 3.2b** ($T \cdot REDUCE1$).

Actually, some nonHermitian matrices can only be reduced to block tridiagonal form, and no algorithm can reduce them to the tridiagonal form.

We also consider the following problem:

Problem 3.2c ($I \cdot EIGENVALUES$), *inverse tridiagonal eigenvalue problem. Given the coefficients (or the zeros) of the characteristic polynomial of an $n \times n$ real symmetric tridiagonal matrix T and the coefficients (or the zeros) of the characteristic polynomial of its $(n-1) \times (n-1)$ leading principal submatrix, compute the entries of T.*

Problems 3.2b ($T \cdot REDUCE1$) and 3.2c ($I \cdot EIGENVALUES$) are related to each other since similar matrices A and T have a common characteristic polynomial. This property, however, does not extend to their principal submatrices.

Tridiagonalization and block tridiagonalization algorithms.

Next, we will consider some algorithms for the tridiagonalization problems [Problems 3.2a ($T \cdot REDUCE$) and 3.2b ($T \cdot REDUCE1$)] and for the inverse tridiagonal eigenvalue problem [Problem 3.2c ($I \cdot EIGENVALUES$)].

The known effective solutions of these computational problems with general matrices reduce these problems to computations with structured matrices and with polynomials (compare [Gut], [Par92]). This will be a feature of our algorithms too.

For simplicity, first let A be a Hermitian matrix. To solve Problem 3.2a ($T \cdot REDUCE$), we shall look for a unitary matrix Q and a real symmetric tridiagonal matrix

$$T = \begin{pmatrix} \alpha_0 & \beta_0 & & O \\ \beta_0 & \alpha_1 & \ddots & \\ & \ddots & \ddots & \beta_{n-2} \\ O & & \beta_{n-2} & \alpha_{n-1} \end{pmatrix} \tag{3.2}$$

such that $Q^H AQ = T$. Rewriting this condition as $AQ = QT$ and letting $\mathbf{q}_i$ denote column i of Q, $i = 0, \ldots, n-1$, we obtain the vector equations

$$A\mathbf{q}_j = \beta_{j-1}\mathbf{q}_{j-1} + \alpha_j \mathbf{q}_j + \beta_j \mathbf{q}_{j+1}, \quad j = 0, \ldots, n-1, \tag{3.3}$$

where we assume that $\beta_{-1} = \beta_{n-1} = 0$. Now, we may fix the vector $\mathbf{q}_0$, recall about the orthonormality of the $\mathbf{q}_i$, and compute the entries α_i, β_i and the vectors $\mathbf{q}_i$ by means of the Lanczos algorithm (compare [GL]):

Algorithm 3.1.

Input: a natural n, the entries of an $n \times n$ Hermitian matrix A and the components of a vector $\mathbf{q}_0 \in \mathbf{C}^n$.

Output: either DEGENERACY or else the entries $\alpha_0, \ldots, \alpha_{n-1}$, β_0, $\ldots$, β_{n-2} of a real symmetric tridiagonal matrix T of (3.2) and the column vectors $\mathbf{q}_1, \ldots, \mathbf{q}_{n-1}$ defining a unitary matrix Q with the first column $\mathbf{q}_0$ and such that $Q^H AQ = T$.

Computation:

1. Set $\beta_{-1} = 0$, $j = 0$.
2. Compute $\alpha_j = \mathbf{q}_j^H A\mathbf{q}_j$, $\mathbf{r}_j = (A - \alpha_j I)\mathbf{q}_j - \beta_{j-1}\mathbf{q}_{j-1}$.
3. If $\mathbf{r}_j \neq 0$, then set $\beta_j = ||\mathbf{r}_j||_2$ and $\mathbf{q}_{j+1} = \mathbf{r}_j/\beta_j$ [compare (3.3)]. Otherwise stop and output: DEGENERACY.
4. Set $j = j+1$; if $j \leq n-1$ go to Stage 2, otherwise stop and output the entries of T and Q.

Observe that the numbers β_j are defined at Stage 3 up to a scaling factor of modulus 1.

The most expensive stage of the Lanczos algorithm is the computation of matrix-by-vector products, which is simplified for sparse and/or structured matrices A. Another interesting feature is that the algorithm can be implemented with the memory space just for a pair of n-dimensional vectors if only the entries of T must be computed. On the other hand, the Lanczos algorithm is numerically unstable and intrinsically sequential. Moreover, the algorithm may have an abnormal interruption if $\mathbf{r}_j = \mathbf{0}$ at Stage 3 for some $j < n - 1$, due to an unsuccessful choice of the vector $\mathbf{q}_0$. In this case it is possible to modify the algorithm by setting $\beta_\ell = 0$ and choosing for $\mathbf{q}_{j+1}$ a random unit vector orthogonal to the vectors $\mathbf{q}_0, \dots, \mathbf{q}_j$. With a high probability, such a process shall compute the desired solution matrices Q and T, which always exist for a real symmetric (or Hermitian) input matrix A.

The equations (3.3), together with the orthonormality of the *Lanczos vectors* $\mathbf{q}_0, \dots, \mathbf{q}_{n-1}$, imply that these vectors constitute an orthonormal basis of the linear space spanned by the Krylov sequence $\{\mathbf{q}_0, A\mathbf{q}_0, \dots, A^{n-1}\mathbf{q}_0\}$. Therefore, such vectors can be alternatively computed by means of any algorithm that computes the QR factorization of the Krylov matrix $K(A, \mathbf{q}_0) = [\mathbf{q}_0, A\mathbf{q}_0, \dots, A^{n-1}\mathbf{q}_0]$, introduced in Definition 1.15. In fact, from the tridiagonalization condition $AQ = QT$ it follows that $A^iQ = QT^i$, $i = 0, 1, \dots$, whence $A^i\mathbf{q}_0 = QT^i\mathbf{e}_0$, $i = 0, \dots, n-1$, that is,

$$K(A, \mathbf{q}_0) = Q[\mathbf{e}^{(0)}, T\mathbf{e}^{(0)}, \dots, T^{n-1}\mathbf{e}^{(0)}] = QR. \tag{3.4}$$

QR factorization is uniquely defined up to scaling the columns of Q by factors having unit moduli. In particular, scaling at Stage 3 of Algorithm 3.1 gives us the QR factorization (3.4).

We now observe that the matrix R can be also obtained from the Cholesky factorization of the matrix $K(A, \mathbf{q}_0)^H K(A, \mathbf{q}_0)$, whose (i, j) entry is $a_{ij} = \mathbf{q}_0^H A^{i+j} \mathbf{q}_0$; indeed, from (3.4) it follows that

$$K(A, \mathbf{q}_0)^H K(A, \mathbf{q}_0) = R^H R.$$

The entries a_{ij} and $a_{i+h,j-h}$ of the matrix $K(A, \mathbf{q}_0)^H K(A, \mathbf{q}_0)$ satisfy the equation $a_{ij} = a_{i+h,j-h}$ for all h, i, j. Matrices with this property are called *Hankel* matrices and will be studied in sections 5, 9, 10 and 11.

Now, to solve Problem 3.2c ($I{\cdot}EIGENVALUES$), let T_i denote the $i \times i$ leading principal submatrix of T and recursively express $p_i(\lambda) = \det(\lambda I - T_i)$, $i = 1, \dots, n$, by means of the Laplace rule applied to the last column of T_i. This yields the following three term recursion:

$$p_0(\lambda) = 1$$
$$p_1(\lambda) = \alpha_0 - \lambda,$$
$$p_{i+1}(\lambda) = (\lambda - \alpha_i)p_i(\lambda) - \beta_{i-1}^2 p_{i-1}(\lambda), \quad i = 1, 2, \dots, n-1.$$

It is easy to prove that the polynomials $p_i(\lambda)$, up to within a scaling factor, constitute the remainder sequence obtained by applying the Euclidean scheme to the pair of polynomials $p_n(\lambda)$, $p_{n-1}(\lambda)$, that is, to the characteristic polynomials of the matrices T_n and T_{n-1}. Thus, the Euclidean scheme gives us the solution of the inverse tridiagonal eigenvalue problem, that is, of Problem 3.2c ($I \cdot EIGENVALUES$), in the case where $\beta_i \neq 0$, $i = 0, \ldots, n-2$. In particular, this is the case where both polynomials $p_n(\lambda)$ and $p_{n-1}(\lambda)$ have only real zeros and the zeros of $p_{n-1}(\lambda)$ interlace the zeros of $p_n(\lambda)$. For other solution algorithms for Problem 3.2c ($I \cdot EIGENVALUES$), we refer the reader to [Hoc], [Ha], [dBG].

Finally, consider tridiagonalization Problems 3.2a ($T \cdot REDUCE$) and 3.2b ($T \cdot REDUCE1$) for a general matrix A. We will first assume that there exists a similarity transformation of A into an *irreducible* tridiagonal matrix T of (3.2), that is, into a matrix of (3.2) whose all codiagonal entries are nonzero [we call the entry (i, j) codiagonal if $|i - j| = 1$]. In this case we will call the matrix A *nonderogatory* (compare [GL]).

Theorem 3.1 ([Par92]). *For an $n \times n$ matrix A, let there exist two nonsingular $n \times n$ matrices P and Q such that $P^H AQ = T$, $P^H Q = D = \mathrm{diag}(d_0, \ldots, d_{n-1})$, and T is an irreducible complex symmetric tridiagonal matrix [compare Problem 3.2a ($T \cdot REDUCE$)]. Then the matrices P, Q, D and T are uniquely determined by the pair of the first (or last) columns of P and Q.*

In the case where A is a Hermitian matrix, Theorem 3.1 follows from the Lanczos algorithm for $D = I$. In the general case it is sufficient to extend the Lanczos algorithm by rewriting the reduction conditions as

$$AQ = QD^{-1}T, \tag{3.5}$$

$$P^H A = TD^{-1}P^H. \tag{3.6}$$

By equating columns j on both sides of (3.5), as well as rows j on both sides of (3.6), we arrive at three term recurrence equations similar to (3.3). Based on these equations, the columns of P and Q and the scalars defining the symmetric tridiagonal matrix T can be computed by means of a two-sided orthogonalization process applied to suitable Krylov matrices.

Denote that $U_\ell = [\mathbf{u}_0, \ldots, \mathbf{u}_{\ell-1}]$, $V_\ell = [\mathbf{v}_0, \ldots, \mathbf{v}_{\ell-1}]$, $Q_\ell = [\mathbf{q}_0, \ldots, \mathbf{q}_{\ell-1}]$, $P_\ell = [\mathbf{p}_0, \ldots, \mathbf{p}_{\ell-1}]$, and recall one of the customary algorithms for the two-sided orthogonalization (compare [Par92]).

Algorithm 3.2, *generalized Gram-Schmidt orthogonalization.*

Input: a natural n and the components of $2n$ (nonzero) vectors $\mathbf{u}_i \in \mathbf{C}_{n,1}$, $\mathbf{v}_i \in \mathbf{C}_{n,1}$, $i = 0, \ldots, n-1$, or, equivalently, the matrices $U = [\mathbf{u}_0, \ldots, \mathbf{u}_{n-1}]$, $V = [\mathbf{v}_0, \ldots, \mathbf{v}_{n-1}]$.

Output: natural $\ell \le n$ and vectors $\mathbf{p}_i$ and $\mathbf{q}_i$, $i = 0, \ldots, \ell-1$, such that $\mathbf{p}_\ell^H \mathbf{q}_\ell = 0$, if $\ell < n$; $\mathbf{p}_i^H \mathbf{q}_j = 0$, $i \neq j$, $\mathbf{p}_i^H \mathbf{q}_i = d_i \neq 0$, $i = 0, \ldots, \ell-1$, that is, $P_\ell^H Q_\ell = D_\ell = \operatorname{diag}(d_0, \ldots, d_{\ell-1})$ and $U_\ell = Q_\ell R_\ell$, $V_\ell = P_\ell L_\ell^H$, where the matrices U_ℓ, V_ℓ have been defined above; the $\ell \times \ell$ lower triangular matrix $L_\ell = [l_{ij}]$ and the $\ell \times \ell$ upper triangular matrix $R_\ell = [r_{ij}]$ have unit diagonal entries; and $r_{ij} = \mathbf{p}_i^H \mathbf{u}_j / d_j$, $l_{ji} = \mathbf{v}_j^H \mathbf{q}_i / d_i$, for $i < j$.

Computation:

1. Set $k = 0$.

2. Compute

$$\mathbf{q}_k = \mathbf{u}_k - \sum_{i=0}^{k-1} \mathbf{q}_i (\mathbf{p}_i^H \mathbf{u}_k) / d_i,$$

$$\mathbf{p}_k^H = \mathbf{v}_k^H - \sum_{i=0}^{k-1} (\mathbf{v}_k^H \mathbf{q}_i) \mathbf{p}_i^H / d_i,$$

where the sums vanish for $k = 0$.

3. Compute $d_k = \mathbf{p}_k^H \mathbf{q}_k$.

4. If $k = n$ or $d_k = 0$, then set $\ell = k - 1$ and stop; otherwise set $k = k + 1$ and go to Stage 2.

Correctness of the algorithm is verified by inspection (compare [Par92]).

Algorithm 3.2 is an auxiliary algorithm for the tridiagonal reduction of a matrix, and we will next show how to apply it. First observe that, under the hypotheses of Theorem 3.1, applying (3.5) and (3.6) inductively yields the matrix equations:

$$A^i Q = Q(D^{-1}T)^i,$$
$$P^H A^i = (TD^{-1})^i P^H,$$

for $i = 0, 1, \ldots$. Whence $A^i \mathbf{q}_0 = Q(D^{-1}T)^i \mathbf{e}^{(0)}$, $\mathbf{e}^{(0)T} P^H A^i = \mathbf{e}^{(0)T} (TD^{-1})^i P^H$, $i = 0, 1, \ldots, n-1$, and we arrive at the following factorizations of the Krylov matrices $K(A, \mathbf{q}_0)$ and $K(A^H, \mathbf{p}_0)$, which generalize (3.4):

$$U = K(A, \mathbf{q}_0) = Q[\mathbf{e}^{(0)}, D^{-1}T\mathbf{e}^{(0)}, \ldots, (D^{-1}T)^{n-1}\mathbf{e}^{(0)}] = QR,$$
$$V = K(A^H, \mathbf{p}_0) = P[\mathbf{e}^{(0)}, (TD^{-1})^H \mathbf{e}^{(0)}, \ldots, ((TD^{-1})^H)^{n-1}\mathbf{e}^{(0)}] = PL^H. \tag{3.7}$$

Observe that, from the condition $P^HQ = D$ and from the symmetry of the matrix $V^HU = (\mathbf{p}_0^H A^{i+j}\mathbf{q}_0)$, it follows that

$$\begin{aligned} V^HU &= LDR, \\ L &= R^T, \end{aligned} \tag{3.8}$$

which gives us the LDL^T factorization of the matrix V^HU.

Now, under the hypotheses of Theorem 3.1, we apply Algorithm 3.2 to the matrices $U = K(A, \mathbf{q}_0)$ and $V = K(A^H, \mathbf{p}_0)$ and arrive at four matrices P_ℓ, Q_ℓ, R_ℓ and L_ℓ such that $V_\ell^H U_\ell = L_\ell D_\ell R_\ell$. Therefore, since the matrix $V_\ell^H U_\ell$ is symmetric, we have that $R_\ell = L_\ell^T$. Moreover, suppose that Algorithm 3.2 is carried out with $\ell = n$. Then, from the uniqueness of the LDL^T factorization (3.8), we have $L_\ell = L$, $R_\ell = R$, $D_\ell = D$, whence $Q_\ell = Q$, $P_\ell = P$, and the matrix $T_\ell = P_\ell^H A Q_\ell$ is tridiagonal. Moreover, from (3.7) and (3.8) it follows that $P_\ell^H A Q_\ell = L^{-1}V^H AUL^{-T}$. Therefore, since the matrix $H^{(1)} = V^H AU = (\mathbf{p}_0^H A^{i+j+1}\mathbf{q}_0)$ is symmetric, then also the tridiagonal matrix T_ℓ is symmetric.

We summarize this result in the following theorem.

Theorem 3.2 ([Par92]). *Under the hypotheses of Theorem 3.1, if $\ell = n$, then the matrix $T_\ell = P_\ell^H A Q_\ell$ satisfies the equation*

$$T_n = L^{-1}H^{(1)}L^{-T}, \tag{3.9}$$

where $L = L_n$, $H^{(1)} = V^T AU$, $L^{-T} = (L^T)^{-1}$.

In the general case, for $\ell \leq n$, the following result holds:

Theorem 3.3 ([Par92]). *$T_\ell = P_\ell^H A Q_\ell$ is a complex tridiagonal symmetric matrix.*

Note that $H = V_\ell^H U_\ell$ is the leading principal nonsingular submatrix of the matrix $H^{(0)} = V^H U$ of the maximum size for which there exists its LDL^T factorization; in particular, $\ell = n$ and $H = H^{(0)}$ if and only if $H^{(0)}$ is nonsingular and allows its LDL^T factorization.

Theorem 3.2 implies the following randomized algorithm, which solves Problems 3.2a ($T \cdot REDUCE$) and 3.2b ($T \cdot REDUCE1$) (excluding the singular case where $\ell < n$):

Algorithm 3.3. *Tridiagonal reduction.*

Input: natural n and the entries of an $n \times n$ matrix A.

Output: either FAILURE or the entries of four matrices D, T, P and Q, all of the size $n \times n$, that together constitute a solution to Problem 3.2a

($T \cdot REDUCE$), where D is a diagonal matrix; T is a complex, symmetric and tridiagonal matrix; and D, P and Q are nonsingular matrices.

Computation:

1. Choose two random vectors $\mathbf{p}, \mathbf{q} \in \mathbf{C}_{n,1}$.
2. Compute the Krylov matrices $U = K(A, \mathbf{q})$ and $V = K(A^H, \mathbf{p})$.
3. Compute $H^{(0)} = VU$.
4. If the LDL^T factorization of $H^{(0)}$ does not exist, output FAILURE and stop. Otherwise compute this factorization.
5. Compute $Q = UR^{-1}$ and $P = VR^{-1}$, where $L = R^T$.
6. Compute $T = P^H AQ$.

The algorithm outputs FAILURE in the case where the LDL^T factorization of $H^{(0)}$ does not exist. This case corresponds to a breakdown of Lanczos algorithm in the symmetric case, i.e., to the case where $\ell < n$. Unlike the symmetric case, however, in the general case the input matrix A may be irreducible to the tridiagonal form, and then the output FAILURE shall occur for any initial choice of the vectors $\mathbf{p}$ and $\mathbf{q}$.

For the solution of Problem 3.2b ($T \cdot REDUCE1$), Stages 5 and 6 can be replaced by the computation of the matrix T_n of (3.9), due to Theorem 3.2.

If the matrix A and the vectors $\mathbf{p}$ and $\mathbf{q}$ are such that $H^{(0)}$ is strongly nonsingular, then there exists the LDL^T factorization of the matrix $H^{(0)}$. We observe that if A has eigenspaces of dimension 1 (each eigenspace of A is defined as $\{\mathbf{x} \in \mathbf{C}_{n,1} : A\mathbf{x} = \lambda\mathbf{x}\}$, where λ is an eigenvalue of A), in particular, if A has pairwise distinct eigenvalues, then for almost any pair of vectors $\mathbf{p}$ and $\mathbf{q}$, the matrix $H^{(0)}$ is strongly nonsingular, in which case the generalized Gram-Schmidt orthogonalization process can be brought to its completion.

Theorems 3.1 and 3.3 reduce Problem 3.2a ($T \cdot REDUCE$) to the computation of Krylov matrices [Problem 2.1a ($KRYLOV$)] and either to the generalized Gram-Schmidt orthogonalization or, alternatively, to the computation of the LDL^T factorization of the symmetric matrix $H^{(0)}$ defined above.

With the latter approach, we may simplify the computations, since we may exploit the Hankel structure of the matrices $H^{(0)}$ and $H^{(1)}$ when we compute their factorizations. We will show this in sections 9 and 10 where we first develop some effective techniques for computations with structured matrices and then use these techniques to devise the desired algorithm (Algorithm 10.2) for tridiagonalization or block tridiagonalization. More specifically, in the general case, we are led to block tridiagonalization, which always turns into tridiagonalization in the regular case, where $\ell = n$.

By using the classical result of [GrLi] on block triangular factorization (see Proposition 10.1), we obtain the following theorem, on which we rely in our Algorithm 10.2:

Theorem 3.4 ([Par92]). *Let $H^{(0)} = LDL^T$ be the block LU factorization of $H^{(0)}$ of Proposition 10.1. Then the matrix T_n of (3.9) is a block tridiagonal matrix. The diagonal blocks are $\delta_s \times \delta_s$ matrices, for appropriate δ_s, $s = 1, \ldots, \ell$, having entries $h_{ij}^{(s)}$, $h_{ij}^{(s)} = h_{i+j}^{(s)}$, $h_r^{(s)} = 0$ if $r < \delta_s - 2$. Each codiagonal block has a single nonzero entry in the lower right position. The value of this entry is equal to the values of all other nonzero entries lying on the same antidiagonal of T_n.*

We rely on this theorem in our Algorithm 10.2 for such a tridiagonalization of an arbitrary $n \times n$ matrix A (including the degeneration case of $\ell < n$). The solution is computed by using a bounded computational precision, which improves the numerical stability and decreases the Boolean complexity of the block tridiagonalization.

Computing SVD, norms and condition numbers of a matrix

Our next computational problem for a matrix A is equivalent to the symmetric eigenvalue problem for $A^H A$ and AA^H (see Fact 1.5), which has no rational solution and whose complexity is estimated in chapter 8. Practical algorithms compute approximate solution without computing $A^H A$ (see [GL]).

Problem 3.3 (SVD), *singular value decomposition (SVD) of a matrix. Given an $m \times n$ matrix A, compute its SVD.*

Remark 3.1. The SVD of A gives generators $G = \tilde{U}\tilde{\Sigma}^{1/2}$, $H^T = \tilde{\Sigma}^{1/2}\tilde{V}^H$ or $G = \tilde{U}$, $H^T = \tilde{\Sigma}\tilde{V}^T$ of length r for A, where $\tilde{U}$ and $\tilde{V}$ are the submatrices made up by the first r columns of U and V, respectively, and $\tilde{\Sigma}$ is the $r \times r$ leading principal submatrix of Σ. Computing the SVD is numerically stable in practice [GL], although this is not a rational algorithm.

In an application in chapter 4, we will need to solve this problem where the input matrix A is represented by its short generator (M, N), $A = MN^T$, $M \in \mathbf{F}_{n,r}$, $N \in \mathbf{F}_{n,r}$. In this case, it is most effective to at first compute the SVD's of M and N^T and then to obtain the desired SVD of A by using the techniques of [EL], [FH] and [HLPW].

Next, we will consider the evaluation of matrix norms and condition numbers.

Problem 3.4 ($M \cdot NORM$). *Compute* $||A||_h$ *for* $h = 1, 2, \infty$.

Proposition 1.1 provides a very simple solution of this problem [by using $O(n^2)$ ops] in the cases where $h = 1$ and $h = \infty$, as well as some upper and lower estimates for $||A||_2$. Proposition 1.1 and Fact 1.5 also reduce the computation of $||A||_2$ to solving Problem 3.1a ($EXTREME \cdot EIGENVALUES$) for $A^H A$. [Both computations for $||A||_2$ and for Problem 3.1a ($EXTREME \cdot EIGENVALUES$) generally have no rational solution algorithms.]

Problem 3.5 ($M \cdot CONDITION$). *Compute* $\mathrm{cond}_h(A)$ *for* $h = 1, 2, \infty$.

For $h = 1$ and $h = \infty$ for a nonsingular matrix A, Problem 3.5 ($M \cdot CONDITION$) immediately reduces to Problems 2.5 ($INVERT$) of matrix inversion and 3.4 ($M \cdot NORM$); for $h = 2$, it reduces to Problem 3.3 (SVD) since $\mathrm{cond}_2(A) = \sigma_1/\sigma_n$ (see Proposition 1.1). Several practically effective algorithms estimate $\mathrm{cond}_h(A)$ rather than compute it. Usually, such algorithms are described in the case of Hermitian or real symmetric A (see [GL], [Di], [Hi86]), but this also handles the general case, since $(\mathrm{cond}_2(A))^2 = \mathrm{cond}_2(A^H A)$ and $1/n \le \mathrm{cond}_h(A)/\mathrm{cond}_2(A) \le n$ for $h = 1$, $h = \infty$.

Approximation by a matrix of a lower rank.

Problem 3.6 ($M \cdot COMPRESS$). *Given an* $n \times n$ *matrix* A *and an integer* d, $0 < d \le r = \mathrm{rank}\ A$, *approximate* A *by an* $n \times n$ *matrix* A_d *of rank at most* d *(representing* A *with its generator of length at most* d*), so as to minimize* $||A - A_d||_2$.

The solution of this problem is immediately reduced to computing the SVD of A, due to the following fact ([GL], p. 73):

Fact 3.1. *Let* $W = U\Sigma V^H$ *be the SVD of* $W \in \mathbf{C}_{m,n}$, Σ *be an* $m \times n$ *diagonal matrix with exactly* r *nonzeros on the diagonal,* $\sigma_1 \ge \sigma_2 \ge \cdots \ge \sigma_r > 0$. *Let* $d < r$ *be a positive integer,* Σ_d *be the* $m \times n$ *diagonal matrix,* $\Sigma_d = \mathrm{diag}(\sigma_1, \sigma_2, \ldots, \sigma_d, 0, \ldots, 0)$, $W_d = U\Sigma_d V^H$. *Then* $||W - W_d||_2 = \sigma_{d+1} \le ||W - Y||_2$ *for all matrices* Y *of rank at most* d.

For $d = r$, this solution of Problem 3.6 ($M \cdot COMPRESS$) gives us a (numerically stable) solution of Problem 2.11a ($G \cdot COMPRESS$). We refer the reader to [PS] and [Chan] on some alternative solutions of Problem 3.6 ($M \cdot COMPRESS$), without computing and truncating the SVD of A.

Matrix computations over any field.

We will conclude this section with a remark on the extension of general matrix computations presented so far in this chapter to the computations over arbitrary fields.

Remark 3.2. Almost all the algorithms of this chapter are rational, that is, they involve only arithmetic operations and comparisons, with the exception of the algorithms for Choleski factorization [Problem 2.7 ($LU \cdot FACTORS$) for an h.p.d. input matrix], for orthogonal factorization [Problem 2.9 ($QR \cdot FACTORS$)] and for Hessenberg reduction [Problem 3.2 ($H \cdot REDUCE$)], in whose solutions computing the square roots is involved. [The solutions to Problems 3.1 ($EIGENVALUES$), 3.1a ($EXTREME \cdot EIGENVALUES$), 3.3 ($SVD$) and 3.6 ($M \cdot COMPRESS$), as well as to Problems 3.4 ($M \cdot NORM$) and 3.5 ($M \cdot CONDITION$) for $h = 2$, are irrational but are not included into this chapter.] In many cases these rational algorithms can be extended to the case of computations over any field (compare section 7 of chapter 1), but we need to guard against involving comparisons of the absolute values of the field elements and to recall that the sums of positive integers may vanish. This means, in particular, that computing least-squares solutions to linear systems [Problem 2.8 ($L \cdot SQUARES$)], matrix norms and condition numbers, as well as the main purpose for introducing the latter tools (the norms and conditioning), that is, bounding the errors of computations, are not defined over an arbitrary field. The definition of A^+ based on (2.9) and on the matrix equation $\begin{pmatrix} O & A \\ A^T & O \end{pmatrix}^+ = \begin{pmatrix} O & (A^T)^+ \\ A^+ & O \end{pmatrix}$ applies over any field, but the properties (1.4) of A^+ do not generally hold anymore (see exercise 14), and the solution to Problem 2.12 ($GEN \cdot INVERSE$) loses its value over the fields of positive characteristics. Similar comments apply to the QR factorization of Problem 2.9 ($QR \cdot FACTORS$) over an arbitrary field $\mathbf{F}$, even in the case where there exists such a factorization over $\mathbf{F}$. The second and the third of the presented solutions of Problem 2.10 ($M \cdot RANK$) of computing the rank of a matrix A and the algorithm that solves Problem 2.3a ($LIN \cdot SOLVE1$) by its reduction to Problem 2.8 ($L \cdot SQUARES$) may give wrong answers over finite fields and thus should not be used. The definitions of the singularity and the rank of a matrix involve comparisons, but only between a field element and 0 or between some pairs of auxiliary integers (associated with the matrix sizes), respectively. Moreover, we may restate Problem 2.10 ($M \cdot RANK$) by only involving comparisons with 0 in its statement [BGH]. Thus, Problems 2.3a ($LIN \cdot SOLVE1$) and 2.10 ($M \cdot RANK$) are well defined over any field, and the solutions using $O(M(m)n/m)$ ops for $m \leq n$, or $O(M(n)m/n)$ for $m > n$ and based on the LSP-factorization can be applied over any field.

As a simple exercise, the reader may trace how the above comments on computation over any field extend to the computations with special matrices in the remainder of this chapter (see also Remarks 5.4, 6.1, 9.5, 11.1 and 11.2).

4. Vandermonde Matrices. Cauchy (Generalized Hilbert) Matrices. Correlation to Polynomial Evaluation and Interpolation.

Matrices encountered in practical computations frequently have good structure, which can be exploited to simplify the computations. The resulting improvement against the computations with dense unstructured matrices is dramatic: the complexity bounds decrease to $O(n \log n)$ or $O(n \log^2 n)$ ops and to $O(n)$ words of storage space with small overhead constants, for the solution of the fundamental Problems 2.1–2.5. In the remainder of this chapter we will demonstrate this improvement for several important classes of structured matrices, exploiting various techniques of independent interest and, in particular, the *correlation to computations with polynomials.* We will start with studying some specific classes of structured matrices (of Vandermonde, Cauchy-Hilbert, Hankel and Toeplitz type) and then will present their *unified treatment* based on the study of the associated operators.

In addition to our previous notation, we will use the following definitions.

Definition 4.1. J is the *reversion matrix* filled with zeros, except for its antidiagonal, filled with ones, and Z is the *down-shift matrix* filled with zeros, except for its first subdiagonal filled with ones, i.e.

$$Z = \begin{pmatrix} 0 & \dots & & \dots & 0 \\ 1 & 0 & & & \vdots \\ 0 & \ddots & \ddots & & \\ \vdots & & \ddots & \ddots & \vdots \\ 0 & \dots & \dots & 1 & 0 \end{pmatrix}, \quad J = \begin{pmatrix} 0 & \dots & 0 & 1 \\ \vdots & & \cdot^{\cdot^{\cdot}} & 0 \\ 0 & \cdot^{\cdot^{\cdot}} & & \vdots \\ 1 & 0 & \dots & 0 \end{pmatrix}.$$

Thus, for a vector $\mathbf{v} = [v_0, v_1, \dots, v_{n-1}]^T$, we have $J\mathbf{v} = [v_{n-1}, \dots, v_1, v_0]^T$, $Z\mathbf{v} = [0, v_0, v_1, \dots, v_{n-2}]^T$, $Z^T\mathbf{v} = [v_1, v_2, .., v_{n-1}, 0]^T$.

Definition 4.2. A matrix $V = V(\mathbf{v}) \in \mathbf{F}_{m+1,n+1}$ is an $(m+1) \times (n+1)$ *Vandermonde matrix* (associated with the vector $\mathbf{v} = [v_0, \ldots, v_m]^T$) if $(V)_{i,j} = v_i^j$, $i = 0, 1, \ldots, m$, $j = 0, 1, \ldots, n$. (Many authors call V^T, rather than V, a Vandermonde matrix.)

For $m = 1$ and $n = 3$ we have

$$V = \begin{pmatrix} 1 & v_0 & v_0^2 & v_0^3 \\ 1 & v_1 & v_1^2 & v_1^3 \end{pmatrix}.$$

Example 4.1. An important example of a Vandermonde matrix is given by the matrix $F_{n+1} = \sqrt{n+1}\,\Omega = (\omega^{ij})$, where $\Omega = (\omega^{ij}/\sqrt{n+1})$ is the *Fourier matrix* introduced in the solution of Problem 1.2.3 (*IDFT*), and ω is a primitive $(n+1)$-st root of 1. It is important to point out (compare [vL]) that different FFT algorithms correspond to different factorizations of the matrix F_{n+1}, and the matrix formulation of FFT algorithms is a tool for unifying and simplifying the literature in this field. We refer the reader to exercise 49 for the factorizations on the base of *Decimation in Time* (DIT) and *Decimation in Frequency* (DIF) algorithms for the *base 2 FFT*, such as Cooley-Tukey and Sande-Tukey algorithms [CT]. More details concerning, for instance, the base 4 and the split-radix algorithm (in both DIF and DIT versions) can be found in [Duh], [vL], [Nus81], and we refer the reader to [Bo], [BBo] for the analysis of Bruun's algorithm ([Bru]) and of other algorithms for different trigonometric transforms, such as Hartley, Sine and Cosine transforms (see also [PFTV], [Hou], [VN]).

Solutions of Problems 2.1 ($M{\cdot}VECTOR$) **and 2.3** ($LIN{\cdot}SOLVE$) (in the case of Vandermonde matrices). A square Vandermonde matrix is nonsingular if and only if $v_i \neq v_j$ for $i \neq j$. For a Vandermonde matrix V and for a fixed vector $\mathbf{p} = [p_0, \ldots, p_n]^T$, the evaluation of the product $V\mathbf{p}$ is equivalent to Problem 1.2.2 ($POL \cdot EVAL$), that is, to the evaluation of the polynomial $p(x) = p_0 + p_1 x + \cdots + p_n x^n$ at the points $x = v_i$ for $i = 0, 1, \ldots, m$, whereas solving the linear system $V\mathbf{x} = \mathbf{p}$, that is, the evaluation of $\mathbf{x} = V^{-1}\mathbf{p}$, is equivalent to interpolation by a polynomial [Problem 1.2.3 ($POL \cdot INTERP$)], provided that $m = n$ and $v_i \neq v_j$ if $i \neq j$. ∎

Due to this *reduction to polynomial computations,* both evaluations of $V\mathbf{p}$ and $V^{-1}\mathbf{p}$ [which are Problems 2.1 ($M \cdot VECTOR$) and 2.3 ($LIN \cdot SOLVE$) for the matrix $A = V$ defined above] cost $O((m+n)\log^2(m+n))$ ops (see section 4 of chapter 1), which is much less than in the case where V

is a general $m \times n$ matrix. Based on the next definition and the next simple fact, we may effectively compute $(\det V)^2$.

Definition 4.3. For a vector $\mathbf{v} = [v_0, \dots, v_m]^T$, let $\Gamma(x) = \Gamma_{\mathbf{v}}(x)$ denote the polynomial $\prod_{i=0}^{m}(x - v_i) = x^{m+1} + \sum_{i=0}^{m} \gamma_i x^i$, and $\gamma = \gamma(\mathbf{v})$ denote the vector $(\gamma_0, \dots, \gamma_m)^T$.

Fact 4.1. *For an $m \times m$ Vandermonde matrix $V = V(\mathbf{v})$ of Definition 4.2,*

$$(\det V)^2 = (-1)^{m(m+1)/2} \prod_{i=0}^{m} \Gamma'(v_i) \,.$$

Definition 4.4. Given two vectors $\mathbf{s} = [s_i]$ and $\mathbf{t} = [t_i]$ such that $s_i \neq t_j$ for any pair $[i, j]$, the $m \times n$ matrix $C(\mathbf{s}, \mathbf{t}) = C = [c_{ij}]$, $c_{ij} = 1/(t_i - s_j)$, $i = 0, 1, \dots, m-1$, $j = 0, 1, \dots, n-1$, is the *Cauchy (generalized Hilbert) matrix* associated to the vectors $\mathbf{s}$ and $\mathbf{t}$.

This class of matrices (associated with the names of Cauchy in [PSz], [GO2] and Hilbert in [Ger]) appears in the study of integral equations, conformal mappings and singular integrals ([Ger], [ODR89], [Rok85]), where, in particular, $C\mathbf{v}$ and $C^{-1}\mathbf{v}$ are sought for given vectors $\mathbf{s}, \mathbf{t}$ and $\mathbf{v}$.

For instance, if $m = 2$, $n = 3$, then

$$C = \begin{pmatrix} (t_0 - s_0)^{-1} & (t_0 - s_1)^{-1} & (t_0 - s_2)^{-1} \\ (t_1 - s_0)^{-1} & (t_1 - s_1)^{-1} & (t_1 - s_2)^{-1} \end{pmatrix}.$$

Observe that a Cauchy (generalized Hilbert) matrix would not change if the vector $\alpha[1, 1, \dots, 1]^T$ is added to both vectors $\mathbf{s}$ and $\mathbf{t}$ for any scalar α.

Cauchy (generalized Hilbert) matrices form a special subclass of *Loewner matrices*, that is, matrices B such that

$$(B)_{i,j} = \frac{u_i - v_j}{s_i - t_j}$$

(see [FPt], [Fi]).

Solution of Problem 2.1 ($M \cdot VECTOR$) [for a Cauchy (generalized Hilbert) matrix]. For a Cauchy (generalized Hilbert) matrix C and for a vector $\mathbf{v} = [v_j]$, the vector $\mathbf{w} = C\mathbf{v}$ is filled up with the values

$$w_i = (C\mathbf{v})_i = \sum_{j=0}^{n-1} v_j/(t_i - s_j) = p(t_i)/\Gamma(t_i)$$

where $\Gamma(y) = \Gamma_{\mathbf{s}}(y)$ is defined in Definition 4.3 for $\mathbf{v} = \mathbf{s}$ and where

$$p(y) = \Gamma(y) \sum_{j=0}^{n-1} v_j/(y - s_j). \tag{4.1}$$

Compare the Lagrange interpolation formulae (1.4.1) and (1.4.2) and deduce that $p(y)$ is the interpolation polynomial for the values $v_j\Gamma'(s_j)$ at s_j, $j = 0, 1, \ldots, n-1$.

Thus, we may effectively compute $C\mathbf{v}$ by using the fan-in method and the polynomial evaluation and interpolation algorithms of section 4 of chapter 1.

Algorithm 4.1 ([Ger]).

Input: natural n, m; the component v_j, $j = 0, \ldots, n-1$, of a vector $\mathbf{v} = [v_i]$; and the entries c_{ij}, $i = 0, \ldots, m-1$, $j = 0, \ldots, n-1$, of the Cauchy (generalized Hilbert) matrix C of Definition 4.4 (or just the entries of its first row and column).

Output: the components of the vector $C\mathbf{v}$.

Computation:

0. Set $s_0 = 0$ and compute $t_i = 1/c_{i0}$, $i = 0, \ldots, m-1$, and $s_j = 1/c_{0j} - t_0$, $j = 0, \ldots, n-1$.

1. Apply the fan-in method to compute the coefficients of the polynomial $\Gamma(y) = \Gamma_{\mathbf{s}}(y)$. Then compute the coefficients of $\Gamma'(y)$.

2. Compute the values $\Gamma'(s_j)$ and then $v_j\Gamma'(s_j)$, for $j = 0, 1, \ldots, n-1$.

3. Compute the coefficients of the polynomial $p(y)$ of (4.1), which interpolates to its values $p(s_j) = v_j\Gamma'(s_j)$, for $j = 0, 1, \ldots, n-1$, on the set of points $\{s_0, \ldots, s_{n-1}\}$.

4. Compute the values of the polynomials $p(y)$ and $\Gamma(y)$ and then their ratios $w_i = p(t_i)/\Gamma(t_i)$ at the points $y = t_i$, $i = 0, 1, \ldots, m-1$.

By using the algorithms of section 4 of chapter 1, we perform Stages 1–3 in $O(n \log^2 n)$ ops and Stage 4 in $O(m \log^2 m + n \log n)$ ops, so their overall cost is $O((m+n)\log^2(m+n))$ ops. Other effective solution algorithms are presented in section 4 of the next chapter and in [An], [Ap], [BHu], [Leo], [OS], [vDR], and [Rok85]. ■

Solution of Problem 2.3 (*LIN·SOLVE*) [for a Cauchy (generalized Hilbert) linear system]. If $m = n$ and if the matrix C is nonsingular, we may similarly solve the linear system $C\mathbf{x} = \mathbf{v}$, that is, we may evaluate $C^{-1}\mathbf{v}$ as follows ([Gast]):

Algorithm 4.2.

Input: natural n, the component v_j, $j = 0, \ldots, n-1$, of a vector $\mathbf{v} = [v_i]$ and the entries c_{ij}, $i, j = 0, \ldots, n-1$, of the Cauchy (generalized Hilbert) matrix C of Definition 4.3 (or just the entries of its first row and column).

Output: the components of the vector $\mathbf{v} = C^{-1}\mathbf{w}$.

Computation (Stages **0** and **1** are the same as in Algorithm 4.1):

0. Set $s_0 = 0$ and compute $t_i = 1/c_{i0}$, $i = 0, \ldots, m-1$, and $s_j = 1/c_{0j} - t_0$, $j = 0, \ldots, n-1$.

1. Compute the coefficients of the polynomials $\Gamma(y) = \Gamma_{\mathbf{s}}(y)$ of Definition 4.3 for $\mathbf{v} = \mathbf{s}$ (by using the fan-in method), and $\Gamma'(y)$.

2. Compute the values $w_i\Gamma(t_i) = (C\mathrm{v})_i\Gamma(t_i) = \Gamma(t_i)\sum_{j=0}^{n-1} v_j/(t_i - s_j)$, for all i.

3. Interpolate to the set of the values $w_i\Gamma(t_i)$ at the points t_i by a polynomial $h(y)$ of degree at most $n-1$. [Compare the Lagrange interpolation formulae (1.4.1) and (1.4.2) and deduce that this polynomial takes the values $v_j\Gamma'(s_j)$ at s_j, for $j = 0, 1, \ldots, n-1$.]

4. Evaluate $h(s_j) = v_j\Gamma'(s_j)$ and $\Gamma'(s_j)$ and then evaluate and output $v_j = h(s_j)/\Gamma'(s_j)$, for all j.

Applying the fast algorithms of section 4 of chapter 1, we will only use $O(n\log^2 n)$ ops in this computation. ∎

Finally, we recall the following matrix equations from [GO2], [FHR]:

$$C = C(\mathbf{t}, \mathbf{s}) = \mathrm{diag}([1/\Gamma_{\mathbf{t}}(s_i)])V(\mathrm{s})V^{-1}(\mathbf{t})\mathrm{diag}([\Gamma'_{\mathbf{t}}(t_i)]), \qquad (4.2)$$

$$\left(C^{-1}\right)^T = -\mathrm{diag}([\Gamma_{\mathbf{s}}(t_i)/\Gamma'_{\mathbf{t}}(s_i)])C\mathrm{diag}([\Gamma_{\mathbf{t}}(s_i)/\Gamma'_{\mathbf{s}}(t_i)]), \qquad (4.3)$$

where $\mathbf{s}$ and $\mathbf{t}$ are two vectors of dimension $n+1$ filled with $2n+2$ distinct values; $V(\mathbf{s})$, $V(\mathbf{t})$ and $\Gamma_{\mathbf{t}}(x)$ are defined in Definitions 4.1 and 4.3, and

$$\mathrm{diag}([w_i]) = \mathrm{diag}(w_0, \ldots, w_n)$$

(compare Definition 1.3). (4.2) follows from (1.4.1) and implies (4.3). The reader may compare the evaluation of $C\mathbf{v}$ and $C^{-1}\mathbf{v}$ based on (4.2) and (4.3) with Algorithms 4.1 and 4.2. (4.2) and (4.3) can also be used in order to rapidly evaluate $(\det C)^2$.

5. Toeplitz and Hankel Matrices. Correlation to Polynomial Multiplication and Division and to Padé Approximation.

In this section we extend our study of dense structured matrices. We introduce the classes of Toeplitz, Hankel and f-circulant matrices and analyze their intimate correlation to certain polynomial computations. We also recall two well-known inversion formulae for Toeplitz matrices.

Definition 5.1. $T = [t_{ij}]$ is a *Toeplitz matrix* if $t_{i,j}= t_{i+k,j+k}$, for all positive k, that is, if all the entries of T are invariant in their shifts in the diagonal direction, so that the matrix T is completely defined by its first row and its first column [compare the matrices of the equations (5.1), (5.3), (5.4) and (5.6), for example]. $H = [h_{i,j}]$ is a *Hankel matrix* if $h_{i,j} = h_{i-k,j+k}$, that is, if all the entries of H are invariant in their shifts in the antidiagonal direction, so that the matrix H is completely defined by its first row and its last column [compare the matrices in (9.2)]. $Z_f(\mathbf{c}) = Z_{f,m,n}(\mathbf{c}) = [z_{i,j}]$, for a vector $\mathbf{c} = [c_0, \dots, c_{m-1}]^T$ and for a scalar $f \neq 0$, is an *f-circulant $m \times n$ matrix* if $z_{i,j} = c_{i-j \ mod\ m}$ for $i \geq j$; $z_{i,j} = f c_{i-j \, mod \, m}$ for $i < j$, so that in this special case the Toeplitz matrix $Z_f(\mathbf{c})$ is completely defined by its first column $\mathbf{c}$ and by the scalar f. For example, for $m = n = 4$ we have

$$Z_f(\mathbf{c}) = \begin{pmatrix} c_0 & fc_3 & fc_2 & fc_1 \\ c_1 & c_0 & fc_3 & fc_2 \\ c_2 & c_1 & c_0 & fc_3 \\ c_3 & c_2 & c_1 & c_0 \end{pmatrix}.$$

(1)- and (−1)-circulant matrices are called *circulant and anticirculant (skew-circulant)*, respectively. Alternatively, we could have interchanged the roles of m and n and of columns and rows, but we will only deal with the square f-circulant matrices.

Remark 5.1. TJ and JT are Hankel matrices for any Toeplitz matrix T and for the reversion matrix J of Definition 4.1, whereas HJ and JH are Toeplitz matrices for any Hankel matrix H. For any pair of distinct scalars f and g, any Toeplitz matrix can be represented as a sum of an f-circulant matrix and a g-circulant matrix.

Definition 5.2. $L(\mathbf{v}) = Z_0(\mathbf{v})$ denotes the lower triangular Toeplitz matrix with the first column $\mathbf{v}$, that is, $L(\mathbf{v}) = \sum_{i=0}^{n-1} v_i Z^i$ where $\mathbf{v} = (v_0, \dots, v_{n-1})^T$ and Z is the down-shift matrix of Definition 4.1.

Now, consider the basic operations with Toeplitz and Hankel matrices. For the following problems, the solutions will exploit their *correlations to polynomial computations:*

Problem 5.1 *(TOEPL · VECTOR). Compute the product* $\mathbf{w} = U\mathbf{v}$ *of a special Toeplitz matrix* U *by a vector* $\mathbf{v}$*:*

$$\begin{pmatrix} w_0 \\ w_1 \\ \vdots \\ \\ \\ \\ \vdots \\ w_{m+n} \end{pmatrix} = \begin{pmatrix} u_0 & & & O \\ u_1 & \ddots & & \\ \vdots & \ddots & \ddots & \\ \vdots & \ddots & \ddots & u_0 \\ \vdots & \ddots & \ddots & u_1 \\ u_m & \ddots & \ddots & \vdots \\ & \ddots & \ddots & \vdots \\ & & \ddots & \vdots \\ O & & & u_m \end{pmatrix} \begin{pmatrix} v_0 \\ v_1 \\ \vdots \\ v_n \end{pmatrix}. \tag{5.1}$$

Solution. The vector equation (5.1) is equivalent to the equation (1.2.2) and, consequently, to the problem of polynomial multiplication (vector convolution), [see Problem 1.2.4a (*POL · MULT*)]. Therefore, Problem 5.1 (*TOEPL · VECTOR*) can be solved in at most $K(2 + 4.5 \log K)$ ops, where $K = 2^k$, k is an integer, and $m + n < K \leq 2m + 2n$. ■

Problem 5.2 *(CTH · VECTOR). Given an* f*-circulant, a Toeplitz or a Hankel matrix* U*, and a vector* $\mathbf{v}$*, compute the product* $\mathbf{w} = U\mathbf{v}$*.*

Solution. Let m, n and K be nonnegative integers such that $K = m - n + 1 > 0$. Delete the first n rows and the last n rows of the matrix U of (5.1), which then becomes a general $K \times (n+1)$ Toeplitz matrix. This extends the solution of Problem 5.1 (*TOEPL · VECTOR*) to Problem 5.2 (*CTH·VECTOR*) for f-circulant and Toeplitz matrices (and, consequently, for Hankel matrices as well, due to Remark 5.1). Problem 5.2 (*CTH · VECTOR*) is, therefore, solved in $O(n \log n)$ ops if $K = O(n)$. ■

About 50% of ops can be saved in the case of solving Problem 5.2 (*CTH· VECTOR*) for a circulant $(n+1) \times (n+1)$ matrix $U = [u_{i-j \bmod (n+1)}]$. Indeed, such a problem amounts to computing the coefficients of the polynomial

$$w(x) \bmod (x^{n+1} - 1) = \sum_{i=0}^{n} w_i x^i =$$
$$u(x)v(x) \bmod (x^{n+1} - 1) = (\sum_i u_i x^i)(\sum_j v_j x^j) \bmod (x^{n+1} - 1),$$

which is the positive wrapped convolution [see Problem 1.2.4b (*±CONVOL*)]. Similar savings are obtained if U is an f-circulant matrix, for any $f \neq 0$ (compare Theorem 5.1). If $f = -1$, Problem 5.2 turns into computing a negative wrapped convolution.

Problem 5.3 (*TTOEPL · INVERT*). *Invert a nonsingular* $(n+1) \times (n+1)$ *lower triangular Toeplitz matrix* T.

Solution. It can be easily proved that the set of lower triangular Toeplitz matrices is the algebra generated by the down-shift matrix Z of Definition 4.1 (see exercise 20), so that T^{-1} is a lower triangular Toeplitz matrix, and it is sufficient to compute its first column or its last row. The problem amounts to solving the linear system in $\mathbf{v}$ formed by the first $n+1$ equations of (5.1) for $\mathbf{w} = [1, 0, \ldots, 0]^T$. This in turn amounts to Problem 1.3.4 (*POL · RECIPR*) of computing $v(x) = (1/\sum_{i=0}^{m} u_i x^i) \bmod x^{n+1}$, the reciprocal of a polynomial. Actually, by writing the relation $v(x) \sum_{i=0}^{m} u_i x^i = 1 \bmod x^{n+1}$ in vector form, we arrive at the linear system (5.1) where $w_0 = 1$, $w_1 = \cdots = w_{m+n} = 0$. The solution of section 3 of chapter 1, using $O(n \log n)$ ops, applies. ∎

Problem 5.4 (*CIRC·INVERT*). *Invert an* $(n+1)\times(n+1)$ *nonsingular* f*-circulant matrix* $C_f = Z_{f,n+1,n+1}(\mathbf{c})$*, for a scalar* $f \neq 0$ *and a vector* $\mathbf{c} = [c_i]$*. (Our notation* C_f *for* f*-circulant matrices should not be confused with* C *for Cauchy matrices.)*

It can be easily proved that the set of f-circulant $(n+1) \times (n+1)$ matrices, for any fixed $f \neq 0$ and n, is a commutative algebra with a generator matrix $Z_f(\mathbf{e}^{(1)})$ and is isomorphic to $\mathbf{F}[x]/(x^{n+1} - f)$ (see the next theorem and/or exercise 21). Therefore, C_f^{-1} is an f-circulant matrix again, so that it is sufficient to compute its first column $\mathbf{v} = [v_i]$, such that

$$(\sum_i v_i x^i)(\sum_j c_j x^j) = 1 \bmod (x^{n+1} - f)$$

provided that $C_f = \sum_{i=0}^{n} c_i Z_f^i(\mathbf{e}^{(1)})$. We will rely on the next well-known result:

Theorem 5.1 ([CPW], [Da]). *For a complex* $f \neq 0$*, let* $D_f = \mathrm{diag}(1, g, g^2, \ldots, g^n)$*,* $g^{n+1} = f$*. Let* $\mathbf{c}^T$ *denote the first row of an* f*-circulant* $(n+1) \times (n+1)$ *matrix* C_f*. Let* Ω *be the* $(n+1) \times (n+1)$ *Fourier matrix,* $(\Omega)_{i,j} = \omega^{ij}/\sqrt{n+1}$*,* $(\Omega^H)_{i,j} = \omega^{-ij}/\sqrt{n+1}$*,* $i, j = 0, 1, \ldots, n$*,* $\omega^{n+1} = 1$*,* $\omega^s \neq 1$ *for* $0 < s \leq n$*. Then* $\Omega^H \Omega = I$*,* $\Omega D_f C_f D_f^{-1} \Omega^H = D$*,* $D = \mathrm{diag}(d_0, \ldots, d_n)$ *where* $d_i = (\mathbf{d})_i$*,* $\mathbf{d} = \sqrt{n+1}\, \Omega D_f \mathbf{c}$*.*

Therefore, $C_f\mathbf{v} = \mathbf{w}$ implies that $\mathbf{v} = D_f^{-1}\Omega^H D^{-1}\Omega D_f \mathbf{w}$, and computing $\mathbf{v} = C_f^{-1}\mathbf{w}$ (and similarly, computing $C_f\mathbf{v}$) for a given f-circulant matrix C_f and for a given vector $\mathbf{w}$ essentially amounts to three DFT's on $n+1$ points, that is, to $O(n \log n)$ ops. This also suggests alternative solutions of Problem 5.2 ($CTH \cdot VECTOR$), by means of embedding a Toeplitz matrix into a circulant matrix and padding the vector $\mathbf{v}$ with zeros or, alternatively, by representing a Toeplitz matrix as a sum of an f-circulant matrix and a g-circulant matrix, for a pair of distinct scalars f and g.

Remark 5.2. Since multiplication and inversion of triangular Toeplitz, f-circulant and diagonal matrices cost so little, we may use such matrices and their products as *preconditioners.* For instance, in order to avoid dealing with singular submatrices of a given matrix A, which we need to invert, we may first invert $B = RAS$, where R and S are appropriate preconditioners, and then obtain $A^{-1} = SB^{-1}R$. Another example of preconditioners is encountered in the implementation of the preconditioned conjugate gradient method: in [ChS], [Ch88], [Ch89], [Ch89a], [Str], the class of circulant matrices is used; in [BDB90], [DiB92], [DiBFS] and in [BF], the classes of matrices of exercises 22 and 23 are respectively used.

Problem 5.5 ($TOEPL \cdot SOLVE$). *Given a vector* $\mathbf{v}$ *and a nonsingular Toeplitz matrix* T, *compute the solution vector* $T^{-1}\mathbf{v}$ *to the Toeplitz linear system* $T\mathbf{x} = \mathbf{v}$.

The solution of this problem is immediately extended to the solution of Hankel linear systems, due to Remark 5.1.

For an $n \times n$ Toeplitz matrix T, the matrix T^{-1} generally has $n(n+1)/2$ distinct entries, so its evaluation must involve at least $n(n+1)/2$ ops. On the other hand, the vector $T^{-1}\mathbf{v}$ for any vector $\mathbf{v}$, in particular, any column of T^{-1}, can be computed by using $O(n \log n)$ ops if we are given a vector $\mathbf{v}$ and a basis set of two vectors consisting of the first column of T^{-1} and of either the last column of T^{-1} or the vector $T^{-1}\mathbf{t}$, where $\mathbf{t}$ is the vector of Theorem 5.3. This follows from the solution of Problem 5.2 ($CTH \cdot VECTOR$), from the next result, known as a *Gohberg-Semencul formula*, and from its extensions in Theorem 5.3:

Theorem 5.2. *Let* T *be an* $n \times n$ *nonsingular Toeplitz matrix,* $X = T^{-1}$, $(\mathbf{x})_i = x_i = (X)_{i,0}$, $(\mathbf{y})_i = (X)_{i,n-1}$, J *and* Z *be the* $n \times n$ *matrices of Definition 4.1, and* $L(\mathbf{v})$ *be the* $n \times n$ *matrices of Definition 5.2. Then*

$$x_0 X = L(\mathbf{x})L^T(J\mathbf{y}) - L(Z\mathbf{y})L^T(ZJ\mathbf{x}). \tag{5.2}$$

Remark 5.3. Theorem 5.2 and its various extensions can be found in [GSe], [GF, p. 90]; [T64], [T65], [BGY, pp. 277-279], [FMKL, p. 34], [Hei], [B-AS85], [B-AS86], [KC], [HR], [Ti81], [BA] and [LDC] [see also Proposition 9.3, Theorems 11.3 and 11.4, equation (11.21) and exercise 47]. This theorem is a matrix analog of the Christoffel-Darboux formula for orthogonal polynomials. [GK], [GF], [HR], [Ti81] and [T90] extend the result to the case where $x_0 = 0$ (see also [Ioh] and the next theorem). [AG89], [AG89a] and [Gad] simplify the solution of Hermitian Toeplitz and real symmetric Toeplitz linear systems by means of replacing two triangular Toeplitz matrices in (5.2) by circulant and/or f-circulant matrices, and similar simplifications for general Toeplitz linear systems have been shown in [GO], [GO1] (compare our Example 11.2).

Theorem 5.3 ([Hei], [T90]). *Let T be a nonsingular $n \times n$ Toeplitz matrix, $(T)_{i,j} = t_{i-j}$, $i, j = 0, 1, \ldots, n-1$; $\mathbf{t} = [s, t_{1-n}, t_{2-n}, \ldots, t_{-1}]^T$, for any fixed scalar s; $\mathbf{x} = [x_0, \ldots, x_{n-1}]^T = T^{-1}\mathbf{t}$; $\mathbf{v} = [-1, x_{n-1}, \ldots, x_1]^T$; $\mathbf{y} = T^{-1}[1, 0, \ldots, 0]^T$; $\mathbf{u} = ZJ\mathbf{y}$. Then $T^{-1} = L(\mathbf{x})L^T(\mathbf{u}) - L(\mathbf{y})L^T(\mathbf{v})$.*

Solutions of Problem 5.5 ($TOEPL \cdot SOLVE$). The first and the last columns of the inverse of an $n \times n$ Toeplitz matrix T, as well as the vector $\mathbf{x}$ of the latter theorem, can be computed in $O(n \log^2 n)$ ops by means of different algorithms, presented in [BGY], in [BA] and [Morf] (with larger overhead constants in both cases), and in [AGr85], [AGr87], [AGr88], [CK], [GG], [Mus] and [dH] (with smaller overhead constants). There are also $O(n^2)$ ops algorithms, which are superior for smaller n ([DG], [T64] and [CB]). The algorithms of [BA] and [Morf], [CK], [Mus] and [dH] require *strong nonsingularity* of T, that is, they require that all the leading principal submatrices of the matrix T be nonsingular. This is ensured, for instance, if the matrix is diagonally dominant and/or h.p.d., and we may shift from a nonsingular to an h.p.d. system by means of symmetrization, that is, of premultiplying the linear system by the matrix T^H. The algorithms of [BA], [Morf], [CK] and [Mus] solve the symmetrized system and also compute the determinants of its coefficient matrix in $O(n \log^2 n)$ ops (compare our sections 7 and 11–13). The algorithm of [BA] and [Morf] will be presented in section 13. The algorithm of [Mus] and [dH] (see also [AGr87], [AGr88], [AGr85], [GG] and [L-AK]) reduces the problem to Schur's transformations of complex functions homomorphic in the unit disc on the complex plane; such transformations are recursively applied to Schur's polynomials, related to the matrix T via the Szegö orthogonal polynomials associated with this matrix T. This algorithm, as well as the $O(n^2)$ algorithms of [DG] and [CB], exploits *correlations between orthogonal polynomials and Toeplitz matrices* (compare [Gen] on this general subject). Finally, there are effective iterative Toeplitz solvers that use various preconditioners (see Remark 5.2), Newton's iteration and the steepest descent method (see [P92c] and [LV]).

Next, we will recall the algorithm of [BGY], which reduces the evaluation of the first and last columns of the inverse of any $n \times n$ Toeplitz matrix to Padé approximation [Problem 1.5.2b ($PADE$)] and always computes the solution in $O(n \log^2 n)$ ops. Let $N = m+n$, let $r_0(x) = x^{N+1}$, and let $r_1(x) = t_N x^N + t_{N-1}x^{N-1} + \cdots + t_1 x + t_0$ denote a fixed N-th degree polynomial. Let $[y_0, \ldots, y_m]^T$, $[z_0, \ldots, z_n]^T$ and $[b_0, \ldots, b_{n-1}]^T$ denote the coefficient vectors of the polynomials $r_i(x)$, $t_i(x)$ and $-s_i(x)$, respectively, defined by the equations (1.5.6)-(1.5.8) for i such that $r_i(x)/t_i(x)$ is the Padé approximation to $r_1(x)$ with the upper bounds m and n on the degrees of $r_i(x)$ and $t_i(x)$, respectively. Then (1.5.9) amounts to the following linear system (for simplicity, written for $m \le n$):

$$\begin{pmatrix} t_0 & & & \\ \vdots & \ddots & O & \\ t_m & & \ddots & \\ \vdots & \ddots & \ddots & \\ t_n & & \ddots & t_0 \\ \vdots & \ddots & \ddots & \vdots \\ t_{m+n} & & \ddots & t_m \\ & \ddots & \ddots & \vdots \\ & & \ddots & t_n \\ & O & \ddots & \vdots \\ & & & t_{m+n} \end{pmatrix} \begin{pmatrix} z_0 \\ \vdots \\ \vdots \\ z_n \end{pmatrix} = \begin{pmatrix} y_0 \\ \vdots \\ y_m \\ 0 \\ \vdots \\ \vdots \\ 0 \\ b_0 \\ \vdots \\ \vdots \\ b_{n-1} \end{pmatrix}. \tag{5.3}$$

Extracting the middle part of the latter linear system, we arrive at the linear system

$$\begin{pmatrix} t_s & \cdots & t_0 & & O \\ \vdots & \ddots & & \ddots & \\ t_n & & \ddots & & t_0 \\ \vdots & \ddots & & \ddots & \vdots \\ \cdots & \cdots & \cdots & \cdots & \cdots \\ t_{m+n} & \cdots & \cdots & \cdots & t_m \end{pmatrix} \begin{pmatrix} z_0 \\ z_1 \\ \vdots \\ \vdots \\ z_n \end{pmatrix} = \begin{pmatrix} y_s \\ 0 \\ \vdots \\ \vdots \\ 0 \end{pmatrix}, \tag{5.4}$$

provided that $y_m = y_{m-1} = \cdots = y_{s+1} = 0$, $y_s \neq 0$. For $s = m = n$, this system turns into the linear system

$$\begin{pmatrix} t_n & t_{n-1} & \cdots & t_0 \\ t_{n+1} & \ddots & \ddots & \vdots \\ \vdots & \ddots & \ddots & t_{n-1} \\ t_{2n} & \cdots & t_{n+1} & t_n \end{pmatrix} \begin{pmatrix} z_0 \\ z_1 \\ \vdots \\ z_n \end{pmatrix} = \begin{pmatrix} y_n \\ 0 \\ \vdots \\ 0 \end{pmatrix}$$

with a general Toeplitz matrix T of coefficients [compare the solutions of Problems 5.1 ($TOEPL \cdot VECTOR$) and 5.2 ($CTH \cdot VECTOR$)]. Given such a linear system, the algorithm of [BGY] computes its solution vector $\mathbf{z}$ (and then the first column $\mathbf{z}/y_n$ of the matrix T^{-1}) by applying the extended Euclidean scheme (1.5.6)-(1.5.9) to the input $r_0(x) = u(x) = x^{2n+1}$, $r_1(x) = v(x) = \sum_{i=0}^{2n} t_i x^i$. Similarly, we may compute the last column of T^{-1}. This gives us an $O(n \log^2 n)$ algorithm for solving a nonsingular linear system with an $n \times n$ Toeplitz matrix T, where $(T^{-1})_{0,0} \neq 0$. [BGY] also shows how to relax the latter assumption. ■

Conversely, we may solve Problem 1.5.2b ($PADE$) (Padé approximation) via Toeplitz computations.

Solutions of Problem 1.5.2b ($PADE$). Let s denote the minimum nonnegative integer such that the system of equations (5.4) is consistent for $y_s = 1$, $s \leq m$. Suppose that we have computed the solution of this system. Then the desired Padé approximation can be obtained simply by means of polynomial multiplications. Surely, the system (5.4) has a solution if and only if its homogeneous subsystem obtained by deleting the first equation of (5.4) has a nontrivial solution or, equivalently, has a coefficient matrix of rank at most n. The search for the desired minimum s is thus equivalent to the search for the minimum s such that removing the first s rows of the coefficient matrix of (5.3) defines a matrix of rank at most n. Therefore, we may apply the binary search (Algorithm 8.2) in order to compute the desired value s in $\lceil \log(n+1) \rceil$ tests, each consisting in the evaluation of the rank of a Toeplitz matrix. Once the desired s is available, it remains to solve the Toeplitz system (5.4) for this s and to arrive at the desired Padé approximation by means of polynomial multiplications.

It is possible, however, to reduce the computation of (m, n) Padé approximation to the solution of a single Toeplitz system [BGY]. We first remove the last n equations of (5.3) (representing the polynomial multiplication steps of the solution) and thus represent the original problem in the form of the following linear system:

$$
\begin{pmatrix}
t_0 & & & & O \\
t_1 & \ddots & & & \\
\vdots & \ddots & \ddots & & \\
t_{m-n-1} & & \ddots & \ddots & \\
t_{m-n} & \ddots & & \ddots & t_0 \\
\vdots & \ddots & \ddots & & t_1 \\
\vdots & & \ddots & \ddots & \vdots \\
t_{m-1} & & & \ddots & t_{m-n-1} \\
t_m & \ddots & & & t_{m-n} \\
\vdots & \ddots & \ddots & & \vdots \\
\vdots & & \ddots & \ddots & \vdots \\
\vdots & & & \ddots & \vdots \\
t_{m+n} & \cdots & \cdots & \cdots & t_m
\end{pmatrix}
\begin{pmatrix} z_0 \\ \vdots \\ \vdots \\ z_n \end{pmatrix}
=
\begin{pmatrix} y_0 \\ \vdots \\ \vdots \\ \vdots \\ \vdots \\ \vdots \\ y_m \\ 0 \\ \vdots \\ \vdots \\ 0 \end{pmatrix}.
$$

Then we successively move the right sides of these equations on the left side, by appending the vector $[y_0, \ldots, y_m]^T$ at the bottom of the vector $[z_0, \ldots, z_n]^T$ and by appending the matrix $\begin{pmatrix} -I_{m+1} \\ O \end{pmatrix}$ on the right of the coefficient matrix. Finally, we multiply the first column $[t_0, \ldots, t_{m+n}]^T$ by z_0 and move the product on the right side of the linear system. This gives us the system of equations

$$
\begin{pmatrix} U & -I_{m+1} \\ T & O \end{pmatrix}
\begin{pmatrix} z_1 \\ \vdots \\ z_n \\ y_0 \\ \vdots \\ y_m \end{pmatrix}
= -z_0 \begin{pmatrix} t_0 \\ t_1 \\ \vdots \\ t_{m+n} \end{pmatrix}
\tag{5.5}
$$

where

$$
T = \begin{pmatrix} t_m & \cdots & t_{m-n+1} \\ \vdots & \ddots & \vdots \\ t_{m+n-1} & \cdots & t_m \end{pmatrix}, \quad
U = \begin{pmatrix} 0 & & & O \\ t_0 & 0 & & \\ \vdots & \ddots & \ddots & \\ t_{m-1} & \cdots & t_0 & 0 \end{pmatrix},
$$

equivalent to the original problem. The solution is immediately defined by the solution to the Toeplitz subsystem formed by the last n equations with

n unknowns (for $z_0 = 1$ or for $z_0 = 0$),

$$\begin{pmatrix} t_m & \cdots & t_{m-n+1} \\ \vdots & \ddots & \vdots \\ t_{m+n-1} & \cdots & t_m \end{pmatrix} \begin{pmatrix} z_1 \\ \vdots \\ z_n \end{pmatrix} = -z_0 \begin{pmatrix} t_{m+1} \\ \vdots \\ t_{m+n} \end{pmatrix}. \tag{5.6}$$

This system (and thus the original problem too) is either nonsingular and then has a solution for $z_0 = 1$, or otherwise, is singular and then has infinitely many solutions for $z_0 = 0$. (The resulting ratio of the polynomials $\sum_i y_i x^i / \sum_j z_j x^j$ is always unique, however [BGY].)

Furthermore, due to the results of [Gr] (cited as Theorem 2 in [BGY]), the $r \times r$ leading principal submatrix of the coefficient matrix JT of the linear system (5.6) is nonsingular for $r = \text{rank}\, T$ (here, J denotes the reversion matrix of Definition 4.1), and we arrive at the following algorithm:

Algorithm 5.1, *Padé approximation.*

Input: two natural numbers m and n and a polynomial $r_1(x) = \sum_{i=0}^{m+n} t_i x^i$.

Output: the (m, n) Padé approximation to $r_1(x)$ by the polynomials $y(x) = \sum_{i=0}^{m} y_i x^i$ and $z(x) = \sum_{i=0}^{n} z_i x^i$.

Computation:

1. Compute $r = \text{rank}\, T$, $T = [t_{m+i-j}]$ being the coefficient matrix of the Toeplitz linear system (5.6), so that $r \times r$ is the maximum size of a nonsingular leading principal submatrix of JT.
2. If $r = n$, set $z_0 = 1$ and compute the solution $z_n, \ldots, z_1$ to (5.6), by using any Toeplitz linear system solver. If $r < n$, set $z_0 = 0$. Stage 1 has given us the $r \times r$ nonsingular submatrix of T, and thus we may immediately solve (5.6).
3. Compute $y_0, \ldots, y_m$ from (5.5). ∎

Since T is a Toeplitz matrix, it suffices to use $O(n^2 \log n)$ ops at Stage 1 (see section 7), $O(n \log^2 n)$ ops at Stage 2 and $O(n \log n)$ ops at Stage 3. In section 13, we will use randomization to decrease the cost bounds of Stage 1 and, therefore, the overall cost of performing Algorithm 5.1, to $O(n \log^3 n)$ ops. This still exceeds the bound $O(n \log^2 n)$ for the solution of chapter 1; however, Algorithm 5.1 runs fast as a parallel algorithm (see chapter 4).

Extension of Algorithm 5.1 to Problem 1.5.3 ($RECUR \cdot SPAN$) (of computing the minimum span for a linear recurrence) is immediate [since the solution of Problem 1.5.3 ($RECUR \cdot SPAN$) has been reduced in chapter 1 to computing Padé approximation].

On the other hand, the reduction to Padé approximation is not actually needed here, since Problem 1.5.3 ($RECUR \cdot SPAN$) has originally been stated as the problem of computing a solution $t_0, \ldots, t_{n-1}$ to the Toeplitz (Hankel) linear system

$$t_{n-1}v_{i-1} + \cdots + t_0 v_{i-n} = v_i, \quad i = n, \ldots, 2n-1, \tag{5.7}$$

for the minimum n for which such a system is consistent, for given $v_0, \ldots,$ v_{2n-1}. Let $n = n^*$ be this minimum value of n. Then the properties of the minimum span imply that this system is nonsingular for $n = n^*$ and is singular if $n^* < n$ (see exercise 18 or [KP]).

Remark 5.4. Over the fields $\mathbf{F}$ of constants that do not support DFT on $O(n)$ points, we have to increase by the factor of $\log \log n$ the estimates for the complexity of computations with Toeplitz and other structured matrices based on fast polynomial multiplication (compare Theorem 1.7.1). Also, we need to have all the $(n+1)$-st roots of 1 in $\mathbf{F}$ in order to state and to apply Theorem 5.1.

(m, n) Padé approximation for a formal power series can be naturally extended and defined for the ratio of two such series. The computation of such a generalized Padé approximation is equivalent to solving a Sylvester linear system. In [CaMe] such correlations have been studied, and weakly numerically stable algorithms running in time $O(n^2)$ have been derived for the solution of Toeplitz and Hankel linear systems (by reducing these computations to computing generalized Padé approximations). On numerical stability of solving Toeplitz, Hankel, Toeplitz-like and Hankel-like linear systems, see [C80], [Bun], [L92].

6. Multiplication of a Vector by a Vandermonde Matrix and by its Inverse.

In this section we will compute the product of a vector by a Vandermonde matrix and by its inverse, by reducing the problems to polynomial and Toeplitz-Hankel computations. We will extend this approach further on.

The following auxiliary result, which can be easily proved by direct inspection, gives a matrix interpretation of Newton's identities (1.4.8) and (1.4.9).

Proposition 6.1. *Given $p_0, p_1, \ldots, p_{n-1}$, Newton's identities (1.4.8) define a lower triangular Toeplitz linear system in $s_1, \ldots, s_m$; its solution* *[see Problem 5.3 ($TTOEPL \cdot INVERT$)] requires $O(n \log n)$ ops. Given $s_1, \ldots, s_m$, (1.4.9) is a Toeplitz linear system in $p_{n-1}, \ldots, p_0$, and (1.4.8) is a triangular linear system in $p_{n-1}, \ldots, p_0$, whose coefficient matrix can be represented as the sum of a diagonal matrix $D = \text{diag}(1, 2, \ldots, n)$ and a lower triangular Toeplitz matrix T with the first column $[0, s_1, s_2, \ldots, s_{n-1}]^T$ (see exercise 4e).*

Remark 6.1. Since trace $A = \sum_{i=1}^n \lambda_i$, where λ_i, $i = 1, \ldots, n$, are the eigenvalues of the $n \times n$ matrix A, we have that the values $s_i = \text{trace } (A^i)$ equal the power sums of the eigenvalues of A for $i = 0, 1, \ldots$ (and if A is nonsingular, then also for $i = -1, -2, \ldots$). Therefore, we may compute the characteristic polynomial of an $n \times n$ matrix A by means of the following old algorithm of Leverrier:

a) compute $s_i = \text{trace}(A^i)$, $i = 0, 1, \ldots, n$;

b) solve Problem 1.4.8 ($I \cdot POWER \cdot SUMS$) for the input set $s_1, \ldots, s_n$; output the coefficients of $c_A(\lambda) = \sum_{i=0}^{n} c_i \lambda^i$, the characteristic polynomial of A.

The algorithm for Problem 1.4.8 (section 4 of chapter 1) performs part b) in $O(n \log n)$ ops, but part a) is much harder: its computational costs exceeds the order of $M(n)$. The algorithm, however, serves as a basis of effective parallel solution to Problem 2.6 ($CHAR \cdot POL$), (see [Cs], [PSa], [GP89]). There is a restriction, however: this algorithm only applies over the fields of constants of characteristic 0 or greater than n, that is, over any field of constants that allows division by $n!$, such as the field of integers modulo any prime $p > n$ (on the extension to an arbitrary field, see section 7 of chapter 1, Remark 3.2 of the present chapter, and section 6 and Appendix C of chapter 4). This restriction on the allowed fields of constants has to be extended to our Algorithms 7.1, 11.1 and 11.2 for computations with Toeplitz and other structured matrices, as well as to further applications of these algorithms. Moreover, in Algorithms 9.1 and 10.1 we will have an additional problem when we compute the rank of structured matrices. Then again, the fastest deterministic algorithms for this problem do not generally extend to arbitrary fields, and alternative algorithms are needed (see Remark 3.2 of this chapter and section 6 and Appendix C of chapter 4).

Next, we will apply Newton's identities in order to multiply a vector $\mathbf{w}^T$ by the inverse of a nonsingular $n \times n$ Vandermonde matrix.

Problem 6.1 ($V^T \cdot SOLVE$). *Given vectors $\mathbf{v} = [v_i]$, $\mathbf{w} = [w_i]$ of* *dimension n such that $v_i \neq v_j$, $i \neq j$, compute the product $(V^T)^{-1}\mathbf{w} = (\mathbf{w}^T V^{-1})^T$ where V is the Vandermonde matrix $V(\mathbf{v})$ (compare Definition 4.2),*

$$V = [v_{ij}], \quad v_{ij} = v_i^j, \quad i, j = 0, 1, \ldots, n-1. \tag{6.1}$$

Solution. Observe that $V^T V$ is a Hankel matrix:

$$V^T V = [h_{ij}], \quad h_{ij} = \sum_{k=0}^{n-1} v_k^{i+j}, \quad i, j = 0, 1, \ldots, n-1. \tag{6.2}$$

The computation of $\mathbf{w}^T V^{-1}$ is performed by using the relation $\mathbf{w}^T V^{-1} = \mathbf{w}^T (V^T V)^{-1} V^T$, by means of the following algorithm:

Algorithm 6.1.

Input: natural n; the components $v_0, v_1, \ldots, v_{n-1}$ of a vector $\mathbf{v}$ such that $v_i \neq v_j$ if $i \neq j$; the components $w_0, w_1, \ldots, w_{n-1}$ of a vector $\mathbf{w}$.

Output: the components of the vector $\mathrm{w}^T V^{-1}$ for the matrix V of (6.1).

Computation:

1. First compute the coefficients of the polynomial $p(x) = \prod_{k=0}^{n-1}(x - v_k) = \sum_i p_i x^i$ by means of the fan-in method (see section 4 of chapter 1). Then compute the entries of the Hankel matrix (6.2) by solving for $s_j = \sum_{i=0}^{n-1} v_i^j$, $j = 0, \ldots, n$, the triangular Toeplitz system of Newton's identities [defined by (1.4.8)] and by applying Proposition 6.1. Substitute the solution into the Toeplitz linear system (1.4.9) for $m = 2n-2$ and then solve the resulting triangular Toeplitz linear system in $s_j = \sum_{i=0}^{n-1} v_i^j$, $j = n+1, \ldots, 2n-2$.
2. Compute $\mathbf{y}^T = \mathbf{w}^T (V^T V)^{-1}$ by solving a Hankel linear system.
3. Compute $\mathbf{w}^T V^{-1} = (V\mathbf{y})^T$ by means of the multipoint polynomial evaluation algorithm of chapter 1.

The cost of Algorithm 6.1 is $O(n \log^2 n)$ ops; the approach used is due to [CKL89]. ∎

Problem 6.2 ($V^T \cdot VECTOR$). *Given vectors $\mathbf{v} = [v_i]$ and $\mathbf{w} = [w_i]$ of dimension n such that $v_i \neq v_j$, $i \neq j$, compute the product $V^T\mathbf{w}$, where V is the Vandermonde matrix defined in Problem 6.1 ($V^T \cdot SOLVE$).*

Solution ([P89a]). Note that $V^T \mathrm{w} = (\mathrm{w}^T V)^T$. Consider the Vandermonde matrix $B = [b_{ij}]$, $b_{ij} = 1/v_i^j$, and observe that

$$B^T V = [t_{ij}], \; t_{ij} = \sum_{k=0}^{n-1} v_k^{j-i}, \; i, j = 0, \ldots, n-1, \tag{6.3}$$

is a Toeplitz matrix and that $\mathbf{w}^T V = (\mathbf{w}^T B^{-T}) B^T V$. The entries of $B^T V$ can be computed from two systems of Newton's identities (1.4.8), where in one of them we have $\sum_{i=0}^{n-1} p_i x^i = \prod_{k=0}^{n-1}(x - v_k)$ and $s_j = \sum_{i=0}^{n-1} v_i^j$ and in another we have $\sum_{i=0}^{n-1}(p_{n-i-1}/p_0)x^i = \prod_{k=0}^{n-1}(x - v_k^{-1})$ and $s_{-j} = \sum_{i=0}^{n-1} v_i^{-j}$, $j = 0, 1, \dots, n-1$ (compare Remark 1.4.3). The subsequent computation is reduced to polynomial interpolation (computation of $\mathbf{w}^T B^{-T}$) and to a Toeplitz matrix-by-vector multiplication. Then again, the cost of this approach is $O(n \log^2 n)$. ■

The technique we used for the solution of Problem 6.2 ($V^T \cdot VECTOR$) can also be applied to solve Problem 6.1 ($V^T \cdot SOLVE$). In fact, $\mathbf{w}^T V^{-1} = \mathbf{w}^T (B^T V)^{-1} B^T$, so that Problem 6.1 ($V^T \cdot SOLVE$) can be reduced to solving a Toeplitz system and to multipoint polynomial evaluation. Conversely, the techniques that we used for Problem 6.1 ($V^T \cdot SOLVE$) can also be applied to solve Problem 6.2 ($V^T \cdot VECTOR$). The cost is $O(n \log^2 n)$ ops for each solution; the latter solution (with the use of Toeplitz computations) involve a slightly smaller overhead constant.

In section 12, we will substantially generalize the above solutions to Problems 6.1 ($V^T \cdot SOLVE$) and 6.2 ($V^T \cdot VECTOR$) by exploiting similar correlations among more general classes of structured matrices.

To conclude, we will point out that the transition between V^T and V^{-1} (and the computation of $|\det V|^2$) can be alternatively based on the following matrix equations from [FHR], [GO2]:

$$VJL(J\gamma)V^T = \mathrm{diag}([\Gamma'(v_i)]),$$

$$V^T = JZ_f^{-1}(\gamma + f\mathbf{e}^{(0)})\, V^{-1}\mathrm{diag}([\Gamma'(v_i)(f - v_i^n)])$$

where $V = V(\mathbf{v})$, $\Gamma(x) = \Gamma_{\mathbf{v}}(x)$ and $\gamma = \gamma(\mathbf{v})$ are defined in Definition 4.3, $Z_f(\gamma + f\mathbf{e}^{(0)})$ is the f-circulant matrix with the first column $\gamma + f\mathbf{e}^{(0)}$, and J is defined in Definition 4.1.

7. Computing the Determinant and the Characteristic Polynomial of a Toeplitz Matrix.

Let us revisit Problems 2.4 ($DETERMINANT$) and 2.6 ($CHAR \cdot POL$) of computing $\det A$ and the coefficients of the characteristic polynomial

$$c_A(\lambda) = \sum_{i=0}^{n} c_i \lambda^i = \det(\lambda I - A)$$

provided that A is an $n \times n$ Toeplitz matrix. Our solution will be given by Algorithm 7.1. It relies on computing the powers of A and thus also solves Problems 2.1a ($KRYLOV$) and 2.2a ($M \cdot POWERS$) for A, which are of independent interest. The techniques of section 2 lead us to the extensions of the solution for $c_A(\lambda)$ and of its computational cost bounds to the case of other computational problems, such as Problems 2.8 ($L \cdot SQUARES$) and 2.10 ($M \cdot RANK$), for Toeplitz matrices.

Any algorithm for $\det A$ also enables us to compute $\det H$ for any $n \times n$ Hankel matrix H, since HJ is a Toeplitz matrix and $\det H = (-1)^{\lfloor n/2 \rfloor}$ $\det(HJ)$.

If the input Toeplitz or Hankel matrix A (or even Toeplitz-like or Hankel-like matrix, according to the definition of section 11) is strongly nonsingular and, therefore, has its balanced recursive factorization (2.1)-(2.3), then $O(n \log^2 n)$ ops suffice in order to compute the diagonal entries of its triangular factorization (see [BA], [dH], [Morf] or section 13), and then we may immediately obtain $\det A$ in $O(n)$ ops.

If we agree to compute $|\det A|^2$ rather than $\det A$ (over any subfield of the complex field $\mathbf{C}$), we may relax the assumption about strong nonsingularity since $|\det A|^2 = \det(A^H A)$ and since the matrix $A^H A$ is either singular if A is singular (and then $\det A = 0$) or a strongly nonsingular Toeplitz-like matrix having its balanced recursive factorization (2.1)-(2.3). Therefore, the above approach applies to computing $|\det A|$ in $O(n \log^2 n)$ ops.

Now, we turn to computing the coefficients of $c_A(\lambda)$ in $O(n^2 \log n)$ ops for any $n \times n$ Toeplitz matrix A [which will also give us $\det A = (-1)^n c_0$, as well as $A^+ \mathbf{v}$, for any vector $\mathbf{v}$; see (2.9)]. We will follow [P92b].

We assume that the field of constants $\mathbf{F}$ supports FFT and allows divisions by $n!$ (see Remarks 5.4 and 6.1). Then the problem is reduced to computing trace (A^i) for $i = 1, \ldots, n$, and we apply the following algorithm:

Algorithm 7.1.

Input: natural n and the entries of an $n \times n$ matrix A.

Output: the entries (as polynomials in λ) of $(I - \lambda A)^{-1} \bmod \lambda^{n+1} = I + \lambda A + \cdots + (\lambda A)^n$.

Computation: Denote $d = \lceil \log(n+1) \rceil$, $X_0 = I$, $B = I - \lambda A$, recursively compute

$$X_{i+1} = X_i(2I - BX_i), \; i = 0, \ldots, d-1, \tag{7.1}$$

and output $X_d \bmod \lambda^{n+1}$.

To verify the correctness of this algorithm over any field of constants, deduce from (7.1) that

$$I - BX_{i+1} = (I - BX_i)^2, \quad i = 0, 1, \ldots,$$

observe that $I - BX_0 = \lambda A = O \bmod \lambda$, and conclude that

$$I - BX_i = O \bmod \lambda^{2^i}, \quad i = 0, 1, \ldots . \tag{7.2}$$

Let us next estimate the computational cost of Algorithm 7.1, assuming that A and therefore B are Toeplitz matrices and that we restrict our task to the computation of the first column and last column of $(I-\lambda A)^{-1} \bmod \lambda^{n+1}$. Then (7.2) implies that X_i are the inverses modulo λ^{2^i} of Toeplitz matrices and have their (0,0) entries equal to 1 mod λ. Note that we consider X_i in the ring of polynomials modulo λ^{2^i}, and in this ring X_i^{-1} turns into the Toeplitz matrix B, due to (7.2). Thus, due to Theorem 5.2, we have $X_i = \tilde{X}_i + O(\lambda^{2^i})$, where

$$\tilde{X}_i = \frac{1}{x_0^{(i)}}[L(\mathbf{x}^{(i)})L^T(J\mathbf{y}^{(i)}) - L(Z\mathbf{y}^{(i)})L^T(ZJ\mathbf{x}^{(i)})],$$
$$\mathbf{x}^{(i)} = X_i\mathbf{e}^{(0)}, \ \mathbf{y}^{(i)} = X_i\mathbf{e}^{(n-1)}.$$

We have reduced the iteration (7.1) to computing

$$\mathbf{x}^{(i+1)} = \mathbf{x}^{(i)} + X_i(I - BX_i)\mathbf{e}^{(0)},$$
$$\mathbf{y}^{(i+1)} = \mathbf{y}^{(i)} + X_i(I - BX_i)\mathbf{e}^{(n-1)}.$$

Moreover, since $I - BX_i = O \bmod \lambda^{2^i}$, it follows that

$$\begin{aligned} \mathbf{x}^{(i+1)} &= \mathbf{x}^{(i)} + \tilde{X}_i(I - B\tilde{X}_i)\mathbf{e}^{(0)} \bmod \lambda^{2^{i+1}}, \\ \mathbf{y}^{(i+1)} &= \mathbf{y}^{(i)} + \tilde{X}_i(I - B\tilde{X}_i)\mathbf{e}^{(n-1)} \bmod \lambda^{2^{i+1}}. \end{aligned} \tag{7.3}$$

Therefore, iteration (7.3) essentially amounts to a finite number of multiplications of Toeplitz matrices by vectors [Problem 5.2 ($CTH \cdot VECTOR$)], where all the entries are polynomials in λ of degrees less than $2^{i+1} = O(n)$. Such multiplications can be reduced to $O(1)$ multiplications of $O(1)$ pairs of polynomials of degrees $O(n)$ whose coefficients are polynomials in λ of degrees less than 2^{i+1}.

From the first column $\mathbf{x} = [x_0, x_1, \ldots, x_{n-1}]^T$ and the last column $\mathbf{y} = [y_{n-1}, \ldots, y_1, y_0]^T$ of $(I - \lambda A)^{-1} \bmod \lambda^{n+1}$, we recover

$$\text{trace } (I - \lambda A)^{-1} \bmod \lambda^{n+1} =$$
$$x_0^{-1}\sum_{j=0}^{n-1}(\sum_{i=0}^{j} x_i y_{n-i-1} - \sum_{i=0}^{j-1} y_i x_{n-i-1}) \bmod \lambda^{n+1},$$

at the cost of $O_A(n^2 \log n)$ ops, and this gives us trace (A^i), $i = 0, \dots, n$.

In chapter 4 we will easily verify the estimates $O_A(\log^2 n, n^2/\log n)$ for the overall parallel computational cost of application of this algorithm, and will also recall (in exercise 4d of chapter 4) its modification that supports the bounds $O_A(n^a \log^2 n, n^{2-2a}/\log n)$ for any nonnegative $a < 1$.

These cost estimates dominate the complexity of further extension of the Algorithm 7.1 to the evaluation of the coefficients of the characteristic polynomial of A [see the solution of Problem 1.4.8 ($I \cdot POWER \cdot SUMS$)].

Remark 7.1. The iteration (7.1) is actually Newton's iteration for solving the matrix equation $BX^{-1} - I = O$, also used in chapter 3.

In section 11 of this chapter and in section 4 of chapter 4, we will extend Algorithm 7.1 to obtain effective randomized parallel algorithms for several fundamental computations with general matrices.

8. Toeplitz-like Matrices and Polynomial Computations: GCD, LCM and the Extended Euclidean Scheme.

In this section we will introduce the resultant (Sylvester) matrix and recall the concept of pseudoresultants and its correlations to the extended Euclidean scheme. We will analyze the polynomial gcd and lcm computations and their correlations to computations with Toeplitz-like matrices. The study of the gcd, the lcm and the structured matrices will be continued in sections 9–12, where we will extend the study of the structured matrices introduced so far and will consider some other major classes of structured matrices. For motivation, let us next examine some structured block matrices (see Definition 1.14).

Remark 8.1. Block Toeplitz and block Hankel matrices, that is, Toeplitz or Hankel matrices whose entries are matrices themselves, can be turned into block matrices with Toeplitz and Hankel blocks, respectively, and vice versa, by means of row and column interchange. In particular, if A is a $p \times p$ block Toeplitz matrix with the blocks of size $r \times r$, it suffices to interchange both rows and columns of A according to the same permutation: $ip + j \to jr + i$, $i = 0, 1, \dots, r-1$, $j = 0, 1, \dots, p-1$.

Block Toeplitz and block Hankel matrices, block matrices with Toeplitz and Hankel blocks, and several other classes of Toeplitz-like and Hankel-like matrices naturally arise in numerous applications, in particular, to control, signal processing, solution of PDE's and algebraic computations. In this section, we will show how some linear systems whose coefficients form 2×1

block matrices with Toeplitz blocks define all the polynomials generated by the extended Euclidean scheme (1.5.6)-(1.5.9), including $\gcd(u(x), v(x))$ ([Col] and [BT]). Solving such systems, we may compute all the entries of the extended Euclidean scheme, including the gcd.

We will also show other reductions of the evaluation of the polynomial gcd, lcm and all the polynomials generated in the extended Euclidean scheme to the solution of Toeplitz, Hankel and/or Toeplitz-like linear systems. In fact, we have already done this for the lcm and gcd, which we may compute simply by combining Algorithms 1.5.3 and 1.5.4 with Algorithm 5.1. This will lead us to one of the current best parallel solution of both problems (see chapter 4). Later on, we will show some alternative approaches. In comparison with the above solution, they are only slightly worse (Algorithm 8.1) or equally good (Algorithm 9.1) for the gcd of two polynomials; moreover, they lead to effective randomized computation of the gcd for several polynomials and of all the polynomials generated in the extended Euclidean algorithm. Besides, this study involves some important concepts [such as ones of the Bezoutians, the resultant, pseudo resultant and subresultant matrices] and some fundamental properties of structured matrices [such as the extensions of Theorem 5.2 to the case of Hankel matrices, based on Proposition 9.3 and the equation (10.3)].

Let us first revisit Problems 1.5.1 ($POL \cdot GCD$), then Problems 1.5.1c ($SEV \cdot POL \cdot GCD$) and 1.5.1d ($SEV \cdot POL \cdot LCM$), and finally, Problem 1.5.2 ($RAT \cdot INTERP$), where we will compute the gcd and the lcm of two polynomials $u(x)$ and $v(x)$ by means of the reduction to (sub)resultant linear systems. Let $r(x) = \gcd(u(x), v(x))$ be the (unknown) monic polynomial of the (unknown) degree k, where $u(x)$, $v(x)$ are polynomials of degrees m and n, respectively; $m \geq n \geq k$. Then

$$r(x) = s(x)u(x) + t(x)v(x) \tag{8.1}$$

for some polynomials $s(x) = \sum_i s_i x^i$ and $t(x) = \sum_i t_i x^i$ of degrees at most $n-k-1$ and $m-k-1$, respectively; furthermore, k is the minimum degree of all polynomials $r(x)$ satisfying (8.1) (see Fact 1.5.2).

Equation (8.1) can be rewritten as a system of linear equations

$$\mathbf{r} = [U, V] \begin{pmatrix} \mathbf{s} \\ \mathbf{t} \end{pmatrix}, \tag{8.2}$$

$\mathbf{r} = [0, \ldots, 0, r_k, \ldots, r_0]^T$, $\mathbf{s} = [s_{n-k-1}, \ldots, s_0]^T$, $\mathbf{t} = [t_{m-k-1}, \ldots, t_0]^T$, and U, V are Toeplitz matrices of sizes $(m+n-k) \times (n-k)$, $(m+n-k) \times (m-$

k), respectively, with the first rows $[u_m, 0, \ldots, 0]$ and $[v_n, 0, \ldots, 0]$ and the first columns $[u_m, u_{m-1}, \ldots, u_0, 0, \ldots, 0]^T$ and $[v_n, v_{n-1}, \ldots, v_0, 0, \ldots, 0]^T$, respectively [compare Problem 5.1 ($TOEPL \cdot VECTOR$) and its solution where the components are in the order reversed to the order used in (8.2)]. The first $m+n-2k$ equations of (8.2) define the $(m+n-2k) \times (m+n-2k)$ linear system,

$$S_k \begin{pmatrix} \mathbf{s} \\ \mathbf{t} \end{pmatrix} = \begin{pmatrix} 0 \\ \vdots \\ 0 \\ 1 \end{pmatrix}, \tag{8.3}$$

whose solution yields the coefficients of $s(x)$ and $t(x)$.

Observe that the matrix S_k is a submatrix of the *Sylvester* or *resultant* matrix S of the polynomials $u(x)$ and $v(x)$, that is, $S = [T_1, T_2]$ where T_1 and T_2 are Toeplitz matrices of sizes $(m+n) \times n$ and $(m+n) \times m$, respectively, with the first rows $[u_m, 0, \ldots, 0]$ and $[v_n, 0, \ldots, 0]$ and the first columns $[u_m, u_{m-1}, \ldots, u_0, 0, \ldots, \;0]^T$ and $[v_n, v_{n-1}, \ldots, v_0, 0, \ldots, 0]^T$, respectively. In other words, the matrix S_k, called the k-th *subresultant* matrix, is obtained by deleting the last $2k$ rows of S and the last k columns of each of the submatrices T_1 and T_2.

For example, for $m = 3$, $n = 2$, we have

$$S = \begin{pmatrix} u_3 & 0 & v_2 & 0 & 0 \\ u_2 & u_3 & v_1 & v_2 & 0 \\ u_1 & u_2 & v_0 & v_1 & v_2 \\ u_0 & u_1 & 0 & v_0 & v_1 \\ 0 & u_0 & 0 & 0 & v_0 \end{pmatrix},$$

$$S_1 = \begin{pmatrix} u_3 & v_2 & 0 \\ u_2 & v_1 & v_2 \\ u_1 & v_0 & v_1 \end{pmatrix}.$$

The resultant matrix S has various further applications, in particular, to the study and computing polynomial zeros, based on the following properties of $\det S$, called the *resultant* of $u(x)$ and $v(x)$:

Theorem 8.1 ([vdW]). *Let $u(x) = u_m \prod_{i=1}^m (x - a_i)$, $v(x) = v_n \prod_{j=1}^n (x - b_j)$. Then $\det S = u_m^n v_n^m \prod_{i=1}^m \prod_{j=1}^n (a_i - b_j)$, so that the resultant of $u(x)$ and $v(x)$ vanishes if and only if $u(x)$ and $v(x)$ have a common zero.*

A modified version of the extended Euclidean algorithm of [Schw] enables us to compute the resultant, $\det S$, in $O(N \log^2 N)$ ops, $N = m + n$.

Now, let k be an integer, $k_0 = \deg \gcd(u(x), v(x))$, and observe that the linear system (8.3) is inconsistent for $k < k_0$ and has a unique solution (and therefore, is nonsingular) for $k = k_0$ since in this case there exists the gcd, unique up to its scaling (also compare Fact 1.5.2). If $r(x) = \hat{r}(x)$, $s(x) = \hat{s}(x)$, and $t(x) = \hat{t}(x)$ satisfy (8.1) for $k = k_0$, then $r(x) = x^h\hat{r}(x)$, $s(x) = x^h\hat{s}(x)$ and $t(x) = x^h\hat{t}(x)$ also satisfy (8.1) for any nonnegative integer h, so that there exists a solution to (8.1)-(8.3) for any $k > k_0$. Moreover, once the coefficients of the polynomials $s(x)$ and $t(x)$ have been obtained by solving the system (8.3), the coefficients of the polynomial $r(x)$ can be computed by using the last $k + 1$ equations of (8.2) or by using (8.1). Fact 2.3 or a least-squares solution to (8.3) can be used to test the consistency of the system (8.3).

Algorithm 8.1.

Input: natural numbers m and n and the coefficients $u_0, \dots, u_m$, $v_0, \dots, v_n$ of two polynomials $u(x) = \sum_{i=0}^{m} u_i x^i$, $v(x) = \sum_{i=0}^{n} v_i x^i$.

Output: The coefficients of $\gcd(u(x), v(x)) = r(x) = \sum_{i=0}^{k_0} r_i x^i$, $r_{k_0} \neq 0$.

Computation:

1. Test the consistency of the system (8.3), for $k = 0, 1, \dots, n$, either by applying Fact 2.3 or by computing the residuals of a least-squares solution to (8.3) for each k, and compute the least integer k_0 such that the system (8.3) is consistent for $k = k_0$.
2. Solve the system (8.3) for $k = k_0$, which is Problem 2.3 ($LIN \cdot SOLVE$) with the subresultant input matrix.
3. Compute the coefficients of $r(x)$ from (8.1) by performing polynomial multiplications.

The above algorithm requires more ops than Algorithm 1.5.1 but runs faster in parallel: it supports the arithmetic parallel time $O(\log^2 n)$ versus the order of $n \log n$ for Algorithm 1.5.1. To reach the bound $O(\log^2 n)$, simply perform Stage 1 by applying Algorithm 11.1 and then use its parallel computational cost estimates (see section 11). Similar comments apply to most of other computations of this and the next sections.

We may dramatically decrease the arithmetic sequential time of Algorithm 8.1 by using the following well-known algorithm:

Algorithm 8.2, *binary search.*

Input: Two integers g and h, $g < h$, and a function $TEST = TEST(k)$ such that, for some unknown integer r, $g \leq r \leq h$, $TEST(k) = TRUE$ if and only if $r \leq k \leq h$.

Output: r.

Computation:

```
a := g, b := h.
WHILE b − a > 1 DO
    BEGIN
    k := ⌈(a + b)/2⌉
    IF TEST(k) = TRUE THEN b := k
    ELSE a := k
    END
IF TEST(b) = TRUE THEN output r := b
ELSE output r := a
END
```

Binary search is a general tool for searching in a finite ordered set. In our particular case, we may apply this tool at Stage 1 of Algorithm 8.1, to compute $r = k_0 = \deg \gcd(u(x), v(x))$, such that the system (8.3), for a fixed k, is consistent if and only if $k \geq k_0$. With the binary search, we need to test the consistency of $\lceil \log n \rceil$ systems (8.3), rather than n such systems, and the arithmetic sequential time respectively decreases.

Solution of Problem 1.5.1c ($SEV \cdot POL \cdot GCD$) (*computing the gcd of several polynomials*). Besides the solution of chapter 1, we may apply the alternative approach of [G84], which reduces this problem to Sylvester-like linear systems, and actually this was the basis for our solution above in the case of two polynomials. The reduction relies on the easily verified fact that there exist polynomials $s_1(x), \ldots, s_m(x)$ such that $\sum_{i=1}^m s_i(x)u_i(x) = d(x)$, $\deg s_i(x) < n$ for all i. Then $\deg d(x) = \min\{\deg \sum_{i=1}^m s_i(x)u_i(x)\}$, where the minimum is over all the polynomials $s_1(x), \ldots, s_m(x)$ of degrees less than n. It remains to compute the coefficients of the polynomials $s_i(x) = \sum_{j=0}^{n-1} s_{ij}x^j$ (for given coefficients of the polynomials $u_i(x) = \sum_{i=0}^n u_{ij}x^j$) such that the polynomial $d_k(x) = \sum_{i=0}^m s_i(x)u_i(x)$ is monic, has degree k, and k is minimum. We will follow the same pattern as in the case where $m = 2$.

Modify the latter problem by lifting the requirement that k be minimum and let this problem have a k-th degree solution $d_k(x)$ (where k is not necessarily minimum). Allow k range from 0 to n and allow $s_i(x)$ have degrees at most $2n-1$ for all i. Then $x^h d_k(x)$ is also a solution of degree $k+h$ for $h = 1, 2, \ldots$, so that the binary search can be applied in order to compute the minimum k. Specifically, if we ignore the minimality requirement and fix k, the problem will be reduced to computing the values s_{ij} that satisfy the Sylvester-like linear system,

$$\sum_{i=1}^{m}\sum_{j=0}^{k} s_{ij}u_{i,h-j} = \delta_{hk}, \quad h = k, \ldots, 2n-1, \tag{8.4}$$

where $\delta_{hk} = 0$ if $h \neq k$, $\delta_{kk} = 1$, $u_{i,q} = 0$ if $q > n$. Applying the binary search of k in the interval from 0 to n (such a search requires at most $\lceil \log(n+1) \rceil$ tests), we compute the minimum k such that the latter system has a solution. ∎

Solution of Problem 1.5.1d ($SEV \cdot POL \cdot LCM$) (*computing the lcm of several polynomials*). In addition to the randomized algorithm of chapter 1, let us now follow [G84] and reduce the deterministic evaluation of $w(x) = \text{lcm}(u_1(x), \ldots, u_m(x))$ to solving linear systems of equations as follows. Assume that all the polynomials $u_i(x)$ are monic for all i [or scale these polynomials $u_i(x)$ to ensure such a property]. Set $d_i = \deg u_i(x)$, $D = \sum_{i=1}^{m} d_i$. Find the minimum k such that there exist monic polynomials $v_j(x)$ of degrees at most $k - d_j$ (for $j = 1, \ldots, m$) satisfying the equations $v_i(x)u_i(x) - v_{i+1}(x)u_{i+1}(x) = 0$ for $i = 1, 2, \ldots, m-1$. [These polynomial equations are equivalent to a nonhomogeneous Sylvester-like linear system of equations in the coefficients of the monic polynomials $v_1(x), \ldots, v_m(x)$. Clearly, these equations are satisfied for $v_j(x) = L(x)/u_j(x)$, $L(x) = \prod_j u_j(x)$, $k = D$, but, generally, this is not the minimum k.] For this minimum k, compute the coefficients of the associated monic polynomial $v_1(x)$. Output $v(x) = u_1(x)v_1(x)$. ∎

The entries of the extended Euclidean scheme, in particular the polynomial remainder sequence $\{r_i(x)\}$ defined by (1.5.6), are closely related to the family of *pseudo resultants.* The pseudo resultant $R_i(x)$ is defined [Col] as the determinant of the $(m+n-2i) \times (m+n-2i)$ matrix obtained from the subresultant matrix S_i by replacing its last row by the row vector

$$[x^{n-i-1}u(x), x^{n-i-2}u(x), \ldots, u(x), x^{m-i-1}v(x), x^{m-i-2}v(x), \ldots, v(x)].$$

The pseudo resultants coincide (within constant factors) with the polynomials $r_i(x)$ of the remainder sequence defined by (1.5.6). Unlike the remainder polynomials $r_i(x)$, the pseudo resultants have all their coefficients lying in the ring formed by the coefficients of $u(x)$ and $v(x)$; in particular, the integrality of the coefficients of $u(x)$ and $v(x)$ is preserved in $R_i(x)$. It is said that the pseudo resultant polynomials $R_i(x)$, $i = 0, 1, \ldots,$ form the *pseudo remainder sequence* for $u(x)$ and $v(x)$ (see [Col], [Kn] and [Ka89] on its evaluation).

We will now compute the remainder sequence and other entries of the extended Euclidean scheme [Problem 1.5.2 ($RAT \cdot INTERP$)] by following [G84]. We need to refine our previous approach to computing the gcd, due to the restrictions on the degrees shown in Facts 1.5.1 and 1.5.2.

An application of (1.5.6)-(1.5.9) leads us to the following result:

Fact 8.1 ([G84]). *There exists an integer i such that $0 \le i \le \ell$, $\deg r_i(x) = j$ for some integer j if and only if $\det S_j \neq 0$; in this case, there exists a number c_i such that $s_i(x) = c_i(y_0 + y_1 x + \cdots + y_{n-j-1}x^{n-j-1})$, $t_i(x) = c_i(z_0 + z_1 x + \cdots + z_{m-j-1}x^{m-j-1})$,*

$$S_j[y_{n-j-1}, \ldots, y_0, z_{m-j-1}, \ldots, z_0]^T = [0, \ldots, 0, 1]^T. \tag{8.5}$$

Proof. For a fixed j, $0 \le j < m \le n$, let i be the largest integer such that $\deg r_{i-1}(x) > j$. Due to Fact 1.5.1,

$$\begin{aligned}
&\deg s_i(x) = \deg v(x) - \deg r_{i-1}(x) < \deg v(x) - j = n - j,\\
&\deg t_i(x) = \deg u(x) - \deg r_{i-1}(x) < \deg u(x) - j = m - j,\\
&\deg\,(s_i(x)u(x) + t_i(x)v(x)) = \deg r_i(x) \le j.
\end{aligned}$$

It follows that $S_j[s_{i,n-j-1}, \ldots, s_{i,0}, t_{i,m-j-1}, \ldots, t_{i,0}]^T = [0, \ldots, 0, r_{i,j}]^T$ provided that $s_i(x) = \sum_{g=0}^{n-j-1} s_{i,g}x^g$, $t_i(x) = \sum_{g=0}^{m-j-1} t_{i,g}x^g$, $r_i(x) = \sum_{g=0}^{j} r_{i,g}x^g$.

If $r_{i,j} = 0$, then this linear system with the coefficient matrix S_j is homogeneous and has a nontrivial solution, and therefore S_j is singular. Otherwise, the linear system is not homogeneous and has a solution $[s_{i,n-j-1}, \ldots, t_{i,0}]^T$, which is unique due to Fact 1.5.2. Therefore, S_j is nonsingular if and only if $r_{i,j} \neq 0$. This linear system turns into (8.5) if we scale it by the factor $1/c_i = 1/r_{i,j}$. ∎

Fact 8.1 enables us to compute (within constant factors) all the entries $s_i(x)$, $t_i(x)$ of the extended Euclidean scheme for the two input polynomials $u(x) = \sum_{g=0}^{m} u_g x^g$ and $v(x) = \sum_{g=0}^{n} v_g x^g$ if we solve the Sylvester-like linear systems (8.5), for all $j = \deg r_i(x)$ and for all i, that is, for all j for which $\det S_j \neq 0$. Then, by using the equations (1.5.9), we may compute the polynomials $r_i(x)$ for all i (within constant factors). By performing polynomial divisions with remainders in the extended Euclidean scheme, we may also recover the polynomials $q_i(x)$ and compute the constant factors c_i

of Fact 8.1. Specifically, let $r_i^*(x)$, $s_i^*(x)$, $t_i^*(x)$ denote the monic associates of $r_i(x)$, $s_i(x)$, $t_i(x)$, respectively [that is, the monic polynomials obtained from $r_i(x)$, $s_i(x)$, $t_i(x)$ by scaling], and let

$$b_i r_i^*(x) = r_{i-2}^*(x) - q_{i-1}^*(x) r_{i-1}^*(x)$$

[compare (1.5.6)]. Then it can be easily verified ([G84], [IKo]) that

$$\begin{aligned}
&c_i = c_0 b_2 b_4 b_6 \cdots b_i \quad \text{if } i \text{ is even,}\\
&c_i = c_1 b_3 b_5 b_7 \cdots b_i \quad \text{if } i \text{ is odd,}\\
&r_i(x) = c_i r_i^*(x),\ s_i(x) = c_i s_i^*(x),\ t_i(x) = c_i t_i^*(x),\\
&q_i(x) = (c_{i-1}/c_i) q_i^*(x) \quad \text{for all } i,
\end{aligned}$$

and we may now completely determine the polynomials $q_i(x), r_i(x), s_i(x)$ and $t_i(x)$ for all i, that is, all the polynomials generated by the extended Euclidean scheme. The overall cost of applying this algorithm is $O(n^2 \log^2 n)$ ops, versus $O(n^2)$ ops used in the extended Euclidean algorithm of chapter 1.

The approach of this section, however, has a substantial advantage in parallel implementation. Indeed, it reduces the computation of the coefficients of all the output polynomials of the extended Euclidean scheme to solving the resultant and subresultant linear systems of k equations for k ranging from n to 1. Such a system can be solved at the cost $O_A(\log^2 k, k^2/\log k)$, (see Algorithm 11.2). However, the very simple algorithm, presented on page 1478 of [P89] and also recalled at the end of our section 11, solves all the systems at the cost $O_A(\log^3 n, n^2/\log^2 n)$, provided that the degrees of the output polynomials have been predetermined cost-free. This assumption holds, in particular, in the usual case where the classical Euclidean scheme is carried out in n steps, which is ensured in the application to the symmetric tridiagonal eigenvalue problem (compare [BP90]). The subsequent distinct and more sophisticated approach of [BG90] reached the same result, but it was modified and improved in [BG92] so as to relax the predetermined degrees assumption, still keeping the same complexity estimates. This was done by reducing the Euclidean scheme to computing a block *LU* factorization of a Hankel-like matrix, and we will present this algorithm in section 10.

9. Bezout, Hankel and Frobenius Matrices and Alternative Algorithms for the Polynomial GCD.

Let us next study some properties relating the polynomials generated in the extended Euclidean scheme to Hankel matrices, to Bezout matrices $B(u, v)$ (to be introduced soon) and to the block triangular factorization of $JB(u, v)J$. The algorithms reveal some interesting correlations between computations with polynomials and structured matrices. We will use these properties and correlations in order to devise effective parallel algorithms for the evaluation of the gcd of two polynomials (Algorithm 9.1), for the evaluation of all the polynomials in the extended Euclidean scheme (Algorithm 10.1), and for the reduction of a general matrix to the tridiagonal form (Algorithm 10.2). Besides having good parallel implementation, these algorithms are also highly attractive as sequential algorithms: they have a lower Boolean computational cost and better numerical stability than Algorithms 5.1 and 1.5.4, and, actually, they dramatically improve both numerical stability and the Boolean computational cost of the customary algorithms for the same computational problems (compare Remarks 9.4, 10.2 and 10.4).

For these problems, the algorithms support the record parallel computational complexity estimates (as does Algorithm 5.1 combined with Algorithm 1.5.4 and an algorithm of [P89], page 1478). Over the fields of characteristic 0, these estimates are $O_A(\log^2 n, n^2/\log n)$ for the gcd and $O_A(\log^3 n, n^2/\log^2 n)$ for the extended Euclidean scheme and for the tridiagonal reduction of Problem 3.2b ($T \cdot REDUCE1$). Unlike the algorithm recalled at the end of section 8, Algorithm 10.1 applies to the general case of the Euclidean scheme computation, that is, it does not require any restriction on the degrees of the polynomials in the remainder sequence.

The correlations between polynomials and structured matrices exploited in sections 9 and 10 can be also applied to computing the inertia of a Hankel matrix ([Gem91]) and lead to new inversion formulae for Hankel and Toeplitz matrices. The latter formulae lead to some acceleration of the well-known algorithm of [BGY] for Toeplitz and Hankel matrix inversion ([Gem92]).

Algorithms 9.1, 10.1 and 10.2 rely on some deep correlations among various structured matrices, polynomials, formal power series and general matrices. We start this section with the *study of this auxiliary but advanced material*, in particular, of the Bezoutians and their properties.

Let $f(z, w) = \sum_{i,j=0}^{+\infty} a_{i,j} z^i w^j$ be a formal power series in two variables, z and w. We associate with $f(z, w)$ the semi-infinite matrix $A = [a_{i,j}]$, $i, j = 0, 1, \ldots$, and, for any natural n, the $n \times n$ matrix $A_n = [a_{i,j}]$, $i, j =$

$0, \dots, n-1$, a finite section of A. We refer to $f(z, w)$ as to the *generating function* of A. It is easy to check that, for any polynomial $u(z)$, the function

$$\frac{u(z)z - u(w)w}{z - w}$$

is the generating function of a Hankel matrix (with zeros below its antidiagonal).

Now consider two polynomials $u(x) = \sum_{i=0}^{n} u_i x^i$ and $v(x) = \sum_{i=0}^{m} v_i x^i$ of degrees n and m, respectively, where $m \leq n$. The expression

$$\frac{u(z)v(w) - v(z)u(w)}{z - w} \tag{9.1}$$

is easily verified to be a polynomial in w and z. The $n \times n$ matrix $B(u, v)$, whose generating function is given by (9.1), is called the *Bezout matrix* or the *Bezoutian* of $u(x)$ and $v(x)$. [In the literature, the polynomial of (9.1) is also called the *Bezoutian* of $u(x)$ and $v(x)$.] The Bezoutians appear in the context of the stability theory and in the classical elimination theory ([FPt], [H70a], [HF], [FD], [GrLi], [LTi], [Pt], [Wim]). In the vast literature on the Bezoutians, we refer the reader to the general references [LTi], [HR], on some specific properties to [H70a], [Pt], [FD], [Wim], and, on the relation to the gcd and the Euclidean scheme for polynomials, to [BG92] and [BG90].

Observe as particular cases that

$$\begin{aligned} B(u, 1) &= \begin{pmatrix} u_1 & \cdots & u_n \\ \vdots & \iddots & \\ u_n & & O \end{pmatrix}, \\ B(u, x^n) &= -\begin{pmatrix} O & & u_0 \\ & \iddots & \vdots \\ u_0 & \cdots & u_{n-1} \end{pmatrix} \end{aligned} \tag{9.2}$$

are special Hankel matrices. Moreover, the following matrix representation of the Bezoutian can be easily proved ([LTi]):

$$\begin{aligned} B(u, v) = &-\begin{pmatrix} v_1 & \cdots & v_n \\ \vdots & \iddots & \\ v_n & & O \end{pmatrix} \begin{pmatrix} u_0 & \cdots & u_{n-1} \\ & \ddots & \vdots \\ O & & u_0 \end{pmatrix} + \\ &\begin{pmatrix} u_1 & \cdots & u_n \\ \vdots & \iddots & \\ u_n & & O \end{pmatrix} \begin{pmatrix} v_0 & \cdots & v_{n-1} \\ & \ddots & \vdots \\ O & & v_0 \end{pmatrix}. \end{aligned} \tag{9.3}$$

Remark 9.1. Given the m-th row and the last row of $B(u,v)$, together with the entry $(0,0)$, lying in the first row and in the first column of $B(u,v)$, it is possible to compute the coefficients of the polynomials $u_n v(x)$ and $u(x)/u_n$. In fact, if $m < n$, the last row of $B(u,v)$ is $u_n[v_0, \dots, v_{n-1}]$, which defines the coefficients of $u_n v(x)$. Moreover, the m-th row $\mathbf{b}^T$ of $B(u,v)$ is

$$\mathbf{b}^T = -v_m[u_0, \dots, u_{n-1}] + [u_m, \dots, u_n, 0, \dots, 0] \begin{pmatrix} v_0 & \dots & v_{n-1} \\ & \ddots & \vdots \\ O & & v_0 \end{pmatrix}, \quad (9.3a)$$

and this relation can be used to express the coefficients of $u(x)/u_n$ as linear functions in u_{n-1}. Indeed, assume for simplicity that $u_n = 1$, so that the last row of $B(u,v)$ gives us the value of m and the coefficients $v_0, \dots, v_m$. Moreover, rewrite (9.3a) as an $n \times n$ system of linear equations

$$[u_0, \dots, u_{n-1}](-v_m I_n + \begin{pmatrix} & & O & & O \\ v_0 & \dots & v_m & & \\ & \ddots & & \ddots & \\ O & & v_0 & \dots & v_m \end{pmatrix}) =$$
$$= \mathbf{b}^T - [0, \dots, 0, v_0, \dots, v_{m-1}].$$

Remove the last equation of this linear system and consider the triangular Toeplitz system consisting of the remaining last $n-m$ equations. From the latter system, express the $u_n, \dots, u_{n-2}$ in terms of u_{n-1}. Substitute $u_m, \dots, u_{n-1}$ in the first m equations and express them via u_{n-1}. Since we are given the entry in the first row and in the first column of $B(u,v)$, we may define the remaining unknown value u_{n-1}.

Now assume that $m < n$, consider the reverse polynomials $\hat{v}(x) = x^m v(x^{-1})$ and $\hat{u}(x) = x^n u(x^{-1})$, and observe that the polynomials $u(x)$ and $v(x)$ define the rational function

$$x^{n-m-1}\frac{\hat{v}(x)}{\hat{u}(x)} = \frac{v(x^{-1})}{xu(x^{-1})}$$

and the formal power series

$$\frac{v(x^{-1})}{xu(x^{-1})} = \sum_{i=0}^{+\infty} h_i x^i.$$

Their coefficients are related to the coefficients of $u(x)$ and $v(x)$ via the triangular Toeplitz system of equations

$$\begin{pmatrix} u_n & & & O \\ u_{n-1} & u_n & & \\ u_{n-2} & u_{n-1} & u_n & \\ \ddots & \ddots & \ddots & \ddots \end{pmatrix} \begin{pmatrix} h_0 \\ h_1 \\ h_2 \\ \vdots \end{pmatrix} = \begin{pmatrix} v_{n-1} \\ v_{n-2} \\ v_{n-3} \\ \vdots \end{pmatrix}, \quad (9.4)$$

where we assume that $u_j = v_j = 0$ for $j < 0$ and $v_i = 0$ for $i > m$. This power series can be viewed as the power series expansion of the function $R(x) = v(x)/u(x)$ at the infinity, i.e.

$$R(x) = \frac{v(x)}{u(x)} = \sum_{i=0}^{+\infty} h_i x^{-i-1}. \tag{9.5}$$

The above power series defines the $n \times n$ Hankel matrix $H(u, v)$ whose (i, j) entry is h_{i+j}, $i, j = 0, 1, \ldots, n-1$. The coefficients of the power series are also known as *Markov parameters*, and the Hankel matrix $H(u, v)$ is said to be the Hankel matrix generated by the Markov parameters (see [HJu]).

Given the coefficients of $u(x)$ and $v(x)$, the entries of the Hankel matrix $H(u, v)$ can be computed by solving [at the cost of $O(n \log n)$ ops] the lower triangular Toeplitz system made up by the first $2n - 1$ equations of (9.4). Moreover, the vector equation (9.4) can be rewritten in the following form:

$$\begin{pmatrix} h_0 & & & O \\ h_1 & h_0 & & \\ h_2 & h_1 & h_0 & \\ \ddots & \ddots & \ddots & \ddots \end{pmatrix} \begin{pmatrix} u_n \\ u_{n-1} \\ u_{n-2} \\ \vdots \end{pmatrix} = \begin{pmatrix} v_{n-1} \\ v_{n-2} \\ v_{n-3} \\ \vdots \end{pmatrix}. \tag{9.6}$$

This fact enables us to prove the following result [compare Problem 1.5.2b ($PADE$)]:

Proposition 9.1. *For any pair of relatively prime polynomials $u(x)$, $v(x)$ of degree n and m, respectively, where $m < n$, the matrix $H(u, v)$ is nonsingular. Moreover, for any nonsingular $n \times n$ Hankel matrix H, there exists a pair of relatively prime polynomials $u(x)$, $v(x)$, of degrees n and m, respectively, such that $u(x)$ is monic and $H = H(u, v)$. The polynomials $u(x)$ and $v(x)$ are related to $H(u, v)$ by the following equations:*

$$H(u, v) \begin{pmatrix} u_0 \\ \vdots \\ u_{n-1} \end{pmatrix} = -u_n \begin{pmatrix} h_n \\ \vdots \\ h_{2n-1} \end{pmatrix}, \tag{9.6a}$$

$$\begin{pmatrix} v_{n-1} \\ \vdots \\ v_0 \end{pmatrix} = \begin{pmatrix} h_0 & & O \\ \vdots & \ddots & \\ h_{n-1} & \ldots & h_0 \end{pmatrix} \begin{pmatrix} u_n \\ \vdots \\ u_1 \end{pmatrix} \tag{9.6b}$$

where any real or complex value can be fixed for h_{2n-1}.

Proof. The first n equations of (9.6) form (9.6b) and express the coefficients of $v(x)$ in terms of the coefficients of $u(x)$. The next n equations can be rewritten as a system in the unknowns $u_0, \ldots, u_{n-1}$, with the matrix $H(u,v)$ and with the known terms $-u_n[h_n, \ldots, h_{2n-1}]^T$ on the right side. Therefore, if $\det H(u,v) \neq 0$, then there exists a unique (up to a scalar factor defined when h_{2n-1} has been fixed) solution $u(x)$ to this system, and consequently, there exists a unique polynomial $v(x)$. In this case $\gcd(u(x), v(x)) = 1$ since otherwise the above linear system would have had more than one solution. Vice versa, if $\det H(u,v) = 0$, the above system (with h_{2n-1} fixed) would have more than one solution, say, $u(x)$, $v(x)$ and $a(x)$, $b(x)$ of degrees t, s, respectively, $s < t$. Therefore, from the relation $\dfrac{v(x^{-1})}{xu(x^{-1})} - \dfrac{b(x^{-1})}{xa(x^{-1})} = 0 \bmod x^{2n}$, it follows that $v(x^{-1})a(x^{-1}) - u(x^{-1})b(x^{-1}) = 0 \bmod x$. Since the left-hand side is a polynomial in x^{-1} of degree at most $2n-1$, it follows that $v(x)a(x) - u(x)b(x) = 0$; hence $u(x)$ and $v(x)$ have a common nonconstant divisor. ∎

Remark 9.2. Let $H = H(u,v)$ be an $n \times n$ nonsingular Hankel matrix and let $u_n \neq 0$. Then (9.6b) implies that $m = \deg v = n - k - 1$ if and only if $h_i = 0$, $i = 0, 1, \ldots, k-1$, $h_k \neq 0$, that is, if and only if $\det H_i = 0$, $i = 1, \ldots, k$, $\det H_{k+1} \neq 0$, where H_i is the $i \times i$ leading principal submatrix of H.

The matrices $H(u,v)$ and $B(u,v)$ are correlated via the following result:

Proposition 9.2 ([HR], [HF]). *Let F_u be the $n \times n$ Frobenius matrix associated with the polynomial $u(x)$ of degree n [see (3.1)]. Let $v(x)$ be a polynomial of degree $m < n$. Then we have*

$$B(u,v) = B(u,1)H(u,v)B(u,1),$$

$$B(u,v) = B(u,1)v(F_u)$$

where $v(F_u) = \sum_{i=0}^{m} v_i F_u^i$.

The second representation of the Bezoutian is also known as the *Barnett factorization* [H70a]. Since $B(u,1)$ is nonsingular [see (9.2)], comparing the two representations of $B(u,v)$ given in Proposition 9.2 implies that

$$H(u,v) = v(F_u)B(u,1)^{-1}. \qquad (9.7)$$

By choosing $v(x) = x^n$ in Proposition 9.2, we arrive at a triangular factorization of the n-th power of a Frobenius matrix in terms of triangular Toeplitz matrices

$$F_u^n = B(u,1)^{-1}B(u,x^n) = -\begin{pmatrix} u_n & & O \\ \vdots & \ddots & \\ u_1 & \ldots & u_n \end{pmatrix}^{-1} \begin{pmatrix} u_0 & \ldots & u_{n-1} \\ & \ddots & \vdots \\ O & & u_0 \end{pmatrix}$$

[compare (9.2)].

The following result, due to Lander (see [Fi]), states that the inverse of any nonsingular Hankel matrix is a Bezoutian and, together with (9.3), extends the Gohberg-Semencul formula to Hankel matrices [compare Theorem 5.2 and recall that if T is a Toeplitz matrix then TJ and JT are Hankel matrices, where J is the reversion matrix of Definition 4.1].

Proposition 9.3. *Let $u(x)$ and $v(x)$ be as in Proposition 9.1, and let $w(x)$ be the polynomial of degree less than n such that $w(x)v(x) = 1 \bmod u(x)$. Then $B(u,w)H(u,v) = I$.*

Proof. From Proposition 9.2 it follows that $B(u,w) = B(u,1)w(F_u)$. Combine this equation and (9.7) and deduce that

$$B(u,w)H(u,v) = B(u,1)w(F_u)v(F_u)B(u,1)^{-1}.$$

Now, since $w(x)v(x) = 1 \bmod u(x)$ and since $u(F_u) = O$, it follows that $w(F_u)v(F_u) = I$, whence $B(u,w)H(u,v) = I$. ∎

The following lemma is needed in order to prove our next result (Proposition 9.4), which characterizes the polynomial gcd in terms of the Hankel matrix $H(u,v)$:

Lemma 9.1. *Let $u(x)$ and $v(x)$ be two polynomials of degrees n and m, respectively, where $m < n$. Then $v(F_u^T)\mathbf{z} = \mathbf{0}$ if and only if $v(x)z(x) = 0 \bmod u(x)$, where $z(x) = \sum_{i=0}^{n-1} z_i x^i$. Therefore, $v(F^T)$ is singular if and only if $s(x) = gcd(u(x), v(x))$ is a nonconstant polynomial. Moreover, if $u(x) = w(x)s(x)$, the null space of $v(F^T)$ is $\{\mathbf{z} \in \mathbf{C}^n : \; z(x) = p(x)w(x)\}$ and has dimension deg s.*

Proof. Since $(F^T)^i\mathbf{e}^{(0)} = \mathbf{e}^{(i)}$, $i = 0,\ldots,n-1$, the first column of $v(F^T)$ is $(v_0, v_1, \ldots, v_m, 0, \ldots, 0)^T$. Therefore, $v(F^T)\mathbf{z} = v(F^T)z(F^T)\mathbf{e}^{(0)} = r(F^T)\mathbf{e}^{(0)}$, where $r(x) = v(x)z(x) \bmod u(x)$ is a polynomial of degree less than n. Whence, $v(F^T)\mathbf{z} = \mathbf{0}$ if and only if $v(x)z(x) = 0 \bmod u(x)$. In particular, $z(x)$ must be a multiple of $w(x)$, i.e., there exists a polynomial $p(x)$ such that $z(x) = p(x)w(x)$. ∎

The next proposition is due to [BG90], but its proof below follows [BG92]:

Proposition 9.4. *Let $u(x)$, $v(x)$ be two monic polynomials of degrees n, m, respectively, $m < n$; let $s(x) = gcd(u(x), v(x))$ be a monic polynomial of degree $n - k$. Then we have:*

a) $\text{rank}(H(u,v)) = k$, $\det H_k \neq 0$, *where* H_i *is the* $i \times i$ *leading principal submatrix of* $H(u,v)$, *so that* $\det H_i = 0$ *for* $i > k$,

b) *if* $H_{k+1}\mathbf{w} = \mathbf{0}$, $\mathbf{w} = (w_i)_{i=0,\ldots,k}$, $w_k = 1$, *that is, if*

$$H_k \begin{pmatrix} w_0 \\ \vdots \\ w_{k-1} \end{pmatrix} = -\begin{pmatrix} h_k \\ \vdots \\ h_{2k-1} \end{pmatrix},$$

then

$$u(x) = s(x) \sum_{i=0}^{k} w_i x^i.$$

Proof. Assume that $u(x) = w(x)s(x)$ and $v(x) = t(x)s(x)$, so that $w(x)$ and $t(x)$ are relatively prime. From (9.7), we obtain that $H(u,v)\mathbf{z} = \mathbf{0}$ if and only if $v(F_u^T)\mathbf{z} = \mathbf{0}$. So, the problem is reduced to analyzing the null space of $v(F_u^T)$. Now, by the virtue of Lemma 9.1, $v(F_u^T)\mathbf{z} = \mathbf{0}$ holds if and only if the polynomial $z(x) = \sum_{i=0}^{n-1} z_i x^i$ can be factorized as $z(x) = p(x)w(x)$ for some polynomial $p(x)$. Since $\{z(x) : \; z(x) = p(x)w(x), \; \deg p(x) \leq n-k-1\}$ is a linear space of dimension $n-k$, we have that $\text{rank}\, H(u,v) = \text{rank}\, v(F^T) = k$. Moreover, by choosing $p(x) = 1$, we satisfy the relations a) and b). ∎

Remark 9.3. Since $JB(u,1)$ and $B(u,1)J$ are lower triangular and upper triangular matrices, respectively, we deduce from the relation $JB(u,v)J = JB(u,1)H(u,v)B(u,1)J$ that $JB(u,v)J\mathbf{y} = \mathbf{0}$, $y_k = 1$, $y_i = 0$, $i > k$, if and only if $\mathbf{w} = B(u,1)J\mathbf{y}$ where $\mathbf{w}$ is the vector of Proposition 9.4. Moreover, $\det(JB(u,v))J)_k \neq 0$, whereas $\det(JB(u,v)J)_i = 0$, $i > k$, where $(JB(u,v)J)_i$ is the $i \times i$ leading principal submatrix of $B(u,v)$.

Proposition 9.4 and Remark 9.3 are the basis for the following algorithm, performed at the cost $O_A(\log^2 n, n^2/\log n)$:

Algorithm 9.1, *computing the gcd of two polynomials.*

Input: natural numbers m and n $(m < n)$ and the coefficients $u_0, \ldots, u_n$, $v_0, \ldots, v_m$ of two polynomials $u(x) = \sum_{i=0}^{n} u_i x^i$, $v(x) = \sum_{i=0}^{m} v_i x^i$.

Output: The coefficients of $\gcd(u(x), v(x)) = s(x) = \sum_{i=0}^{k} s_i x^i$.

Computation:

1. Compute rank $B(u,v)$ (see Problem 2.10 $(M \cdot RANK)$ and Remarks 11.1, 11.2):

   i) Compute the coefficients c_i, $i = 0, \ldots, n-1$, of the characteristic polynomial of $B(u,v)$ by applying Algorithm 11.1.

ii) Set k such that $c_{n-k} \neq 0$, $c_i = 0$, $i = 0, \ldots, n-k-1$ [observe that $B(u,v)$ is symmetric; if the coefficients of u and v were complex, replace u and v by $\bar{u}u$ and $\bar{v}v$, respectively, so that $B(u,v)$ is still real symmetric, and replace k by $k/2$].

2. Solve the $k \times k$ system $C\mathbf{y} = \mathbf{b}$, $\mathbf{y} = (y_{k-1}, \ldots, y_0)^T$, where

$$C = (JB(u,v)J)_k = -\begin{pmatrix} v_n & & O \\ \vdots & \ddots & \\ v_{n-k+1} & \cdots & v_n \end{pmatrix} \begin{pmatrix} u_{n-1} & \cdots & u_{n-k} \\ \vdots & \cdot^{\cdot^{\cdot}} & \vdots \\ u_{n-k} & \cdots & u_{n-2k+1} \end{pmatrix} + \begin{pmatrix} u_n & & O \\ \vdots & \ddots & \\ u_{n-k+1} & \cdots & u_n \end{pmatrix} \begin{pmatrix} v_{n-1} & \cdots & v_{n-k} \\ \vdots & \cdot^{\cdot^{\cdot}} & \vdots \\ v_{n-k} & \cdots & v_{n-2k+1} \end{pmatrix};$$

$$\mathbf{b} = \begin{pmatrix} v_n & & O \\ \vdots & \ddots & \\ v_{n-k+1} & \cdots & v_n \end{pmatrix} \begin{pmatrix} u_{n-k+1} \\ \vdots \\ u_{n-2k} \end{pmatrix} - \begin{pmatrix} u_n & & O \\ \vdots & \ddots & \\ u_{n-k+1} & \cdots & u_n \end{pmatrix} \begin{pmatrix} v_{n-k+1} \\ \vdots \\ v_{n-2k} \end{pmatrix}.$$

3. Compute $\mathbf{w} = (w_i)_{i=0,\ldots,k} = (B(u,1)J)_{k+1} \begin{pmatrix} y_0 \\ \vdots \\ y_k \end{pmatrix}$, $y_k = 1$.

4. Compute $s(x)$ such that $u(x) = s(x)w(x)$, where $w(x) = \sum_{i=0}^{k} w_i x^i$.

Remark 9.4. In view of Proposition 9.4, the above algorithm could be rewritten with replacing the matrix $B(u,v)$ by the Hankel matrix $H(u,v) = [h_{i+j}]$, and with replacing the matrix C and the vector $\mathbf{b}$ at Stage 2, by the leading principal $k \times k$ submatrix $(H(u,v))_k$ of $H(u,v)$ and the vector $(h_k, \ldots, h_{2k-1})^T$, respectively. In this way we would arrive at Algorithm 5.2. In this case, however, the entries of $H(u,v)$ may grow exponentially with n, whereas the entries of $B(u,v)$ have moduli bounded by $\phi = 2n\mu\nu$ where $|u_i| \leq \mu$, $|v_i| \leq \nu$. This gives a substantial advantage to the $B(u,v)$ approach in terms of its numerical stability (numerical stability issues are discussed in chapter 3). Moreover, if u and v have integer coefficients, then the matrix $B(u,v)$ has integer entries, and the computation at Stage 1 can be performed over the integers. Since $|c_i| \leq (\phi n)^n$, due to Hadamard's well-known inequality (see chapter 3), the number of bits needed at Stage 1 is $O(n \log(n\mu\nu))$. If we choose $y_k = \det C$, the vector $\mathbf{w}$ at Stage 3 will have only integer components, each represented with $O(n \log(n\mu\nu))$ bits. The computation at Stage 2 can be still kept within the integers, by applying the algorithms for Toeplitz-like and Hankel-like systems of section 11. In this way, it is possible to prove that (compare exercise 34 in chapter 3) if the polynomials u and v have integer coefficients, all represented with at most L bits, then $O(n(L+\log n))$ bits are sufficient to compute $\gcd(u,v)$ by using integer arithmetic. If u and v have rational coefficients represented as pairs of L-bit integers, then $O(n^2(L+\log n))$ bits are sufficient to compute $\gcd(u,v)$ by using integer arithmetic. This fact substantially improves the previous record bound of $O(n^4L)$ bits (due to the algorithm of [SL]) for computing the gcd of two polynomials.

Remark 9.5. For computations over an arbitrary field $\mathbf{F}$, Stage 1 of Algorithm 9.1 in its present form may fail, because of the use of Newton's identities and rank computation, and we refer the readers to Remarks 6.1 of this chapter and to section 6 of chapter 4 on this problem and on some ways of solving it.

10. Block LU Factorization. Application to the Extended Euclidean Computation Scheme and to the Reduction of a Matrix to the Tridiagonal Form.

Next, we will analyze correlations between the matrices $H(u,v)$, $B(u,v)$ and the coefficients of the polynomials generated by the Euclidean scheme applied to $u(x)$ and $v(x)$. Specifically, we will prove that the entries of the triangular matrices of the block triangular factorization of $H(u,v)$ and of the associated matrix $JB(u,v)J$ yield the coefficients of all polynomials generated by the Euclidean scheme (compare Remarks 10.1 and 10.2). Such a reduction to computing a block triangular factorization allows us to compute these coefficients with a record parallel complexity bound (see Algorithm 10.1 and its parallel cost estimates in chapter 4).

Then, we will apply Theorem 3.2 and the properties of Bezoutians to devise an algorithm for reducing a matrix to the tridiagonal or block tridiagonal form [compare Problem 3.2b ($T \cdot REDUCE$)].

We first recall a result of [GrLi] on the block LU factorization of a Hankel matrix and on the reduction into tridiagonal form of a Frobenius matrix (Proposition 10.1). We relate the matrices of this factorization to the polynomials generated by the Euclidean scheme (Proposition 10.2 and Corollary 10.1). Then we prove some properties of the Schur complements of a Bezout matrix by using some of their correlations to the polynomial remainder sequence (Proposition 10.4), and we apply these results to devise the basic relations for computing the block LU factorization of a Bezout matrix

(Proposition 10.6 and Algorithm 10.1) and thus for the Euclidean scheme computation. Then we follow [Gem92] and, as a byproduct of our study, arrive at the inversion formula (10.3) for Hankel matrices. This formula (of [Gem92]) enables us to accelerate the algorithm of [BGY] for Hankel and Toeplitz matrix inversion. We end the section by presenting a solution of Problem 3.2a ($T \cdot REDUCE$) of the reduction of a general matrix to (block) tridiagonal form.

Proposition 10.1 ([GrLi]). *Let $H = [h_{i+j}]$ be a semi-infinite Hankel matrix and H_n be its $n \times n$ leading principal submatrix. Suppose that H_n is nonsingular and let $0 = m_0 < m_1 < \cdots < m_\ell = n$ be integers such that* $\det H_{m_i} \neq 0$, $i = 1, \ldots, \ell$, $\det H_k = 0$, *otherwise. Let $\delta_i = m_i - m_{i-1}$. Let $\mathcal{D}_n$ be the class of block diagonal matrices* $\operatorname{diag}(T_1, \ldots, T_\ell)$ *where T_i is a $\delta_i \times \delta_i$ lower triangular Hankel matrix. Then we have the following properties:*

1) *There exists an upper triangular matrix R having unit diagonal entries, such that $R^T H R = D$, $D \in \mathcal{D}_n$. In particular, one can choose*

$$R = \tilde{R} =$$
$$[\mathbf{q}_0, Z\mathbf{q}_0, \ldots, Z^{\delta_1 - 1}\mathbf{q}_0, \mathbf{q}_1, Z\mathbf{q}_1, \ldots, Z^{\delta_2 - 1}\mathbf{q}_1, \ldots, Z^{\delta_\ell - 1}\mathbf{q}_{\ell-1}]$$

where Z is the down-shift matrix of Definition 4.1, $\mathbf{q}_0 = \mathbf{e}^{(0)}$,

$$\mathbf{q}_i = \begin{pmatrix} \mathbf{s}_i \\ 1 \\ \mathbf{0} \end{pmatrix}, \quad \mathbf{s}_i = -H_{m_i}^{-1} \begin{pmatrix} h_{m_i} \\ \vdots \\ h_{2m_i - 1} \end{pmatrix}, \quad i = 1, \ldots, \ell - 1.$$

2) *For any upper triangular matrix $R = [\mathbf{r}_0, \mathbf{r}_1, \ldots, \mathbf{r}_{n-1}]$ such that $R^T H R \in \mathcal{D}_n$, we have $\mathbf{r}_{\delta_i} = \mathbf{q}_i$, $i = 0, \ldots, \ell - 1$, where we assume that $\delta_0 = 0$.*
3) *Setting $Q_i(x) = [1, x, \ldots, x^{n-1}]\mathbf{q}_i$, $i = 0, 1, \ldots, \ell$, where $\mathbf{q}_\ell = -H_n^{-1} \begin{pmatrix} h_{n+1} \\ \vdots \\ h_{2n} \end{pmatrix}$, we obtain that*

$$Q_i(x) = c_i(x) Q_{i-1}(x) - \theta_{i-1} Q_{i-2}(x), \; i = 1, 2, \ldots, \ell,$$
$$Q_0(x) = 1, \; Q_{-1}(x) = 0,$$

where $c_i(x)$ are monic polynomials of degree δ_i, θ_{i-1} are constants, $i = 1, \ldots, \ell$.

4) If F_{Q_ℓ} is the Frobenius matrix associated with $Q_\ell(x)$, then

$$F_{Q_\ell}^T R = RA, \quad A = \begin{pmatrix} A_{11} & A_{12} & & O \\ A_{21} & \ddots & \ddots & \\ & \ddots & \ddots & A_{\ell-1,\ell} \\ O & & A_{\ell,\ell-1} & A_{\ell\ell} \end{pmatrix},$$

and for every i, $A_{ii} = F_{c_i}^T$, $A_{i,i+1}$ has θ_i in the upper right corner and has zero entries elsewhere, $A_{i+1,i}$ has 1 in the upper right corner and zero entries elsewhere.

Note that this block triangular factorization is not generally unique, unless the matrix H is strongly nonsingular (see exercise 27).

Applying Proposition 10.1 to the matrix $H = H(u,v)$ we may define the vectors $\mathbf{q}_i$ and, consequently, the polynomials $Q_i(x)$.

Proposition 10.2. *Let $u(x)$ and $v(x)$ be monic polynomials of degrees n and m, respectively, such that $n > m$, $\gcd(u(x), v(x)) = 1$. Then, for the monic polynomials $Q_i(x)$ defined in Proposition 10.1 for $H_n = H(u,v)$, we have $Q_\ell(x) = u(x)$, $Q_{\ell-1}(x) = w^*(x)$, where $w^*(x)$ is the monic polynomial associated with $w(x)$ [i.e. $w^*(x) = cw(x)$ for a nonzero constant c] and $w(x)v(x) = 1 \bmod u(x)$.*

Proof. From Proposition 9.3 and Remark 9.1, we have $H(u,v)\mathbf{w} = \mathbf{e}^{(n-1)}$, where $\mathbf{w} = [w_0, \ldots, w_{n-1}]^T$ and $w_i = 0$ for $i > \deg w(x)$. Since

$$H_i \begin{pmatrix} w_0 \\ \vdots \\ w_{i-1} \end{pmatrix} = 0 \quad \text{if } n > i \geq 1 + \deg w(x)$$

[this condition does not apply if $\deg w(x) = n-1$], we have $\det H_i = 0$, $i = 1 + \deg w(x), \ldots, n-1$, whereas $\det H_i \neq 0$ for $i = \deg w(x)$ [due to the uniqueness of $w(x)$], whence $\deg w(x) = m_{\ell-1}$; moreover,

$$H_{m_{\ell-1}} \begin{pmatrix} w_0 \\ \vdots \\ w_{m_{\ell-1}-1} \end{pmatrix} = -w_{m_{\ell-1}} \begin{pmatrix} h_{m_{\ell-1}} \\ \vdots \\ h_{2m_{\ell-1}-1} \end{pmatrix}.$$

From parts 1) and 3) of Proposition 10.1, it follows that $Q_{\ell-1}(x) = w^*(x)$. Now, let $\tilde{u}(x) = (x-\alpha)u(x)$ and $\tilde{v}(x) = (x-\alpha)v(x)$ for a scalar α. Due to (9.7) and to the symmetry of $H(\tilde{u},\tilde{v})$, the null space of $H(\tilde{u},\tilde{v})$ coincides with the null space of $\tilde{v}(F_{\tilde{u}}^T)$. Therefore, due to Lemma 9.1, the vector $\mathbf{u} = [u_0, u_1, \ldots u_{n-1}, 1]^T$ generates the null space of the matrix $H(\tilde{u},\tilde{v})$.

Thus, from the equation $H(\tilde{u},\tilde{v})\mathbf{u}=\mathbf{0}$, we obtain that $H(u,v)\begin{pmatrix} u_0 \\ \vdots \\ u_{n-1}\end{pmatrix} = -\begin{pmatrix} h_{n+1} \\ \vdots \\ h_{2n}\end{pmatrix}$, that is, $u(x)=Q_\ell(x)$. ■

From Propositions 10.1 and 10.2, we immediately obtain the following corollary that relates the factorization $R^THR=D$ of Proposition 10.1 to the polynomials generated by the extended Euclidean scheme:

Corollary 10.1. *Let $q_i(x)$ and $t_{i+1}(x)$, for $i=1,\ldots,\ell$, denote the polynomials generated in the extended Euclidean scheme (1.5.6)-(1.5.8). Then we have*

$$\begin{aligned} Q_i(x) &= t^*_{i+1}(x), \\ c_i(x) &= q^*_i(x), \end{aligned} \qquad i=1,\ldots,\ell,$$

where $q^*_i(x)$ and $t^*_{i+1}(x)$, for $i=1,\ldots,\ell$, are the monic polynomials associated with the polynomials $q_i(x)$ and $t_{i+1}(x)$, respectively.

Proof. Both sequences of polynomials $Q_i(x)$ and $t^*_{i+1}(x)$ can be viewed as the polynomial remainder sequences obtained by applying the Euclidean scheme to $Q_\ell(x)$, $Q_{\ell-1}(x)$ and to $t^*_{\ell+1}(x)$, $t^*_\ell(x)$, respectively. The corollary follows from Proposition 10.2 since the polynomial remainder sequence is uniquely determined by its first two polynomials. ■

Remark 10.1. The above result can be restated in the following way: let $u(x)$ and $v(x)$ denote two monic and relatively prime polynomials. Let the monic polynomial $w(x)$ solve the congruence $v(x)w(x)=c \bmod u(x)$ for a constant $c\neq 0$. Let the Euclidean scheme applied to $u(x)$ and $v(x)$ generate the sequence of polynomials $t_{i+1}(x)$ such that $t_n(x)=u(x)$, $t_{n-1}(x)=v(x)$, and $t_{i-1}(x)$ is the quotient of the division of $t_{i+1}(x)$ by $t_i(x)$. Then the coefficients of the polynomials of this sequence are given by suitable columns of the matrix $(\tilde{L})^{-1}$ where $H(u,w)=\tilde{L}\tilde{D}\tilde{L}^T$ is the block triangular factorization of $H(u,w)$.

The following result relates the block triangular factorizations of the matrices $H(u,v)$ and $H(u,w)$:

Fact 10.1. *Let $u(x)$, $v(x)$ and $w(x)$ be three polynomials such that $u(x)$ and $v(x)$ are relatively prime and $w(x)v(x)=1 \bmod u(x)$. If $H(u,v)=LDL^T$ is a block triangular factorization of $H(u,v)$, then $H(u,w)=\tilde{L}\tilde{D}\tilde{L}^T$ is a block triangular factorization of $H(u,w)$, where $(\tilde{L}^T)^{-1}=B(u,1)LJ$, $\tilde{D}^{-1}=JDJ$.*

Proof. $\tilde{L}\tilde{D}\tilde{L}^T = (JL^{-1}B(u,1)^{-1})^T JD^{-1}J(JL^{-1}B(u,1)^{-1}) = B(u,1)^{-1}\ (L^{-1})^T D^{-1}L^{-1}B(u,1)^{-1} = B(u,1)^{-1}H^{-1}(u,v)B(u,1)^{-1} = B(u,v)^{-1} = H(u,w)$. ∎

Remark 10.2. The computation of the coefficients of all the polynomials generated in the Euclidean scheme, applied to the polynomials $u(x)$ and $v(x)$, has been reduced to computing the block triangular factorization $H(u,v) = LDL^T$ and to polynomial multiplication; in fact, $\tilde{L}^{-1} = JL^T B(u,1)$. Moreover, since $JB(u,v)J = JB(u,1)H(u,v)B(u,1)J$, the block triangular factorization of $H(u,v)$ can be obtained from the block triangular factorization of $JB(u,v)J = \hat{L}\hat{D}\hat{L}^T$, and in fact, $L = B(u,1)^{-1}J\hat{L}$, so that

$$\tilde{L}^{-1} = J\hat{L}^T J.$$

Computing the block triangular factorization of $JB(u,v)J$ has a substantial advantage, in terms of the numerical stability (compare section 1 of chapter 3), over computing the block triangular factorization of $H(u,v)$. Namely, the entries of $H(u,v)$ may grow exponentially with n, whereas the moduli of the entries of $JB(u,v)J$ are bounded by $2n\mu\nu$, where μ and ν are upper bounds on the moduli of the coefficients of $u(x)$ and $v(x)$, respectively. Moreover, if $u(x)$ and $v(x)$ have integer coefficients, then also $JB(u,v)J$ has integer coefficients.

Remark 10.3. Under the assumptions of Proposition 10.2, Remark 9.2 implies that $\det(H(u,v))_i = 0$, $i = 1,\ldots,n-m-1$, and $\det(H(u,v))_{n-m} \neq 0$; therefore, from Remark 9.3 it follows that $\det(JB(u,v)J)_i = 0$, $i = 1,\ldots,n-m-1$ and $\det(JB(u,v)J)_{n-m} \neq 0$.

Our next objective is the design of an algorithm for computing the block triangular factors of $JB(u,v)J$, which should give us all the polynomials $t_i(x)$ generated in the Euclidean scheme. We need some propositions characterizing the Schur complements of $(JB(u,v)J)_{n-m}$ and of some other principal submatrices of $JB(u,v)J$. We first recall the following auxiliary result [PK]:

Lemma 10.1. *If Q is an $n \times n$ matrix having the generating function $(a(z)b(w) - b(z)a(w))/(z-w)$ and if $\det Q_i = 0$, $i = 1,\ldots,t-1$, $\det Q_t \neq 0$, then the Schur complement of Q_t in Q has the generating function $(a_t(z)b_t(w) - b_t(z)a_t(w))/(z-w)$, where $a_t(z) = z^{-t}a(z)$ and $b_t(z)$ is* defined by the recurrence

$$zb_{j+1}(z) = b_j(z) - a_t(z)\frac{b_j(0)}{a_t(0)},\ j = 0, \dots, t-1,$$
$$b_0(z) = b(z).$$

■

Proposition 10.3. *Let S be the Schur complement of $(JB(u,v)J)_{n-m}$ in $JB(u,v)J$, then $S = -JB(v,r)J$ where $u(x) = q(x)v(x) + r(x)$, deg $r(x)$ $< m$.*

Proof. The generating function of $JB(u,v)J$ is easily verified to be

$$(-\hat{u}(z)\hat{v}(w)w^{n-m} + \hat{u}(w)\hat{v}(z)z^{n-m})/(z-w),$$

where $\hat{u}(z)$ and $\hat{v}(z)$ are the reverse polynomials of $u(z)$, $v(z)$, respectively. Therefore, due to Lemma 10.1, the generating function of the Schur complement S is given by $(v_{n-m}(z)u_{n-m}(w) - u_{n-m}(z)v_{n-m}(w))/(z-w)$ where $v_{n-m}(z) = \hat{v}(z)$, and

$$zu_{j+1}(z) = u_j(z) - \hat{v}(z)\frac{u_j(0)}{\hat{v}(0)}, \quad j = 0, \dots, n-m-1, \qquad (10.1)$$
$$u_0(z) = \hat{u}(z).$$

Now observe that the polynomials $u_j(z)$ in (10.1) have degrees $n-j$, $j = 0, \dots, n-m$, and that $z^{n-j}u_j(z^{-1})$ is the polynomial obtained at stage j of the synthetic polynomial division algorithm applied to $u(x)$ and $v(x)$, so that $u_{n-m+1}(z) = \hat{r}(z)z^{m-\deg r-1}$, where $u(z) = q(z)v(z) + r(z)$, deg $r(z) < m$. Therefore, $u_{n-m}(z) = z^{m-\deg r}\hat{r}(z) + \beta\hat{v}(z)$, $\beta = \hat{u}_{n-m}(0)/\hat{v}(0)$. Hence, the generating function of S is

$$(\hat{v}(z)\hat{r}(w)w^{m-\deg r} - \hat{v}(w)\hat{r}(z)z^{m-\deg r})/(z-w)$$

so that $S = -JB(v,r)J$. ■

Fact 10.2 (compare Proposition 2.3). *Let h and k be integers, $0 < h < k$; A_i denote the $i \times i$ leading principal submatrix of A; and A, A_k, A_{h+k} be nonsingular matrices. Consider the Schur complements S_k and S_{h+k} of A_k, A_{h+k} in A, respectively. Let $(S_k)_h$ denote the $h \times h$ leading principal submatrix of S_h. Then S_{h+k} is the Schur complement of $(S_k)_h$ in S_k.*

Proposition 10.4. *Let $u(x)$ and $v(x)$ be as in Proposition 10.2. Let m_i, $i = 1, \dots, \ell$, be as in Proposition 10.1, where the semi-infinite Hankel* *matrix H is generated by the power series (9.5). Then* $\det(JB(u,v)J)_{m_i} \neq 0$, $i = 1,\ldots,\ell$, $\det(JB(u,v)J)_j = 0$, *for* $j \neq m_i$, $i = 1,\ldots,\ell$. *Moreover, for the Schur complement* S_{m_i} *of* $(JB(u,v)J)_{m_i}$ *in* $JB(u,v)J$, *the following relation holds:*

$$S_{m_i} = (-1)^i JB(r_i, r_{i+1})J,$$

where $\{r_i(x),\ i = 1,\ldots,\ell\}$ *is the polynomial remainder sequence (1.5.6) obtained by applying the Euclidean scheme to* $u(x)$, $v(x)$.

Proof. The relations for the determinants follow from Remark 10.3. To prove the relation $S_{m_i} = (-1)^i JB(r_i, r_{i+1})J$, inductively apply Proposition 10.3 and take into account Fact 10.2. ∎

Thus, the polynomials r_i, generated in the Euclidean scheme (1.5.6), can be obtained from the Schur complements S_{m_i}.

Now observe that, from the definition of the Bezoutian, it follows that the $(n-m)\times(n-m)$ leading principal submatrix of $JB(u,v)J$ can be factorized as

$$(JB(u,v)J)_{n-m} = \begin{pmatrix} u_n & & O \\ \vdots & \ddots & \\ u_{m+1} & \cdots & u_n \end{pmatrix} \begin{pmatrix} O & & v_m \\ & \iddots & \vdots \\ v_m & \cdots & v_{2m-n+1} \end{pmatrix}. \tag{10.2}$$

Let us next factorize the $m_i \times m_i$ leading principal submatrices of $JB(u,v)J$, for $i = 1,\ldots,\ell$.

Proposition 10.5. *Under the hypotheses of Proposition 10.4,*

$$(JB(u,v)J)_{m_i} = \begin{pmatrix} u_n & & O \\ \vdots & \ddots & \\ u_{n-m_i+1} & \cdots & u_n \end{pmatrix} \begin{pmatrix} u^{(i)}_{m_i} & & O \\ \vdots & \ddots & \\ u^{(i)}_1 & \cdots & u^{(i)}_{m_i} \end{pmatrix}^{-1}$$
$$JB(u^{(i)}, v^{(i)})J \begin{pmatrix} u^{(i)}_{m_i} & \cdots & u^{(i)}_1 \\ & \ddots & \vdots \\ O & & u^{(i)}_{m_i} \end{pmatrix}^{-1} \begin{pmatrix} u_n & \cdots & u_{n-m_i+1} \\ & \ddots & \vdots \\ O & & u_n \end{pmatrix},$$

where the polynomials $u^{(i)}(z)$, $v^{(i)}(z)$ *have the coefficients* $u^{(i)}_r$, $v^{(i)}_s$, *respectively,* $r = 0,\ldots,m_i$; $s = 0,\ldots,m_i - 1$, *and are defined by the relation* $H(u^{(i)}, v^{(i)}) = (H(u,v))_{m_i}$.

Proof. The proposition follows from the relation

$$(JB(u,v)J)_{m_i} = \begin{pmatrix} u_n & & O \\ \vdots & \ddots & \\ u_{n-m_i+1} & \cdots & u_n \end{pmatrix} (H(u,v))_{m_i} \begin{pmatrix} u_n & \cdots & u_{n-m_i+1} \\ & \ddots & \vdots \\ O & & u_n \end{pmatrix},$$

in view of Proposition 9.2. ∎

From the above results, we immediately obtain the following block factorization of $JB(u,v)J$:

Proposition 10.6. *Under the hypothesis of Proposition 10.4, set* $Y_{m_i} = (JB(u,v)J)_{m_i}$. *Then*

$$JB(u,v)J = \begin{pmatrix} Y_{m_i} & G^T \\ G & W \end{pmatrix} = \begin{pmatrix} I & O \\ GY_{m_i}^{-1} & I \end{pmatrix} \begin{pmatrix} Y_{m_i} & O \\ O & S_{m_i} \end{pmatrix} \begin{pmatrix} I & Y_{m_i}^{-1}G^T \\ O & I \end{pmatrix} =$$

$$\begin{pmatrix} T & O \\ GY_{m_i}^{-1}T & I \end{pmatrix} \begin{pmatrix} JB(u^{(i)},v^{(i)})J & O \\ O & (-1)^i JB(r_i, r_{i+1})J \end{pmatrix} \begin{pmatrix} T^T & T^T Y_{m_i}^{-1} G^T \\ O & I \end{pmatrix}$$

where

$$S_{m_i} = W - GY_{m_i}^{-1}G^T = (-1)^i JB(r_i, r_{i+1})J,$$

$$T = \begin{pmatrix} u_n & & O \\ \vdots & \ddots & \\ u_{n-m_i+1} & \dots & u_n \end{pmatrix} (B(u^{(i)},1))^{-1})J.$$

Remark 10.4. Proposition 10.4 enables us to express the coefficient vector $\mathbf{r}_{i+1}$ of the $(i+1)$-st remainder $r_{i+1}(x)$ in the extended Euclidean scheme as $\mathbf{r}_{i+1} = J(W - GY_{m_i}^{-1}G^T)J\mathbf{e}^{(n-m_i)}$, that is, we may compute the vector $\mathbf{r}_{i+1}$ by solving a Hankel-like $m_i \times m_i$ linear system with the matrix Y_{m_i}, and the computation can be performed over the integers if the input values are integers and if we multiply the above relation by $\det Y_{m_i}$. It follows that the polynomial $\det Y_{m_i} r_{i+1}(x)$ has integer coefficients with their moduli being $O((2n\mu\nu)^{m_i})$. These properties can be used in order to ensure numerical stability of these computations (see chapter 3).

The following algorithm computes the block triangular factorization of the matrix $JB(u,v)J$ and, therefore, the coefficients of the polynomials generated in the Euclidean scheme [at the cost $O_A(\log^3 n,\ n^2/\log^2 n)$]:

Algorithm 10.1, *computing the block triangular factorization of* $JB(u,v)J$.

Input: natural m and n and the coefficients $u_0, \dots u_n$, $v_0, \dots, v_m$ of two polynomials $u(x) = \sum_{i=0}^{n} u_i x^i$, $v(x) = \sum_{i=0}^{m} v_i x^i$ (assume, without loss of generality, that $u(x)$ and $v(x)$ are relatively prime, otherwise compute $s(x)$ =gcd($u(x)$, $v(x)$) and replace $u(x)$, $v(x)$, n and m with $u(x)/s(x)$, $v(x)/s(x)$, $n - \deg s(x)$ and $m - \deg s(x)$, respectively).

Output: The entries of the lower triangular matrix $\hat{L}$ and of the matrix $\hat{D} \in \mathcal{D}_n$, such that $JB(u,v)J = \hat{L}\hat{D}\hat{L}^T$.

Computation:

1. If $n - m > n/2$, set $k = n - m$ [in this case $(B(u,v))_k$ is a nonsingular matrix whereas $\det(B(u,v))_i = 0$, for $i = 1, \ldots, k-1$ (compare Remark 9.2)]; otherwise, compute rank $((B(u,v))_{n/2})$ and set $k =$ rank $((B(u,v))_{n/2})$.

2. Let i be such that $k = m_i$ and factorize the matrix $JB(u,v)J$ as in Proposition 10.6. Compute the matrices

$$Y_{m_i}^{-1},\ GY_{m_i}^{-1},\ S_{mi} = W - GY_{m_i}^{-1}G^T$$

by applying any algorithm for Hankel-like matrix inversion (see Definition 11.3). Compute the coefficients of $r_i(x)$, $r_{i+1}(x)$, by using Propositions 10.6 and Remark 9.1. Compute the coefficients of the polynomials $u^{(i)}(x)$, $v^{(i)}(x)$ such that $H(u^{(i)}, v^{(i)}) = (H(u,v))_{m_i}$, that is, apply (9.6a) and (9.6b) to the nonsingular matrix $(H(u,v))_{m_i}$ (compare Proposition 10.5).

3. Recursively apply Algorithm 10.1 to $JB(u^{(i)}, v^{(i)})J$, $JB(r_i, r_{i+1})J$, obtaining the factorization $JB(u^{(i)}, v^{(i)})J = L'D'L'^T$, $JB(r_i, r_{i+1})J = L''D''L''^T$.

4. Compute and output

$$\hat{L} = \begin{pmatrix} I & O \\ GY_{m_i}^{-1} & I \end{pmatrix} \begin{pmatrix} T & O \\ O & I \end{pmatrix} \begin{pmatrix} L' & O \\ O & L'' \end{pmatrix}, \quad \hat{D} = \begin{pmatrix} D' & O \\ O & D'' \end{pmatrix},$$

where T is the matrix defined in Proposition 10.6, thus obtaining the coefficients of all the polynomials $r_i(x)$ (see Remark 10.2).

It is not hard to prove that $O(\log n)$ recursive steps are sufficient in this algorithm (for more details see [BG92]). If $H(u,v)$ is strongly nonsingular, that is, if the Euclidean scheme applied to $u(x)$ and $v(x)$ is carried out in n steps, then also $JB(u,v)J$ is strongly nonsingular, and Stage 1 can be eliminated.

Once the polynomials $r_i(x)$ have been computed, the quotients $q_i(x)$ may be computed by performing $O(n)$ polynomial divisions at the cost $O_A(\log^2 n, n^2)$ (see chapter 4). The polynomials $s_i(x)$, $t_i(x)$ generated by the extended Euclidean scheme (1.5.6)-(1.5.8) can be computed by means of the relation

$$\begin{pmatrix} \mathbf{s}_i \\ \mathbf{t}_i \end{pmatrix} = R^{-1}\mathbf{r}_i, \quad i = 1, \ldots, \ell,$$

where R is the $(n+m) \times (n+m)$ resultant matrix of $u(x)$ and $v(x)$ (see section 8 for the definition), and $\mathbf{s}_i$, $\mathbf{t}_i$, $\mathbf{r}_i$ are the coefficient vectors of the polynomials $s_i(x)$, $t_i(x)$, $r_i(x)$ filled with zeros up to dimensions m, n and $m+n$, respectively.

We will next apply the above techniques for block LU factorization to the solution of the block tridiagonalization problem, which in the strongly nonsingular case (with no degeneration) defines tridiagonalization. We have the following algorithm, solving Problem 3.2b ($T \cdot REDUCE1$), which extends Algorithm 3.3 by specifying its Stage 4:

Algorithm 10.2, *block tridiagonal reduction.*

Input: a natural n, the entries of $A \in \mathbf{C}_{n,n}$.

Output: either FAILURE or the entries of the block tridiagonal matrix T_n and of the block diagonal matrix D of Theorem 3.4.

Computation:

1. Choose two random vectors $\mathbf{p},\mathbf{q} \in \mathbf{C}_{n,1}$.
2. Compute the Krylov matrix $K(A, \mathbf{q}, 2n)$ [Problem 2.1a ($KRYLOV$)].
3. Compute the scalars $\mathbf{p}^T A^i \mathbf{q}$, $i = 0, \ldots, 2n$, that is, the entries of the Hankel matrices $H^{(0)}$ and $H^{(1)}$ of Algorithm 3.3 and Theorem 3.2.
4. Compute the lower triangular matrix L^{-1} and compute and output a block diagonal matrix D such that $H^{(0)} = LDL^T$. Proceed by performing the following stages:
   i) by solving two Hankel systems, compute the coefficients and the degrees n and m of two polynomials $u(x)$ and $v(x)$ defining the Bezout matrix $B(u,v) = H^{-1}$ where $H = H^{(0)}$ (compare Remark 9.1);
   ii) compute the matrices $\hat{L}$ and $\hat{D}$ such that $JB(u,v)J = \hat{L}\hat{D}\hat{L}^T$, by applying Algorithm 10.1;
   iii) set $L^{-1} = J\hat{L}^T J$, $D = J\hat{D}^{-1}J$.
5. Compute the entries along the three main block diagonals of the matrix $L^{-1}H^{(1)}L^{-T}$; output these entries, thus defining the matrix T_n.

The correctness of the above algorithm follows from Theorems 3.2 and 3.4, since $\hat{L}D\hat{L}^T = JB(u,v)J = JH^{-1}J = (JL^{-T}J)(JD^{-1}J)(JL^{-1}J)$ where $JL^{-T}J$ is a lower triangular matrix.

Remark 10.5. In the case where the matrix A is sparse or structured such that the matrix-by-vector multiplication costs, say, $O_A(\log n, n)$ (as in the case of Toeplitz matrices), the overall computational cost is dominated by the cost of performing Stages 2–4.

Algorithm 10.2 can be adapted to the computation of the entries of the matrices P and Q of Problem 3.2a ($T \cdot REDUCE$).

In the case $\ell = n$, (3.7) and (3.8) imply that $A^{-1} = UH^{-1}V$, where $H = H^{(0)}$. Therefore, in the case where the matrix H is strongly nonsingular, the solution of a linear system can be reduced to computing a Krylov matrix and to solving a Hankel linear system.

We also observe that the columns of the matrix $R = L^{-1}$ of Theorem 3.4 are fully defined by the coefficients of the polynomials generated by the Euclidean scheme applied to the polynomials that define the rational function associated with the Hankel matrix $H^{(0)}$.

In the particular case where A is a Frobenius matrix, Problem 3.2a ($T \cdot REDUCE$) turns into Problem 3.2b ($T \cdot REDUCE1$), where the choice of the vectors $\mathbf{p}$ and $\mathbf{q}$ determines the characteristic polynomial of the $(n-1) \times (n-1)$ leading principal submatrix of T. Specifically, if we choose $\mathbf{p} = \mathbf{e}^{(0)}$, then the vectors $\mathbf{q}_i$ defining L^{-1} give us the coefficients of the remainder sequence generated by the characteristic polynomial of A and the polynomial defined by the vector $\mathbf{q}$.

To conclude this section we follow [Gem92] and will briefly recall two other applications of Proposition 10.4. First consider the $2n \times 2n$ lower triangular Hankel matrix

$$H = \begin{pmatrix} h_1 & h_2 & \dots & h_{2n} \\ h_2 & \dots & h_{2n} & \\ \vdots & \cdot^{\cdot^{\cdot}} & & \\ h_{2n} & & & \end{pmatrix} = \begin{pmatrix} H_n & \hat{H} \\ \hat{H} & O \end{pmatrix},$$

where H_n is the $n \times n$ Hankel matrix defined by $h_1, h_2, \dots, h_{2n-1}$ and $h_{2n} = 1$. The Schur complement of H_n in H is

$$S_n = -\hat{H} H_n^{-1} \hat{H}.$$

Therefore, from Proposition 10.4 it follows that

$$H_n^{-1} = (-1)^k \hat{H}^{-1} J B(r_k(x), r_{k+1}(x)) J \hat{H}^{-1}, \tag{10.3}$$

where $r_k(x)$ and $r_{k+1}(x)$ are the polynomials generated by the Euclidean scheme (1.5.6)-(1.5.7) applied to $r_0(x) = x^{2n}$, $r_1(x) = \sum_{i=0}^{2n} h_i x^{2n-i}$, such that $\deg r_k(x) = n$.

Thus, in the view of the above formula, the solution of a Hankel linear system can be computed by performing convolutions and by means of a single call to the EMGCD (Extended Middle GCD) algorithm of [BGY] for the computation of the polynomials $r_k(x)$ and $r_{k+1}(x)$, whereas the algorithm of [BGY] for solving Hankel systems requires at least two calls to EMGCD.

Proposition 10.4 is also applied in [Gem91] to computing the inertia of a Hankel matrix in $O_A(n \log^3 n)$ ops.

11. Operators Associated with Dense Structured Matrices. Extension of Toeplitz Matrix Computations.

In this section and in section 12, we will follow [P89a] and [P92c] in order to extend effective algorithms for Toeplitz matrix computations to computations with more general classes of dense structured matrices. In this way we will unify efficient computations with various dense structured matrices (with further extensions to computations with polynomials, rational functions and general matrices). We will introduce some techniques for dealing with the class of matrices that are each representable as the sum of a pair of Toeplitz and Hankel matrices or of a pair of matrices of Toeplitz and Hankel types.

We will first recall the representation of an $m \times n$ matrix $A = GH^T$ of a low rank r by means of its generator G, H of length r with using only $r(m+n)$ (rather than mn) words of storage space. Furthermore, to compute the vector $A\mathbf{v}$, for a given pair of matrices G, H and a vector $\mathbf{v}$, we only need $r(2n-1)+m(2r-1) < 2r(m+n)$ ops, rather than $(2n-1)m$ ops. If, in addition, we are given a generator K, L of length s for an $n \times p$ matrix B, then we may compute [in less than $2(m+n)rs$ ops] the generator GH^TK, L of length s for AB and [in less than $2(n+p)rs$ ops] the generator $G, (H^TKL)^T$ of length r for AB [to be compared with $mp(2n-1)$ ops in the straightforward evaluation of AB for given entries of A and B].

Given a dense structured matrix A, we look for an operator F that transforms A into a low rank matrix $F(A)$ such that we could easily recover A from its image $F(A)$, and then we may take all the advantages of operating with low rank matrices, even though the structured input matrix may have the full rank. The operators F that shift and scale the entries of the matrices turn out to be appropriate tools for defining the matrices of Toeplitz, Hankel, Vandermonde and Cauchy (Hilbert) types, which generalize Toeplitz, Hankel, Vandermonde and Cauchy (generalized Hilbert) matrices.

Next, we will specify this approach, due to [KVM], [KKM] (also see [B83], [K87], [AG89], [BP93], [GO], [GO1]), that is, we will specify such operators F and certain operations with the F-images of structured matrices and will demonstrate the resulting unification and substantial simplification of the computations with structured matrices. We will stay with the operators F of scaling and displacement in this section and will follow [P89a], to show a more general class of operators, in the next section.

Definition 11.1. Let $F : \mathbf{F}_{m,n} \to \mathbf{F}_{m,n}$ be an operator, let $A \in \mathbf{F}_{m,n}$ and let $G \in \mathbf{F}_{m,d}$ and $H \in \mathbf{F}_{n,d}$ denote two matrices such that $F(A) = GH^T$. Then $r = \text{rank}\ (F(A))$, the rank of the matrix $F(A)$, is called the *F-rank* of A, and the pair of the matrices G and H is called an *F-generator* of A of length d.

Observe that if F is the identity operator, the F-rank of A coincides with the rank of A and an F-generator of A coincides with a generator of A.

For a fixed operator F and a fixed matrix A, computing an F-generator of the minimum length r for A can be reduced to Problems 2.7c ($LSP \cdot FACTORS$) or 3.3 (SVD) for the matrix $F(A)$; if we are given a generator of length $R \geq r$ for $F(A)$ (with R much smaller than n), it is most effective to apply the reduction to Problem 2.11b ($G \cdot COMPRESS$).

Hereafter, we will use the matrices J and Z of Definition 4.1 and $L(\mathbf{v})$ of Definition 5.2. Given an $m \times m$ matrix X and an $n \times n$ matrix Y, define the operator $F_{(X,Y)}(A) = A - XAY$, and consider the special cases of the displacement operators of *Hankel-Toeplitz type*:

$$F(A) = F_{(Z,Z^T)}(A) = A - ZAZ^T, \tag{11.1}$$

$$F(A) = F_{(Z^T,Z)}(A) = A - Z^T AZ, \tag{11.2}$$

$$F(A) = F_{(Z,Z)}(A) = A - ZAZ, \tag{11.3}$$

$$F(A) = F_{(Z^T,Z^T)}(A) = A - Z^T AZ^T. \tag{11.4}$$

We will also denote the operators of (11.1) and (11.2) as

$$F_{(Z,Z^T)}(A) = F_+(A),$$
$$F_{(Z^T,Z)}(A) = F_-(A).$$

At the end of this section (Examples 11.1–11.4), we will show some other important operators of this kind.

Due to the shifting properties of multiplications by Z and Z^T, the operators F of (11.1) and (11.2) turn into zero all the entries of a Toeplitz matrix A, except for the first row and column under (11.1) and for the last row and column under (11.2), which are invariant in the transition from A to $F(A)$ (see Table 11.1). This defines F-generators of A of lengths at most

2 in such cases and similarly in the cases where A is a Hankel matrix and F is an operator of (11.3) or (11.4). The lengths only increase to $p+q$ for $p \times q$ block matrices with Toeplitz or Hankel blocks.

The matrix equations $JJ = I$, $JZJ = Z^T$ imply the following simple relations:

Proposition 11.1. *$F(A^T) = [F(A)]^T$ for the operators F of (11.1) and (11.2); $F_{(Z,Z)}(A^T) = [F_{(Z^T,Z^T)}(A)]^T$. Moreover, the operators (11.1) and (11.2) are related to the operators (11.3) and (11.4) by the following equations:*

$$F_{(Z^T,Z)}(A) = JF_{(Z,Z)}(JA) = F_{(Z^T,Z^T)}(AJ)J,$$
$$F_{(Z,Z^T)}(A) = F_{(Z,Z)}(AJ)J = JF_{(Z^T,Z^T)}(JA).$$

The next two (rather simple but fundamental) theorems (due to or implicit in [KKM]) relate the F-ranks of A and A^{-1} to each other and express the matrix A via its F-generator in the case of the operators F of (11.1)-(11.4):

Theorem 11.1. *For any nonsingular matrix A,*

$$\text{rank } F_{(Z,Z^T)}(A) = \text{rank } F_{(Z^T,Z)}(A^{-1}),$$
$$\text{rank } F_{(Z,Z)}(A) = \text{rank } F_{(Z,Z)}(A^{-1}),$$
$$\text{rank } F_{(Z^T,Z^T)}(A) = \text{rank } F_{(Z^T,Z^T)}(A^{-1}).$$

Proof. rank $F_{(Z,Z^T)}(A) =$ rank $(A - ZAZ^T) =$ rank $((A-ZAZ^T)A^{-1})$ $=$ rank $(I - ZAZ^TA^{-1})$, rank $F_{(Z^T,Z)}(A^{-1}) =$ rank $(A^{-1} - Z^TA^{-1}Z) =$ rank $(A(A^{-1} - Z^TA^{-1}Z)) =$ rank $(I - AZ^TA^{-1}Z)$. This implies that rank $F_{(Z,Z^T)}(A) =$ rank $F_{(Z^T,Z)}(A^{-1})$ since rank $(I_n - WZ) =$ rank $(I_n - ZW) = 1 +$ rank $(I_{n-1} - W_{1,n-1})$ for any $n \times n$ matrix W and its $(n-1) \times (n-1)$ northeastern submatrix $W_{1,n-1}$ (in our case, let $W = AZ^TA^{-1}$). Similarly, the two other equations of Theorem 11.1 are proved. ∎

Theorem 11.2. *For the operators F of (11.1)-(11.4),*

$$F(A) = GH^T = \sum_{i=1}^{d} \mathbf{g}_i \mathbf{h}_i^T, \; G = [\mathbf{g}_1, \ldots, \mathbf{g}_d], \; H = [\mathbf{h}_1, \ldots, \mathbf{h}_d]$$

if and only if

$$A = \sum_{i=1}^{d} L(\mathbf{g}_i)L^T(\mathbf{h}_i), \text{ under (11.1)},$$

$$A = \sum_{i=1}^{d} L^T(J\mathbf{g}_i)L(J\mathbf{h}_i), \text{ under (11.2)},$$

$$A = \sum_{i=1}^{d} L(\mathbf{g}_i)L^T(J\mathbf{h}_i)J, \text{ under (11.3)},$$

$$A = \sum_{i=1}^{d} JL(J\mathbf{g}_i)L^T(\mathbf{h}_i), \text{ under (11.4)}.$$

Proof. Verify the result by direct calculations for $d = 1$; then immediately extend to any d. For an alternative proof, see exercise 38. ∎

Theorem 11.2 enables us to recover a matrix A from its F-image for the operators F of (11.1)-(11.4). Therefore, in the Hankel-Toeplitz case, we may shift to operating with $F(A)$, rather than with A, and thus save computational resources (of time and space); we will specify the details later on, when we also introduce the operators of Cauchy (Hilbert) and Vandermonde types. Let us, however, observe an immediate application of Theorem 11.2 to approximating an $n \times n$ matrix W having F-rank r by a matrix having F-rank at most d for a fixed $d < r$ and for an operator F of (11.1)-(11.4). To arrive at such an approximation, we may first compute the matrix $F(W)$. Then we compute its SVD and a rank d approximation F_d to $F(W)$, by relying on Fact 3.1 with W and W_d replaced by $F(W)$ and F_d, respectively. Finally, we compute the desired approximation to W by a matrix of F-rank d, by applying Theorem 11.2 for F_d replacing $F(A)$. We may modify this approach by relying on the algorithms of [Chan] or [PS] instead of computing and truncating the SVD. In all the 3 modifications (based on the SVD, [Chan] and [PS]), the cost of these computations is low for smaller r (we leave the estimates to the reader). However, compared with the evaluation of $W_d = U\Sigma_d V^H$ of Fact 3.1, the latter algorithm needs two more stages for the back and forth transition from W to $F(W)$ and from F_d to $W_d(F)$. This requires us to increase the error bound of Fact 3.1 as follows:

Lemma 11.1 ([P93]). *Let W be an $n \times n$ matrix, F any operator of (11.1)-(11.4),* rank $F(W) = r$, $0 < d \leq r$, *$W_d(F)$ an approximation to W having F-rank d and obtained via truncating the SVD of $F(W)$, and Y*

any $n \times n$ matrix having F-rank d. Then $||W - W_d(F)||_2 \leq (1 + 2n(r - d))||W - Y||_2$.

We may immediately extend this approach to approximating the structured matrices of Vandermonde and Cauchy (Hilbert) types by relying on Facts 11.2 and 11.5, and we may also compress an F-generator of length ℓ for a structured matrix having F-rank $r < \ell$, simply by extending our solution of Problem 2.11b ($G \cdot COMPRESS$).

The next result has a corollary (Corollary 11.1) that we will use in chapter 4 (compare [KKM] and [BA]):

Proposition 11.2. *Let $\mathbf{x}^T = [x_0, \dots, x_{n-1}]$, $\mathbf{y}^T = [y_0, \dots, y_{n-1}]$ be a pair of n-dimensional vectors, let $\mathbf{e}^{(k)}$, $k = 0, \dots, n-1$, denote the k-th unit coordinate vector of Definition 1.2, J the reversion matrix of Definition 4.1, $\mathbf{v}^T = [0, x_{n-1}, \dots, x_1]$, $\mathbf{w}^T = [0, y_{n-1}, \dots, y_1]$, $\mathbf{p} = \text{diag}(0, 1, \dots, 1)L^T(\mathbf{y})\mathbf{x}$, $\mathbf{q} = L^T(\mathbf{x})\mathbf{y}$. Then $L(\mathbf{x})L^T(\mathbf{y}) = L(\mathbf{p}) + L^T(\mathbf{q}) - L^T(\mathbf{v})L(\mathbf{w})$, $L^T(\mathbf{x})L(\mathbf{y}) = L^T(\mathbf{p}) + L(\mathbf{q}) - L(\mathbf{v})L^T(\mathbf{w})$.*

Proof. To prove the two latter matrix identities, first verify their respective images in the application of the operators F of (11.2) and (11.1) to both sides of these identities and then apply Theorem 11.2. ∎

Corollary 11.1. *For any matrix A, its $F_{(Z,Z^T)}$-generator of length at most $r + 2$ can be computed by using $O(rn \log n)$ ops provided that an $F_{(Z^T,Z)}$-generator of A of length r is available, whereas an $F_{(Z^T,Z)}$-generator of length at most $s + 2$ can be computed by using $O(sn \log n)$ ops provided that an $F_{(Z,Z^T)}$-generator of A of length s is available.*

Next, we will follow [HR], [GKK] and [GKKL] and, for a pair of n-dimensional vectors $\mathbf{s}$ and $\mathbf{t}$, will define the operators of *Cauchy (Hilbert) type*,

$$F(A) = \widetilde{F}_{\mathbf{s},\mathbf{t}}(A) = D(\mathbf{s})A - AD(\mathbf{t}), \tag{11.5}$$

where $D(\mathbf{v}) = \text{diag}(v_0, v_1, \dots, v_{n-1})$ if $\mathbf{v} = [v_0, v_1, \dots, v_{n-1}]^T$.

An operator of the Cauchy (Hilbert) type is nonsingular (that is, represented by a nonsingular matrix) if and only if $s_i \neq t_j$, $i, j = 0, \dots, n-1$.

Let us apply such an operator to a Cauchy (generalized Hilbert) matrix

$$C = [c_{ij}], \quad 1/c_{ij} = s_i - t_j \quad \text{for all } i \text{ and } j \tag{11.6}$$

(compare Definition 4.4). Then

$$\widetilde{F}_{\mathbf{s},\mathbf{t}}(C) = \mathbf{e}\mathbf{e}^T, \tag{11.7}$$

where $\mathbf{e} = [1, \ldots, 1]^T$ (see Definition 1.2).

Definition 11.2. An operator $\hat{F}$ is called *dual* to an operator F if $\hat{F}(A^T) = -F(A)^T$ and if, for every nonsingular matrix A, $\hat{F}(A^{-1}) = -A^{-1}F(A)A^{-1}$.

If $\hat{F}$ is dual to F, then also F is dual to $\hat{F}$, rank $F(A) =$ rank $\hat{F}(A^T)$ for any matrix A, and furthermore, rank $\hat{F}(A^{-1}) =$ rank $F(A^{-T}) =$ rank $F(A)$, $F(A^{-T}) = A^{-T}HG^TA^{-T}$, provided that A is a nonsingular matrix, $F(A) = GH^T$, and A^{-T} denotes $(A^{-1})^T$.

Theorem 11.2 can be extended to the case of the operators of the Cauchy (Hilbert) type of (11.5) by using the two following simple facts:

Fact 11.1. *The operators $\widetilde{F}_{\mathbf{s},\mathbf{t}}$ of (11.5) and $\widetilde{F}_{\mathbf{t},\mathbf{s}}$ are dual to each other for any fixed pair of vectors* $\mathbf{s}$ *and* $\mathbf{t}$ *of the same dimension.*

Fact 11.2. *Let $A = [a_{ij}]$, $F = \widetilde{F}_{\mathbf{s},\mathbf{t}}(A) = [f_{ij}]$. Then $f_{ij} = (s_i - t_j)a_{ij}$ for all i and j.*

Combining the equation (11.7) and Facts 11.1 and 11.2 implies the following results:

Corollary 11.2. *Let A denote the $n \times n$ nonsingular Cauchy (generalized Hilbert) matrix of (11.6), $B = [b_{ij}] = A^{-1}$, $\widetilde{F}_{\mathbf{t},\mathbf{s}}(B) = [f_{ij}]$. Then the vector $B\mathbf{e}$ has no zero components, the matrix $\widetilde{F}_{\mathbf{t},\mathbf{s}}(B) = -B\mathbf{e}\mathbf{e}^TB$ has rank 1, and $b_{ij} = f_{ij}/(t_i - s_j)$ for all i and j; furthermore, every fixed pair of a row and a column of the matrix B can be recovered from its first row and column by using $O(n)$ ops.*

Next, assume that the vectors $\mathbf{s}$ and/or $\mathbf{t}$ have no zero components, which is no loss of generality (in a field of a large cardinality), since $\widetilde{F}_{\mathbf{s},\mathbf{t}}(A) = \widetilde{F}_{\mathbf{s}+a\mathbf{e},\,\mathbf{t}+a\mathbf{e}}(A)$, for any scalar a, for $\mathbf{e} = (1, \ldots, 1)^T$, and for any matrix A. Apply *scaling* to reduce the operator $\widetilde{F}_{\mathbf{s},\mathbf{t}}$ to the format

$$F_{\mathbf{s},\mathbf{t}} = F_{(K,L)}, \ F_{(K,L)}(A) = A - KAL \tag{11.8}$$

for appropriate matrices K and L. Specifically, consider the operators

$$F_{(D^{-1}(\mathbf{s}),D(\mathbf{t}))} \text{ and } F_{(D(\mathbf{s}),D^{-1}(\mathbf{t}))},$$

which have the format (11.8) and are defined by scaling the equation (11.5) and the respective matrices of the F-generators of A by the factors $D^{-1}(\mathbf{s})$ and $D^{-1}(\mathbf{t})$. Such scaling does not change the F-rank of A.

Finally, for a vector $\mathbf{v} = [v_j]$, consider the four following operators of *Vandermonde type* (compare [HR], [GKK] and [GKKL]):

$$F(A) = \widetilde{F}_{\mathbf{v},Z}(A) = D(\mathbf{v})A - AZ, \tag{11.9}$$

$$F(A) = \widetilde{F}_{\mathbf{v},Z^T}(A) = D(\mathbf{v})A - AZ^T, \tag{11.10}$$

$$F(A) = \widetilde{F}_{Z,\mathbf{v}}(A) = AD(\mathbf{v}) - ZA, \tag{11.11}$$

$$F(A) = \widetilde{F}_{Z^T,\mathbf{v}}(A) = AD(\mathbf{v}) - Z^T A. \tag{11.12}$$

Hereafter, $\mathbf{v}^{-1} = [1/v_i]$ provided that $\mathbf{v} = [v_i]$ and $v_i \neq 0$ for all i. Let $A = V$ be the $(n+1) \times (n+1)$ Vandermonde matrix of Definition 4.2 (for $m = n$) associated to the vector $\mathbf{v}$. Then we have:

$$\widetilde{F}_{\mathbf{v},Z}(A) = D^{n-1}(\mathbf{v})\mathbf{e}\mathbf{e}^{(n)T}; \tag{11.13}$$

if, in addition, $v_i \neq 0$ for all i, then

$$\widetilde{F}_{\mathbf{v}^{-1},Z^T}(A) = D^{-1}(\mathbf{v})\mathbf{e}\mathbf{e}^{(0)T}. \tag{11.14}$$

Fact 11.1 is immediately extended as follows:

Fact 11.3. *The two following equations hold (for any fixed vector* $\mathbf{v}$*):* $\widetilde{F}_{Z,\mathbf{v}}(A^{-1}) = A^{-1}\widetilde{F}_{\mathbf{v},Z}(A)A^{-1}$, *if* A *is a nonsingular matrix, and* $\widetilde{F}_{\mathbf{v},Z^T}(A^T)$ $= \widetilde{F}_{Z,\mathbf{v}}(A)^T$ *for any matrix* A.

In particular, (11.13), (11.14) and the first equation of Fact 11.3 imply that

$$\widetilde{F}_{Z,\mathbf{v}}(A^{-1}) = A^{-1}D^n(\mathbf{v})\mathbf{e}\mathbf{e}^{(n)T}A^{-1}, \tag{11.15}$$

and similarly,

$$\widetilde{F}_{Z^T,\mathbf{v}^{-1}}(A^{-1}) = A^{-1}D^{-1}(\mathbf{v})\mathbf{e}\mathbf{e}^{(0)T}A^{-1} \tag{11.16}$$

for any nonsingular Vandermonde matrix $A = V(\mathbf{v})$.

Definition 11.3. An $m \times n$ matrix that has F-rank bounded from above by a constant independent of m and n is called *Toeplitz-like* if F is one of the two operators defined in (11.1) or (11.2), *Hankel-like* if F is one of the two operators defined in (11.3) or (11.4), *Cauchy (Hilbert)-like* if F is the operator defined in (11.5), and *Vandermonde-like* if F is any of the operators defined in (11.9)-(11.12).

Observe that a Loewner matrix is Cauchy (Hilbert)-like (see exercise 28).

Next, we will extend Fact 11.2.

Fact 11.4. *Let* $A = [a_{ij}]$, $F(A) = [f_{ij}]$, $\mathbf{v} = [v_i]$. *Then*

$$\begin{aligned} f_{ij} &= v_i a_{ij} - a_{i,j+1}, &\quad \text{for } F = \widetilde{F}_{\mathbf{v},Z}, \\ f_{ij} &= v_i a_{ij} - a_{i,j-1}, &\quad \text{for } F = \widetilde{F}_{\mathbf{v},Z^T}, \\ f_{ij} &= a_{i-1,j} - v_j a_{ij}, &\quad \text{for } F = \widetilde{F}_{Z,\mathbf{v}}, \\ f_{ij} &= a_{i+1,j} - v_j a_{ij}, &\quad \text{for } F = \widetilde{F}_{Z^T,\mathbf{v}}, \end{aligned}$$

where $a_{pq} = 0$ *if* p *and/or* q *are out of the range.*

Fact 11.5 ([GKKL]). *Let* $F = \widetilde{F}_{Z,\mathbf{v}}$ *denote the operator of (11.11), let* $\widetilde{F}_{Z,\mathbf{v}}(A) = -GH^T$, $G = [\mathbf{g}_1, \ldots, \mathbf{g}_r]$, $H = [\mathbf{h}_1, \ldots, \mathbf{h}_r]$; *and let* $V = [v_i^{-j}]$. *Then*

$$A = \sum_{i=1}^{r} L(\mathbf{g}_i) D^{-1}(\mathbf{v}) V^T D(\mathbf{h}_i).$$

There also exist expressions similar to the latter one [defining the reversion of the operator F of (11.11)] for the reversion of the operators F of (11.9), (11.10) and (11.12); this extension immediately follows from Fact 11.4 and enables us to recover matrices A from their images $F(A)$ for such operators F. We will omit these explicit formulae, as well as the respective extension of Corollary 11.2. Note that the second equation of Fact 11.3 reduces the reversion of the operators of (11.9), (11.10) to the reversion of the operators of (11.12), (11.11), respectively (and vice versa).

Finally, the operators of (11.9)-(11.12) can be immediately reduced to the form (11.8) by means of their scaling by the matrix $D^{-1}(\mathbf{v})$.

The reader is encouraged to look for other useful operators F associated with structured matrices. We will next recall some examples of such operators (also see exercises 35 and 39).

Example 11.1 ([CK]). Associate with block Toeplitz and block Hankel matrices the operators F of (11.1)–(11.4) where Z is replaced by the similar matrices with the one-entries replaced by the blocks I of appropriate sizes.

Example 11.2 ([GO], [GO1], see also [B83], [AG89], [AG89a], [Gad]). Associate with Toeplitz matrices A the operator

$$F(A) = A - Z_f A Z_{1/f}^T \tag{11.17}$$

where $f \neq 0$ and $Z_g = Z_g(\mathbf{e}^{(1)})$ is the $n \times n$ *unit* g*-circulant matrix* whose first row is $g\mathbf{e}^{(n-1)T}$ and which otherwise coincides with $Z = Z_0$ [(11.17) is (11.1) with $Z = Z_0$ replaced by Z_f and with $Z^T = Z_0^T$ replaced by $Z_{1/f}^T$]. Let $Z_{f,lr}(\mathbf{u}^T)$ and $Z_{f,lc}(\mathbf{v})$ in the next two propositions denote the pair of f-circulant matrices with the last row $\mathbf{u}^T$ or the last column $\mathbf{v}$, respectively. Then it is not hard to verify the following properties (by using commutativity of pairwise multiplication of the g-circulant matrices and the matrix equation $Z_f Z_{1/f}^T = I$ for $f \neq 0$).

Proposition 11.3 ([GO], [GO1]). *Let e and f be two nonzero constants, $e \neq f$. Let A be an $n \times n$ matrix,* rank $F(A) \leq d$, $F(A) = \sum_{i=1}^d \mathbf{g}_i \mathbf{h}_i^T$ *for the operator F of (11.17). Then*

a) $F(A) = 0$ *if and only if A is an f-circulant matrix;*

b) $\sum_{i=1}^{d} Z_f(\mathbf{g}_i) Z_{1/f}^T(\mathbf{h}_i) = O,$

c) $A = Z_{f,\ell r}((\mathbf{e}^{(n-1)})^T) + \frac{f}{f-e}\sum_{i=1}^{d} Z_e(\mathbf{g}_i) Z_{1/f}^T(\mathbf{h}_i)$, *if* $f \neq 0$;

d) $A = Z_{f,\ell c}(A\mathbf{e}^{(n-1)}) + \frac{e}{e-f}\sum_{i=1}^{d} Z_f(\mathbf{g}_i) Z_{1/e}^T(\mathbf{h}_i)$ *if* $e \neq 0$.

In [B83], [AG89], [AG89a], [Gad], Proposition 11.3 a)–c) appeared in the special case where $f = 1$, $e = 0$, $Z_e(\mathbf{v}) = L(\mathbf{v})$, for any vector $\mathbf{v}$.

Now combine Proposition 11.3 c) with the matrix equation $F(A^{-1}) = -A^{-1} F(A) Z_f A^{-1} Z_{1/f}^T$, which follows from (11.17) since $Z_f Z_{1/f}^T = I$, and deduce

Proposition 11.4 ([GO], [GO1]). *Under the assumptions of Proposition 11.3, let $f \neq 0$, A be a nonsingular matrix, $\mathbf{q}_i = A^{-1}\mathbf{g}_i$,*

$$\mathbf{r}_i^T = -\mathbf{h}_i^T (Z_f A Z_{1/f}^T)^{-1}\ ,\ i = 1, \ldots, d,$$

$\mathbf{x}^T = (\mathbf{e}^{(n-1)})^T A^{-1}$, *$r$ be a constant. Then*

$$A^{-1} = Z_{f,\ell r}(\mathbf{x}^T) + \frac{f}{f-e}\sum_{i=1}^{d} Z_e(\mathbf{q}_i)\, Z_{1/f}^T(\mathbf{r}_i)\ .$$

Furthermore, if A is a Toeplitz matrix, $(A)_{i,j} = a_{i-j}$, $i, j = 0, 1, \ldots, n-1$,

$$\mathbf{u} = A^{-1}\mathbf{e}^{(0)}\ ,\quad \mathbf{z} = A^{-1}[r, a_1 - (1/f)a_{1-n}, \ldots, a_{n-1} - (1/f)a_{-1}]^T\ ,$$

then

$$A^{-1} = \frac{f}{f-e}\left(Z_e(\mathbf{u})Z_f(\mathbf{z}) - Z_e(\mathbf{z} - \frac{f-e}{f}\mathbf{e}^{(0)})Z_f(\mathbf{u})\right)\ .$$

Moreover, if, in addition, $\mathbf{v} = A^{-1}\mathbf{e}^{(n-1)}$ *and* $u_0 \neq 0$*, then*

$$u_0 A^{-1} = Z_f(\mathbf{u}) Z_{1/f}(ZJ\mathbf{v} + x_0\mathbf{e}^{(0)}) - Z_f(Z\mathbf{v} + f x_0 \mathbf{e}^{(0)}) Z_{1/f}(Z_{1/f}J\mathbf{u}) \,.$$

Similar results, for Toeplitz matrices and for $e = 0$, $f = 1$, have been obtained in [B83]; whereas [AG89], [AG89a], [Gad] contain similar formulae for A^{-1} in the case where A is a Hermitian Toeplitz matrix.

The extension of Theorem 5.2 given by Proposition 11.4 turns out to be a slightly more effective basis for the actual evaluation of the vectors $A^{-1}\mathbf{v}$ (after preprocessing A) than Theorem 5.2 itself, due to the lower complexity of multiplying an f-circulant matrix by a vector.

Example 11.3 ([B83], [P92b]). Consider the operators $F_+ = F_{(Z,Z^T)}$ of (11.1) and $F_- = F_{(Z^T,Z)}$ of (11.2) and define two similar operators:

$$F^+(A) = AZ - ZA, \tag{11.18}$$

$$F^-(A) = AZ^T - Z^T A = -(F^+(A^T))^T. \tag{11.19}$$

The operators F_+, F_-, F^+ and F^- are closely related to each other.

Fact 11.6 ([P92b]). *Let A be an $n \times n$ matrix. Then*

$$\begin{aligned}
&F^+(A)Z^T = F_+(A) - A\mathbf{e}^{(0)}\mathbf{e}^{(0)T},\\
&Z^T F^+(A) = \mathbf{e}^{(n-1)}\mathbf{e}^{(n-1)T}A - F_-(A),\\
&F^-(A)Z = F_-(A) - A\mathbf{e}^{(n-1)}\mathbf{e}^{(n-1)T},\\
&ZF^-(A) = \mathbf{e}^{(0)}\mathbf{e}^{(0)T}A - F_+(A),\\
&F_+(A)Z = F^+(A) + ZA\mathbf{e}^{(n-1)}\mathbf{e}^{(n-1)T},\\
&Z^T F_+(A) = \mathbf{e}^{(n-1)}\mathbf{e}^{(n-1)T}AZ^T - F^-(A),\\
&F_-(A)Z^T = F^-(A) + Z^T A\mathbf{e}^{(0)}\mathbf{e}^{(0)T},\\
&ZF_-(A) = \mathbf{e}^{(0)}\mathbf{e}^{(0)T}AZ - F^+(A),\\
&F^-(A) = -(F^+(A^T))^T,\\
&F^-(A) = JF^+(JAJ)J.
\end{aligned}$$

Proof. The last two identities of Fact 11.6 are immediate; the remaining identities are implied by the matrix equations (11.1), (11.2), (11.18), (11.19) and by the following easily verified matrix equations:

$$Z^T Z = I - \mathbf{e}^{(n-1)}\mathbf{e}^{(n-1)T}, \quad ZZ^T = I - \mathbf{e}^{(0)}\mathbf{e}^{(0)T}. \tag{11.20}$$

■

Fact 11.6 implies, in particular, that the F-ranks of Toeplitz-like matrices are $O(1)$ also for $F = F^+$ and $F = F^-$ (compare exercise 36).

Theorem 11.2 can be extended to the operators F^+ and F^- as follows:

Theorem 11.3 ([B83]). *a) For F denoting the operator F^+ of (11.18), we have*

$$F(A) = GH^T = \sum_{i=1}^{d} \mathbf{g}_i \mathbf{h}_i^T, \quad G = [\mathbf{g}_1, \dots, \mathbf{g}_d], \quad H = [\mathbf{h}_1, \dots, \mathbf{h}_d]$$

if and only if simultaneously

$$\sum_{i=1}^{d} \sum_{j=0}^{n-1} g_j^{(i)} (Z^T)^j \mathbf{h}_i = \sum_{i=1}^{d} \sum_{j=0}^{n-1} h_j^{(i)} Z^{n-j-1} \mathbf{g}_i = \mathbf{0},$$

where $\mathbf{g}_i = [g_j^{(i)}]$, $\mathbf{h}_i = [h_j^{(i)}]$ and either or both of the two following matrix equations holds:

$$A = L(A\mathbf{e}^{(0)}) + \sum_{i=1}^{d} L(\mathbf{g}_i) L^T(Z\mathbf{h}_i),$$

$$A = L(JA^T\mathbf{e}^{(n-1)}) - \sum_{i=1}^{d} L^T(ZJ\mathbf{g}_i) L(J\mathbf{h}_i).$$

b) Similarly, for F denoting the operator F^- of (11.9), we have $F(A) = GH^T$ if and only if simultaneously

$$\sum_{i=1}^{d} \sum_{j=0}^{n-1} h_j^{(i)} (Z^T)^j \mathbf{g}_i = \sum_{i=1}^{d} \sum_{j=0}^{n-1} g_j^{(i)} Z^{n-j-1} \mathbf{h}_i = \mathbf{0}$$

and any of the two following matrix equations holds:

$$A = L^T(A^T\mathbf{e}^{(0)}) - \sum_{i=1}^{d} L(Z\mathbf{g}_i) L^T(\mathbf{h}_i),$$

$$A = L^T(JA\mathbf{e}^{(n-1)}) + \sum_{i=1}^{d} L^T(J\mathbf{g}_i) L(ZJ\mathbf{h}_i).$$

Proof. Apply the techniques of [B83], that is, rewrite $F^+(A) = \sum_{i=1}^d \mathbf{g}_i \mathbf{h}_i^T$ in matrix form, where the entries of the matrix $A = [\mathbf{a}_0, \dots, \mathbf{a}_{n-1}]$ have been arranged column-wise, to form the vector $[\mathbf{a}_0^T, \dots, \mathbf{a}_{n-1}^T]^T$. Obtain a block bidiagonal system of linear equations with the diagonal blocks $-Z$, the superdiagonal blocks I, and the right-hand side vector $\sum_{i=1}^{d} \mathbf{h}_i \otimes \mathbf{g}_i$, where $\mathbf{u} \otimes \mathbf{v}$ denotes the vector $(u_i \mathbf{v})$, which is the *tensor product* (the *Kronecker product*) of $\mathbf{u}$ and $\mathbf{v}$ (see exercise 49). Fix the first column $\mathbf{a}_0$ and express the remaining columns by using back substitution. This proves the first matrix equation of part a). Make similar substitutions in the last block equation, recall that $Z^n = O$ and obtain the second vector equation of part a). Verify the first vector equation of part a) by simple inspection. To yield the extensions to the other equations of parts a) and b), combine together the relations $(F^-(A))^T = -F^+(A^T)$, $F^-(A) = JF^+(JAJ)J$ of Fact 11.6. (For further details, see [B83] or [BP93]). ∎

Theorem 11.1 can be easily extended to F^+ and F^-. Indeed, pre- and postmultiply the matrix equations (11.18) and (11.19) by A^{-1} and obtain that

$$F(A^{-1}) = -A^{-1}F(A)A^{-1}, \quad \text{for} \quad F = F^+ \text{ and } F = F^-, \tag{11.21}$$

and consequently,

$$\text{rank } F(A) = \text{rank } F(A^{-1}), \quad \text{for} \quad F = F^+, \ F = F^-,$$

and for any nonsingular matrix A.

Note that F^+ and F^- are self-dual operators. Moreover, combining Theorem 11.3 and the equation (11.21) immediately yields new inversion formulae for Toeplitz-like matrices.

Another interesting property of the operators $F = F^+$ of (11.18) and $F = F^-$ of (11.19), obviously shared by the operator $F = F^{\pm}(A)$ of (11.23) below, is that an F-generator of the product of two matrices can be easily expressed in terms of the F-generators of these two matrices:

$$F(AB) = AF(B) + F(A)B, \quad \text{for } F = F^+, \ F = F^-, \text{ and } F = F^{\pm}. \tag{11.22}$$

The next three examples and exercises 35 and 39 show some operators that can be appropriately associated with matrices, each of which is represented as the sum of a Toeplitz-like matrix and a Hankel-like matrix.

Example 11.4 ([B83]). Consider the self dual operator

$$F^{\pm}(A) = A(Z + Z^T) - (Z + Z^T)A = F^+(A) + F_-(A), \tag{11.23}$$

introduced and analyzed in [B83] (compare Examples 11.5 and 11.6 and exercises 35 and 39). Observe that (11.21) and (11.22) hold for $F = F^{\pm}$ and that rank $F^{\pm}(A) \leq 4$ for any Toeplitz+Hankel matrix A, that is, for any sum of Toeplitz and Hankel matrices (see also exercises 35 and 39). Moreover, rank $F^{\pm}(A)$ is bounded from above by a constant for any Toeplitz-like + Hankel-like marix A, and the following extension of Theorem 11.3 holds:

Theorem 11.4 ([B83], [BP93]). *Let $Q_j(x)$ denote the Chebyshev-like polynomials defined by the equations*

$$Q_{j+1}(x) = xQ_j(x) - Q_{j-1}(x),$$
$$Q_1(x) = x, \; Q_0(x) = 1.$$

For the operator $F^{\pm}$ of (11.23) and for any matrix A, we have

$$F^{\pm}(JAJ) = JF^{\pm}(A)J$$
$$F^{\pm}(A^T) = -(F^{\pm}(A))^T.$$

Moreover, $Q_n(Z + Z^T) = O$, and for any matrix A, we have

$$F^{\pm}(A) = GH^T = \sum_{i=1}^{d} \mathbf{g}_i \mathbf{h}_i^T, \; H = [\mathbf{h}_1, \ldots, \mathbf{h}_d]$$

if and only if $\sum_{i=1}^{d} \sum_{j=0}^{n-1} h_j^{(i)} Q_{n-j-1}(Z + Z^T)\mathbf{g}_i = \mathbf{0}$, and if simultaneously any of the four following matrix equations holds:

$$A = \tau(A\mathbf{e}^{(0)}) + \sum_{i=1}^{d} \tau(\mathbf{g}_i)L^T(Z\mathbf{h}_i),$$

$$A = \tau(JA\mathbf{e}^{(n-1)}) + \sum_{i=1}^{d} \tau(J\mathbf{g}_i)L(ZJ\mathbf{h}_i),$$

$$A = \tau(A^T\mathbf{e}^{(0)}) - \sum_{i=1}^{d} L(Z\mathbf{g}_i)\tau(\mathbf{h}_i),$$

$$A = \tau(JA^T\mathbf{e}^{(n-1)}) - \sum_{i=1}^{d} L^T(ZJ\mathbf{g}_i)\tau(J\mathbf{h}_i).$$

Here $U = \tau(\mathbf{u})$ is the $n \times n$ matrix that belongs to the (commutative) matrix algebra τ generated by the matrix $Z + Z^T$ and such that $U\mathbf{e}^{(0)} = \mathbf{u}$, that is, its entries are defined by the equations

$$u_{i,j-1} + u_{i,j+1} = u_{i-1,j} + u_{i+1,j}, \quad i, j = 0, 1, \ldots, n-1;$$
$$u_{i,j} = 0 \text{ if } i \in \{-1, n\}, \text{ or } j \in \{-1, n\}.$$

Note that for $F = F^+$, $F = F^-$ and $F = F^{\pm}$, we need to know an F-generator of A and one of the vectors $A^T\mathbf{e}^{(n-1)}$, $A\mathbf{e}^{(n-1)}$, $A^T\mathbf{e}^{(0)}$ or $A\mathbf{e}^{(0)}$ in order to recover A, whereas in the case of the operators $F = F_+$ and $F = F_-$, we may recover A just from its F-generator (see Theorem 11.2).

The matrix algebra τ, introduced in [BC83a], has some interesting computational properties, strongly related to the sine transform (exercise 22). In particular, we recall the following result from [BC83a], immediately implied by exercise 22:

Fact 11.7. *Given two vectors* $\mathbf{u}$, $\mathbf{v}$, *each having* n *components, the following vectors can be computed at the cost* $O_A(\log n, n)$ *by means of a constant number of sine transforms:* $\tau(\mathbf{u})\mathbf{v}$, $\tau(\mathbf{u})^{-1}\mathbf{v}$ *[if the matrix* $\tau(\mathbf{u})$ *is nonsingular] and* $\tau(\mathbf{u})\tau(\mathbf{v})\mathbf{e}^{(0)}$.

Here is some bibliography on further applications of this matrix algebra: in [BC83a], [B88a], this algebra applies to the analysis of the spectral properties of banded Toeplitz matrices; in [BC87], to computing the tensor rank of band Toeplitz matrices; in [BDB90], [Se], [FS], [DiB93], [DiBFS] and [Se93], to the solution of Toeplitz and block Toeplitz linear systems, by means of the preconditioned conjugate gradient and multigrid methods, in the specific but important case where the matrix is generated by a nonnegative analytic function defined on the unit circle in the complex plane.

Exercise 39 shows that several properties of the operator $F^{\pm}$ are shared by the operator

$$F^{\mp}(A) = A(Z - Z^T) - (Z - Z^T)A.$$

In Table 11.1 we display the structures of the matrices $F(A)$ for the operators $F = F^+$, $F = F^-$, $F = F_+$, $F = F_-$, $F = F^{\pm}$ in the case where A is a Toeplitz matrix (or a Hankel+Toeplitz matrix for $F = F^{\pm}$ and $F = F^{\mp}$).

Example 11.5 ([BDiF]). Consider the operator

$$F_\alpha(A) = AS_\alpha - S_\alpha A,$$

where

$$S_\alpha = \begin{pmatrix} \alpha & 1 & & & O \\ 1 & 0 & \ddots & & \\ & \ddots & \ddots & 1 & \\ & & \ddots & 0 & 1 \\ O & & & 1 & \alpha \end{pmatrix},$$

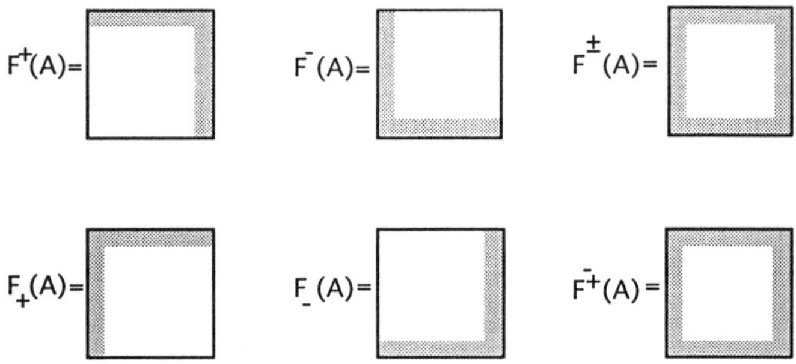

Table 11.1

so that $F_0 = F^{\pm}$. It is easy to prove that rank $F_\alpha(A) \le 4$ for any Toeplitz+ Hankel matrix A. Moreover, if $k = \text{rank } F_1(A)$, there exist matrices τ_i^+, τ_i^-, $i = 1, \ldots, k$, belonging to the algebra generated by S_1 and S_{-1}, respectively, such that $A = \sum_{i=1}^k \tau_i^+ \tau_i^-$. Some inversion formulae based on this property are given in [BDiF]. A similar result can be obtained for the operator F_{-1}. Each of the matrices S_1 and S_{-1} can be diagonalized by means of a similarity transformation associated with a fast discrete transform. In this way, one may multiply each of the matrices τ_i^+, τ_i^- by a vector at the cost of $O(n \log n)$ ops, and in fact, for a large class of matrices, this cost is lower than in the case of the representations of Examples 11.1–11.5.

The paper [DiFZ] analyzes the operator $F_S = AS - SA$, where S is a general matrix in Hessenberg form.

Example 11.6 ([Bo93]). Consider the operator $F^{(\alpha)}(A) = AS^{(\alpha)} - S^{(\alpha)}A$, where

$$S^{(\alpha)} = Z_\alpha + Z_\alpha^T = \begin{pmatrix} 0 & 1 & 0 & \ldots & 0 & \alpha \\ 1 & 0 & 1 & \ddots & \ddots & 0 \\ 0 & 1 & \ddots & \ddots & \ddots & \vdots \\ \vdots & \ddots & \ddots & \ddots & 1 & 0 \\ 0 & \ddots & \ddots & 1 & 0 & 1 \\ \alpha & 0 & \ldots & 0 & 1 & 0 \end{pmatrix}.$$

It is easy to prove that $F^{(\alpha)}(A) \le 4$ for any Hankel+Toeplitz matrix A. Moreover, if rank $F^{(1)}(A) = k$, then there exist $h < k$ ($h \approx k/2$) and matrices C_i^+, C_i^-, $i = 1, \ldots, h$, belonging to the null spaces (algebras) of the operators $F^{(1)}$ and $F^{(2)}$, respectively, such that $A = \sum_{i=1}^{h} C_i^+ C_i^-$. Again, for a large class of matrices A, the multiplication of A by a vector based on this formula is simpler than by using other approaches of this section.

Next, we will present two important applications of the operators F^+, F^-, and $F^{\pm}$ (and of the matrices of the algebra τ) in the form of two algorithms. Both use $O(n^2 \log n)$ ops and have parallel cost $O_A(\log^2 n, n^2/\log n)$. The first of them computes the coefficients of the characteristic polynomial, and the second computes a short F-generator of the inverse of an $n \times n$ Toeplitz-like + Hankel-like matrix. The algorithms are based on the technique used in Algorithm 7.1 and involve FFT's and divisions by $2, 3, \ldots, n$ (compare section 7 of chapter 1). A short generator of the inverse can be computed in fewer ops (see section 13), but the algorithms of this section support the record parallel complexity bounds (see chapter 4).

First recall Newton's iteration

$$X_{i+1} = 2X_i - X_i B X_i \bmod \lambda^{2^{i+1}}, \quad i = 0, 1, \ldots, \tag{11.24}$$

$B = I - \lambda A$, $X_0 = I$. As in section 7, we perform the i-th step of Newton's iteration in the ring of polynomials modulo $\lambda^{2^{i+1}}$ and observe that $X_{i+1} = B^{-1} \bmod \lambda^{2^{i+1}}$. Therefore, by applying (11.21) with $F = F^+$, $F = F^-$, $F = F^{\pm}$, we yield $F(X_i) = -X_i F(B) X_i \bmod \lambda^{2^i}$. Now, by using (11.24), we obtain

$$\begin{aligned} F(X_{i+1}) &= U_{i+1} V_{i+1}^T \bmod \lambda^{2^{i+1}}, \\ U_{i+1} &= -(2X_i - X_i B X_i) U, \\ V_{i+1}^T &= V^T (2X_i - X_i B X_i) \end{aligned} \tag{11.25}$$

where $F(B) = UV^T$ and U and V are $n \times k$ matrices. Thus, in view of Theorems 11.3 and 11.4, the F-generator U_{i+1}, V_{i+1}^T of X_{i+1} can be computed, together with the first column of X_{i+1}, from the F-generator U_i, V_i^T and the first column of X_i by performing relatively few multiplications of triangular Toeplitz matrices and/or of algebra τ matrices by vectors, that is, by multiplying the right sides of the expressions (11.25) by vectors from a fixed set of $2k + 1$ vectors (i.e. by the columns of U_i and V_i and the first column of X_i).

Algorithm 11.1.

Input: an F-generator of length k for an $n \times n$ matrix A (having F-rank at most k), where F denotes F^+, F^-, or $F^{\pm}$, that is, two $n \times k$ matrices U, V such that $F(A) = UV^T$.

Output: the coefficients of the characteristic polynomial of A.

Computation:

1. Set $B = I - \lambda A$, $X_0 = I$ and compute

$$U_{i+1} = -(2X_i - X_i B X_i)U \bmod \lambda^{2^{i+1}},$$
$$V_{i+1}^T = V^T(2X_i - X_i B X_i) \bmod \lambda^{2^{i+1}},$$
$$\mathbf{x}_{i+1} = 2\mathbf{x}_i - X_i B \mathbf{x}_i,$$

where $i = 1, \ldots, h-1$; $h = \lceil \log(n+1) \rceil$; $\mathbf{x}_0 = \mathbf{e}^{(0)}$ for the operators F^+ and $F^\pm$ and $\mathbf{x}_0 = \mathbf{e}^{(n-1)}$ for the operator F^-. Performing matrix multiplications, apply the representations of B and X_i given in Theorems 11.3 and 11.4 and operate with F-generators rather than with the matrices (compare Fact 12.1 of the next section). Note that $U_h V_h^T = F(I + \lambda A + \cdots + \lambda^n A^n) \bmod \lambda^{n+1}$, due to the first matrix equation of (11.25) for $i+1 = h$ and since $X_h = B^{-1} \bmod \lambda^{2^h}$, $B = I - \lambda A$, and $2^h \geq n+1$.

2. By using Theorems 11.3 or 11.4, recover from $U_h V_h$ the diagonal entries of the matrix $X_h \bmod \lambda^{n+1}$ and their sum, which is a polynomial in λ of degree at most n, whose coefficients are the power sums of the eigenvalues of this matrix.

3. Apply the algorithm for Problem 1.4.7 (*POWER·SUMS*) (see Remark 6.1) in order to compute the coefficients of the characteristic polynomial of A.

The most expensive computation in the above algorithm is the performance of $O(k)$ multiplications of matrices given with their F-generators of length $O(k)$ over the ring of polynomials in λ modulo $\lambda^{2^{i+1}}$. This stage can be reduced to $O(k^2)$ multiplications of the bivariate polynomials, and for $k = O(1)$, can be performed in $O(n^2 \log n)$ ops implemented at the computational cost $O_A(\log^2 n, n^2/\log n)$, under parallel RAM models of computation (see chapter 4).

Remark 11.1 (compare section 6 of chapter 4). If the field $\mathbf{F}$ is any field of characteristic 0, then Algorithm 11.1 can be immediately extended to computing rank A at the same asymptotic cost. Indeed, we may apply the solution of Problem 2.10 ($M \cdot RANK$) based on the equation

$$\text{rank } A = \frac{1}{2}\text{rank } S, \quad S = \begin{pmatrix} O & A^H \\ A & O \end{pmatrix}$$

and on the observation that the value $2n - \text{rank } S$ coincides with the lowest degree of a nonzero monomial in the characteristic polynomial of S (see section 2); here we also observe that the F-rank of S is not greater than $4 + 2$ rank $F(A)$ for F denoting F^+, F^- or $F^\pm$ (over any field). This approach leads to effective parallel computation of rank A. On sequential computation of rank A, see Remark 13.1 in section 13.

The technique used in Algorithm 11.1 and the resulting computational cost estimates can be applied to solve a Toeplitz-like + Hankel-like linear system of equations [with no increase of the asymptotic computational cost bounds $O_A(\log^2 n, n^2/\log n)$].

Algorithm 11.2.

Input: a vector $\mathbf{b}$ and an F-generator of length k for an $n \times n$ nonsingular matrix A (having F-rank at most k) for F being F^+, F^- or $F^\pm$.

Output: the vector $\mathbf{x} = A^{-1}\mathbf{b}$.

Computation:

1. Compute the coefficients c_i, $i = 0, \ldots, n-1$, of the characteristic polynomial of A, an F-generator modulo λ^{n+1} of the matrix $\sum_{i=0}^{n} \lambda^i A^i$ and the first column (if $F = F^+$ or $F = F^\pm$) or the last column (if $F = F^-$) of this matrix by using Algorithm 11.1 and by extending it to the first (or last) column computation.
2. Compute $\mathbf{b}_i$ such that $\sum_{i=0}^{n} \lambda^i A^i \mathbf{b} = \sum_{i=0}^{n} \mathbf{b}_i \lambda^i$, $i = 0, \ldots, n$.
3. Compute and output the vector $\mathbf{x} = -(1/c_0) \sum_{i=0}^{n-1} c_{i+1} \mathbf{b}_i$, where $c_n = 1$. (This vector equals $A^{-1}\mathbf{b}$ due to the Cayley-Hamilton theorem.)

The matrix equations (2.9) enable us to extend Algorithm 11.2 to computing a least-squares solution to $A\mathbf{x} = \mathbf{b}$, whereas the inversion formulae [based on (11.21) and Theorems 11.3 and 11.4] immediately enable us to extend Algorithm 11.2 to the evaluation of a short generator of the inverse matrix A^{-1} (see Algorithm 4.2.2). As an exercise, the reader may elaborate the extensions to some other computational problems too.

If F is one of the operators $F_{(Z,Z^T)} = F_+$ or $F_{(Z^T,Z)} = F_-$, we may make transition to F^+ or F^- based on Fact 11.6. An alternative "direct" algorithm of [P92b] for $F = F_+$ and $F = F_-$ supports the same asymptotic cost estimates for the solutions but in a more complicated way.

Remark 11.2. We refer the readers to Remarks 3.2, 5.4, 6.1 and 9.5 and to section 6 of chapter 4 on the extension of the above results to any field of constants.

Remark 11.3. If the input matrix A for Problem 2.5a (*ALL · INVERT*) (of inverting all the leading submatrices of A) is an $n \times n$ strongly nonsingular Toeplitz-like+Hankel-like matrix represented with its short F-generator, then we may follow [P89], p. 1478, and solve this problem (representing the output matrices also by their short F-generators) in almost optimum time of $O(n^2 \log n)$ ops and at the parallel cost $O_A(\log^3 n, n^2/\log n)$. Indeed, for simplicity, assuming that $n = 2^h - 1$, h integer, for $i = 1, 2, \ldots, h$, we recursively invert the $s_i \times s_i$ leading principal submatrices for all s_i of the form $(n+1)(2k+1)/2^i$, $k = 0, 1, \ldots, 2^{i-1} - 1$. For each $i > 1$, we apply Theorem 1.2 in order to reduce the computation essentially to 2^{i-1} concurrent inversions of $[(n+1)/2^i] \times [(n+1)/2^i]$ matrices given with their short F-generators, and this immediately gives us the solution at the claimed cost. Similarly, we solve Problem 2.5b $(ALL \cdot INVERT1)$ within the same cost bounds, and as a special case (due to the results of section 8), we may compute all the remainders in the extended Euclidean scheme, if their degrees have been precomputed.

Remark 11.4 ([KP,a]). In section 6 of chapter 4, we will need to compute $c_H(\lambda)$ and $K(H, \mathbf{v}, 2n-1)$ for an $n \times n$ Hankel or Hankel-like matrix H. We may apply Algorithms 11.1 and 11.2 with $F = F^{\pm}$ or $F = F^{\mp}$. Let us, however, show that in this case we may stay with the operators $F = F^{+}$ or $F = F^{-}$, which will decrease the overhead constants hidden in the "O" of the resulting asymptotic complexity estimates. First suppose that the Toeplitz or Toeplitz-like matrix $T = HJ$ is symmetric. Then

$$T^{2i} = H^{2i}, \quad H^{2i+1} = T^{2i+1}J ,$$

so that column j of $K(H, \mathbf{v}, 2n-1)$ coincides with column i of $K(T^2, \mathbf{v}, n-1)$ for $j = 2i$ and with column i of $K(T^2, H\mathbf{v}, n-1)$ for $j = 2i+1$, $i = 0, 1, \ldots, n-1$. Thus, it suffices to apply Algorithm 11.2 with $F = F^{+}$ or $F = F^{-}$ to the Toeplitz-like matrix T^2 and to two vectors $\mathbf{b} = \mathbf{v}$ and $\mathbf{b} = H\mathbf{v}$, or we may even use a simple extension of Algorithm 7.1 for $A = T$. Similarly, we compute trace(H^j) = trace(T^j) for $j = 2i$, trace(H^j) = trace$(T^j J)$ for $j = 2i+1$, $i = 0, \ldots, n-1$. If $T = HJ$ is not symmetric, we simply shift from H to $\widetilde{H} = \operatorname{diag}(H, J_n H^T J_n)$ and from $T = HJ_n$ to $\widetilde{T} = \widetilde{H} J_{2n}$ where J_k denotes the $k \times k$ reversion matrix for $k = n$ and $k = 2n$. $\widetilde{H}$ is a symmetric Hankel-like matrix, and $c_{\widetilde{H}}(\lambda) = (c_H(\lambda))^2$ can be computed by using the above construction; then $c_H(\lambda)$ can be immediately recovered by using Newton's iteration [see Problem 1.6.2 $(POL \cdot ROOT)$].

12. Further Extensions of Computations with Structured Matrices by Using Associated Operators.

We know how to recover a matrix from its F-generators for the operators F of the previous sections (here we exclude $F = F^{+}$, $F = F^{-}$ and $F = F^{\pm}$, to simplify the argument). Thus, it suffices for us to save and to store such F-generators rather than the matrix itself; this is economical if the lengths of the F-generators are small.

Suppose that we need to perform some arithmetic operations with matrices stored in this way. We do not necessarily have to recover the entries of the input matrices. In many cases, we will be better off if we operate with F-generators to the very end of the computations, and we will assume doing this hereafter, as we did in Algorithm 11.1. Additions and subtractions of matrices will be then reduced to the union operations with their F-generators. For multiplication of matrices, we will use (11.22) or the following extension of a more special result from [CKL-A]:

Proposition 12.1. *Let K, L, M and N be four fixed matrices, $F = F_{(K,N)}$, $F_1 = F_{(K,L)}$ and $F_2 = F_{(M,N)}$ be the three operators defined by the equations*

$$F_1(A) = A - KAL, \quad F_2(B) = B - MBN, \tag{12.1}$$

$$F(C) = C - KCN. \tag{12.2}$$

Then the following matrix equations hold, where $\Delta = LM - I$:

$$F(AB) = F_1(A)B + KALF_2(B) + \Delta_0, \tag{12.3}$$

$$\Delta_0 = KA\Delta BN. \tag{12.4}$$

Proof. Since $F(AB) = AB - KAIBN$ and $I = LM - \Delta$, we have that

$$F(AB) = (A - KAL)B + KAL(B - MBN) + \Delta_0.$$

Substitute the equations (12.1) and deduce (12.3). ∎

Given the matrices A and B and the operators F_1 and F_2 of (12.1), we may compute the image $F(AB)$ of the operator F of (12.2) by applying the equations (12.3) and (12.4). We will refer to this computation as to **Algorithm 12.1.**

To estimate its cost, let $r = \text{rank } F(AB)$, $r_0 = \text{rank } \Delta$, $r_1 = \text{rank } F_1(A)$, $r_2 = \text{rank } F_2(B)$. Then it follows that $r \leq r_0 + r_1 + r_2$.

For smaller r_0, r_1 and r_2, Algorithm 12.1 is essentially reduced to computing the generators of the matrix products $F_1(A)B$ and $KALF_2(B)$, which amounts to r_1 multiplications of vectors by the matrix B and r_2 multiplications of the matrix KAL by vectors.

To apply Proposition 12.1 and Algorithm 12.1, we need to represent the operators F_1 and F_2 in the form (11.8) [we will use scaling in the cases (11.5), (11.9)-(11.12)] and to choose matrices L and M for F_1 and F_2, respectively, so as to keep the rank of the matrix $\Delta = LM - I$ smaller. Here are three relevant choices:

$$L = Z,\ M = Z^T,\ \Delta = -\mathbf{e}^{(0)}\mathbf{e}^{(0)T}, r_0 = 1, \tag{12.5}$$

$$L = Z^T,\ M = Z,\ \Delta = -\mathbf{e}^{(n-1)}\mathbf{e}^{(n-1)T}, r_0 = 1, \tag{12.6}$$

$$L = D(\mathbf{v}),\ M = D^{-1}(\mathbf{v}),\ \Delta = O, r_0 = 0. \tag{12.7}$$

Corollary 12.1. *Let $n \times n$ matrices A, B, K, L, M, N and Δ satisfy the equations (12.5), (12.6) and the assumptions of Proposition 12.1, where F, F_1 and F_2 are the displacement operators of Hankel and/or Toeplitz type (11.1)-(11.4); let the matrices A and B be given with their F_1- and F_2-generators of lengths d_1 and d_2, respectively. Then an F-generator of length at most $d_1 + d_2 + 1$ for the matrix AB can be computed by using $O(d_1 d_2)$ applications of FFT at $O(n)$ points and $O(d_1+d_2)$ summations of $O(d_1+d_2)$ vectors of dimension n, at the overall cost of $O(n(d_1 + d_2)^2 \log(n(d_1 + d_2)))$ ops.*

Remark 12.1. Corollary 12.1 and Theorem 11.1 imply, for instance, that rank $(F(A^+)) \leq 5$ for a Toeplitz matrix A of full rank and for the operator $F = F_1 = F_2$ of (11.1)-(11.2) [compare the solutions to Problem 2.12 ($GEN \cdot INVERSE$) of computing A^+], and similar bounds can be deduced for the F-ranks of the generalized inverses and Schur complements of other structured matrices that we study.

Other pairs of matrices L and M, in addition to ones of (12.5)-(12.7), may suggest further interesting applications of Proposition 12.1. Also, its extensions, in particular, to the operators F_1 and F_2 having the commutant form $KA - AL$ with singular matrices K and L, are useful in some applications (see [BDB90], [B83], [P92b], our section 11, and exercises 35, 36).

In particular, for the operators F^+, F^- of (11.18) and (11.19) and $F^\pm = F^+ + F^-$ of (11.23), we have the following extension of Corollary 12.1:

Fact 12.1. *Given an operator F of (11.18) ($F = F^+$), of (11.19) ($F = F^-$) or of (11.23) ($F = F^\pm$), and F-generators of lengths d_1 and d_2 for two matrices A and B, respectively, $O(nd_1d_2 \log n)$ ops suffice in order to compute an F-generator of length at most $d_1 + d_2$ for AB.*

Proof. From (11.22) it follows that an F-generator of AB, for $F = F^+$, $F = F^-$ and $F = F^\pm$, is given by $G_{AB} = [AG_B, G_A]$, $H_{AB} = [H_B, B^T H_A]$, where G_A, H_A and G_B, H_B are F-generators of A and B, respectively. To ensure the desired complexity bound $O(nd_1d_2 \log n)$ ops for computing G_{AB} and H_{AB}, apply Theorem 11.3 (for the operators $F = F^+$ and $F = F^-$), Theorem 11.4 and Fact 11.7 (for $F = F^\pm$), and recall that the product of a triangular Toeplitz matrix by a vector can be computed in $O(n \log n)$ ops [see Problem 5.1 ($TOEPL \cdot VECTOR$)]. ∎

Following [P89a], we may regard Proposition 12.1 as a general tool, which, in particular, reduces computations for one class of matrices to similar computations for another class [an alternative reduction relies on the specific transition formulae given in (12.9)–(12.12)]. For example, let HT stand for a Hankel-like or Toeplitz-like matrix, C for a Cauchy (Hilbert)-like matrix and V for a Vandermonde-like matrix. Then Proposition 12.1 and the equations (12.5)-(12.7) enable us to make transition between the classes of matrices HT, C and V according to Table 12.1.

A	B	AB	F-rank of AB
HT	V	V	$r \le r_1 + r_2 + 1$
V	V	C	$r \le r_1 + r_2 + 1$
V	V	HT	$r \le r_1 + r_2$
C	V	V	$r \le r_1 + r_2$
HT	HT	HT	$r \le r_1 + r_2 + 1$
C	C	C	$r \le r_1 + r_2 + 1$

Table 12.1

This transition requires, of course, that for the operators F_1 and F_2 reduced to the format (12.1), the matrices L and M be reconciled with each other so as to ensure (12.5), (12.6) or (12.7).

Using these transitions, we may reduce the original problem to other problems for which we have effective algorithms, such as ones of sections 4–6 and 13. In particular, we may bound the complexity of the Cauchy (Hilbert) and Vandermonde type computations by means of their reduction to computations of the Hankel-Toeplitz type, for which some effective algorithms are known (see sections 11 and 13).

Some interesting extensions of this approach of [P89a] appeared in [Wo], [Wo1]. In particular, Proposition 12.1 has been extended to generate various other formulae for the product of two structured matrices. Application of the original formula of Proposition 12.1, as well as of its extensions, to the product AA^{-1} was exploited to generate the expressions of the Gohberg-Semencul type for A^{-1} (compare Theorems 5.2 and 5.3). Furthermore, some applications of the operator $F_{(M,N)}$ [such that $F_{(M,N)}(K) = K - MKN$] to the computations with Krylov matrices $K = K(A, \mathbf{v})$ (see Definition 1.15) have been shown in [Wo].

Next, let us demonstrate some applications of Algorithm 12.1. Let us first be given an $F_{\mathbf{s},\mathbf{t}}$-generator of length r for an $n \times n$ matrix A of the Cauchy (Hilbert) type (and of $F_{\mathbf{s},\mathbf{t}}$-rank r) and for the (nonsingular) operator $F_{\mathbf{s},\mathbf{t}}$ of (11.5), with the vectors $\mathbf{s}$ and $\mathbf{t}$ having no components 0, and let us seek $\det A$ and/or (if $\det A \neq 0$) an $F_{\mathbf{t},\mathbf{s}}$-generator of length r for A^{-1}. To solve this problem, we may choose the Vandermonde matrix $B = [1/t_i^j]$, such that $\det B \neq 0$ and AB is a matrix of F-rank at most $r + 1$, for the operator F of (11.9) with $\mathbf{v} = [v_i] = \mathbf{s}^{-1} = [1/s_i]$ where $\mathbf{s} = [s_i]$. Then compute an F-generator of AB of length at most $r + 1$, by using Algorithm 12.1. [We will scale the F-generators when we need to reduce them to the form (11.8).] Then invert the matrix AB and/or compute $\det(AB)$ and $\det B$, and finally, compute $\det A = \det(AB)/\det B$ and/or $\hat{F}_{\mathbf{t},\mathbf{s}}$-generators of $A^{-1} = B(AB)^{-1}$, first of length $r_1 \geq r$, and then of length r (see Lemma 11.1). Since B is a Vandermonde matrix and AB is a matrix of the Vandermonde type, we reduced a Cauchy (Hilbert) type computation to a Vandermonde type computation.

A similar transition is possible from any of the classes HT, C and V to any other of these classes, by extending our solution to Problem 2.11b ($G \cdot COMPRESS$). The last two lines of Table 12.1 also indicate that we may make transitions within each of the classes HT and C, respectively, changing the associated operators F. Furthermore, we may extend this approach, say, to computing the F-generators of the Moore-Penrose generalized inverses of the structured matrices of full rank (compare Remark 12.1) and thus to computing least-squares solutions to linear systems with such matrices, which gives us better algorithms than other known approaches, in particular, in the cases of rectangular Vandermonde and Cauchy (generalized Hilbert) matrices.

Similarly, to compute the rank and/or the null space of a Cauchy (Hilbert)-like or Vandermonde-like matrix, we may pre- and postmultiply the input matrix by nonsingular Vandermonde and/or transposed Vandermonde matrices and thus reduce the problem to computing the rank and/or the null space of the resulting matrix of the class HT.

Next, we will specify a reduction of the class C to HT, which will enable us to extend effective matrix algorithms available for the class HT to the class C. To simplify our presentation we will only show such an extension from HT to the Cauchy (generalized Hilbert) matrices. An extension to the more general classes of Cauchy (Hilbert)-like and Vandermonde-like matrices is immediate, due to Facts 11.2, 11.4 and 11.5 [also see the equations (12.9)–(12.12) and exercise 32].

Thus let $C = C(\mathbf{s}, \mathbf{t})$ denote a nonsingular Cauchy (generalized Hilbert) matrix of (11.6) and of Definition 4.4, and let us compute $\det C$ (where C is real) and the vector $C^{-1}\mathbf{v}$ for a given Cauchy matrix C and a vector $\mathbf{v}$. Let us first compute an F-generator of the matrix

$$S = V^T(\mathbf{t}^{-1})CV(\mathbf{s}) \tag{12.8}$$

where $V(\mathbf{w})$, for a vector $\mathbf{w}$, denotes the Vandermonde matrix $[w_i^j]$ (compare Definition 4.2); $\mathbf{t}^{-1}$ denotes the vector $[1/t_i]$. Without loss of generality, we choose the vectors $\mathbf{s}$ and $\mathbf{t}$ [which define $H = A$ via (11.6)] having no zero coordinates. We will twice apply Algorithm 12.1 (for appropriate operators F) in order to compute:

1) at first, an F-generator of the matrix $V^T(\mathbf{t}^{-1})C$ of length 2 or less, and then
2) an F-generator of S of length 3 or less.

Let us specify the input to Algorithm 12.1 in these applications:

1) $K = Z$, $L = D^{-1}(\mathbf{t})$, $M = L^{-1}$, $N = D(\mathbf{s})$, $A = V^T(\mathbf{t}^{-1})$, $B = C$;
2) $K = Z$, $L = D(\mathbf{s})$, $M = L^{-1}$, $N = Z^T$, $A = V^T(\mathbf{t}^{-1})C$, $B = V(\mathbf{s})$.

Here $D(\mathbf{v})$ is defined as in (11.5).

Our estimates for the F-ranks of Vandermonde and Cauchy (generalized Hilbert) matrices, for the operators F of (11.5), (11.9)-(11.12) (valid also after scaling the operator F), together with Proposition 12.1, imply the following bounds at the Stages 1) and 2):

1) $\Delta = O$, $r_0 = 0$, $r_1 = r_2 = 1$, $r \le 2$;

2) $\Delta = O$, $r_0 = 0$, $r_1 \le 2$, $r_2 = 1$, $r \le 3$.

In the second application of Algorithm 12.1, an F-generator of length 3 (or less) for S is output where F is the operator of (11.1), so that by applying the known algorithms for Toeplitz-Hankel computations, we will immediately compute $\det S = \det V^T(\mathbf{t}^{-1}) \det H \det V(\mathbf{s})$.

Then we may compute the determinants of the two Vandermonde matrices $V(\mathbf{t})$ and $V(\mathbf{s})$ by using Fact 4.1, and this will give us $\det C = \det S \det V(\mathbf{t}) / \det V(\mathbf{s})$.

Likewise, (12.8) implies that $C^{-1}\mathbf{v} = V(\mathbf{s}) S^{-1} V^T(\mathbf{t}^{-1})\mathbf{v}$, and the evaluation of $C^{-1}\mathbf{v}$ is reduced to solving a Toeplitz-like linear system and to multiplication of $V(\mathbf{s})$ and $V^T(\mathbf{t}^{-1})$ by vectors (section 4 and Algorithms 6.1, 6.2).

Now let us estimate the cost of these applications of Algorithm 12.1. Due to Proposition 12.1, this cost is dominated by the cost of premultiplying each of the matrices $V(\mathbf{s})$ (once) and $AV(\mathbf{s})$ (once) and of postmultiplying the matrix A (once) and the matrix $V^T(\mathbf{t}^{-1})$ (twice) by vectors. This amounts

a) to two multiplications of vectors by the Cauchy (generalized Hilbert) matrices A and A^T (by means of Algorithm 4.1) and

b) to four multiplications of vectors by the Vandermonde matrices $V(\mathbf{s})$ and $V(\mathbf{t}^{-1})$.

The overall cost of this algorithm is $O(n \log^2 n)$ ops. Actually, in the case of a Cauchy (rather than a Cauchy-like) matrix $C = C(\mathbf{s}, \mathbf{t})$, we may compute $\det C$ at the same asymptotic cost but with a smaller overhead constant, based on the known equation

$$\det C = (-1)^{n(n-1)/2} \det V(\mathbf{s}) \det V(\mathbf{t}) \,/\, \prod_{i=0}^{n-1} \Gamma_{\mathbf{t}}(s_i).$$

Let us also demonstrate the same approach by computing an F-generator of length 1 for the inverse of a nonsingular Cauchy (generalized Hilbert) matrix A of (11.6). We will first compute an F-generator of length at most 3 for the matrix S of (12.8). Then we will compute an F-generator of length at most 3 for S^{-1} by means of the known efficient algorithms. We have the equation $A^{-1} = V(\mathbf{s}) S^{-1} V^T(\mathbf{t}^{-1})$, so that Corollary 11.2 implies that it suffices to compute the vectors $A^{-1}\mathbf{e}^{(0)}$ and $\mathbf{e}^{(0)T} A^{-1}$. Due to Theorem 11.2, six multiplications of triangular Toeplitz matrices by vectors suffice in order to compute each of the vectors $\mathbf{c} = S^{-1} V^T(\mathbf{t}^{-1})\mathbf{e}^{(0)}$ and

$\mathbf{d}^T = \mathbf{e}^{(0)T}V(\mathbf{s})S^{-1}$. Then it will remain to compute the vectors $V(\mathbf{s})\mathbf{c}$ and $\mathbf{d}^T V^T(\mathbf{t}^{-1}) = (V(\mathbf{t}^{-1})\mathbf{d})^T$, amounting to the evaluation at n points of two polynomials of degrees at most $n-1$. The overall asymptotic cost of computing the F-generator for A^{-1} is $O(n \log^2 n)$ ops.

The correlation between the classes HT, C and V of structured matrices, shown in [P89a] and in this section, have been more recently further specified in [GO1]. Next, we will show such specific correlations.

Recall the operator $F_{\mathbf{s},\mathbf{t}}(A) = A - \text{diag}(\mathbf{s})A\ \text{diag}(\mathbf{t})$ of (11.8), of the Cauchy (Hilbert) type. Assume that $s_i \neq 0$ for all i, denote that $\mathbf{s}^{-1} = (1/s_0, \ldots, 1/s_{n-1})^T$, $\mathbf{e} = (1, \ldots, 1)^T$, and observe that

$$F_{\mathbf{s},\mathbf{t}}(C(\mathbf{s},\mathbf{t})) = \mathbf{s}^{-1}\mathbf{e}^T\ , \quad F_{\mathbf{s},\mathbf{t}}(A) = \text{diag}(\mathbf{s})\, \widetilde{F}_{\mathbf{s}^{-1},\mathbf{t}}(A)\ ,$$

for any matrix A [compare (11.5)–(11.7)]. Associate (with a matrix A and with the two vectors $\mathbf{s}^{-1}$ and $\mathbf{t}$, $\mathbf{s}^{-1}$ having no components 0) a generator (G, H), such that

$$F_{\mathbf{s},\mathbf{t}}^{-1}(A) = GH^T\ ,\ G = (\mathbf{g}^{(k)})\ ,\ H = (\mathbf{h}^{(k)})\ , \qquad k = 1, \ldots, d,$$

for an appropriate integer d. In [GO1], it has been proved that the latter matrix equations hold if and only if

$$A = \text{diag}(\mathbf{s}) \sum_{k=1}^{d} \text{diag}(\mathbf{g}^{(k)})\, C(\mathbf{s},\mathbf{t})\text{diag}(\mathbf{h}^{(k)})\ , \tag{12.9}$$

and also if and only if

$$\widetilde{A} - Z_f \widetilde{A} Z_{1/f}^T = \widetilde{G}\widetilde{H}^T \tag{12.10}$$

where f is a scalar, $f \neq 0$, $V(\mathbf{s}^{-1})\widetilde{A}V^T(\mathbf{t}) = A$, $\widetilde{G} = (\tilde{\mathbf{g}}^{(k)})$, $\widetilde{H} = (\tilde{\mathbf{h}}^{(k)})$, $k = 1, \ldots, d+2$, $\tilde{\mathbf{g}}^{(k)} = V^{-1}(\mathbf{s}^{-1})\mathbf{g}^{(k)}$, $\tilde{\mathbf{h}}^{(k)} = V^{-1}(\mathbf{t})\mathbf{h}^{(k)}$, $k = 1, \ldots, d$;

$\tilde{\mathbf{g}}^{(d+1)} = -\mathbf{q} - f\mathbf{e}^{(0)}$, $\tilde{\mathbf{h}}^{(d+1)} = (Z_{1/f} - (\mathbf{p} + (1/f)\mathbf{e}^{(0)})(\mathbf{e}^{(n-1)})^T)\widetilde{A}^T\mathbf{e}^{(n-1)}$,
$\tilde{\mathbf{g}}^{(d+2)} = Z_f\widetilde{A}\mathbf{e}^{(n-1)}$, $\tilde{\mathbf{h}}^{(d+2)} = -\mathbf{p} - (1/f)\mathbf{e}^{(0)}$,

$$\mathbf{q} = (q_i) = \gamma(\mathbf{s}^{-1}),\ x^n + \sum_{i=0}^{n-1} q_i x^i = \prod_{i=0}^{n-1}(x - 1/s_i),$$

$$\mathbf{p} = (p_i) = \gamma(\mathbf{t}),\ x^n + \sum_{i=0}^{n-1} p_i x^i = \prod_{i=0}^{n-1}(x - t_i)$$

(compare Definition 4.3).

In particular, (12.10) relates to each other the operator representations of the matrices of the Cauchy (Hilbert), Vandermonde and Toeplitz types.

Extending (11.9), we consider the operator

$$F_{\mathbf{v},Z_{1/f}}(A) = A - \operatorname{diag}(\mathbf{v})AZ_{1/f}^T ,$$

of the Vandermonde type. We observe that

$$F_{\mathbf{v},Z_{1/f}}(V(\mathbf{v})) = (1 - v_0^n/f, \ldots, 1 - v_{n-1}^n/f)^T(\mathbf{e}^{(0)})^T .$$

Furthermore, due to the close relation between $Z_{1/f}$ and $Z = Z_0$, the matrix $F_{\mathbf{v},Z_{1/f}}(A)$ has a short generator if and only if A is a Vandermonde-like matrix. In [GO1] it has been proved that

$$F_{\mathbf{v},Z_{1/f}}(A) = GH^T , \ G = (\mathbf{g}^{(k)}), \ H = (\mathbf{h}^{(k)}) , \quad k = 1, \ldots, d ,$$

if and only if

$$\operatorname{diag}([1 - v_i^n/f]_{i=0}^{n-1})A = \sum_{k=1}^{d} \operatorname{diag}(\mathbf{g}^{(k)})V(\mathbf{v})Z_{1/f}^T(\mathbf{h}^{(k)}) , \qquad (12.11)$$

and also if and only if

$$\widehat{A} - Z_f\widehat{A}Z_{1/f} = \widehat{G}\widehat{H}^T , \qquad (12.12)$$

where

$$\widehat{A} = V^{-1}(\mathbf{v})A , \ \widehat{G} = (\hat{\mathbf{g}}^{(k)}) , \ \widehat{H} = (\mathbf{h}^{(k)}) , \quad k = 1, \ldots, d+1 ,$$
$$\hat{\mathbf{g}}^{(k)} = V^{-1}(\mathbf{v})\mathbf{g}^{(k)} , \quad k = 1, \ldots, d ,$$
$$\hat{\mathbf{g}}^{(k+1)} = -\gamma(\mathbf{v}) - f\mathbf{e}^{(0)} , \ \mathbf{h}^{(k+1)} = Z_{1/f}\widehat{A}^T\mathbf{e}^{(n-1)} ,$$
$$\gamma(\mathbf{v}) = (\gamma_i), \ x^n + \sum_{i=0}^{n-1} \gamma_i x^i = \prod_{i=0}^{n-1} (x - v_i)$$

(compare Definition 4.3). The latter equations relate to each other the operator representations of structured matrices of Toeplitz and Vandermonde types.

13. Computations of Toeplitz Type. Regularization of a Matrix via Preconditioning with Randomization.

In section 5 we cited the known effective algorithms of [BA] and [Morf] (see our Theorem 13.1), [Mus] and [AGr88], for Toeplitz matrix inversion.

These algorithms also apply to computing F-generators of length d for the inverses and for the recursive triangular factors of the strongly nonsingular matrices given with their F^*-generators of length d, for the pairs of operators F and F^* where F and F^* are related through Theorem 11.1: $F^* = F$, for F of (11.3) and (11.4), $F^* = F_{(Z^T,Z)}$ if $F = F_{(Z,Z^T)}$, $F^* = F_{(Z,Z^T)}$ if $F = F_{(Z^T,Z)}$ [see (11.1), (11.2)]; the recursive factorization gives us the determinants of such matrices too. The overall cost of these computations is $O(d^2 n \log^2 n)$ ops, where $d = 2$ for Hankel and Toeplitz matrices. Given a Toeplitz-like or a Hankel-like matrix A (in a field of characteristic 0), we may compute $|\det A|^2 = \det(A^H A)$ at the above cost since $A^H A$ is strongly nonsingular (unless $\det A = 0$) and, therefore, has the (balanced) recursive factorization (2.1)-(2.3) (compare exercise 4c and section 7). This approach can be extended to a randomized algorithm for computing rank A in $O(n \log^3 n)$ ops if we apply the binary search (Algorithm 8.2) to find the nonsingular leading principal submatrix of maximum size for the Toeplitz-like matrix RAS where R^T and S are two random unit lower triangular Toeplitz matrices. This implies that, with a high probability, the matrix RAS is strongly nonsingular if A is nonsingular (compare Lemma 13.1 and Procedure 1.7.1).

If we seek $\det T$ (rather than $|\det T|$) for any $n \times n$ matrix $T = [t_{ij}]$, we may just compute the $n+1$ determinants of the $n \times n$ matrices $T_k = T + t\omega^k I$ for $k = 0, 1, \dots, n$ (where t is a real number, ω is a primitive $(n+1)$-st root of 1), because $(n+1)\det T = \sum_{i=0}^{n} \det T_i$. Choosing $t > n \max_{i,j} |t_{ij}|$, we may ensure that the matrices T_k are diagonally dominant and therefore strongly nonsingular. This deterministic solution costs $O(d^2 n^2 \log^2 n)$ ops and applies over any field of characteristic 0. To extend the solution to the case of any field, we pre- and postmultiply the matrices A or JA by random unit upper and lower triangular Toeplitz matrices, respectively. This does not change the determinants of the matrices A and/or JA. Choosing the entries of the multipliers at random from a sufficiently large set ensures strong nonsingularity of the product with a high probability (see Procedure 1.7.1 and Lemma 13.1), and then with a high probability we arrive at the determinant of A in $O(d^2 n \log^2 n)$ ops. These and other techniques of *regularization via randomization* will be studied in the second part of this section.

Due to the fundamental role of Toeplitz-like computations, we will next present the algorithm of [BA], [Morf] for computing both the inverse A^{-1} of a Toeplitz-like matrix A and the recursive triangular factorizations of A and A^{-1} based on (2.1)-(2.3) (provided A and A^{-1} have such factorizations). In this presentation we will assume that $F = F_{(Z,Z^T)}$ is the operator of (11.1) and will a little simplify and clarify the algorithm of [BA], [Morf], by including the recursive compression of the computed displacement generators of the auxiliary matrices [see Problem 2.11b ($G \cdot COMPRESS$)].

The computation is simple to define and to perform, since we leave all the recursive blocks DB^{-1} and $B^{-1}C$ in factorized form and represent all the matrices by their F-generators. We only need to keep an appropriate order of operations. Specifically, to factorize and even to compute the Schur complement S, we need to have B^{-1}. We observe, however, that computing the recursive factorization (2.2) is very simple if we know the recursive factorization (2.1), for it is enough to invert the 1×1 diagonal blocks in this case. Thus, in every recursive factorization step, *we will delay the inversion of B and S* until we compute their factorization, so that the actual evaluation of the recursive factorization starts with the inversion of the 1×1 leading principal submatrix of A, then its Schur complement is computed and inverted, then the 2×2 leading principal submatrix of A is inverted, its Schur complement is computed and inverted, and so on.

Hereafter, we will assume for simplicity that $n = 2^h$ is an integer power of 2 and that the recursive factorization is balanced, so that, in particular, B, C, D and E are $(n/2) \times (n/2)$ matrices; then we only need $h = \log n$ recursive steps (2.1), (2.2).

Theorem 13.1. *Given an $n \times n$ strongly nonsingular matrix A with its F-generator of length r for the operator $F = F_{(Z,Z^T)}$ of (11.1), the F-generators of all the matrices encountered in the balanced recursive factorizations (2.1) of A and (2.2) of A^{-1} can be computed in $O(nr^2 \log^2 n)$ ops and can be stored by using $O(nr \log n)$ words of storage space; furthermore, $O(nr^2 \log^2 n)$ ops suffice in this case in order to compute $\det A$ and an $F_{(Z^T,Z)}$-generator of length r for A^{-1} and for the operator $F_{(Z^T,Z)}$ of (11.2).*

Proof. We recall that S^{-1} is the southeastern principal submatrix of A^{-1} (Proposition 2.3), so that the results of section 11 imply that

$$\begin{aligned}
&\text{rank } F_{(Z^T,Z)}(S^{-1}) = \text{rank } F_{(Z,Z^T)}(S) \le r,\\
&\text{rank } F_{(Z,Z^T)}(B) = \text{rank } F_{(Z^T,Z)}(B^{-1}) \le r,\\
&\text{rank } F_{(Z,Z^T)}(C) \le r, \text{rank } F_{(Z,Z^T)}(D) \le r
\end{aligned}$$

where $r = \text{rank } F_{(Z,Z^T)}(A) = \text{rank } F_{(Z^T,Z)}(A^{-1})$.

The latter bounds are recursively extended further on, throughout the recursive factorizations (2.1) and (2.2). Applied together with the results of section 11, these bounds imply that all the matrices encountered in the recursive factorizations (2.1) and (2.2) have their F-ranks of the order of $O(r)$, for $F = F_{(Z,Z^T)}$. Moreover, we will keep representation of these matrices by their F-generators of lengths $O(r)$ by applying our solution of Problem 2.11a ($G \cdot COMPRESS$) [in $O(r^2n)$ ops].

Hereafter, let $\mu_r(k)$, $\sigma_r(k)$ and $\phi_r(k)$ denote the numbers of ops required for multiplication, subtraction and computing $F_{(ZZ^T)}$-generators for the recursive factorization (2.1) of $k \times k$ matrices A given with their $F_{(ZZ^T)}$ -generators of length at most r, respectively. In the above construction, the subtractions are only needed for computing the F-generator of length r for the Schur complement S of (2.3) or, more precisely, for a reduction to r of the length of a readily available longer F-generator, for $F = F_{(Z,Z^T)}$. The cost of such a reduction is $\sigma_r(k) = O(r^2k)$. Since we only need to compute F-generators of length $O(r)$ for the products and differences of matrices, we also obtain that $\mu_r(k) = O(r^2k \log k)$ (see Corollary 12.1).

Now, computing (2.1) amounts to computing the matrices B^{-1} and S, since we keep DB^{-1} and $B^{-1}C$ in factorized form. Since we delay computing the inverse of B and at the end have the inversion cost negligible, we will ignore this cost and will write:

$$\phi_r(n) \le 2\mu_r(n/2) + 2\phi_r(n/2) + \sigma_r(n/2) \le \sum_{i=1}^{h} 2^i(\mu_r(n/2^i) + \frac{1}{2}\sigma_r(n/2^i)),$$

$h = \log n$.

Substitute the above bounds on $\mu_r(k)$ and $\sigma_r(k)$ for $k = n/2^i$ and obtain that $\phi_r(n) = O(r^2 n \log^2 n)$, which is immediately extended to bound the cost of computing $\det A$ and F-generators for the recursive factorization (2.2) too, and thus also the cost of computing F-generators for A^{-1}.

To prove the storage space bound, assume that we are given F-generators of lengths at most r for A and A^{-1} and keep all the matrices in (2.1) and (2.2) in factorized form, that is, store only F-generators of the matrices B, B^{-1}, C, D, S, and S^{-1} of (2.1) and (2.2), which have lengths at most $r+1$. Of these F-generators for C and D, only the entries of two pairs of $(n/2)$-dimensional vectors are not the entries of the matrices of the F-generator for A, and two of these vectors are unit coordinate vectors, which we do not need to store. The entries of the matrices of the F-generators for B, B^{-1}, S and S^{-1} are from the matrices of the F-generators of A and A^{-1}, except for at most $2rn$ additional entries, so that we may recursively apply the same argument to the recursive factorization of the matrices B, B^{-1}, S and S^{-1} and thus arrive at the desired storage space bound. ∎

Remark 13.1. The reader may specify the time bound of Theorem 13.1 based on the proof of Theorem 13.1, together with the proofs of the auxiliary results from the previous section; the bound on the storage space has been specified to be $n(4r + (2r + 1)\log n)$ in [P88]. Note that the algorithm supporting the latter proof can be extended to the evaluation of the rank of a tame matrix A given with its (short) F-generator. [On taming a matrix A, see the solution of Problem 2.10b ($PRECNDTN$), which keeps the length of an F-generator bounded.] Indeed, since A is tame, so are the matrices S (provided that B is nonsingular, so that S is defined) and B; furthermore, rank A = rank B if B is singular, rank A = rank B + rank S otherwise. Thus, we may bound rank A from above or below whenever we solve Problem 2.4a ($SINGULAR?$) for B and, recursively, for any of the diagonal northwestern blocks encountered in the recursive factorization of B and S (compare [Ka93]). The size of a block in this factorization should be each time chosen so as to halve the remaining interval for the value of rank A. We also recall that the solution of Problems 2.3a ($LIN \cdot SOLVE1$) and 2.3b ($NULL \cdot SPACE$) can be reduced to $SUBMATRIX$, which for a tame matrix is equivalent to $M \cdot RANK$.

In the remainder of this section, we will study regularization via randomization. Specifically, for a given nonsingular $n \times n$ matrix A, we will fill two auxiliary $n \times n$ matrices B and C with constants and with k indeterminates, $x_0, \ldots, x_{k-1}$, so as to ensure strong nonsingularity of the matrix $V = BAC$, which is equivalent to the inequalities

$$\det V_{(s,s)} \neq 0\ , \quad s = 1, \ldots, n\ , \tag{13.1}$$

or to ensure the equation

$$m_V(\lambda) = c_V(\lambda)\ . \tag{13.2}$$

We will also expect to have the matrix V tame if A is tame.

(13.2) is equivalent to the inequality $\det T_n \neq 0$ where T_n is the $n \times n$ Toeplitz matrix, $(T_n)_{i,j} = \mathbf{u}^T V^{n-1+i-j}\mathbf{v}$, $\mathbf{u} = [u_0, \ldots, u_{n-1}]^T$, $\mathbf{v} = [v_0, \ldots, v_{n-1}]^T$, u_i and v_j are indeterminates, $i, j = 0, \ldots, n-1$ (see exercise 44), so that (13.2) does not hold if and only if $\det T_n \equiv 0$ identically in $u_0, \ldots, u_{n-1}$.

Therefore, we may ensure the relations (13.1) and/or (13.2) with a high probability for a random choice of $x_0, \ldots, x_{k-1}$ if we ensure that these relations hold for at least one specific choice of $x_0, \ldots, x_{k-1}$ (see Procedure 1.7.1 and Lemma 1.5.1). Pursuing this goal, we will prefer to use fewer basic indeterminates $x_0, \ldots, x_{k-1}$ and to choose B and C having Toeplitz-like structure if A has it.

In particular, Procedure 1.7.1, together with Lemma 13.1 (to be presented later on), implies that strong nonsingularity of A and, therefore, the existence of the balanced recursive factorization (2.1) can be ensured with a high probability for the matrix UAL where L and U^T are random lower unit triangular Toeplitz matrices. Then Theorem 13.1 will lead to a fast randomized evaluation of A^{-1} and $\det A$, over any field of constants.

Since our objective is to satisfy some algebraic (polynomial) inequalities in $x_0, \ldots, x_{k-1}$, we may apply the *assignment techniques*, that is, just choose B and C such that some assignment of values to $x_0, \ldots, x_{k-1}$ satisfies one or both of the equations (say)

$$V = BAC = I + Z \ , \tag{13.3}$$

$$V = BAC = I + Z^T \ . \tag{13.4}$$

Each of these equations implies both (13.1) (that is, strong nonsingularity of V) and (13.2). [As a simple exercise, verify (13.2) for $V = Z + aI$ for any scalar a.] Hereafter, we refer to this solution as **Regularization 13.1**. In particular, if for a nonsingular matrix A we know that, say, rank $F_{(Z^T,Z)}(A) \leq r$, then we may recall Theorem 11.1 and set either

$$B = I + Z^T \ , \qquad C = \sum_{i=0}^{r-1} L_i U_i$$

or

$$B = \sum_{i=0}^{r-1} L_i U_i \ , \qquad C = I + Z$$

where L_i and U_i^T, for $i = 0, \ldots, r-1$, are $n \times n$ *generic* lower triangular Toeplitz matrices (whose first columns are filled with indeterminates), so that $k = 2rn$ (in particular, $k = 4n$ if A is a Toeplitz matrix).

In the next approach, we follow [KS91] and make the matrix V strongly nonsingular by ensuring algebraic independence of the terms in the Cauchy-Binet well-known expansion of $\det V_{(s,s)}$ for every s. This *algebraic independence approach* decreases k to $2n-1$ or $2n-2$ for any input matrix A. We will refer to this regularization technique as **Regularization 13.2.**

Let B^T and C be generic unit lower triangular Toeplitz matrices with the first columns filled with indeterminates below their diagonals, so that $k = 2n - 2$. The next lemma implies strong nonsingularity of BAC for a nonsingular matrix A.

Lemma 13.1 ([KS91]). *Let A be an $n \times n$ matrix of rank r and let U^T and L be generic unit lower triangular Toeplitz matrices. Then the $i \times i$ leading principal submatrices of UAL are nonsingular for $i = 1, \dots, r$. In particular, UAL is strongly nonsingular if $r = n$.*

Proof. For a matrix C and for two sets of positive integers, S and T, both having the same cardinality $|S| = |T|$, let $C_{S,T}$ denote the determinant of the submatrix of C formed by the entries (i, j) of C such that $i \in S$, $j \in T$. Denote $K = \{1, \dots, k\}$, $I = \{i_1, \dots, i_k\}$, $J = \{j_1, \dots, j_k\}$, $k = 1, \dots, r$, observe that $U_{K,I} L_{J,K} \neq 0$, and apply the well-known Cauchy-Binet formula,

$$\widetilde{A}_{K,K} = (UAL)_{K,K} = \sum_I \sum_J U_{K,I} A_{I,J} L_{J,K} \,. \tag{13.5}$$

We need to prove that $\widetilde{A}_{K,K} \neq 0$. Expand the subdeterminant $U_{K,I}$ as the sum of monomials in the entries of $U = [u_{i,j}]$ where $u_{i,j} = u_{j-i}$, $u_s = 0$ if $s < 0$, $u_0 = 1$, $u_1, \dots, u_{n-1}$ are indeterminates; write the terms of $U_{K,I}$ in the descending order of the variables, assuming the order $u_1 < u_2 < \cdots < u_{n-1}$. Define the lexicographically strictly lowest term of such a sum for a fixed I and observe that this term uniquely identifies the set I. Similarly proceed with the subdeterminants $L_{J,K}$, and then uniquely identify the product $U_{K,I} A_{I,J} L_{J,K}$ by its lexicographically strictly lowest term, for each pair (I, J) such that $A_{I,J} \neq 0$. Such terms cannot cancel each other in (13.5). On the other hand, since rank $A = r$, $A_{I,J} \neq 0$ for some pair (I, J), as long as $k \leq r$. Therefore, $\widetilde{A}_{K,K} \neq 0$. ∎

Another type of regularization, **Regularization 13.3**, can be performed analogously. Let A be a nonsingular $n \times n$ matrix, $B = I$, C be the generic Toeplitz or the generic Hankel matrix, $k = 2n - 1$. Then strong nonsingularity of BAC follows similarly to the proof of Lemma 13.1 (examine either the diagonal terms of the expansions of the minors of C, as in [KP,a], or the antidiagonal terms depending on whether C is a Hankel or a Toeplitz matrix).

Regularizations 13.2 and 13.3 imply (13.1) but not (13.2). We may, however, ensure (13.2) by scaling A by a generic diagonal matrix, provided that A is a nonsingular matrix. We need the following simple auxiliary results ([vdW]):

Fact 13.1. *$m_A(\lambda) = c_A(\lambda)$ if $c_A(\lambda)$ is square-free (has no multiple zeros).*

Fact 13.2. *A polynomial $p(x)$ is square-free if and only if it has non-vanishing discriminant, that is, if and only if the (scaled) resultant of $c_A(\lambda)$ and of its derivative does not vanish (compare Theorem 8.1).*

Furthermore, in [Wied] the following lemma is proved by induction on n:

Lemma 13.2. *If an $n \times n$ matrix A is strongly nonsingular over a field $\mathbf{F}$, then the discriminant of the characteristic polynomial of the matrix* $\mathrm{diag}(y_1, \ldots, y_n)A$ *does not vanish identically in $y_1, \ldots, y_n$, and thus the characteristic polynomial of this matrix is square-free over the ring $\mathbf{F}[y_1, \ldots, y_n]$ of polynomials in $y_1, \ldots, y_n$ over $\mathbf{F}$ and over the field $\mathbf{F}(y_1, \ldots, y_n)$ of rational functions in $y_1, \ldots, y_n$ over $\mathbf{F}$.*

Corollary 13.1. *If A is a strongly nonsingular matrix and if $Y = \mathrm{diag}(y_1, \ldots, y_n)$, $y_1, \ldots, y_n$ being indeterminates, then $m_{YA}(\lambda) = c_{YA}(\lambda)$, $m_{AY}(\lambda) = c_{AY}(\lambda)$.*

Thus, for a strongly nonsingular $n \times n$ matrix A, we may ensure (13.2) by choosing $B = I$ and $C = \mathrm{diag}(x_0, x_1, \ldots, x_{n-1})$ or vice versa, due to Corollary 13.1. This *scaling procedure* will be referred to as **Regularization 13.4.** In such a regularization (for a strongly nonsingular matrix A), $k = n$, and the Toeplitz-like structure of A is not preserved in V.

An alternative way, with $k = 2n - 1$, is to apply **Regularization 13.5** ([KS91], [KP,a]). Set $B = U$, $C = L$, $V = UAL$, where L and U^T are lower triangular Toeplitz matrices, L is generic and U^T is unit generic, that is, the diagonal entries of U equal 1, and all other entries of the first row of U and all the entries of the first column of L are indeterminates. By combining the techniques of the proofs of Lemmas 13.1 and 13.2, we may deduce that this procedure ensures both (13.1) and (13.2) for any nonsingular matrix A. If A has all its trailing principal submatrices nonsingular or, equivalently, if JAJ is strongly nonsingular, and if we replace the unit diagonal entries of U by an indeterminate, then we immediately extend (13.2) to $m_{V_i}(\lambda) = c_{V_i}(\lambda)$ for all leading principal blocks V_i of V [the extension proof is via substituting zeros for the first i (sub)diagonals of U^T and L, $i = 0, 1, \ldots, n-1$].

Remark 13.2. Since $L(\mathbf{v}) = Z_0(\mathbf{v})$, $L^T(\mathbf{v}) = Z_0^T(\mathbf{v})$, we may replace every triangular Toeplitz matrix by an f-circulant matrix with a random f in all the regularization procedures of this section.

If we have an $n \times n$ matrix W of rank r, with the strongly nonsingular leading principal submatrix of size $r \times r$, we may compute r by solving Problem 2.4a ($SINGULAR?$) for an appropriate set $\{W_s\}$ of leading principal submatrices of W. Specifically, with the binary search, we need $\lceil \log n \rceil$ tests for nonsingularity of the $\lceil \log n \rceil$ matrices W_s (of sizes $s \times s$); the search proposed in [ChR] requires at most $2\lfloor \log r \rfloor + 1$ tests: indeed, first set $s = 2^i$ and perform successive tests for $i = 0, 1, \dots, k$, until the matrix W_s (for $s = 2^k$, $k = \lfloor \log r \rfloor + 1$) turns out to be singular. Then it will remain to perform $k - 1$ steps of the binary search for the largest nonsingular matrix W_s for s ranging from 2^{k-1} to $2^k - 1$.

This algorithm gives us the following randomized and *output sensitive* upper bound on the complexity of Problem 2.10 ($M \cdot RANK$):

$$c_r(A) = (2\lfloor \log r \rfloor + 1)c_r + c_r^*(A) \tag{13.6}$$

where $c_r^*(A)$ denotes the complexity of Regularization 13.2 applied to an input matrix A of rank r (which is justified by Lemma 13.1), and c_r denotes the complexity of solving Problem 2.4a ($SINGULAR?$) for a matrix of size at most $(2r) \times (2r)$.

We recall that $DETERMINANT \succeq SINGULAR?$ and also observe that the randomization of the input (right-hand side) vector $\mathbf{b}$ of $LIN \cdot SOLVE$ leads to the implication:

$$LIN \cdot SOLVE \succeq SINGULAR?.$$

Then we arrive at the following estimates for the number of ops in $c_r^*(A)$ and c_r:

$$\begin{aligned} c_r^*(A) &= O(n^2 \log n), \\ c_r &= O(M(r)), \end{aligned} \tag{13.7}$$

for a general $n \times n$ matrix A of rank r ([ChR]).

14. Computations with Band Structured Matrices.

There are other important classes of structured matrices that we have not covered so far, in particular, banded matrices. A matrix $A = (a_{ij})$ is *banded, with the band* (h, k), provided that $a_{ij} = 0$ unless $h > i - j > k$ for fixed positive h and k,

$$h < n-1\,, \quad k < n-1\,. \tag{14.1}$$

The computations with banded matrices can be greatly simplified compared with the case of general matrices, in particular, a band linear system $A\mathbf{x} = \mathbf{b}$ can be solved in $O((k+h)^2 n)$ ops (see [GL]) or even in $O((k+h)^{\omega-1}n)$ ops for $\omega < 2.376$ (see [E92], [PSA], [PSA,a]). For further results on banded matrices, we refer the reader to [GL], [Sw], [BaF], [Hal], [BeC], [BeC], [BeCR], [BeLR], [Ike], [Riz], [Ro86], [RBFR], [E92], [PSA], [PSA,a].

In this section we will briefly review the important case of solving a more special linear system $A\mathbf{x} = \mathbf{b}$ where A is an $n \times n$ nonsingular banded Toeplitz matrix. Such systems arise, for instance, in many PDE and signal processing applications.

We will first recall some auxiliary formulae for the solution of such systems; for more details we refer the reader to the papers [GrS], [BC83], [BC83a], [B84].

The techniques of [GrS] are based on the following:

Fact 14.1([GrS]). *Let $A = (a_{ij})$ be a $2k+1$ diagonal Toeplitz matrix, that is, $a_{ij} = a_{j-i}$, $a_s = 0$ for $|s| > k$. Then, $A = Z_1^T(\mathbf{a}) + UV^T$ where, as before, $Z_1(\mathbf{a})$ denotes the circulant matrix whose first column is $\mathbf{a} = (a_0, a_{-1}, \ldots, a_{-k}, 0, \ldots, 0, a_k \ldots, a_1)^T$ and where U, V are $n \times k$ matrices.*

Applying the Sherman-Morrison-Woodbury formula (Theorem 1.2) to the above relation, we reduce the solution of the linear system $A\mathbf{x} = \mathbf{b}$ to solving an $n \times n$ circulant system (that is, to application of three DFT's) and to solving an $(2k) \times (2k)$ unstructured linear system. However, this approach requires nonsingularity of the $(2k) \times (2k)$ matrix that appears in the Sherman-Morrison-Woodbury formula, and this may lead to numerical stability problems. In [GrS] some modifications of this method are presented in order to improve its numerical stability.

A different approach, introduced in [BC83], [BC83a], mainly for the eigenvalue problem, consists in considering a band symmetric Toeplitz matrix as a submatrix of a matrix belonging to a suitable algebra.

For instance, it is possible to embed a symmetric $(2k+1)$-diagonal $n \times n$ Toeplitz matrix into an $(n+2\lfloor k/2 \rfloor) \times (n+2\lfloor k/2 \rfloor)$ matrix of the τ algebra, defined in Theorem 11.4. Here is an example with $k = 2$ and $n = 4$:

$$\begin{pmatrix} a_0 - a_2 & a_1 & a_2 & 0 & 0 & 0 \\ a_1 & a_0 & a_1 & a_2 & 0 & 0 \\ a_2 & a_1 & a_0 & a_1 & a_2 & 0 \\ 0 & a_2 & a_1 & a_0 & a_1 & a_2 \\ 0 & 0 & a_2 & a_1 & a_0 & a_1 \\ 0 & 0 & 0 & a_2 & a_1 & a_0 - a_2 \end{pmatrix} \in \tau.$$

This property allows us to reduce the solution of a banded Toeplitz linear system to performing the sine transform and to solving a $k \times k$ linear system.

An alternative way is to embed A into a circulant matrix or into a triangular Toeplitz matrix (in the case where A is not symmetric). In this way, it is also possible to deal with the case of Hessenberg Toeplitz matrices [satisfying (14.1), except that, in this case, we should allow that $h = n - 1$ or $k = n - 1$].

Below, we give more details on the technique of embedding a banded Toeplitz matrix into a triangular Toeplitz matrix (compare [B84]).

We will first consider the auxiliary linear system

$$B\begin{pmatrix} \mathbf{z} \\ \mathbf{w} \end{pmatrix} = \begin{pmatrix} \mathbf{b} \\ \mathbf{y} \end{pmatrix}, \quad B = \begin{pmatrix} C & A \\ O & E \end{pmatrix}, \quad C = \begin{pmatrix} C_k \\ O \end{pmatrix}, \quad E = (\,O, E_k\,) \qquad (14.2)$$

where C_k and E_k are $k \times k$ upper triangular Toeplitz matrices, B is an $(n+k)\times(n+k)$ upper triangular Toeplitz matrix, $\mathbf{z}$ and $\mathbf{y}$ have k components and $\mathbf{w}$ has n components. Let

$$B^{-1} = \begin{pmatrix} S & R \\ U & T \end{pmatrix}, \quad \begin{pmatrix} \mathbf{z} \\ \mathbf{w} \end{pmatrix} = \begin{pmatrix} S & R \\ U & T \end{pmatrix}\begin{pmatrix} \mathbf{b} \\ \mathbf{y} \end{pmatrix}, \qquad (14.3)$$

where R is a $k \times k$ matrix. Since B^{-1} is an upper triangular Toeplitz matrix, R is a Toeplitz matrix. Moreover, $\mathbf{z} = S\mathbf{b} + R\mathbf{y}$, and R is a nonsingular matrix since A is nonsingular (see exercise 47). We compute $\mathbf{y} = -R^{-1}S\mathbf{b}$, set $\mathbf{z} = \mathbf{0}$, and then obtain that $\mathbf{x} = \mathbf{w}$ provided that (14.3) holds. Thus, at the overall cost of $O(n \log n + k \log^2 k)$ ops, we reduce the solution of the linear system with a banded Toeplitz matrix A essentially to solving two Toeplitz linear systems, one with the $(n + k) \times (n + k)$ triangular Toeplitz matrix B and another with the $k \times k$ Toeplitz matrix R. This is substantially less than in the case of the solution of a general Toeplitz system and, if $k > \sqrt{\log n}$, is less than in the solution by means of Gaussian elimination. We refer the reader to [T85] and [Lin] for some alternative approaches to solving band Toeplitz linear systems; in particular, Linzer proves a weak numerical stability of his algorithm; we also refer the reader to [BP91a], [BC83], [BC83a], [BP88] and our volume 2 for some further examples of efficient computations with (block) banded Toeplitz matrices, in particular, for computing their characteristic polynomials.

Exercises to Chapter 2.

1. Draw the reducibility digraph for the problems of matrix computations showing also the techniques involved.

2. Find the threshold value n for which the algorithm multiplying $n \times n$ matrices in $\lceil 4.54n^{2.81} \rceil$ ops improves the straightforward algorithm; then do the same for the algorithm using $\lceil 3.92n^{2.81} \rceil$ ops and cited in [BM], p. 45; finally, compute similar threshold values n for other problems of section 2 that are reduced to matrix multiplication.

3. Prove that a row- (column-) diagonally dominant matrix A is nonsingular. [First note that the matrix $D = \text{diag}(A)$ is nonsingular. Then write $A = D(I + (D^{-1}A - I))$ and use the fact that $d_\infty(A) = ||D^{-1}A - I||_\infty < 1$.]

4. a) Verify the matrix equations (2.1)-(2.3) and the fact that S^{-1} is a submatrix of A^{-1}.

   b) Prove that if there exists the LU factorization of a nonsingular matrix A, then for any k the Schur complement S of the $k \times k$ leading principal submatrix B of A is the southeastern $(n-k) \times (n-k)$ principal submatrix of the matrix obtained from A in k steps of Gaussian elimination (see [GL]).

   c) Deduce from b) that the recursive factorization (2.1)-(2.3) can be carried out if and only if there exists the LU factorization of A; moreover, prove that if A is strongly nonsingular, then also B and its Schur complement S are strongly nonsingular. [It follows that for every strongly nonsingular matrix A, its recursive factorization (2.1)-(2.3) can be carried out.] Also prove that the strong nonsingularity of A is equivalent to the existence of its LU factorization with nonsingular factors L and U.

   d) By using b) and induction on k, prove that if A is diagonally dominant, then also the $k \times k$ Schur complement of the $(n-k) \times (n-k)$ leading principal submatrix of A is diagonally dominant. Moreover, prove that $d_h(A) \geq d_h(S)$, $d_h(A) \geq d_h(B)$, for $h = 1$ or $h = \infty$, for B and S of (2.3) and for $d_h(A) = \|D^{-1}A - I\|_h$, $D = \text{diag}(A)$.

   e) Apply the formula of Remark 2.1 to deduce the complexity bound of $O(n \log^2 n)$ ops for computing the solution $p_0, p_1, \ldots, p_n$ to the linear system (1.4.8) of n Newton identities. Extend this bound to any nonsingular linear system $A\mathbf{x} = \mathbf{b}$ where $A = D + T$, D is a diagonal matrix, T is a triangular Toeplitz matrix.

5. Let $I(k)$, $\det(k)$, $S(k)$ and $M(k)$ denote the numbers of ops required in order to compute the inverse of a $k \times k$ h.p.d. matrix, its determinant, its square and the product of a pair of $k \times k$ matrices, respectively.
   a) Supply the details in order to deduce that $I(n) = O(M(n))$, that is, deduce from the equations (2.1)-(2.3) that $I(2n) \leq 2I(n) + 6M(n) + n^2 + n$, and recursively apply these bounds to show that $I(n) = O(M(n))$.
   b) Extend this approach to the solution (with pivoting) of Problem 2.7 ($LU \cdot FACTORS$) in $O(M(n))$ ops (see [AHU]) and then deduce that $\det(n) = O(M(n))$, $I(n) = O(M(n))$ over any field of constants.
   c) (Schönhage) Show that $M(2k) \leq 4S(2k)+12k^2$ by using the matrix equation

$$\begin{pmatrix} A_3 - R & B_1 \\ A_1 & R \end{pmatrix}^2 - \begin{pmatrix} A_4 - R & -B_2 \\ A_2 & R \end{pmatrix}^2 = \begin{pmatrix} X_1 & X_3 \\ X_2 & X_4 \end{pmatrix}$$

   where $R = B_1 + B_2$, $X_3 = A_3B_1 + A_4B_2$, $X_4 = A_1B_1 + A_2B_2$.
   d) Show that $S(k) = O(I(k))$ by using the identity $P^2 = (P^{-1} - (I + P)^{-1})^{-1} - P$ [compare (1.3.4)] where P is the input matrix A (to be squared), if A and $I+A$ are nonsingular matrices, or $P = I+cA$ for a scalar c such that P and $P+I$ are nonsingular matrices.
   e) Prove that Problem 2.6 ($CHAR \cdot POL$) for a $k \times k$ matrix can be solved at a cost $O(\min\{M(k)\log^2 k, k^\omega \log k\})$, (see [K-G], [FF]).

6. Verify that Theorem 1.3 (of [Li76], [BSt]) indeed implies the reduction of Problem 2.5 ($INVERT$) to Problem 2.4 ($DETERMINANT$). Furthermore, observe that for the characteristic polynomial $\det(\lambda I - A) = \sum_{i=0}^{n} c_i\lambda^i$, $c_n = 1$, we have $c_{n-1} = -\text{trace } A$, $c_{n-2} = c_{n-1}^2 - s_2$, $s_2 = \text{trace } (A^2)$, so that $-c_{n-1}$ can be computed by using n ops and c_{n-2} by using $O(n^2)$ ops. Finally, use Theorem 1.3 to reduce $n \times n$ matrix multiplication to computing c_{n-3}.

7. a) Given a square matrix A and the coefficients of a polynomial $p(\lambda)$ such that $p(A) = O$, extend (2.9).
   b) Use the reduction of a matrix A to the Jordan canonical form ([GL]) to verify that $m_A(\lambda) \neq c_A(\lambda)$ implies that $m_{A^k}(\lambda) \neq c_{A^k}(\lambda)$, for $k = 2, 3, \ldots$, and, if $\det A \neq 0$, then also for $k = -1, -2, \ldots$.

8. Prove that $\det(\lambda I - AB) = \det(\lambda I - BA)$ if $A \in \mathbf{C}_{n,n}$, $B \in \mathbf{C}_{n,n}$, and A and/or B are nonsingular. Show with a counterexample that this

property does not hold if A and B are both singular. Note that the result extends to matrices A and B over the ring $\mathbf{Z}_m$, for any m.

9. Verify that $\det(\lambda I - F) = \sum_{i=0}^{n} p_i \lambda^i$, $p_n = 1$, for the matrix F of (3.1).

10. Prove that $c'(\lambda)/c(\lambda) = \text{trace}\ ((\lambda I - A)^{-1})$ where $c(\lambda) = \det(\lambda I - A)$.

11. Estimate how many ops are used in Gaussian elimination if the coefficient matrix is a) tridiagonal and b) in the Hessenberg form.

12. Given a Hermitian tridiagonal matrix T, find a diagonal matrix $D = \text{diag}(d_1, \ldots, d_n)$ such that $|d_i| = 1$ for all i and DTD^{-1} is a real symmetric matrix.

13 ([Ike]). Let T be a nonsingular tridiagonal matrix. Then there exist vectors $\mathbf{u}, \mathbf{v}, \mathbf{w}$ and $\mathbf{z}$ such that the matrices $T^{-1} - \mathbf{u}\mathbf{v}^T$ and $T^{-1} - \mathbf{w}\mathbf{z}^T$ are lower and upper triangular matrices, respectively, whose diagonals are filled with zeros. Prove this property and its appropriate extension to banded matrices. The inversion formula obtained in this way is numerically unstable (compare exercise 4 of chapter 3).

14. Show that there is no matrix A^+ satisfying (1.4) for $A = \begin{pmatrix} 1 & 1 \\ 1 & 1 \end{pmatrix}$ in $\mathbf{Z}_2$.

15 ([BT]). Given an increasing sequence $e_0, e_1, \ldots, e_k$ of nonnegative integers and the values of a polynomial $p_{e_k}(x) = \sum_{i=0}^{k} p_{e_i} x^{e_i}$ on a fixed set of $k+1$ distinct complex points $x_0, \ldots, x_k$, compute the coefficients $p_{e_0}, \ldots, p_{e_k}$ of $p_{e_k}(x)$. (Note that this problem involves submatrices of a Vandermonde matrix.)

16. Compare the cost of computing $\mathbf{w}^T A$ in the two algorithms of section 6 [based on (6.2) and (6.3)].

17. Let T be an $n \times n$ Toeplitz matrix whose first column and first row are filled with independent random entries chosen from a set S. Let A be an $n \times n$ nonsingular matrix. Estimate the probability, in terms of n and $|S|$, that the matrix TA:
    a) has nonzero entry (0,0);
    b) is strongly nonsingular.

    Do the same exercise assuming that T is a random Hankel matrix, random circulant matrix, or $T = UL$, U^T and L are random lower triangular Toeplitz matrices.

18. Show that the linear system (5.7) is singular if $n^* < n$ and nonsingular if $n = n^*$.

19. Specify the evaluation of the coefficients of all the pseudo resultants $R_i(x)$ for a given pair of polynomials, so as to minimize the number of ops used.

20. a) Show that the set of lower triangular $n \times n$ Toeplitz matrices is the (commutative) matrix algebra generated by the matrix Z of Definition 4.1, i.e., the set of all the matrices $\sum_{i=0}^{n-1} a_i Z^i$.

    b) Prove that this algebra over the field $\mathbf{F}$ is isomorphic to the algebra $\mathbf{F}[x]/x^n$ of all the polynomials over $\mathbf{F}$ modulo x^n.

21. a) Show that the set of all $n \times n$ circulant matrices is a (commutative) algebra generated by the circulant matrix $Z_1^{(n)} = Z_{1,n,n}(\mathbf{e}^{(1)}) = [c_{ij}]$, $c_{ij} = c_{i-j}$, $c_{i-j} = 1$ if $i - j = 1 \bmod n$, $c_{i-j} = 0$ otherwise.

    b) ([AB]). Show that A^+ is a circulant matrix if and only if A is a circulant matrix.

    c) Show that the set of $n \times n$ anticirculant matrices is the algebra generated by $Z_{-1}^{(n)} = Z_{-1,n,n}(\mathbf{e}^{(1)})$. Prove that $Z_{-1}^{(n)}$ is similar to $\theta Z_1^{(n)}$ where θ is an n-th root of -1.

    d) If n is even, prove the similarity of the matrices $Z_1^{(n)}$ and $\text{diag}(Z_1^{(n/2)}, Z_{-1}^{(n/2)})$. Deduce that the class of circulant matrices of the size $n \times n$ is the direct sum of the classes of circulant and anticirculant matrices of the size $(n/2) \times (n/2)$.

    e) Prove that the set of $n \times n$ (anti)circulant matrices is isomorphic to the algebra $\mathbf{F}[x]/(x^n - 1)$ [respectively, $\mathbf{F}[x]/(x^n + 1)$]. Interpret property d) by means of the relation $x^n - 1 = (x^{n/2} - 1)(x^{n/2} + 1)$.

    f) Extend the properties c) and e) to the class of f-circulant $n \times n$ matrices, for any fixed $f \neq 0$.

22. a) Consider the class τ of $n \times n$ matrices $A = [a_{ij}]$ such that $a_{i,j+1} + a_{i,j-1} = a_{i+1,j} + a_{i-1,j}$, where $a_{ij} = 0$ if a subscript is out of the range. Prove that this class is the algebra generated by the matrix $Z + Z^T$ for the matrix Z of Definition 4.1. Moreover, let $A \in \tau$, $S = [\sqrt{\frac{2}{n+1}} \sin \frac{\pi i j}{n+1}]$, $i, j = 1, \dots, n$. Prove that $S^T = S^{-1} = S$, $SAS = D$, $D = \text{diag}(d_1, \dots, d_n)$, $d_i = (S\mathbf{a})_i / (\sqrt{\dfrac{2}{n+1}} \sin \dfrac{\pi i}{n+1})$, where $\mathbf{a}$ denotes the first column of the matrix A. The vector $\sqrt{\frac{n+1}{2}} S\mathbf{v}$ is said to be the *sine transform of* $\mathbf{v}$ (compare exercise 5 of chapter 1).

    b) Prove that the class τ over $\mathbf{F}$ is isomorphic to the algebra $\mathbf{F}[x]/Q_n(x)$, of polynomials modulo $Q_n(x)$, where $Q_n(x)$ is the

Chebyshev-like polynomial of degree n defined in Theorem 11.4.

23 ([BF]). Let $H = (h_{ij})$ be the $n \times n$ Hartley matrix defined by $h_{ij} = \sin(\frac{2\pi}{n}ij) + \cos(\frac{2\pi}{n}ij)$, $i, j = 0, \ldots, n-1$, and associated with the Hartley transform $\mathbf{u} \to H\mathbf{u}$. Prove that $H^T H = I$. Show that each matrix of the class $\mathcal{H} = \{A \in \mathbf{R}_{n,n} : \quad A = H\mathrm{diag}(d_0, \ldots, d_{n-1})H,\ d_0, \ldots, d_{n-1} \in \mathbf{R}\}$ can be represented as the sum of a symmetric circulant matrix and of a suitable Hankel matrix.

24. a) Interpret the evaluation of the rational function $w(t) = \sum_{j=0}^{n-1} \frac{v_j}{t-s_j}$ on the set $\{t_0, \ldots, t_{n-1},\}$ and interpolation to $w(t)$, for given values $s_0, \ldots, s_{n-1}$, $t_0, \ldots t_{n-1}$, $w(t_0), \ldots, w(t_{n-1})$, as Problems 2.1 ($M \cdot VECTOR$) and 2.3 ($LIN \cdot SOLVE$) for a Cauchy matrix $C = C(\mathbf{s}, \mathbf{t})$.

b) Reduce the Cauchy problem of rational interpolation [Problem 1.5.2 ($RAT \cdot INTERP$)] to solving a homogeneous system of linear equations with a Vandermonde type matrix of coefficients.

25. Let $u(x) = a(x)b(x)$, where $a(x)$ and $b(x)$ are polynomials of degrees h and k, respectively, and let F_u be the $(h+k) \times (h+k)$ Frobenius matrix associated with $u(x)$. Prove that

a) the first h columns of $a(F_u^T)$ are linearly independent, and therefore rank $a(F_u^T) \geq h$;

b) the first k columns of $b(F_u^T)$ are linearly independent, and therefore rank $b(F_u^T) \geq k$.

From the relation $O = u(F_u^T) = a(F_u^T)b(F_u^T)$ deduce that

c) $a(F_u)$ maps into $\mathbf{0}$ at least k linearly independent vectors, and therefore $\mathcal{N}(a(F_u)) \geq k$, rank $a(F_u^T) \leq h$;

d) $b(F_u)$ maps into $\mathbf{0}$ at least h linearly independent vectors, and therefore $\mathcal{N}(b(F_u)) \geq h$, rank $b(F_u^T) \leq k$.

Conclude that rank $a(F_u^T) = h$ and rank $b(F_u^T) = k$.

26. Analyze the algebra generated by the matrix

$$S_\alpha = \begin{pmatrix} \alpha & 1 & & & O \\ 1 & 0 & \ddots & & \\ & \ddots & \ddots & 1 & \\ & & \ddots & 0 & 1 \\ O & & & 1 & \alpha \end{pmatrix},$$

and find the similarity transformations that diagonalize S_α for $\alpha \in \{-1, 1\}$.

27. Verify that

$$\begin{pmatrix} 0 & 0 & 1 \\ 0 & 1 & 1 \\ 1 & 1 & 2 \end{pmatrix} = \begin{pmatrix} 1 & 0 & 0 \\ 0 & 1 & 0 \\ \frac{1}{2} & 1 & 1 \end{pmatrix} J \begin{pmatrix} 1 & 0 & \frac{1}{2} \\ 0 & 1 & 1 \\ 0 & 0 & 1 \end{pmatrix}$$
$$= \begin{pmatrix} 1 & 0 & 0 \\ 1 & 1 & 0 \\ 1 & 0 & 1 \end{pmatrix} J \begin{pmatrix} 1 & 1 & 1 \\ 0 & 1 & 0 \\ 0 & 0 & 1 \end{pmatrix}$$

and conclude that the block triangular factorization of Proposition 10.1 is not unique.

28. Let $(A)_{i,j} = \frac{u_i - v_j}{t_i - s_j}$, so that A is a Loewner matrix. Let $F(A) = D(\mathbf{s})A - AD(\mathbf{t})$ be a Cauchy (Hilbert) type operator. Then $F(A) = -\mathbf{u}\mathbf{e}^T + \mathbf{e}\mathbf{v}^T$.

29 ([Fi]). Let $A = (\frac{c_i - d_j}{y_i - z_j})$ be a Loewner matrix and $u(x), v(x)$ be two polynomials of degrees at most n such that $u(y_i)u(z_j) \neq 0$, $c_i = v(y_i)/u(y_i)$, $d_j = v(z_j)/u(z_j)$, $i, j = 1, \ldots, n$. Then prove that $A = D_1^{-1} V(\mathbf{y}) B(u, v) V^T(\mathbf{z}) D_2^{-1}$, where $D_1 = \operatorname{diag}(u(y_1), \ldots, u(y_n))$, $D_2 = \operatorname{diag}(u(z_1), \ldots, u(z_n))$, $V(\mathbf{z})$ and $V(\mathbf{y})$ are two Vandermonde matrices associated with $z_1, \ldots, z_n$ and with $y_1, \ldots, y_n$, respectively.

30. Elaborate the details and the complexity estimates for the tridiagonalization of a general matrix [Problems 3.2a ($T \cdot REDUCE$) and 3.2b ($T \cdot REDUCE1$)], based on Algorithms 3.3 and 10.2, assuming that no degeneracy occurs (compare Proposition 4.5.3).

31. Extend the complexity estimates for Problem 2.5a ($ALL \cdot INVERT$) (compare the end of section 11) to the estimates for Problem 2.5b ($ALL \cdot INVERT1$), assuming nonsingularity of all the submatrices involved. How will your estimates change if you relax the nonsingularity assumptions?

32. Generalize Fact 11.5 to the cases of all the operators of (11.5), (11.8)–(11.12) of the Vandermonde and Cauchy (Hilbert) types.

33. Estimate the F-rank of A where F is defined by (11.1) or (11.2), and
    a) A is a Sylvester matrix;
    b) $A = T^H T$, T is a Toeplitz matrix;
    c) $A = T^{-1}$ where T is a nonsingular Toeplitz matrix;
    d) $A = S^{-1}$ where S is a nonsingular Sylvester matrix;
    e) A is the matrix obtained by deleting the third row and the third column of a $k \times k$ Toeplitz matrix for $k > 3$;
    f) A is the matrix obtained by replacing the i-th row of a Sylvester matrix by a random vector [do for $i = 1$ and $i = 2$ and separately for F of (11.1) and F of (11.2)];

g) A is the Schur complement of a northwestern block of a Toeplitz matrix.

34. Specify some appropriate operators F in order to bound the F-ranks of block Toeplitz and block Hankel matrices.

35 ([B83]). Prove that for the operator $F = F^{\pm}$ of (11.23) the F-rank of any matrix A belonging to the linear space spanned by all Toeplitz and Hankel matrices is at most 4. Determine an inversion formula for such matrices based on the operator $F^{\pm}$.

36. Prove that the F-rank of a Toeplitz matrix is at most 2 and that a $g \times h$ block matrix with Toeplitz blocks has F-rank at most $g + h$, for $F = F^{+}$, $F = F^{-}$, $F = F_{+}$, and $F = F_{-}$. Moreover, for the above F, prove that the F-rank of the product of two Toeplitz matrices is at most 4. Determine some inversion formulae for Toeplitz matrices based on the above operators, in addition to the formulae in the text. Follow the pattern of Example 11.2 to generalize the operators F^{+}, F^{-}, F_{+} and F_{-} by replacing (in their definitions) the matrix $Z = Z_0$ by the unit f-circulant matrix Z_f (nonsingular for $f \neq 0$). Extend the known properties of the operators F^{+}, F^{-}, F_{+}, F_{-} to the case of such new operators.

37. Obtain the asymptotic estimates of Corollary 12.1 for the number of ops required in order to multiply an $n \times n$ matrix A of F_1-rank r by a vector and to compute an F-generator of the product AB of the pair of $n \times n$ matrices A and B satisfying the assumptions of Proposition 12.1 for L and M of (12.5)–(12.7), provided that we are given r_1, r_2 and the number of ops required for computing the products of A by a vector and of a vector by B. Then specify the overhead constants of these asymptotic estimates. Consider the special case where A and B are Hermitian Toeplitz matrices and L and M satisfy (12.5). (Use Ammar-Gader's formula of [AG89].)

38. Prove Theorem 11.2 by extending the technique used in the proof of Theorem 11.3.

39. Prove that

    a) $Z - Z^T = \mathbf{i}D(Z + Z^T)D^{-1}$, where $D = \text{diag}(1, \mathbf{i}, \mathbf{i}^2, \ldots, \mathbf{i}^{n-1})$ and $\mathbf{i}$ is the imaginary unit, $\mathbf{i}^2 = -1$.

    b) Use a) to prove that rank $F^{\mp}(A) \leq 4$ for any Toeplitz+Hankel matrix A, where the operator $F^{\mp}$ is defined by $F^{\mp} = A(Z - Z^T) - (Z - Z^T)A$.

    c) Extend Theorem 11.4 and Algorithms 11.1 and 11.2 to $F^{\mp}$.

d) ([KKM]). For the matrices A of the form $T+H$, where T is a Toeplitz-like matrix and H is a Hankel-like matrix, and for the operator $F_{\pm}(A) = A - ZAZ - ZAZ^T + Z^2AZZ^T$, prove that rank $F(A) = O(1)$. Show some similar operators with the same property. Consider two classes of candidate operators: $F_f^{\pm}(A) = A(Z_f + Z_f^T) - (Z_f + Z_f^T)A$ and $F_f^{\mp}(A) = A(Z_f - Z_f^T) - (Z_f - Z_f^T)A$ such that $F_0^{\pm}(A) = F^{\pm}(A)$, $F_0^{\mp}(A) = F^{\mp}(A)$, and observe that $F_1^{\pm}(A)$ is the operator of Example 11.6. Extend the results of sections 11 and 12 to the case of these operators.

e) Let T and HJ be $n \times n$ Toeplitz-like matrices, and assume that $T+H$ and $T-H$ are nonsingular matrices. Show that the block matrix

$$A = \begin{bmatrix} T & HJ \\ JH & JTJ \end{bmatrix}$$

is nonsingular, $F(A)$ has a rank $O(1)$ for F being any of the operators F_+, F_-, F^+, F^-, $F^{\pm}$, $F_{\pm}$, $F^{\mp}$, and devise an algorithm for computing $(T+H)^{-1}$ and $(T-H)^{-1}$ from A^{-1}. (This reduces solving linear systems with Toeplitz-like + Hankel-like coefficient matrices $T+H$ and $T-H$ to the one with the Toeplitz-like matrix A, although the approach using the operators $F^{\pm}$ and $F^{\mp}$ of Example 11.4 leads to a slighter faster solution.) *Hint:* $A = Q\,\mathrm{diag}(T+H, T-H)\,Q^T$, where $\sqrt{2}Q = \begin{pmatrix} I & I \\ J & -J \end{pmatrix}$.

40. a) Prove inversion formulae for Toeplitz matrices based on Theorem 11.3.

b) ([GO]) Deduce Proposition 11.4 by using Proposition 11.3.

c) Extend Algorithms 11.1 and 11.2 by using the operators $F_f^{\pm}(A)$ of exercise 39 instead of the operator $F^{\pm}(A)$.

41. Let 1/2 be an upper bound on the probability that the randomized algorithm of section 13 for computing the rank of a Toeplitz matrix, with $TEST(k)$ given by Algorithm 8.2, outputs $r \leq \mathrm{rank}\, A$. How many times do we need to repeat calls to $TEST(k)$ in order to ensure the upper bound 1/1000 on the error probability?

42. Prove that $||A^HA||_2 = ||AA^H||_2 = ||A||_2^2$.

43. Indicate over which fields of constants each reduction of Proposition 2.2 holds.

44. Let A be any $n \times n$ matrix.

a) Prove that $m_A(\lambda) = c_A(\lambda)$ if and only if $\det T_n \neq 0$, where $(T_n)_{ij} = \mathbf{u}^T A^{i+j}\mathbf{w}$, for two vectors of indeterminates $\mathbf{u}$ and $\mathbf{v}$. *Hint:* observe that $m_A(\lambda) \neq c_A(\lambda)$ if and only if the Krylov matrices $K(A, \mathbf{u}, n)$ and $K(A^T, \mathbf{w}, n)$ are singular for any pair of vectors $\mathbf{u}$ and $\mathbf{w}$ or, over a field of characteristic zero, if and only if the matrix $(K(A^T, \mathbf{w}, n))^T K(A, \mathbf{u}, n)$ is singular.

b) Answer whether deg($m_A(\lambda)$)=rank(T_n) for any $n \times n$ matrix A, or, equivalently, whether the leading principal $k \times k$ submatrix of T_n is nonsingular if $k =$ deg $(m_A(\lambda))$, and is singular if $k >$ deg $(m_A(\lambda))$.

45. Compute the product of the pair of $n \times n$ Toeplitz matrices by using $4.5n^2 + O(n \log n)$ ops ([B83]). Then improve the bound to $4n^2 + O(n \log n)$ ([Wo1]).

46. Given n scalars $x_0, \ldots, x_{n-1}$ and the coefficients $a_0, \ldots, a_{n-1}$ of the $(n-1)$-st degree polynomial $a(x) = \sum_{i=0}^{n-1} a_i x^i$, we may compute the coefficients of the n-th degree polynomial $\Gamma(x) = \prod_{j=0}^{n-1}(x - x_j)$ in $O(n \log^2 n)$ ops, and then define the $n \times n$ matrix $A = A(a(x), \Gamma(x))$ whose i-th column is obtained as the coefficient vector of the polynomial $x^i a(x)$ mod $\Gamma(x)$, $i = 0, 1, \ldots, n-1$, so that the first column of A is given by $(a_0, a_1, \ldots, a_{n-1})^T$.

a) For any fixed vector $\mathbf{v}$ of dimension n, compute $A\mathbf{v}$ in $O(n \log^2 n)$ ops, and if A is nonsingular, compute $A^{-1}\mathbf{v}$ in $O(n \log^2 n)$ ops.

b) Decrease the above bounds to $O(n \log n)$ in the case where $\Gamma(x)$ divides $x^N - c^N$ for a constant c and for $N = O(n)$.

c) Let $A(\Gamma(x))$ denote the class of all matrices $A(a(x), \Gamma(x))$ for a fixed $\Gamma(x)$ and for all $a(x)$. Prove that this class $A(\Gamma(x))$ is the algebra generated by the Frobenius matrix associated with the polynomial $\Gamma(x)$. In particular, if $\Gamma(x) = x^n$, obtain the algebra of lower triangular Toeplitz matrices; if $\Gamma(x) = x^n - f$, obtain the algebra of f-circulant matrices.

47. Prove that the matrix R of (14.3) is nonsingular if the matrix A of (14.2) is nonsingular. *Hint* ([PSA]): Examine the factorization (2.1) of the matrix $\begin{pmatrix} A & C \\ E & O \end{pmatrix}$ for the matrices A, C, E of (14.2).

48 ([XZ]). Extend the Gohberg-Semencul formulae to represent the Moore-Penrose generalized inverse of a Toeplitz matrix A as $\sum_{i=1}^{8} L_i U_i$ where L_i, U_i^T are lower triangular Toeplitz matrices.

49. Let $F_k = (\omega^{ij}) \in \mathbf{C}_{k,k}$, where ω is a primitive k-th root of 1, so that $\Omega = F_k/\sqrt{k}$ is the Fourier matrix of Theorem 5.1, for $k = n+1$. For any pair of matrices $A \in \mathbf{C}_{n,m}$, $B \in \mathbf{C}_{p,q}$, let $A \otimes B \in \mathbf{C}_{np,mq}$ denote the block matrix having blocks $(a_{ij}B)$. $A \otimes B$ is called the *tensor product*

or the *Kronecker product* of A and B (compare section 11).

a) Prove the following matrix factorization:

$$F_{2k} = (F_2 \otimes I_k) D_{2k} (I_2 \otimes F_k) P_{2k}^T =$$
$$\begin{pmatrix} I_k & I_k \\ I_k & -I_k \end{pmatrix} D_{2k} \begin{pmatrix} F_k & O \\ O & F_k \end{pmatrix} P_{2k}^T,$$

where P_{2k} denotes the permutation matrix such that $P\mathbf{v} = \mathbf{w}$, $w_i = v_{2i}$, for $i = 0, \ldots, k-1$, $w_{i+k} = v_{2i+1}$, for $i = 0, \ldots, k-1$, and $D_{2k} = \mathrm{diag}(I_k, \mathrm{diag}(1, \omega, \ldots, \omega^{k-1}))$. Assume that k is an integer power of 2 and recursively apply this factorization in order to compute the product $F_k\mathbf{v}$, thus obtaining a *Decimation in Time (DIT) base 2 FFT algorithm* (compare Example 2.1). Observe that this algorithm is the matrix version of the algorithm described in the solution of Problem 1.2.2a (DFT).

b) By using the property $F_{2k}^T = F_{2k}$, first prove that

$$F_{2k} = P_{2k} (I_2 \otimes F_k) D_{2k} (F_2 \otimes I_k) =$$
$$P_{2k} \begin{pmatrix} F_k & O \\ O & F_k \end{pmatrix} D_{2k} \begin{pmatrix} I_k & I_k \\ I_k & -I_k \end{pmatrix},$$

then derive a *Decimation in Frequency (DIF) base 2 FFT algorithm* relying on the recursive application of the latter formula. Try to construct an analogous factorization in the case where $k = pq$ for two integers $p, q \geq 2$.

c) Exploit the matrix factorization shown in a) and b) in order to compute the convolution $\mathbf{z}$ of two vectors $\mathbf{x}$ and $\mathbf{y}$, avoiding permutations of the components of $\mathbf{x}$ and $\mathbf{y}$. *Hint:* Use the property $\mathbf{z}$=IDFT(DFT($\mathbf{x}$)$*$DFT($\mathbf{y}$)), where $*$ denotes componentwise multiplication (see the solution of Problem 1.2.4a), and apply a DIF algorithm for DFT and a DIT algorithm for IDFT.

d) Revisit the base 2 FFT algorithm of chapter 1 in the view of properties c), d) and e) of exercise 21.

50. Revise the algorithm of [BA], [Morf]; apply it by using $n-1$ recursive steps, with the matrices B of size 1×1, and assuming that the input Toeplitz-like matrix A is strongly nonsingular. Show that this algorithm solves Problem 2.5a ($ALL \cdot INVERT$) for the matrix A in $O(n^2)$ ops.

51. Extend the algorithm supporting Theorem 13.1 to the solution of Problem 2.3a ($LIN \cdot SOLVE1$) in the case of a Toeplitz-like input matrix. *Hint:* Solve the auxiliary Problem 2.11b ($G \cdot COMPRESS$) by using the algorithms of section 2 (compare [P92b], [Ka93]).

Appendix A. All-Pairs-Shortest-Distances Computation in a Graph by Means of Matrix Multiplication.

Here, we recall an algorithm from [Sei] that computes the shortest distances between all pairs of vertices of undirected and unweighted graph G with n vertices by using $O(M(n)\log n)$ ops performed with integers from $-n$ to n, provided that $M(n)$ ops suffice for $n \times n$ matrix multiplication (section 2). This result (not used in our book) demonstrates the power of matrix multiplication in applications to graph algorithms.

The all-pairs-shortest-distances (APSD) problem generalizes the transitive closure problem (of computing all pairs of vertices of a graph connected by paths), equivalent to Boolean matrix multiplication ([AHU]). Generally, the APSD problem is stated for a graph (or a digraph) with weighted edges. If the graph is undirected and if the weights are integers between 0 and B, then the problem has been solved in $O(B^2 M(n)\log n)$ ops and comparisons (see [GM], [Mar], and [AGM]). The next algorithm works for a less general (although still highly important) problem and has a very simple description.

Algorithm (APSD) (see [Sei] for proofs).

Input: $n \times n$ matrix A filled with 0s and 1s, the adjacency matrix of undirected, connected graph G.

Output: $n \times n$ integer matrix $D = (d_{ij})$, with d_{ij} denoting the length of a shortest path joining vertices i and j in G.

function APSD (A: $n \times n$ matrix filled with 0s and 1s): $n \times n$ integer matrix.

Computations:

- compute $Z = A \cdot A$
- compute $n \times n$ matrix B, where $b_{ij} = 1$ if and only if $i \neq j$ and ($a_{ij} = 1$ or $z_{ij} > 0$)
- if $b_{ij} = 1$ for all $i \neq j$ then return $(2B - A)$
- compute $T = APSD(B)$
- compute $X = T \cdot A$
- return $n \times n$ matrix D, where $d_{ij} = \begin{cases} 2t_{ij} & \text{if} \quad x_{ij} \geq t_{ij} \cdot \text{degree}(j) \\ 2t_{ij} - 1 & \text{if} \quad x_{ij} < t_{ij} \cdot \text{degree}(j) \end{cases}$

Appendix B. The Problems and Their Reductions to Each Other.

We will next list the problems of this chapter together with their reductions to each other and computational complexity estimates. (char 0) will mean the solution over all the fields of characteristic 0.

– Problem 2.1 ($M \cdot VECTOR$), $n \times n$ matrix-by-vector multiplication, $O(n^2)$.

– Problem 2.1a ($KRYLOV$), compute the $n \times m$ Krylov matrix, $O(M(n) \log n)$, where $m = O(n)$.

– Problem 2.2 ($M \cdot MULTIPLY$), $n \times n$ matrix multiplication, $M(n) = O(n^{\omega})$, $2 \leq \omega < 2.376$.

– Problem 2.2a ($M \cdot POWERS$), compute the first m powers of an $n \times n$ matrix, $O(mM(n))$.

$$M \cdot POWERS \succeq CHAR \cdot POL.$$

– Problem 2.3 ($LIN \cdot SOLVE$), compute a solution of an $n \times n$ linear system if the matrix is nonsingular, otherwise output SINGULAR, $O(M(n))$.

$$LIN \cdot SOLVE \preceq LSP \cdot FACTORS$$
$$LIN \cdot SOLVE \preceq (M \cdot VECTOR, INVERT)$$
$$LIN \cdot SOLVE \preceq (CHAR \cdot POL, KRYLOV).$$

– Problem 2.3a ($LIN \cdot SOLVE1$), solve an $m \times n$ linear system if it is consistent, otherwise output INCONSISTENT, $O(M(m)n/m)$ if $m \leq n$, $O(M(n)m/n)$, otherwise.

$$LIN \cdot SOLVE1 \preceq LSP \cdot FACTORS$$
$$LIN \cdot SOLVE1 \preceq (NULL \cdot SPACE, LIN \cdot SOLVE, M \cdot VECTOR)$$
$$LIN \cdot SOLVE1 \preceq (M \cdot VECTOR, L \cdot SQUARES) \quad \text{(char 0)}$$
$$LIN \cdot SOLVE1 \preceq (PRCNDTN, M \cdot MULTIPLY, LIN \cdot SOLVE).$$

– Problem 2.3b ($NULL \cdot SPACE$), compute a basis for the null space of an $m \times n$ matrix, $O(M(m)n/m)$ if $m \leq n$; $O(M(n)m/n)$ otherwise.

$$NULL \cdot SPACE \preceq (LSP \cdot FACTORS, M \cdot MULTIPLY, INVERT)$$
$$NULL \cdot SPACE \preceq (M \cdot PRCNDTN, M \cdot MULTIPLY, INVERT).$$

– Problem 2.4 ($DETERMINANT$), compute the determinant of an $n \times n$ matrix, $O(M(n))$.

$$INVERT \succeq DETERMINANT \succeq INVERT$$

$$DETERMINANT \preceq LSP \cdot FACTORS.$$

– Problem 2.4a ($SINGULAR?$), test if a given $n \times n$ matrix is singular, $O(M(n))$.

– Problem 2.5 ($INVERT$), compute the inverse of an $n \times n$ matrix if it is nonsingular, otherwise output SINGULAR, $O(M(n))$.

– Problem 2.5a ($ALL \cdot INVERT$), compute the inverses of all the leading principal submatrices of a strongly nonsingular $n \times n$ matrix, $O(n^3)$.

– Problem 2.5b ($ALL \cdot INVERT1$), compute the inverses of a fixed sequence of nested nonsingular submatrices of a nonsingular $n \times n$ matrix, $O(n^3)$.

– Problem 2.6 ($CHAR \cdot POL$), compute the coefficients of the characteristic polynomial of an $n \times n$ matrix, $O(\min\{M(n) \log^2 n, n^\omega \log n\})$.

$$CHAR \cdot POL \succeq DETERMINANT.$$

– Problem 2.6a ($MIN \cdot POL$), compute the coefficients of the minimum polynomial of an $n \times n$ matrix, $O(M(n) \log n)$ (randomized).

$$MIN \cdot POL \preceq (RECUR \cdot SPAN, M \cdot VECTOR, KRYLOV) \text{ (randomized).}$$

– Problem 2.7 ($LU \cdot FACTORS$), compute, if it exists, the LU factorization of an $n \times n$ matrix, $O(M(n))$.

$$LU \cdot FACTORS \succeq DETERMINANT.$$

– Problem 2.7a ($PLU\tilde{P} \cdot FACTORS$), compute the $PLU\tilde{P}$ factorization of an $m \times n$ matrix of rank r, $O(mnr)$.

$$PLU\tilde{P} \cdot FACTORS \succeq SUBMATRIX,$$
$$PLU\tilde{P} \cdot FACTORS \succeq M \cdot RANK.$$

– Problem 2.7b ($PLU \cdot FACTORS$), compute the PLU factorization of an $n \times n$ matrix, $O(M(n))$.

– Problem 2.7c ($LSP \cdot FACTORS$), compute the LSP factorization of an $m \times n$ matrix, $m \leq n$, $O(M(m)n/m)$.

$$LSP \cdot FACTORS \preceq M \cdot MULTIPLY$$
$$LSP \cdot FACTORS \preceq LIN \cdot SOLVE1$$

– Problem 2.8 ($L \cdot SQUARES$), least-squares solution of a linear system of m equations with n unknowns, $m \leq n$, $O(M(m)n/m)$.

$$L \cdot SQUARES \preceq (CHAR \cdot POL, KRYLOV)$$
$$L \cdot SQUARES \preceq (GEN \cdot INVERSE, M \cdot VECTOR).$$

– Problem 2.9 ($QR \cdot FACTORS$), compute a QR factorization of an $m \times n$ matrix of full rank n, $O(M(n)m/n)$.

$$QR \cdot FACTORS \preceq (M \cdot MULTIPLY, LU \cdot FACTORS)$$

– Problem 2.9a ($QRP \cdot FACTORS$), compute a QRP factorization of an $m \times n$ matrix of rank r, $O(mnr)$.
– Problem 2.10 ($M \cdot RANK$), compute the rank of an $m \times n$ matrix, $m \leq n$, $O(M(m)n/m)$.

$$M \cdot RANK \preceq LSP \cdot FACTORS$$
$$M \cdot RANK \preceq SVD \quad \text{(char 0)}$$
$$M \cdot RANK \preceq CHAR \cdot POL \text{ (char 0).}$$

– Problem 2.10a ($SUBMATRIX$), compute the nonsingular submatrix of the maximum size of an $m \times n$ matrix of rank r, $m \leq n$, $O(M(m)n/m)$.

$$SUBMATRIX \preceq LSP \cdot FACTORS$$
$$SUBMATRIX \succeq M \cdot PRECNDTN.$$

– Problem 2.10b ($M \cdot PRECNDTN$), precondition an $m \times n$ matrix A of rank r, $m \leq n$, that is, compress A into an $r \times r$ nonsingular matrix RAS, $O(M(m)n/m)$.

$$M \cdot PRECNDTN \succeq M \cdot RANK.$$

– Problem 2.11 ($GENERATOR$), compute a generator of the minimum length for an $m \times n$ matrix, $O(M(n))$, $m = O(n)$.

$$GENERATOR \preceq LSP \cdot FACTORS.$$

– Problem 2.11a ($GENERATOR1$), compute a generator of the minimum length r for an $m \times n$ matrix A, given with the compressors R and S, $O(M(r)(m+n)/r)$.

$$GENERATOR1 \preceq (M \cdot MULTIPLY, INVERT).$$

– Problem 2.11b ($G \cdot COMPRESS$), compression of a generator of length $\hat{r}$ of an $n \times n$ matrix of rank r, $O(M(\hat{r})n/\hat{r})$.

$$G \cdot COMPRESS \preceq (SUBMATRIX, M \cdot MULTIPLY, INVERT).$$

– Problem 2.12 ($GEN \cdot INVERSE$), compute the generalized inverse of an $m \times n$ matrix, $O(M(m)n/m)$ if $m \le n$, $O(M(n)m/n)$ otherwise.

$$GEN \cdot INVERSE \preceq (LSP \cdot FACTORS, M \cdot MULTIPLY, INVERT)$$
$$GEN \cdot INVERSE \preceq (CHAR \cdot POL, M \cdot POWERS).$$

– Problem 3.1 ($EIGENVALUES$), compute the eigenvalues of a matrix.
– Problem 3.1a ($EXTREME \cdot EIGENVALUES$), compute the extreme eigenvalues of a matrix.
– Problem 3.2 ($H \cdot REDUCE$), reduce an $n \times n$ matrix to Hessenberg form, $O(M(n) \log n)$.

$$H \cdot REDUCE \preceq (M \cdot MULTIPLY, QR \cdot FACTORS, KRYLOV).$$

– Problem 3.2a ($T \cdot REDUCE$), reduce a matrix to the tridiagonal form by a similarity transformation and compute the similarity transformation matrix.
– Problem 3.2b ($T \cdot REDUCE1$), reduce a matrix to the tridiagonal form.
– Problem 3.2c ($I \cdot EIGENVALUES$), solve the inverse eigenvalue problem.
– Problem 3.3 (SVD), compute the singular value decomposition.
– Problem 3.4 ($M \cdot NORM$), compute the norm of a given matrix.
– Problem 3.5 ($M \cdot CONDITION$), compute the condition number of a given matrix.

$$M \cdot CONDITION \preceq (INVERT, M \cdot NORM).$$

– Problem 3.6 ($M \cdot COMPRESS$), compute a lower rank matrix approximating a given matrix.

$$M \cdot COMPRESSSS \preceq (SVD, M \cdot MULTIPLY).$$

– Problem 5.1 ($TOEPL \cdot VECTOR$), compute the product of an $m \times n$ Toeplitz matrix and a vector, $O(n \log n)$, $m = O(n)$.
– Problem 5.2 ($CTH \cdot VECTOR$), compute the product of an f-circulant, a Toeplitz or a Hankel $n \times n$ matrix and a vector, $O(n \log n)$.
– Problem 5.3 ($TTOEPL \cdot INVERT$), invert an $n \times n$ triangular Toeplitz matrix, $O(n \log n)$.
– Problem 5.4 ($CIRC \cdot INVERT$), invert an $n \times n$ nonsingular f-circulant matrix, $O(n \log n)$.
– Problem 5.5 ($TOEPL \cdot SOLVE$), solve an $n \times n$ Toeplitz system, $O(n \log^2 n)$.

– Problem 6.1 ($V^T \cdot SOLVE$), solve an $n \times n$ transposed Vandermonde system, $O(n \log^2 n)$.
– Problem 6.2 ($V^T \cdot VECTOR$), compute the product of an $n \times n$ transposed Vandermonde matrix and a vector, $O(n \log^2 n)$.

3

Bit-Operation (Boolean) Cost of Arithmetic Computations

Summary.

We define the Boolean cost (or the bit-cost) of arithmetic computations, relate it to numerical stability of computations with finite precision and survey the major techniques available for estimating and decreasing it. We also study the related problem of reducing the arithmetic complexity (represented by the number of ops involved) of computing approximation to the output. We include some techniques from recent journal articles and technical reports and a new result on the numerical stability of FFT. Some of the results and the techniques have been or promise to be useful in practice of computing.

1. Contents and Introduction. Bit-Operation Count. Numerical Stability and Condition.

Contents.

After this introductory section, we recall the bounds on the values output in the solution of a linear system and in the Euclidian scheme for polynomials, as well as the estimates on the probability that a linear system, nonsingular in the field of rationals, becomes singular in the field $GF(p)=\mathbf{Z}_p$, of integer modulo a random prime p (section 2). In section 3 we study various aspects and techniques of decreasing the precision of computation by means of the reduction modulo an integer. In section 4 we first present a new result on estimating roundoff errors of FFT and then turn to the study of the arithmetic complexity of multipoint polynomial approximation, where we report the results of the recent progress of 1988-1993. In sections 5–7, we recall some data compression techniques, including the compact multigrid, for obtaining high precision solutions to linear systems by means of lower precision computations. Section 8 is devoted to decreasing the arithmetic time by means of allowing to output an arbitrarily close approximation, rather than the exact solution. In section 9, we exploit the correlation between multiplication of integers and polynomials to demonstrate a special data compression technique (called binary segmentation), which also enables us to substantially accelerate some fundamental computations with matrices and polynomials whose input parameters (entries, coefficients) are represented by sufficiently short binary integers. In Appendix A we review the fundamentals of the floating point representation of real numbers and of the error analysis.

Boolean (bit-) complexity: motivation, definition and some general means of estimating and decreasing it.

The arithmetic models of computing, adopted in our first chapters, do not reflect the dependence of the computational complexity on the precision of computing. In practice, this dependence is substantial, and in this chapter we will extend our previous study from the infinite to the finite precision computations and from the arithmetic to the Boolean (RAM or circuit) models of computing [see chapter 1 (section 1 and Appendix A) on these models]. Our main objective in this chapter is to demonstrate some major techniques for establishing upper bounds on the bit-complexity of polynomial and matrix computational problems. We assume the Boolean RAM model, but similar techniques apply under the Boolean circuit model.

The simplest but in many cases the most effective approach is to apply the algorithms with a lower arithmetic cost and to assign an appropriate precision value to both operands and to the output of every arithmetic operation of an algorithm. The Boolean cost of such an operation decreases with the precision, but the user must keep the precision high enough to obtain a meaningful output. We define the Boolean cost of the entire arithmetic computation as the sum of the Boolean cost values of all its arithmetic operations; to decrease the overall Boolean cost, we shall care about decreasing both the arithmetic cost and the precision of the computations.

Two convenient frameworks for our study are suggested by the practice of scientific and engineering computation, where matrix and polynomial computations are performed by means of either algebraic and symbolic error-free computations or numerical computations with roundoff errors. The former (error-free) approach applies to computations with integers and/or rationals, which can be performed in the ring $\mathbf{Z}_M$ of integers modulo a fixed integer $M > 1$ [recall that this ring is a field, denoted $GF(M)$, if M is a prime]. The operands and the output of any operation in $\mathbf{Z}_M$ can be represented with the precision of $\lceil \log M \rceil$ bits (and with no errors), and there are some canonical tools (such as the Chinese remainder algorithm for integers and p-adic lifting) for decreasing the computational precision even further (see section 3). Rational numbers can be represented as pairs of integers (a, b), with the arithmetic operations defined by $(a, b) * (c, d) = (ac, bd)$, $(a, b)/(c, d) = (ad, bc)$, $(a, b) \pm (c, d) = (ad \pm bc, bd)$.

If all or some of the input and output values are (complex or real) irrational numbers, the users may try to replace them by their finite binary (or decimal, ternary, octal) floating point (or fixed point) approximations, thus shifting to numerical computations with roundoff errors and to the approximation models of computing [see (1.A.3), (1.A.4) and Appendix A to this chapter on the definition and the properties of floating point numbers and on some fundamentals of the error analysis]. The computations with roundoff approximate the output values within a fixed upper bound $\epsilon > 0$ on the errors, depending on the precision u of the computations (some comments on the correlation between the parameters u and ϵ will be given in the subsections of this section on numerical stability and condition and on the error analysis). As a rule, numerically stable algorithms with roundoff and with smaller output errors are preferred by the users to the infinite precision algebraic computations, because *the latter computations currently require much more computer time and memory space* (see Remark 3.3, however). For instance, a typical user would prefer to reduce a computational problem to nonrational Problem 2.3.3 (SVD), for which numerically stable approximation algorithms are available, rather than to Problem 2.2.7c ($LSP \cdot FACTORS$), for which we have a rational exact solution algorithm but generally have no numerically stable algorithm in the presence of roundoff errors.

If the precision u of a certain computation and/or the tolerance ϵ to the output errors are fixed, then the Boolean complexity of this computation is roughly proportional to the number of ops involved. This motivates our study (in sections 4, 5 and also 8) of the arithmetic complexity of approximation algorithms, whose class was defined in Appendix A to chapter 1 [compare (1.A.3), (1.A.4)]. We note that, *for several important computational problems, the approximate solutions* (with small output errors) *require substantially fewer ops than the exact solutions.* The two fields (of symbolic and algebraic computing and of numerical computing) have been developed by two independent groups of people and so far had little interaction. We believe that interaction between these two fields is of benefit to both of them. Application of the algebraic computation techniques to the design of effective approximation algorithms (such as the algorithms for multipoint polynomial approximation and for linear systems of equations) can be an example of such benefit (compare [P92e], [Riv], [P93d] and our Remark 4.2).

The choice among numerical approach with roundoff, algebraic approach with the infinite precision and their combination is in fact quite delicate (and should rely on the careful error and complexity analysis). Moreover, there are other competitive approaches, exploiting various *data compression techniques.* Since we mostly deal with the computations where the output values are rational functions in the input values, we could have assumed that any successful solution only involves arithmetic operations and that any decrease of the overall estimated bit-complexity of the computations should rely on using fewer single precision ops. In sections 6, 7 and 9, however, we show that, on the contrary, substantial progress in the solution of some fundamental rational computational problems can be based just on using low precision computation (in sections 6 and 7) or nonarithmetic operations of segmentation (in section 9): the approaches of sections 6, 7 and 9 do not fit the arithmetic (RAM and circuit) models of computations; they improve the previously known Boolean complexity estimates by improving the finite precision representation of the input, output and (in some cases) auxiliary values and thus achieving substantial data compression, which leads to economization of both computational time and memory space and may have a certain practical impact (compare Remark 9.2 and exercises 1 and 14).

For more information on algebraic and symbolic computations, we refer the reader to the books [BCL], [GCL], [Z93] (containing further bibliography), to manuals on the symbolic computation libraries and packages of subroutines, and to periodicals, in particular, to the Journal of Symbolic Computation, as well as to the Proceedings of the ACM-SIGSAM Annual International Symposia on Algebraic and Symbolic Computations (ISSAC). The developed area of numerical approximation is well-represented in various texts on numerical analysis (such as [A], [CdB], [DB], [Ster]), containing further bibliography; the volume [GL] is an outstanding source of material on numerical matrix computations. Abundant material on numerical computing can also be found in periodicals, such as Mathematics of Computation, Numerische Mathematik, Computing, and SIAM Journals on Numerical Analysis and Matrix Analysis.

In the remainder of this section and in section 2, we will recall some basic definitions and auxiliary results and will give some general comments on the

Boolean models of computing (formally defined in chapter 1), on estimating the bit-complexity and on the related topics of numerical analysis.

Remark 1.1. *How realistic are the Boolean models of computing?* As a rule, the time of performing each fixed arithmetic operation is fixed for each computer, together with the precision of computation (which rather suggests using the arithmetic models). The precision, however, can be doubled at the cost of increasing (by more than twice) the performance time per operation. Moreover, the precision varies for various computers, and it generally costs more to build a computer allowing computation with a higher precision. In some cases, it is efficient to partition the computation into stages and to vary the precision from stage to stage, which is a typical feature of p-adic (Newton-Hensel's) lifting and the Chinese remainder algorithm in the area of algebraic computing (see section 3); in practice of numerical computing with roundoff, variation of the precision also takes place, although most frequently amounts to jumping from the single to the double precision mode and back. These observations support using the Boolean models of computation, widely accepted in the literature. Moreover, some modern supercomputers, such as Connection Machines CM1 and MASPAR, have many bit-serial processors and a very limited primary storage memory, generally accessed bit-serially. For such computers, the time for executing an arithmetic operation crucially depends on its precision, which may be considered variable. So, the Boolean complexity (the bit-complexity) is the most appropriate measure for the complexity of the computation on such machines, which perform computations faster when their precision is lower. In section 9, we will show some nontraditional ways of taking similar advantages of low precision computations on conventional computers, by means of binary segmentation (in particular, see Remark 9.3).

Estimating the bit-complexity of arithmetic computations: general comments.

As we have already said, we will estimate the bit-cost of our arithmetic computations by summing the estimates for the bit-cost of all arithmetic operations involved, assuming certain precision bound for each of these operations and assuming that the cost of all nonarithmetic operations is small enough to be ignored.

To be specific, let $\alpha(h)$ and $\mu(h)$ denote the numbers of bit-operations required in order to perform (with an h-bit precision) an addition/subtraction and a multiplication/division, respectively, for a given pair of binary floating point numbers. We assume that fewer than h bits are needed to express the exponent in the floating point representation of such a number. Then we arrive at the estimates

$$\alpha(h) = O(h),\ \mu(h) = O(h \log h \log\log h) \text{ as } h \to \infty \tag{1.1}$$

extending the bound (1.2.4) (from the case of bounded integers). For moderate h, one may consider an alternative extension of the bounds (1.2.5) (from the case of bounded integers) by letting $\mu(h)$ be of the order h^s where $s = 2$ or $s = \log 3 = 1.5849\ldots$. [Indeed, in practice of numerical computing, a very high precision is rarely needed; on the other hand, even (1.2.5) is still not realistic: many pipelined computers add and multiply integers equally fast (as we have already mentioned in Remark 1.1.2)].

Now, let $\alpha(h)$ and $\mu(h)$ be defined, let us be given an algorithm consisting of A additions/subtractions and M multiplications/divisions, and let $h = h(g)$ bits be the minimum upper bound on the precision of the computation (including the input precision) that ensures the g-bit absolute or relative precision of the output for any input [or let it ensure the absolute or relative output error bound $\epsilon = 2^{-g}$ in (1.A.3) or (1.A.4)]. Then we immediately define the sequential time of the solution as $A\alpha(h) + M\mu(h)$. [Note that in this approach, we rely on the Boolean RAM model of computation (chapter 1) and estimate the overall bit-complexity for a fixed h.] Alternatively, we may let h vary throughout the computations, so as to ensure computing the output within the desired error bound at a lower computational cost. Since we rely on the bounds (1.1) and (1.2.5), the decrease by k times of the computational precision is even more desirable than the decrease by k times of the number of ops involved.

A quantitative definition of numerical stability and condition.

In practical numerical computations on a computer with a fixed precision, one needs to ensure that the output precision would not be much less than the precision of the computations and of the input data, and if a solution algorithm does ensure this, then it is customary to call this algorithm numerically stable.

The error affecting the result of a floating point computation has two main components. The first component collects the errors generated in each arithmetic operation performed with floating point arithmetic (these errors are due to the roundoff in the computation); the second component collects the errors induced by the approximation of the input data by means of floating point numbers (these errors are due to the roundoff in the input).

The first kind of errors depends on the specific algorithm used in the computation. For this reason, they are customarily called the *algorithmic errors.* If such an error is "large" with respect to the precision of computation, then numerical analysts informally call the algorithm *numerically unstable.*

The second kind of errors is independent of the solution algorithm; they only depend on the computational problem (that is, on the functions in the input variables that ought to be computed) and are called *the inherent errors.* If "small" changes of the input produce "large" changes in the output, then the inherent error is large and the computational problem is called *ill-conditioned.*

If we consider relative errors, then it is easy to prove that our upper bound on the total relative error of a floating point computation is related to the upper bounds on the algorithmic errors and the inherent errors in the following way (compare Appendix A):

$$\text{total}= \text{inherent}+\text{algorithmic} +(\text{inherent})(\text{algorithmic}).$$

This means that, no matter which solution algorithm is used, the output error bound is always large for an ill-conditioned problem, unless the input data can be represented with no errors by floating point numbers. Moreover, we may find numerically stable algorithms even for ill-conditioned problems (see exercise 30).

We may formally measure the condition of a computational problem and numerical stability of a solution algorithm in terms of the following functions:

Define $c(g)$, the minimum upper bound on the input precision that would guarantee the g-bit (absolute or relative) output precision for any input, assuming the computation with the infinite precision.

Define $s(g)$, the minimum upper bound on the precision of computation that would guarantee the g-bit (absolute or relative) output precision for any floating point input, i.e. for any input with no errors.

Then the quantities $s(g)/g$ and $c(g)/g$ formally measure numerical stability of the algorithm and the condition of the computational problem, respectively.

In some cases $s(g)$ and $c(g)$ can be infinitely large; for instance, under the relative error bound model, this situation occurs for any problem whose solution may vanish (that is, may have some output value equal to zero) for some input values (see exercise 31). Otherwise, there is no formal and commonly accepted specification of the word "large" in these definitions, but a natural idea is to minimize the precision required in the computations and in the input, provided that the computations are kept feasible for the actual computers and for a given tolerance to the error defining the value g.

We may formalize the word "large" by relating the functions $s(g)$ and $c(g)$ to the size of the problem. Assume for simplicity that n is a parameter measuring the (input or output) size of the problem under our consideration (for instance, the degree of a polynomial or the dimension of a vector). In this way, the functions $s(g)$ and $c(g)$ depend also on n. There are situations, customarily considered ill-conditioned, where $c(g,n)$ is of the order ng, as in the case of approximating the zeros of a polynomial, given its coefficients (see [A], pp. 82-83; [H70], pp. 74-77; [Mi] and [Sc82]).

In other cases, such as the summation of n positive numbers $\sum_{i=1}^{n} x_i$, the situation is quite different: we have $c(g,n) = O(g)$ and $s(g,n) = O(g+\log n)$, if we consider the algorithm $(\cdots((x_1+x_2)+x_3)+\cdots+x_{n-1})+x_n$, whereas $s(g,n) = O(g + \log\log n)$ for the fan-in summation algorithm (see exercise 32). Along this line, we could have given an asymptotic definition of ill-conditioning (and of numerical stability) by calling a problem ill-conditioned (or an algorithm numerically unstable) if we have $c(g,n)$ [respectively, $s(g,n)$] of the order gn or, more generally, $gf(n)$, where $f(n)$ is a fixed increasing function, say, $\log n$. (Note a limitation: in the actual computations, we deal with bounded precision, so that their actual numerical properties of stability and conditioning may be quite different from their asymptotic estimates.)

We will now consider the relative errors. In this case from the equation relating the inherent, the algorithmic and the total error, it follows that $h(g) \leq \max\{s(g+1), c(g+1)\}$, where the function $h(g)$ is the minimum upper bound on the precision of computation (both in the input and in the arithmetic operations) that ensures the g-bit relative precision of the output for any input.

Some comments on the error analysis.

The analysis of the dependences of $c(g)$, $s(g)$ and $h(g)$ on g is quite involved for some problems and algorithms but is not hard for other ones. In Appendix A we will recall some basic techniques and results for such an analysis; in chapters 6 and 8 we will show some examples of its application to some major problems of numerical computations with matrices and polynomials.

Next, we will briefly recall some simple expressions for the output errors through the perturbation of the input:

$$\Delta_E(AB) = (A+E)B - AB = EB,$$

$$\Delta_E(A^{-1}) = (A-E)^{-1} - A^{-1} \ ,$$

where E represents the matrix of perturbation of the entries of A by the input errors, the left sides represent the matrices of the resulting absolute output errors, and it is assumed that A and $A-E$ are nonsingular matrices. Therefore, for any fixed matrix norm, induced by (subordinate to) a vector norm (see Definitions 2.1.11 and 2.1.12), we have

$$\frac{||\Delta_E(AB)||}{||AB||} \le \frac{||E||}{||A||}\left(\frac{||A||\ ||B||}{||AB||}\right), \tag{1.2}$$

$$\frac{||\Delta_E(A^{-1})||}{||A^{-1}||} \le \frac{||E||}{||A||}\ \frac{\text{cond}(A)}{1-||EA^{-1}||} \le \frac{||E||}{||A||}\ \frac{\text{cond}(A)}{1-\text{cond}(A)||E||/||A||}, \tag{1.3}$$

provided that $||EA^{-1}|| < 1$. Relations (1.2) and (1.3) bound the relative error norms of the matrix product and of the matrix inverse due to the perturbation of the input matrices; in particular, the norm bound for the perturbation of the inverse of A turns out to be almost proportional to cond(A) if $||E||$ is small, and actually, such an upper bound is quite sharp. This suggests avoiding the inversion of the auxiliary matrices that may have large condition numbers. For instance, we need the bounds (2.1.5) or similar bounds to ensure safe numerical computation of the triangular factorization (2.2.1)-(2.2.3) in the presence of small roundoff and input errors. Numerical analysts call the matrix *ill-conditioned* if cond(A) is "large" and call it *well-conditioned* otherwise.

In addition to the bounds (1.2) and (1.3), we recall the following results on the sensitivity of linear systems of equations to perturbation of the input ([GL]):

Theorem 1.1. *Let* $A\mathbf{x} = \mathbf{b}$, $(A+E)\mathbf{y} = \mathbf{b}+\mathbf{e}$, *$A$ and E being $n \times n$ matrices, A being a nonsingular matrix.*

a) If for a fixed vector norm and for the subordinate matrix norm, we have

$$\|E\| \le \delta\|A\| \ , \quad \|\mathbf{e}\| \le \delta\|\mathbf{b}\| \ ,$$

$r = \delta$ *cond*$(A) < 1$, *then the matrix $A+E$ is nonsingular, and*

$$(1-r)\|\mathbf{x}-\mathbf{y}\| \le 2\delta\|\mathbf{x}\|\ cond(A) = 2r\|\mathbf{x}\|.$$

b) If $|W|$ denotes the matrix such that $(|W|)_{i,j} = |(W)_{i,j}|$ for all i and j, if $|E| \le \delta|A|$, $|\mathbf{e}| \le \delta|\mathbf{b}|$, and if $\delta\ \mathrm{cond}_\infty(A) = r < 1$, then $A + E$ is a nonsingular matrix, and

$$(1 - r)\|\mathbf{y} - \mathbf{x}\|_\infty \le 2\delta\|\mathbf{x}\|_\infty\| \, (|A| \, |A^{-1}|) \, \|_\infty \, .$$

For any h.p.d. and/or diagonally dominant matrix A, there exists its unique LU factorization, $A = LU$. The factors L and U can be computed by means of Gaussian elimination or by means of the recursive factorization (2.2.1)–(2.2.3). Both methods are numerically stable. Indeed, if A is h.p.d., the condition number of the intermediate matrices to be factorized in the recursive factorization (2.2.1)–(2.2.3) is not greater than the condition number of A. This follows from the bounds

$$||A||_2 \ge \max\{||B||_2, ||S||_2\}, \quad ||A^{-1}||_2 \ge \max\{||B^{-1}||_2, ||S^{-1}||_2\},$$

implied by (2.1.5) and Proposition 2.2.3. A similar result can be proved for Gaussian elimination (compare exercise 4.b of chapter 2). If A is a row- (or column-) diagonally dominant matrix, then also B and S are diagonally dominant, with the dominance coefficients not less than ones for A. This can be shown based on the inequality

$$||I - A||_h \ge \max\{||I - B||_h, ||I - S||_h\}$$

where $h = \infty$ (or $h = 1$, respectively) and where we assume for simplicity that $\mathrm{diag}(A) = I$; otherwise, replace A by $(\mathrm{diag}(A))^{-1}A$ or by $A(\mathrm{diag}(A))^{-1}$ (compare exercise 4d of chapter 2). A similar result holds for Gaussian elimination.

Remark 1.2 *(on the uniformity).* Using the models of chapter 1, in particular the models of arithmetic and Boolean RAM's and circuits, involves the important problem of *uniformity*, so that under the uniform models, the overall complexity estimates must include the cost of indexing the operands and selecting the constants of the algorithm from a generally infinite input set (see [Bor77], [Cook85], [Ruz], [G86], and [E] on this interesting and important subject). In the algorithms that we will study, however, the cost of such indexing and a selection of the constants is small enough to be ignored.

2. Some Auxiliary Results.

In this section, we will recall some auxiliary results. The first three facts give us some a priori bounds on the magnitudes of the solutions to Problems 2.2.3–2.2.6 ($LIN \cdot SOLVE$), ($DETERMINANT$), ($INVERT$), ($CHAR \cdot POL$), 2.2.8 ($L \cdot SQUARES$), 2.2.12 ($GEN \cdot INVERSE$) of matrix computations and to the related problems, such as 1.5.1 ($POL \cdot GCD$) (computing the polynomial gcd).

A priori bounds on the magnitude of the output values.

Fact 2.1, *Cramer's rule* ([Str80]). *Given a nonsingular matrix A and a vector $\mathbf{u}$ defining a linear system $A\mathbf{x} = \mathbf{u}$, the entry x_i of the vector $\mathbf{x}$ is the ratio $x_i = \det B_i(A, \mathbf{u}) / \det A$ where the matrix $B_i(A, \mathbf{u})$ is obtained by replacing the i-th column of the matrix A by the vector $\mathbf{u}$.*

Fact 2.2, *Hadamard's inequality* ([Kn], p. 414; and [MM]). *For an $n \times n$ complex matrix $A = [a_{ij}]$, $|\det A|^2 \le \prod_j \sum_i |a_{ij}|^2$, and there are matrices for which an equality holds in the above. Moreover, for any operator matrix norm $||.||$, induced by (subordinate to) a vector norm,*

$$|\det A| \le ||A||^n. \tag{2.1}$$

In particular, for $||A|| = ||A||_1 = \max_j \sum_i |a_{ij}|$, (2.1) implies that $|\det A| \le (\max_j \sum_i |a_{ij}|)^n$.

The inequality (2.1) follows because $\det A = \prod_{j=1}^n \lambda_j$, where $\lambda_1, \ldots, \lambda_n$ denote the n zeros of $\det(\lambda I - A)$ (that is, the n eigenvalues of A counted with their multiplicities), and because, surely, $||A|| \ge |\lambda_j|$ for any operator norm of A and for any eigenvalue λ_j of A.

Combining Facts 2.1 and 2.2 and applying them to the equations (2.8.3), we arrive at the useful a priori bounds on the coefficients of the polynomial entries $s_i(x)$ and $t_i(x)$ of the extended Euclidean scheme (1.5.6)-(1.5.8), in terms of the coefficients of the input polynomials $u(x)$ and $v(x)$ of (1.5.1). In particular, if these input coefficients are integers, then all the polynomial entries of the scheme (1.5.6)-(1.5.8) [including $\gcd(u(x), v(x))$] can be scaled, so that their coefficients become integers in the range from $-B$ to B where

$$B = K^{m+n}(m+1)^{n/2}(n+1)^{m/2}, \quad K = \max\{\max_i |u_i|, \max_j |v_j|\} \tag{2.2}$$

(see [Kn], p. 414).

These bounds also hold for the integer coefficients of the scaled polynomials $a_i r_i(x)$ and $b_i R_i(x)$ of the remainder and pseudo remainder sequences assuming an appropriate choice of scaling constants a_i and b_i that simultaneously ensures the integrality of the coefficients and keeps their magnitude bounded. As a result of such scaling, the overall Boolean cost of computing the coefficients of these polynomials can be substantially decreased (see [BT], [Kn]).

For the gcd itself and for each divisor of a given polynomial, we have stronger bounds (see [Kn], p. 438, and [BCL], pp. 259-263).

Fact 2.3. *Let* $u(x) = v(x)q(x)$ *where* $u(x) = \sum_{i=0}^{m} u_i x^i$, $v(x) = \sum_{i=0}^{n} v_i x^i$ *and* $q(x) = \sum_{i=0}^{m-n} q_i x^i$ *are any polynomials with complex coefficients,* $u_m v_n \neq 0$. *Let* μ *denote* $(\sum_{i=0}^{m}(u_i^2))^{1/2}$. *Then* $|v_j| \leq \binom{n-1}{j}\mu + \binom{n-1}{j-1}|u_m|$ *for all* j, *and if all the coefficients of* $u(x), v(x)$ *and* $q(x)$ *are integers, then also* $\sum_{i=0}^{n} |v_i| \leq 2^n \mu |u_m/v_n|$.

Some further bounds useful for the error analysis and the bit-complexity estimates can be found in chapter 6, including the estimates for the magnitudes of the coefficients of the quotient of polynomial division.

Probability of a singularity.

Next, we will shift to estimating the probability that a matrix B depending on a random parameter p is singular, specifically where $B = A \mod p$ for a fixed nonsingular matrix A filled with integers. We will start with a purely auxiliary result, only needed in order to deduce Corollary 2.1.

Fact 2.4 ([IR]). *Let* $f(n)$ *be a function defined on the set of positive integers such that* $f(n) > 0$ *and* $\lim_{n\to\infty} f(n) = \infty$. *Then there exist a positive constant* C *and an integer* n_0 *such that, for any* $n > n_0$, *the interval*

$$\{p\colon f(n)/n < p < f(n)\} \tag{2.3}$$

contains at least $f(n)/(C \log f(n))$ *distinct primes.*

Proposition 2.1. *Let* $f(n), h(n)$ *and* $k(n)$ *be three functions in* n *such that* $h(n)$ *is integer valued,* $h(n) \neq 0$,

$$0 < (h(n))^{1/k(n)} < f(n)/n,\ k(n) > 0,\ \lim_{n\to\infty} f(n) = \infty. \tag{2.4}$$

Let p *be a random prime in the interval (2.3). Then there exist a positive constant* C *and an integer* n_0 *such that, for any* $n > n_0$, $h(n) = 0 \bmod p$ *with a probability at most* $(Ck(n)\log f(n))/f(n)$.

Proof. Proposition 2.1 is implied by Fact 2.4. Indeed, fewer than $k(n)$ primes in the interval (2.3) may divide $h(n)$ due to (2.4). ∎

Next, we will apply Proposition 2.1 assuming some specific choices of $f(n)$, $h(n)$ and $k(n)$. We will first fix a consistent operator norm of vectors and matrices such that $||\mathbf{e}^{(i)}|| = 1$ for all the unit coordinate vectors $\mathbf{e}^{(i)}$ (for instance, this holds for $||.||_g$, $g = 1, 2, \infty$). Now let A denote an $n \times n$ matrix filled with integers; $\mathbf{v}$ denote an n-dimensional integer vector such that $\|\mathbf{v}\| \le n\|A\|$; $N = N(n) \ge 1$ denote a function in n; b, r and q denote three positive constants such that

$$\|A\|^{-1/N}(Nr\log n + \log\|A\|) = O(n^{r-b-q}), \qquad r - b - q > 0. \tag{2.5}$$

At this point, we let $f(n) = n^r||A||^{1/N}$ and set either $b = 1$, $k(n) = nN$, $h(n) = |\det A|$, or $b = 2$, $k(n) = Nn^2$, $h(n) = |\det K(A, \mathbf{v})|$ [where $K(A, \mathbf{v})$ is the Krylov matrix of Definition 2.1.15]. We assume that $h(n) \ne 0$ in both cases. Then in both cases, (2.4) holds, due to (2.1) and to the following rough but simple bound: $||K(A, \mathbf{v})|| \le n||A||^{n-1}||\mathbf{v}||$ (which holds if $||\mathbf{e}^{(i)}|| = 1$ for all i). Thus, we apply Proposition 2.1 and arrive at the following result:

Corollary 2.1 ([P87a]). *Let $N = N(n) \ge 1$ be a function in n, A an $n \times n$ integer matrix, $\mathbf{v}$ an integer vector; let a consistent operator norm for vectors and matrices be fixed such that $||\mathbf{e}^{(i)}|| = 1$ for all i, $||\mathbf{v}|| \le n||A||$. Let p be a random prime in the interval (2.3) with $f(n) = n^r||A||^{1/N}$. Let r and q be two positive constants, $b = 1$ or $b = 2$, and let (2.5) hold. If $b = 1$ and $\det A \ne 0$, then $\det A \ne 0 \bmod p$ with a probability $1 - O(1/n^q)$ as $n \to \infty$. If $b = 2$ and $\det K(A, \mathbf{v}) \ne 0$, where $K(A, \mathbf{v})$ is the Krylov matrix of Definition 2.1.15, then $\det K(A, \mathbf{v}) \ne 0 \bmod p$ with a probability $1 - O(1/n^q)$ as $n \to \infty$.*

In order to apply randomization, one needs to generate (pseudo) random numbers (see [Kn]), and in many applications we need (pseudo) random primes exceeding a fixed value n. Due to Fact 2.4, applied for $f(x) = n^2$, the probability that a random number in the interval from n to n^2 is prime is at least $\dfrac{1}{2C(1 - 1/n)\log n}$; therefore, we just need to perform a primality test for $\lceil 2C(1 - 1/n)(\log n)\log(1/\epsilon)\rceil$ (pseudo) random numbers in the interval from n to n^2 (say), in order to arrive at the desired prime with a probability at least $1 - \epsilon$ provided that n is sufficiently large. Here C is the positive constant from Proposition 2.1. The Monte Carlo randomized algorithms of [Mil] and [Rab] perform such a primality test at the cost of $O(s\log^3 n)$ bit-operations and have the error probability at most 2^{-s}, for any fixed positive s. [Recall from chapter 1 that the Monte Carlo randomization, unlike the Las Vegas randomization, does not ensure detecting the output randomization errors; even in the Monte Carlo case, however, the probability of the occurrence of such errors is bounded (and in our case can be made as small as desired, by repeating the primality test)]. It follows that the overall complexity of randomized Monte Carlo algorithms for the generation of a (pseudo) random prime is $O(s\log^4 n \log(1/\epsilon))$ Boolean operations. In many applications, the Monte Carlo randomization suffices. For instance, if a linear system of equations that we need to solve turns out to be singular, then the solution algorithm performed with no roundoff errors will usually lead us to division by 0, which will detect the singularity.

If, however, we need to turn the above Monte Carlo algorithm into a Las Vegas algorithm, we may apply additional tests, and this can be done at the cost of $(\log n)^{O(1)}$ ops in a Las Vegas randomized algorithm (see [GKi], [AH]). Thus, it is quite unlikely that a deterministic algorithm for verification of the primality of the candidate integer is required, but such a verification is possible in $O(\,(\log n)^{\log\log n})$ sequential Boolean time ([APR]).

3. Integer and Rational Arithmetic.

Modular arithmetic (some general comments).

In this section, we will study the finite precision rational computations, which rely on performing ops modulo some integers, bounded so as to decrease the precision and, thus, the Boolean cost of the computations. This approach, already cited in section 7 of chapter 1 and widely used in computer algebra (compare, for instance, [Gre], [YG]), applies where the computations are rational, that is, only involves the arithmetic operations: $+, -, *$ and $\div$ (compare Remark 3.4). If all the output values are known to be integers lying between $(1-M)/2$ and $(M-1)/2$ for some natural M and if no divisions are involved, or if each divisor has its inverse modulo M, we may perform all the computations modulo M, that is, with the precision of $\lceil \log M \rceil$ bits, arrive at the output values reduced modulo M, and then immediately recover the exact output. Indeed, $h = h \bmod M$ if $h \bmod M < M/2$; $h = (h \bmod M) - M$ otherwise. We will refer to this simple rule as to *modular rounding-off*, which can be extended to the case of integer output values from any fixed real interval of length M. Note that a much higher precision of the computation may be required without such a reduction modulo M (see exercise 15).

Not only may we bound the precision of computation at the level of the output precision in this way, but we may also decrease it much further if we choose natural numbers $m_1, m_2, \cdots, m_k$, pairwise mutually prime (that is, pairwise having their greatest common divisors, gcd's, equal to 1) and such that $m_1 m_2 \cdots m_k > M$, perform all the computations modulo m_i for each i from 1 to k (thus using by factor k more ops but about k times smaller precision of computation if logarithms of all the moduli $m_1, \dots, m_k$ are of the same order of magnitude), and finally, recover the output values in $\mathbf{Z}_M$ by applying the Chinese remainder computation for integers (see Remark 1.4.2 and exercise 29). Only the latter stage requires the full precision computation, but in many cases this stage is a relatively small part of the algorithms in terms of the number of ops used.

The approach can be extended to the case where the computations involve divisions; in this case, we may either compute modulo p provided that $p \geq M$ and is mutually prime with every divisor, or we may apply the Chinese remainder computations where all the moduli m_i should be pairwise mutually prime with each other and with all the divisors. The latter property can be ensured with a high probability if we choose the values m_i as random primes lying in appropriately fixed intervals (compare Fact 2.4).

If the output values are not necessarily integers but are rational numbers, then a further complication arises at the stage of the recovery of the rationals from their modular representations. In many cases, we may a priori bound the numerators and the denominators of the rational output values (see Facts 2.1, 2.2 and 2.3); in such cases we may first compute these values modulo a large integer K and then recover the output. Specifically, suppose that we seek an output value y/x and let $a = (y/x) \bmod K$, that is, $y = ax + bK$, where x, y and b are three unknown integers, a and K are two fixed precomputed integers. [If $K = m_0 m_1 \cdots m_k$ and if the values $(y/x) \bmod m_i$ are known for all i, we may compute $(y/x) \bmod K$ via the Chinese remainder algorithm for integers.] Let K be chosen such that the values $|y|$ and $|x|$ are bounded from above by $\sqrt{K/2}$. Then x and y can be computed by means of the continued fraction approximation algorithm applied to the pair of integers a and K (see [Wang], [WGD], [Ab], [KRo]). The computation by this algorithm involves $O(k \log^2 k \log\log k)$ bit-operations where $k = \log K$ ([AHU], [Kn]).

As we already pointed out in Remark 1.5.2, the continued fraction approximation algorithm is the customary name for the extended Euclidian algorithm applied to the pair of integers. The name of continued fraction approximation should not mislead: this is an algebraic algorithm, not a numerical approximation algorithm; we deal with numerical approximation algorithms in sections 4 and 5 (also see exercise 37).

The area of the application of the above approach is actually even wider than it may seem. In particular, recall that FFT, our basic tool, can be performed modulo an integer, over a finite field or ring (see the end of this section).

Newton-Hensel's (p-adic) lifting.

Next, let us recall a general, powerful and popular alternative method for the reduction of computations modulo a large integer to computations modulo smaller integers. (The method is called *Hensel's, Newton-Hensel's* or *p-adic lifting.*) Let us apply this method to matrix inversion and solving linear systems of equations.

Technically, the idea is to exploit the algorithms of numerical linear algebra, such as Newton's iteration algorithm for matrix inversion, that is, for the solution of the matrix equation $I - AX^{-1} = O$ [compare (2.7.1), (5.4) and section 6 of chapter 1], and the customary iterative algorithms for the solution of linear systems (see section 5). Instead of obtaining numerical approximations to an integer or rational number Q, we now compute its s-th order *p-adic approximations*, $Q \bmod p^s$ for a fixed natural p. The exact rational Q will then be recovered from its higher order p-adic approximation. There is a striking similarity between the numerical and algebraic (p-adic) versions of some algorithms for matrix computations (compare section 5). We refer the reader to [MC] for further details.

Algorithm 3.1, *Newton-Hensel's lifting process for matrix inversion* ([MC]).

Input: natural numbers $H = 2^h$, p, n and an $n \times n$ matrix A having integer entries such that $\det A \neq 0 \bmod p$ and $\sqrt{p^H/2}$ is an upper bound on the integers $|x_{ij}|$, $|y_{ij}|$, $i, j = 0, \ldots, n-1$, such that $A^{-1} = (x_{ij}/y_{ij})$.

Output: the n^2 pairs of integer numbers x_{ij}, y_{ij}, $i, j = 1, \ldots, n$, such that $A^{-1} = (x_{ij}/y_{ij})$ (or alternatively, the n^2 entries of $S_h = A^{-1} \bmod p^H$).

Computation:

0. (Initialization.) Compute the matrix $S_0 = A^{-1} \bmod p$.

1. For $j = 1, \ldots, h$, compute the matrix S_j given by the equations

$$S_j = S_{j-1}(2I - AS_{j-1}) \bmod p^J, \quad J = 2^j.$$

2. Recover the matrix A^{-1} from $S_h = A^{-1} \bmod p^H$, $H = 2^h$, by applying the continued fraction approximation (extended Euclidean) algorithm (cited above and in Remark 1.5.2) to each entry of S_h. Output A^{-1}. (In some applications, S_h is the output, and then Stage 2 is skipped).

Fact 3.1. *We have:* $S_j = A^{-1} \bmod p^J$, *for* $J = 2^j$ *and for all* j.

Proof (by induction on j). The result has already been assumed for $j = 0$. To extend it from any $j-1$ to j, it remains to observe that $I - S_j A = (I - S_{j-1}A)^2 \bmod p^J$. ∎

Suppose that $\det A \neq 0$. Then Corollary 2.1 for $b = 1$ implies that $\det A \neq 0 \bmod p$ with a (high) probability, $1 - O(1/n^q)$ as $n \to \infty$, for a random choice of a prime p in the interval

$$n^{r-1}||A||^{1/N} < p < n^r||A||^{1/N}, \tag{3.1}$$

provided that (2.5) holds, $b = 1$, q and r are two positive constants, $N = N(n)$ is a function in n, $N \geq 1$, $||A|| = \max_j \sum_i |a_{ij}|$ is the 1-norm of the matrix $A = [a_{ij}]$. Due to Facts 2.1 and 2.2, it is sufficient to choose

$$H = O(n \log ||A||/ \log p) \tag{3.2}$$

in order for us to be able to recover the exact value of any entry of A^{-1} from its value $\bmod p^H$; the cost of the recovery (by means of the continued fraction approximation algorithm, cited above and in Remark 1.5.2) is $O(k \log^2 k \log\log k)$ bit-operations for $k = H \log p$. The denominator of every entry divides det A, and in many cases this observation can be exploited to decrease the worst case upper estimate $O(n^2 k \log^2 k \log\log k)$ on the overall cost of the recovery of A^{-1} from $A^{-1} \bmod p^H$.

Remark 3.1. We may simplify the recovery of A^{-1} from $A^{-1} \bmod p^H$ if A has integer entries and if we know $\det A$, for in this case $A^{-1} \det A =$ adj A is a matrix with integer entries, which we may immediately recover if we know adj $A \bmod p^H$ for $p^H > 2\max_{i,j} |(\text{adj } A)_{ij}|$ (see Fact 2.1). Similar techniques can be applied to simplify the recovery of the exact rational solution $A^{-1}\mathbf{b}$ of a linear system from its solution modulo a prime power and even for the recovery of the rational matrix A^{-1} and/or vector $A^{-1}\mathbf{b}$ from their numerical binary approximations (compare Algorithm 3.3).

The key-idea of Algorithm 3.1 is that the relatively *harder stage of matrix inversion* (Stage 0) *is performed modulo* p, *that is, with a (lower) precision* of at most $\lceil \log p \rceil$ bits, and then the precision gradually increases to $\lceil H \log p \rceil$ at the easier stages of matrix multiplication.

This approach is particularly effective where we need to invert certain structured matrices. For instance, we invert a triangular Toeplitz matrix to perform polynomial division, and we invert Sylvester-like, Toeplitz or Hankel matrices to compute a polynomial gcd (see sections 8 and 9 of chapter 2). In these applications, we may simplify the recovery of the output from its value in $\mathbf{Z}_M$ if the input values are integers: by means of their scaling, we may turn all the output values into integers. For the latter group of input matrices A (that is, of Hankel and Toeplitz matrices), the inverses $S_j = A^{-1} \bmod p^J$ are also well-structured and can be represented by their F-generators of lengths at most 2 for the operators F of (2.11.1) and (2.11.2) (due to Theorem 2.11.1), as well as (2.11.18) and (2.11.19) (due to Theorem 2.11.3), so that performing Stage 1 of Algorithm 3.1, for all j, $j = 1, \ldots, h$, amounts to $O(h)$ multiplications of Toeplitz matrices by vectors and costs $O(hn \log n)$ ops. We may also apply this argument in order to devise effective algorithms for the p-adic inversion of Toeplitz-like and Hankel-like matrices.

Furthermore, we may reduce polynomial interpolation to solving a Vandermonde linear system, where we may apply Algorithm 3.1 too. Here, and similarly in many other cases, however, we may prefer the simpler p-adic algorithms that just solve such a system, rather than invert its coefficient matrix. Such algorithms rely on splitting the input matrix A as follows: $A = N - pL$ (where $N = A \bmod p$ is assumed to be nonsingular). Thus, we rewrite the equation $A\mathbf{x} = \mathbf{b}$ as $N\mathbf{x} = pL\mathbf{x} + \mathbf{b}$, compute an initial approximation vector $\mathbf{x}_1$ such that $A\mathbf{x}_1 = \mathbf{b} \bmod p$ and successively improve it as follows:

$$\mathbf{x}_{i+1} = pN^{-1}L\mathbf{x}_i + N^{-1}\mathbf{b}. \tag{3.3}$$

Then $\mathbf{x} - \mathbf{x}_{i+1} = pN^{-1}L(\mathbf{x} - \mathbf{x}_i)$. Consequently, by induction, $\mathbf{x} - \mathbf{x}_i = \mathbf{0} \bmod p^{i+1}$, and therefore, $A\mathbf{x}_i = \mathbf{b} \bmod p^i$ for all i. We may decrease the Boolean cost of each iteration by performing all the matrix-by-vector multiplications modulo p^2 or p in the following adaptation of this method (where again we choose a random prime p according to Corollary 2.1 applied for $b = 1$):

Algorithm 3.2, *Hensel's lifting process for linear systems* (compare [MC]).

Input: natural numbers H, p, n; a vector $\mathbf{b}$ of dimension n; and an $n \times n$ matrix A, both filled with integers, such that $\det A \neq 0 \bmod p$ and $\sqrt{p^H/2}$ is an upper bound on the integers $|\xi_i|$, $|\eta_i|$, such that $(A^{-1}\mathbf{b})_i = \xi_i/\eta_i$, $i = 0, \ldots, n-1$.

Output: $2n$ integers ξ_i, η_i, $i = 0, \ldots, n-1$, such that $(A^{-1}\mathbf{b})_i = \xi_i/\eta_i$, $i = 1, \ldots, n$ (or, alternatively, the vector $\mathbf{x}_H = A^{-1}\mathbf{b} \bmod p^H$).

Computation:

0. (Initialization.) Compute

$$S_0 = A^{-1} \bmod p, \quad \mathbf{x}_1 = S_0\mathbf{b} \bmod p, \quad \mathbf{v}_1 = A\mathbf{x}_1.$$

1. For $i = 1, \ldots, H-1$, compute the vectors

$$\begin{aligned}
\mathbf{w}_i &= \mathbf{b} - \mathbf{v}_i \bmod p^{i+1}, \ \mathbf{y}_i = (S_0\mathbf{w}_i/p^i) \bmod p^2, \\
\mathbf{v}_{i+1} &= \mathbf{v}_i + p^i A\mathbf{y}_i \bmod p^{i+1} = \quad (\mathbf{v}_i \bmod p^{i+1} + \\
&\quad + p^i(A\mathbf{y}_i \bmod p^2)) \bmod p^{i+1}, \\
\mathbf{x}_{i+1} &= \mathbf{x}_i + p^i\mathbf{y}_i \bmod p^{i+1}.
\end{aligned}$$

2. Recover the vector $A^{-1}\mathbf{b}$ from $\mathbf{x}_H = A^{-1}\mathbf{b} \bmod p^H$ by applying the continued fraction approximation algorithm of [Wang] to each component of $\mathbf{x}_H$. Output $A^{-1}\mathbf{b}$ (or in some applications skip Stage 2 and output $\mathbf{x}_H$).

To prove the correctness of Algorithm 3.2, first combine the above expressions for $\mathbf{y}_i$ and $\mathbf{v}_{i+1}$ and deduce, by induction on i, that $\mathbf{v}_i = A\mathbf{x}_i \bmod p^i$ and, consequently, $\mathbf{w}_i = \mathbf{b} - A\mathbf{x}_i \bmod p^{i+1}$. Then observe that $\mathbf{x} - \mathbf{x}_{i+1} = \mathbf{x} - \mathbf{x}_i - (\mathbf{x}_{i+1} - \mathbf{x}_i) = \mathbf{x} - \mathbf{x}_i - p^i\mathbf{y}_i \bmod p^{i+1} = \mathbf{x} - \mathbf{x}_i - S_0\mathbf{w}_i \bmod p^{i+1} = \mathbf{x} - \mathbf{x}_i - S_0(\mathbf{b} - A\mathbf{x}_i) \bmod p^{i+1}$. Substitute $\mathbf{b} = A\mathbf{x}$ and deduce that $\mathbf{x} - \mathbf{x}_{i+1} = \mathbf{x} - \mathbf{x}_i - S_0A(\mathbf{x} - \mathbf{x}_i) \bmod p^{i+1} = (I - S_0A)(\mathbf{x} - \mathbf{x}_i) \bmod p^{i+1}$. Since $I - S_0A = O \bmod p$, we now deduce, by induction on i, that $\mathbf{x} - \mathbf{x}_i = \mathbf{0} \bmod p^i$. ∎

Stage 1 involves $O(n)$ ops, needed in order to perform vector additions and to multiply three vectors by constants, and also involves two multiplications of matrices A and S_0 by vectors reduced modulo p or p^2, so that these two matrix-by-vector multiplications are performed with a lower precision. We may avoid computing the matrix S_0 and find the vectors $\mathbf{x}_1$ and $\mathbf{y}_i$ for all i by solving linear systems of equations $A\mathbf{x}_1 = \mathbf{b} \bmod p$, $A\mathbf{y}_i = (\mathbf{w}_i/p^i) \bmod p^2$. Stage 2 (performed by means of Wang's algorithm) involves $O(kn \log^2 k \log\log k)$ bit-operations where $k = O(H \log p)$.

For demonstration, let us now apply this algorithm in order to solve a Vandermonde linear system (that is, to perform interpolation) modulo p^H assuming that $\mathbf{Z}_{p^H}$ supports FFT on $O(n)$ points. The solution from chapter 1 involves $O(n \log^2 n)$ ops modulo p^H. With Algorithm 3.2, we reduce the problem (apart from Stages 0 and 2) to performing $O(H)$ linear operations (that is, additions and scalar multiplications) with n-dimensional vectors [this stage is equivalent to $O(Hn)$ arithmetic operations with scalars], to the solution of H Vandermonde systems modulo p (if we avoid computing S_0) and to H multiplications of the Vandermonde input matrix by a vector (performed modulo p^2). Each such a multiplication amounts to solving Problem 1.2.2 ($POL \cdot EVAL$) of polynomial evaluation modulo p^2 at $n+1$ points, that is, to $O(n \log^2 n)$ ops modulo p^2, provided that $\mathbf{Z}_{p^2}$ supports FFT on $O(n)$ points.

Rational roundoff.

To compute the matrix S_0 at the initialization stage of Newton-Hensel's or Hensel's lifting algorithms, we may proceed as follows (compare [P85], [P87]). First compute the matrix $N = (A \bmod p) + 2pnI$. (Note that $S_0 = N^{-1} \bmod p$.) Then compute a high precision numerical approximation $\tilde{N}^{-1}$ to the inverse matrix N^{-1}. We may do this by applying a rapidly converging iterative algorithm [it is sufficient to apply Newton's iteration for matrix inversion by setting $X_0 = I/(2pn)$, $X_{j+1} = X_j(2I - NX_j)$, $j = 0, 1, \ldots, k$, $X_k = \tilde{N}^{-1}$ for $k = O(\log n)$]. Finally, recover the matrix N^{-1} from its approximation $\tilde{N}^{-1}$ and compute $S_0 = N^{-1} \bmod p$. To recover the exact values of the entries of the matrix N^{-1} from their numerical approximations (we will call such a recovery *rational roundoff*), apply the continued fraction approximation algorithm (that is, the extended Euclidean algorithm for integers) as follows (compare [Sch86], p. 66, and [UP]):

Algorithm 3.3, *rational roundoff.*
Input: four natural numbers k, p, u and D where $p > 1$.
Output: the unique pair of natural numbers x and y such that

$$\gcd(x, y) = 1, \ |(u/p^k) - (y/x)| < 1/(2D^2), \ x \le D. \tag{3.4}$$

Computation:
The solution is computed by means of the continued fraction approximation (extended Euclidean) algorithm applied to the input integers u and $v = p^k$ [replacing the input polynomials $u(x)$ and $v(x)$ of Algorithm 1.5.1b]. The output is obtained as $x = s_i$, $y = t_i$ for the smallest integer i such that $r_i < D$ [use (1.5.6)-(1.5.8) with s_i, t_i and r_i replacing $s_i(x), t_i(x)$ and $r_i(x)$, respectively].

We will apply Algorithm 3.3 in order to recover the bounded and relatively prime integers y and x from the known high precision numerical approximation to y/x; in such cases, by assumption, there exists a solution (and its uniqueness can be immediately verified). If we allow arbitrary input values k, p, u and D, there can be no solution in some cases, and a test verifying (3.4) for a candidate pair x, y should be included ([UP]). Algorithms 3.1 and 3.3 can be combined together for an effective inversion of Toeplitz and other structured matrices.

Remark 3.2. Due to Algorithm 3.3, any sufficiently close approximation to the rational output can be converted into the exact solution.

Another application of p-adic lifting

The presented matrix versions of p-adic lifting unify several applications, and Algorithm 3.1 is very convenient under the parallel models of computation. There are also several other versions of Hensel's and Newton-Hensel's algorithm ([Y], [Kn], [G84a] and [Ka87]), customarily applied to polynomial factorization over finite fields and over rational numbers and to many computations for multivariate polynomials (without using any reduction to matrix computations).

Here is a simple example:

Algorithm 3.4, *evaluation of all the integer zeros of a polynomial having only integer zeros* ([Loos]).

Input: natural n and N and the coefficients of a monic polynomial $q(x) = \sum_{i=1}^{n} q_i x^i$, $q_n = 1$, that has only integer zeros, which all lie between $-N$ and N.

Output: the set of all the zeros of $q(x)$.

Computation:

1. Compute the discriminant d of $q(x)$ [that is, the resultant of $q(x)$ and of its derivative $q'(x)$ times $(-1)^{n(n-1)/2}$].
2. Compute the smallest prime p that does not divide d.
3. By means of the exhaustive search, compute modulo p the set $\mathcal{S}_1$ of all the zeros of $q(x)$.
4. For $j = 1, 2, \ldots, k-1$, recursively compute the sets $\mathcal{S}_{j+1} = \{(x_j + p^J r_j) \bmod p^{2J} : x_j \in \mathcal{S}_j, r_j = -u_j v_j \bmod p^J, u_j = q(x_j)/p^J, v_j = 1/(q'(x_j)) \bmod p^J, J = 2^j\}$, choosing k to be the smallest integer such that p^K for $K = 2^k$ exceeds $2N$. (This process is actually Newton's iteration, and S_j is the set of the zeros of $q(x)$ defined up to within their reduction modulo p^J.)
5. Output the set of the zeros of $q(x)$ recovered from the set S_k by means of modular rounding-off.

Remark 3.3. Modular arithmetic (together with the rational roundoff and the Chinese remaindering techniques) can be used as a highly effective means of avoiding numerical stability problems in matrix computations. Creating simple microcodes for modular arithmetic is not a very difficult problem and actually is quite simple for computation modulo $2^m \pm 1$, for a natural m. In fact, this problem is essentially reduced to Problem 1.3.6 of integer division, and we could see some great advantages of using such microcodes for *large scale numerical computing.* Moreover, simple codes for performing modular arithmetic have been created and applied by some present day users and are also available as subroutines of the "C" computer language. So far, however, the users routinely perform computations in numerical linear algebra without taking such advantages, because subroutines for modular arithmetic are usually available only as a small part of large packages of highly sophisticated subroutines for symbolic computation. Presently, such packages and, consequently, modular arithmetic are too expensive (in terms of computer resources of time and memory space involved) in order to be used for large scale numerical computing. We believe that such a situation will soon change, and modular arithmetic will be widely used as a major tool for numerical computing.

Remark 3.4. Effective application of modular arithmetic can be extended to those nonrational computations whose main computational task can be simplified by applying rational algorithms. An interesting example is given by the eigenvalue computation for a general (unsymmetric) matrix. This computation can be greatly simplified by reducing the input matrix to the block tridiagonal form, which generally is a numerically ill-conditioned but rational process [see chapter 2, Problems 2.3.2a ($T \cdot REDUCE$) and 2.3.2b ($T \cdot REDUCE1$)], so that the rational arithmetic seems to be an appropriate means for this reduction step.

(Generalized) DFT and polynomial multiplication in the rings $\mathbf{Z}$ *and* $\mathbf{Z}_M$.

Since the operations of polynomial multiplication and DFT are so fundamental to algebraic computations, we will comment on performing these operations in the rings $\mathbf{Z}$ of integers and $\mathbf{Z}_M$ of integers modulo $M \geq 2$ [where $\mathbf{Z}_M$ is a field if M is a prime]. As before, we will focus our study on the time-complexity, leaving the space-complexity estimates to the reader. By means of modular rounding-off, we may easily recover integers lying between $-H$ and H (for $H < M/2$) from their values reduced modulo M; in particular, we may easily recover the coefficients of $u(x)v(x)$ from $u(x)v(x) \bmod M$ if $u(x)$ and $v(x)$ are polynomials of degrees at most n with integer coefficients in the range from $-L$ to L, where $nL^2 < M/2$. In this way, we may avoid both problems of rounding off and of numerical precomputation of complex roots of unity. A customary choice of M is of the form $2^m \pm 1$ for an integer m, and then the reduction modulo M turns out to be a simple operation.

To apply FFT on K points in $\mathbf{Z}_M$, we need to have a principal and invertible K-th root of 1 in $\mathbf{Z}_M$ (invertible means having a reciprocal in $\mathbf{Z}_M$). It is easy to prove that ω is a primitive (and therefore, principal and invertible) K-th root of 1 in $\mathbf{Z}_M$ if ω and K are some powers of 2 and if $M = \omega^{K/2} + 1$ (see [AHU], p. 288). Based on this observation, we may compute modulo M the DFT of any integer vector $\mathbf{v}$ of dimension K, by using $O(K^2 \log K \log \omega)$ bit-operations. This is immediately extended to computing the product of two polynomials (of degrees at most K with integer coefficients in the range from $-L$ to L) at the cost of $O(\max\{K^2 \log K \log \omega,\ K\mu(K \log \omega)\})$ bit-operations provided that $L \le \omega^{K/2}\sqrt{K}$ and that $\mu(h)$ bit-operations suffice to multiply two integers modulo 2^h (see [AHU], pp. 266–269).

This approach involves a precision of computation of about $K\lceil \log \omega \rceil / 2$ bits, which is rather high if K is large. We may dramatically decrease this precision, together with the resulting upper bounds on the bit-complexity of multiplication of polynomials of degrees at most K, in the case where their coefficients are integers that lie in a smaller range, also from $-L$ to L, but for $L = K^{O(1)}$, for example. Indeed, in this case, we may reduce the problem to the computation in $\mathbf{Z}_M$ for $M = 2KL^2 + 1$ and then apply one of the algorithms of [Nus], [Sc77], [CKa], [CKa87], which all support Theorem 1.7.1 and, in particular, solve our problem at the cost of $O(K \log K \log\log K)$ ops in $\mathbf{Z}_M$. This implies the bit-complexity bound

$$O\big(K \log K \log\log K\, \mu(\log(2KL^2))\big) \tag{3.5}$$

provided that any op in $\mathbf{Z}_M$ can be performed in $\mu(\log(2KL^2))$ bit-operations for $M = 2KL^2 + 1$. If $L = K^{O(1)}$, this bound turns into

$$O(K \log K \log\log K\, \mu(\log K))\ ,$$

and we may substitute (1.2.4) and obtain the bound of

$$O(K(\log K \log\log K)^2 \log\log\log K)$$

bit-operations. Presently, this is the record asymptotic estimate. The algorithm of [Nus], supporting the latter estimates in $\mathbf{Z}$ and $\mathbf{Z}_M$ (for all odd

M), is a good practical algorithm for polynomial multiplication and, therefore, also for the generalized DFT, Problem 1.2.5 ($GEN \cdot DFT$). Inspection shows that the associated overhead constants hidden in the "O" notation of the estimate (3.5) are not large. Potentially competitive are the algorithms of [CKa], of [Can] (both of which can be applied in $\mathbf{Z}$ and $\mathbf{Z}_M$ for all M), and, in the field $\mathbf{Z}_2 = GF(2)$, of [Sc77], not counting the algorithms based on approximating the complex roots of unity.

In practice, the most customary policy (when the computation of $u(x)\,v(x)$ is reduced to performing FFT's in $\mathbf{Z}_M$) is to choose any prime M such that K divides $M-1$ provided that $2K < M$, K is a power of 2, K exceeds the degree of the product $u(x)\ v(x)$, and the ratio M/K is not too large (the latter condition ensures the practical efficacy of this approach). In this case, $\mathbf{Z}_M$ is a field having a readily available primitive K-th root of 1, and the desired primes M are readily available for K in the range from 1 to 2^{30} (say), (see [BM], pp. 86–87). Moreover, we do not have to increase M when the range $(-L,\ L)$ for the input integer coefficients grows. Indeed, we may partition the binary string representing each input coefficient into s shorter strings, for $s > 1$, and then s times apply a forward FFT. Similarly, we may replace the inverse FFT by several inverse FFT's.

The latter customary approach, however, generally requires precomputation of a primitive root of 1 and may involve M not of the form $2^m \pm 1$, which would make it disturbing to do the reduction modulo M generally required after each (or almost each) operation. The algorithms of [Nus], [Can], [CKa87], [CKa] all have the advantage of allowing M in the desired form $2^m + 1$ for an integer $m = kn$, since in this case $\omega = 4^m$ is an n-th principal root of 1 in $\mathbf{Z}_M$ (see [AHU] p. 266). Actually, we may apply the algorithms of [Nus], [CKa], [CKa87] and [Can] for several smaller M's, thus using computations with a low precision, and then we may recover the output in $\mathbf{Z}$ by means of the Chinese remainder algorithm. The main (although a rather mild) disadvantage of these approaches is, apparently, an extra $\log\log n$ factor in the ops count.

Finally, we may solve the problem by applying Remark 1.7.1 or reduce the size of FFT's involved, by following Remark 1.7.2 or exercise 12.

4. Finite Precision Computations and Some Approximation Algorithms.

As we noted in Remark 3.3, modular arithmetic is still a tool too expensive for many users; also this tool cannot handle irrational computations that require *approximation algorithms* (see exercise 37). Thus, in practice, integer and rational computations do not replace numerical computations with a fixed finite precision, which output approximations to the solutions. For many numerical computational problems, such an output can be rapidly refined further. We will show some important refinement schemes in the next sections. We also note that close approximations to rationals can be refined to the exact values due to the rational roundoff by means of Algorithm 3.3. Moreover, approximations with any absolute errors less than 0.5 turn into exact solutions via roundoff if the output values are integers. We will call this technique a *refinement by integer rounding-off.*

Normally, the users expect to obtain the desired output approximations to the solution (within a fixed error bound) by means of (low precision—low cost) numerical computations. Whether this is indeed possible for a given problem depends on the condition of the problem and on the numerical stability of the algorithm (compare the end of section 1). The analysis of numerical stability and condition enjoys a great attention from the designers of numerical algorithms, who gladly exploit the simple bounds such as (1.2) and (1.3), as well as the more sophisticated tools, like Theorem 1.1 and the techniques of backward error analysis (see [GL], [W65] and our Appendix A). Let us next briefly review some polynomial computations from this point of view.

It is fortunate for us that the discrete Fourier transform (DFT) is a well-conditioned problem and that our basis tool, FFT, is a numerically stable algorithm if we consider both input and output as vectors and measure the errors in terms of vector norms.

Proposition 4.1. *Let d and k be positive integers, $K = 2^k$. Let $\mathbf{x}$ and $\mathbf{y}$ be a pair of K-dimensional complex vectors such that $\mathbf{y} = DFT(\mathbf{x})$ is the forward DFT of $\mathbf{x}$. Let $FFT_d(\mathbf{x})$ denote the vector computed by applying the FFT algorithm of base 2 to $\mathbf{x}$ (see exercise 49 in chapter 2) and by using a floating point arithmetic with d bits, $d \geq 10$. Then we have*

$$FFT_d(\mathbf{x}) = DFT(\mathbf{x} + \mathbf{e}_K(\mathbf{x}));$$

$$||\mathbf{e}_K(\mathbf{x})|| \leq ((1 + \rho 2^{-d})^k - 1)||\mathbf{x}||, \ 0 < \rho < 4.83$$

where $||\cdot|| = ||\cdot||_2$ denotes the Euclidean norm. Moreover,

$$||\mathbf{e}_K(\mathbf{x})|| \leq 5k\ 2^{-d}||\mathbf{x}|| \quad \text{for any} \quad K \leq 2^{2^{d-6}}.$$

Proof. We prove this result by performing a *backward rounding error analysis* (compare Appendix A). First, let us fix some notation and assumptions. Consider the $K \times K$ matrix $F_K = (\omega^{ij})$ defined by ω, the primitive

K-th root of the unity, so that $\mathbf{y} = F_K\mathbf{x}$. In this way, since the matrix $F_K/\sqrt{K}$ is orthogonal, we obtain that

$$||F_K\mathbf{x}|| = \sqrt{K}||\mathbf{x}||. \tag{4.1}$$

Let $fl(\cdot)$ denote the result obtained by computing the expression between the parentheses by using floating point arithmetic. For any $a, b \in \mathbf{C}$ such that their real and imaginary parts are floating point numbers, we can easily prove that

$$\begin{aligned}&fl(a \pm b) = (a \pm b)(1+\epsilon),\ fl(a \times b) = (a \times b)(1+\delta),\ |\epsilon| < u = 2^{-d},\\ &|\delta| < 2\sqrt{2}u(1+u),\end{aligned} \tag{4.2a}$$

provided that overflow and underflow conditions do not occur (compare Appendix A and exercise 36) and that complex multiplication is performed by means of the customary algorithm requiring 4 real multiplications and 2 real additions. Moreover, we assume that the floating point representation $fl(\omega^{ij})$ of the complex number ω^{ij} is such that

$$fl(\omega^{ij}) = \omega^{ij}(1+\delta_{ij}), \quad |\delta_{ij}| < u. \tag{4.2b}$$

Now, let us analyze the customary version of the FFT algorithm of base 2 relying on recursive application of the matrix factorization

$$\begin{aligned}&F_{2K} = P_{2K}\begin{pmatrix} F_K & O \\ O & F_K \end{pmatrix} D_{2K}\begin{pmatrix} I_K & I_K \\ I_K & -I_K \end{pmatrix},\\ &F_2 = \begin{pmatrix} 1 & 1 \\ 1 & -1 \end{pmatrix},\end{aligned} \tag{4.3}$$

where P_{2K} denotes the odd-even permutation matrix and where we set that $D_{2K} = \text{diag}(I_K, \text{diag}(1, \omega, \ldots, \omega^{K-1}))$ (compare exercise 49 of chapter 2). (For other base 2 FFT algorithms, relying on different matrix factorizations, the analysis is performed analogously.) Applying (4.2a), we yield the following relations:

$$\begin{aligned}&fl(F_2\mathbf{x}) = F_2(\mathbf{x} + \mathbf{e}_2(\mathbf{x})),\\ &||\mathbf{e}_2(\mathbf{x})|| \le u||\mathbf{x}||.\end{aligned} \tag{4.4}$$

We will deduce Proposition 4.1 from the next relations:

$$\begin{aligned}&fl(F_H\mathbf{x}) = F_H(\mathbf{x} + \mathbf{e}_H(\mathbf{x})),\ ||\mathbf{e}_H(\mathbf{x})|| \le \sigma_h||\mathbf{x}||,\ H = 2^h,\ h = 0, \ldots, k-1,\\ &\sigma_1 = u,\ \sigma_{h+1} = (1+\rho u)(\sigma_h + 1) - 1,\ h = 0, \ldots, k-1,\ \rho < 4.83,\end{aligned} \tag{4.5}$$

which we will prove by induction on h.

We have $\sigma_1 = u$, due to (4.4), and we will inductively extend the claimed result from the case $h = n$, $H{=}N{=}2^n$ to the case $h = n{+}1$, $H{=}2N{=}2^{n+1}$, to arrive at the relations $fl(F_{2N}\mathbf{x}) = F_{2N}(\mathbf{x} + \mathbf{e}_{2N}(\mathbf{x}))$, $||\mathbf{e}_{2N}(\mathbf{x})|| \le \sigma_{n+1}||\mathbf{x}||$.

By using (4.3) rewrite the vector $fl(F_{2N}\mathbf{x})$ as

$$fl(F_{2N}\mathbf{x}) = P_{2N}fl(\begin{pmatrix} F_N\mathbf{z}_1 \\ F_N\mathbf{z}_2 \end{pmatrix}), \tag{4.6}$$

where $\mathbf{z} = \begin{pmatrix} \mathbf{z}_1 \\ \mathbf{z}_2 \end{pmatrix} = fl(D_{2N}\begin{pmatrix} I_N & I_N \\ I_N & -I_N \end{pmatrix}\mathbf{x})$. By applying (4.2a), (4.2b) and (4.4) and observing that the product $\begin{pmatrix} I_N & I_N \\ I_N & -I_N \end{pmatrix}\mathbf{x}$ corresponds to performing N disjointed DFT's of vectors of dimension 2, we obtain that

$$\begin{aligned} &\mathbf{z} = \mathrm{diag}([1+\epsilon_i])D_{2N}\begin{pmatrix} I_N & I_N \\ I_N & -I_N \end{pmatrix}\hat{\mathbf{x}},\ ||\hat{\mathbf{x}} - \mathbf{x}|| \le u||\mathbf{x}||, \\ &\epsilon_i = 0, i = 0, \dots, N-1, \\ &|\epsilon_i| < (1+u)(1+2\sqrt{2}u(1+u)) - 1,\ i = N, \dots, 2N-1. \end{aligned} \tag{4.7}$$

Rewrite $\mathbf{z}$ as

$$\mathbf{z} = D_{2N}\begin{pmatrix} I_N & I_N \\ I_N & -I_N \end{pmatrix}(\mathbf{x} + \mathbf{f}), \tag{4.8}$$

and replace $D_{2N}^{-1}\mathrm{diag}([\epsilon_i])D_{2N}$ by $\mathrm{diag}([\epsilon_i])$, which is justified since D_{2N} is a diagonal matrix. Next obtain from (4.7) and (4.8) that

$$\begin{pmatrix} I_N & I_N \\ I_N & -I_N \end{pmatrix}(\mathbf{x} + \mathbf{f}) = \mathrm{diag}([1+\epsilon_i])\begin{pmatrix} I_N & I_N \\ I_N & -I_N \end{pmatrix}\hat{\mathbf{x}}$$

and, therefore,

$$\mathbf{f} = \hat{\mathbf{x}} - \mathbf{x} + \begin{pmatrix} I_N & I_N \\ I_N & -I_N \end{pmatrix}^{-1}\mathrm{diag}([\epsilon_i])\begin{pmatrix} I_N & I_N \\ I_N & -I_N \end{pmatrix}\hat{\mathbf{x}}.$$

Recall that $\frac{1}{\sqrt{2}}\begin{pmatrix} I_N & I_N \\ I_N & -I_N \end{pmatrix}$ is an orthogonal matrix and, from the latter vector equation and from (4.7), obtain the following bound, which complements (4.8):

$$||\mathbf{f}|| \le ||\mathbf{x}||((1+u)^2(1+2\sqrt{2}u(1+u)) - 1) \le ||x||\rho u,\ \rho < 4.83. \tag{4.9}$$

Now, we rewrite (4.6) as

$$F_{2N}(\mathbf{x} + \mathbf{e}_{2N}(\mathbf{x})) = fl(F_{2N}\mathbf{x}) = P_{2N}\begin{pmatrix} F_N(\mathbf{z}_1 + \mathbf{e}_N(\mathbf{z}_1)) \\ F_N(\mathbf{z}_2 + \mathbf{e}_N(\mathbf{z}_2)) \end{pmatrix}, \tag{4.10}$$

and deduce from (4.3) and (4.8) that

$$F_{2N}(\mathbf{x}+\mathbf{f}) = P_{2N}\begin{pmatrix} F_N & O \\ O & F_N \end{pmatrix}\mathbf{z} = P_{2N}\begin{pmatrix} F_N\mathbf{z}_1 \\ F_N\mathbf{z}_2 \end{pmatrix}. \tag{4.11}$$

Subtract the left-hand sides and the right-hand sides of the two equations (4.10) and (4.11) from each other and obtain that $F_{2N}(\mathbf{e}_{2N} - \mathbf{f}) = P_{2N}\begin{pmatrix} F_N\mathbf{e}_N(\mathbf{z}_1) \\ F_N\mathbf{e}_N(\mathbf{z}_2) \end{pmatrix}$. Recall (4.3) and deduce that

$$\mathbf{e}_{2N}(\mathbf{x}) = \mathbf{f} + \begin{pmatrix} I_N & I_N \\ I_N & -I_N \end{pmatrix}^{-1} D_{2N}^{-1}\begin{pmatrix} \mathbf{e}_N(\mathbf{z}_1) \\ \mathbf{e}_N(\mathbf{z}_2) \end{pmatrix}.$$

Consequently,

$$||\mathbf{e}_{2N}(\mathbf{x})|| \le ||\mathbf{f}|| + \frac{1}{\sqrt{2}}\sqrt{||\mathbf{e}_N(\mathbf{z}_1)||^2 + ||\mathbf{e}_N(\mathbf{z}_2)||^2}.$$

Applying the inductive hypothesis, $||\mathbf{e}_N(\mathbf{z}_i)|| \le \sigma_n||\mathbf{z}_i||$, $i = 1, 2$, we now obtain that

$$||\mathbf{e}_{2N}(\mathbf{x})|| \le ||\mathbf{f}|| + \frac{\sigma_n}{\sqrt{2}}||\mathbf{z}||. \tag{4.12}$$

Due to (4.8), we have $||\mathbf{z}|| \le \sqrt{2}(||\mathbf{x}|| + ||\mathbf{f}||)$, and we obtain from (4.9) and (4.12) that

$$||\mathbf{e}_{2N}(\mathbf{x})|| \le ||\mathbf{x}||((1+\rho u)(\sigma_n + 1) - 1).$$

Therefore, we may now complete the induction proof of (4.5) by setting $\sigma_{n+1} = (1+\rho u)(\sigma_n + 1) - 1$.

We next recall the recursive definition of σ_k (based on the latter equation) and deduce that $\sigma_k < (1+\rho u)^k - 1$. Substitute $u = 2^d$ and obtain the first part of Proposition 4.1. Moreover, since $(1+\rho u)^k - 1 = \sum_{i=1}^k \binom{k}{i}(\rho u)^i < \sum_{i=1}^k \frac{(\rho k u)^i}{2^{i-1}}$, we obtain that $\sigma_k < 5ku$ for $u = 2^{-d}$ and for any $k \le 2^{d-6}, K \le 2^{2^{d-6}}$. ∎

The condition $k \le 2^{2^{d-6}}$ is actually very mild. For instance, to be more specific, assume that we perform floating point computations with the 64-bit arithmetic. Then the inequality $\sigma_k < 5ku$ holds for any $k = \log K < 2^{58} \approx 2.9 \times 10^{17}$. This covers a much wider range of values for K than the range customarily used in the applications.

From (4.1) we obtain the following corollary of Proposition 4.1:

Corollary 4.1. *Under the notations of Proposition 4.1, we have*

$$||FFT_d(\mathbf{x}) - DFT(\mathbf{x})|| \le 5\sqrt{K}(\log K)2^{-d}||\mathbf{x}||$$

for any $K = 2^k$ *such that* $K < 2^{2^{d-6}}$.

A similar result holds for the inverse DFT (see exercise 30).

The above analysis can be extended to the solution of several computational problems reducible to DFT. In particular, we may perform FFT with a relatively low precision when we compute the product $c(x)$ of two polynomials $a(x)$ and $b(x)$ with integer coefficients, and then rounding off will still give us the output with no error. Specifically, if the polynomials $a(x)$ and $b(x)$ have m-bit integer coefficients, $c(x)$ has degree less than $K = 2^k$, and $d \geq 2m + 1.5 \log K + \log\log K + 5$, then it is possible to prove that the d-bit precision computation (with the subsequent integer rounding-off) yields the correct result (see Corollary 4.1 and exercise 30). This bound only slightly exceeds the number of bits generally needed to represent some output values, i.e. $2m + \log K - 1$.

On the other hand, polynomial division and computing the gcd and the zeros of polynomials are generally ill-conditioned problems [see the estimates in chapter 6 for the division; in (2.2), in Fact 2.3, in [Kn, p.414], and in [Sc85] for the gcd, and in [Hen74] and [H70] for polynomial zeros]. Consequently, the fast algorithms of chapter 1 for polynomial interpolation and evaluation on an arbitrary set of points are numerically unstable (at least, in their straightforward implementation) due to the recursive use of polynomial division in these algorithms, although these computations are numerically stable in the specific cases of the Fourier input sets.

Rational arithmetic of the previous section is an effective general way out of this difficulty, but for various reasons, partly explained in Remark 3.3, the present-day user generally cares less about the economization of ops in such algorithms than about the possibility to stay with lower precision numerical computations, provided, of course, that the output precision is satisfactory. The latter property can be ensured by applying numerically stable algorithms to well-conditioned computational problems.

Similar questions arise in the area of matrix computations. In particular, as we explained in the beginning of section 3 of the previous chapter, distinct algorithms are applied in practice for numerical computation of matrix eigenvalues and polynomial zeros: although the eigenvalues are the zeros of the characteristic polynomial, their dependence on the coefficients of this polynomial is much worse conditioned than their dependence on the matrix entries. We refer the reader to [GL], [W65] and also to [Bun], [C80] and to our section 1, on the condition and numerical stability issues in matrix computations.

After this brief review of the fundamental issues, we will next present some sample nonorthodox techniques for devising effective lower precision algorithms. These will be approximation algorithms [compare our section 1, Remark 1.1.1 and (1.A.3), (1.A.4)], which do not generally compute the solutions exactly even if all the operations are performed with the infinite precision. (Again, recall that one may, in principle, recover the exact solutions by means of the integer or rational roundoff by applying Algorithm 3.3, if the output values are integers or rationals and are approximated sufficiently closely.)

We will first recall that the most popular algorithms for polynomial evaluation at a single point are a little slower than Horner's rule but are numerically stable (see [Bak71] and [Ne]). Next, we will consider the problem of approximating a polynomial $p(y) = \sum_{i=0}^{n} p_i y^i$ on a fixed set of N real points $y_0, y_1, \ldots, y_{N-1}$ lying between 0 and 1. We require that the absolute output errors be at most $\epsilon \sum_{i=0}^{n} |p_i|$ (for a fixed positive ϵ). For this problem, we will next describe the numerically stable algorithm of Rokhlin [Rok88], which requires $O((N+n)\log(1/\epsilon) + n\log^3(1/\epsilon))$ ops assuming that all the points $y_0, \ldots, y_N$ lie between 0 and 1 (and we refer to Remark 4.2 on a further substantial improvement of this algorithm). The restriction to the real interval [0,1] can be lifted by means of shifting and scaling the variable (the error bound should change respectively as the value $\sum_i |p_i|$ changes). Numerical stability is the major advantage of the algorithm of [Rok88], for its asymptotic arithmetic cost estimates are inferior to ones of chapter 1 for Problem 1.2.2 ($POL \cdot EVAL$) of multipoint polynomial evaluation, even if we let $N = O(n)$ and $\log(1/\epsilon)$ be of the order $\log n$, say. Due to the numerical stability of Rokhlin's algorithm, its Boolean cost is relatively low, compared to the cost of the fast solution shown in chapter 1, provided that the user accepts a rough (low precision) approximation of the output values (compare Remark 4.2).

The algorithm of [Rok88] relies on the techniques of independent interest. The problem is reduced in [Rok88] to the evaluation of the expression $\sum_{i=0}^{n} p_i \exp(-s_i t)$ for nonnegative s_i at the points $t_k = -\ln y_k$, $k = 0, 1, \ldots, N-1$ (where ln denotes the natural logarithm, to the exponential base). Then $\exp(-s_i t) = y^{s_i}$. This technique brings us to $p(y) = \sum_{i=0}^{n} p_i y^i$ if we set $s_i = i$, $i = 0, 1, \ldots, n$, and also suggests using this method for the evaluation of *sparse polynomials* with $n+1$ nonzero coefficients.

To explain the construction of [Rok88], assume the infinite precision computations of $p(y)$ for the values t_k and s_i ranging in the intervals from

t_0 to t_{N-1} and from s_0 to s_n, respectively, such that

$$0 \le 2t_0 \le t_{N-1}, \quad 0 \le 2s_0 \le s_n, \tag{4.13}$$

and recall the definitions of the Chebyshev polynomials (see exercise 17 of chapter 1) and of the adjusted Chebyshev sets (see section 2 of chapter 1), as well as the following fact from the *approximation theory:*

Fact 4.1 ([DB]). *Let $f(s)$ be a function having K continuous derivatives on the interval $a \le s \le b$. Let $L_h(s) = L(s)/(s - x_h)$ be a polynomial in s of degree $K-1$ where $L(s) = \prod_{h=0}^{K-1}(s - x_h) = 2^{1-K}T_K(\frac{b-a}{2}(s+1)+a)$, $T_K(x)$ is the Chebyshev polynomial of degree K, $L_h(s)/L_h(x_h)$ is the h-th Lagrange interpolation polynomial for the Chebyshev node set $\{x_0, \dots, x_{K-1}\}$ (adjusted to the interval $a \le s \le b$). Let*

$$g(s, f, K) = \sum_{h=0}^{K-1} f(x_h)L_h(s)/L_h(x_h) \tag{4.14}$$

denote the polynomial of degree at most $K-1$ interpolating to the function $f(s)$ on the set $\{x_0, \dots, x_{K-1}\}$. Let $a \le s \le b$. Then

$$|f(s) - g(s, f, K)| \le 2((b-a)/4)^K \max_{a \le s \le b} |f^{(K)}(s)|/K!.$$

In particular, if $b = 2a > 0$, $f(s) = \exp(-st)$, $K \ge 2$, then $|f(s) - g(s, f, K)| \le 2(at)^K/(K!4^K \exp(at)) < 2/4^K$ for $a \le s \le b$, $t \ge 0$.

Recall (4.13), apply Fact 4.1 to the function $f(s) = \exp(-st)$ for $a = s_0 \le s \le s_n = b$, $t \ge 0$, and obtain that

$$|\exp(-st) - g(s, f, K)| < 2/4^K \text{ for } s = s_i,\ i = 0, 1, \dots, n.$$

[Note that $f(s)$, and consequently $g(s, f, K)$, are also functions in t.] Therefore,

$$|\sum_{i=0}^{n} p_i \exp(-s_i t) - G(t)| \le 2\sum_{i=0}^{n} |p_i|/4^K \text{ for all } t \ge 0$$

provided that $G(t) = \sum_{i=0}^{n} p_i g(s_i, f, K)$. Let $2K = \lceil \log(2/\epsilon) \rceil$. Then $2/4^K \le \epsilon$, and the original problem is reduced to the evaluation of $G(t)$ at the points $t = t_k$, $k = 0, 1, \dots, N-1$.

Now, apply (4.14), replace $f(x_h)$ by $\exp(-x_h t)$, and deduce that

$$G(t) = \sum_{i=0}^{n} \sum_{h=0}^{K-1} p_i L_h(s_i) \exp(-x_h t)/L_h(x_h).$$

Set $g_h = \sum_{i=0}^{n} p_i L_h(s_i)/L_h(x_h)$, $h = 0, 1, \ldots, K-1$, and deduce that $G(t) = \sum_{h=0}^{K-1} g_h \exp(-x_h t)$. Now, we can see the power of this simple techniques of *summation reordering*. First evaluate $L(s_i)$ for all i, then divide $L(s_i)$ by $s_i - x_h$ for all h and i, thereby obtaining the values $L_h(s_i)$, and then compute g_h for all h at the overall cost of $O(nK)$ ops. This does not include the cost of computing $L_h(x_h)$ for all h. If x_h are as in Fact 4.1 for $b = -a = 1$, then $L_h(x_h) = (-1)^h \sin(\frac{2h+1}{2K}\pi)/(K2^{K-1})$ ([Hen82], p. 249). We may reduce any pair of a and b to the case $b = -a = 1$ by shifting and scaling the variable. The values $\sin(\frac{2h+1}{2K}\pi)$ can be either precomputed or evaluated by special fast methods with sufficiently small errors. We may need to increase K in order to accommodate the errors of computing $\sin(\frac{2h+1}{K}\pi)$.

Finally, use the values g_h in order to compute $G(t_k)$ for $k = 0, 1, \ldots, N-1$ in time $O(NK)$. Since $2K = \lceil \log(2/\epsilon) \rceil$, the overall cost of the solution is $O(\,(N+n)\log(1/\epsilon)\,)$ ops.

For the general set of nonnegative values s_i and t_k, the same approach is applied in [Rok88] on several subsets of this set, each subset meeting the requirement (4.13) (with the adjusted notation). This results in the overall cost bound of $O((n+N)\log(1/\epsilon) + N(\log(1/\epsilon))^3)$ ops.

Remark 4.1. We may easily extend the algorithm to the evaluation of an n-th degree polynomial $p(z)$ on any set S of points on the unit circle. Indeed, set $x = x(z) = (1/2)(z + 1/z)$, so that $-1 \le x \le 1$ as $|z| = 1$. Substitute $z = x + \sqrt{x^2-1}$ and obtain $p(z) = q_+(x) + s_+(x)\sqrt{x^2-1}$ where $\operatorname{Im} z \ge 0$; substitute $z = x - \sqrt{x^2-1}$ where $\operatorname{Im} z \le 0$ and obtain $p(z) = q_-(x) + s_-(x)\sqrt{x^2-1}$. Finally, evaluate $q_+(x)$, $s_+(x), \sqrt{x^2-1}$ and (if needed) $q_-(x)$ and $s_-(x)$ on a set of real points $x = x(z)$, $z \in S$.

Remark 4.2. The power of Rokhlin's algorithm is quite restricted: its computational cost grows too rapidly with the growth of $1/\epsilon$. Let us sketch a substantial recent improvement (due to [P92e]), which decreases the computational cost to the level of $O(\log^2(\log(1/\epsilon)))$. The algorithm of [P92e] starts with approximating $p(x)$ on the interval from $-a$ to a by a polynomial $v(x)$ of degree at most $d-1$ that interpolates to $p(x)$ on the K point adjusted Chebyshev node set. Due to Fact 4.1, the approximation error E is exponentially small as K grows, provided that $a < 2/3$. [For $a = 1/2$ we have that $E = O(p2^{-K})$, $p = \sum_{i=K}^{n} |p_i|$.] Thus, it remains to choose K, so as to ensure the desired upper bound on E and to evaluate $v(x)$, rather than $p(x)$, on the input set of points. Choosing K of the order $\log n$ suffices to ensure the error bound of the order $p/n^{O(1)}$, and then it is surely simpler to evaluate $v(x)$ than $p(x)$. The coefficients of $v(x)$ are obtained based on the equation $v(x) = p(x) \bmod C_K(2x/a)$ where $C_K(x) = \prod_{k=0}^{K-1}(x - 2x_k) = \sum_{h=0}^{H} c_h x^{K-2h}$ and where $x_k = \cos(\frac{2k+1}{2K}\pi)$, $k = 0, 1, \ldots, K-1$, denote Chebyshev's nodes,

$$c_h = (-1)^h \binom{d-h}{h} d/(d-h) , \quad h = 0, 1, \ldots, H , \ H = \lfloor K/2 \rfloor .$$

[$C_K(x) = 2T_K(x/2)$, $T_K(y)$ denoting the Chebyshev polynomial of degree K (see exercise 17 of chapter 1), and hence $C_K(x) = xC_{K-1}(x) - C_{K-2}(x)$ for $K = 3, 4, \ldots$, but it may be more effective to use the above explicit expressions for the coefficients c_h, rather than the latter recursive equations for $C_K(x)$.] We may evaluate the coefficients of $v(x)$ by means of polynomial division and then evaluate $v(x)$ on the given set $\{x = y_k , \ k = 0, 1, \ldots, N-1\}$ by using the algorithm of our chapter 1. Such a *combination of the idea of numerical approximation* of $p(x)$ [within the error $p\epsilon$] *with the error-free algebraic computation* yields the solution in $O(m \log^2(\log(1/\epsilon)) + n \min\{\log n , \ \log(1/\epsilon)\})$ ops. This is substantially less than in [Rok88] and [already for ϵ as small as $1/n^{O(1)}$] is even below the lower bound on the complexity of the *exact* multipoint evaluation of $p(x)$. Moreover, the evaluation of the coefficients of $v(x)$ can be performed in any ring containing $p_0, \ldots, p_n$, and the algebraic computation techniques, based on the Chinese remainder algorithm for integers (see section 3) and on Kaminski's algorithms for residue computation ([Kam87]), can be applied to achieve a further substantial improvement of the computations (see [P92e]). The user should not, however, discard the fast $O(n \log^2 n)$ algorithms too easily, since the input set (that is, the node and the coefficient set) can be approximated by the set of rationals, and then the evaluation can be performed in the finite fields with the use of the Chinese remainder theorem and the rational roundoff. Carefull error and complexity analysis, where one should also take into account the option of using the binary segmentation (see Remark 9.2), should determine when this approach is superior to one of [P92e].

The next algorithm from [Rok85] is again a fast and numerically stable approximation algorithm. It outperforms the known exact solution algorithm in the case of lower precision computations, and it exploits the techniques of *expansion of partial fractions into Laurent series* and of summation reordering. The algorithm has several important applications to particle simulation ([AGR] and [CGR]) and to the computation of conformal mapping ([ODR89]).

Problem 4.1 *($C(G \cdot H) \cdot VECTOR$), multiplication of a vector by a Cauchy (generalized Hilbert) matrix. Given a natural K and $2K$ complex numbers a_k, w_k, $k = 0, 1, \ldots, K-1$, and defining the partial fraction decomposition of a rational function,*

$$R(x) = \sum_{k=0}^{K-1} \frac{a_k}{x - w_k},$$

compute the values $R(x)$ for $x = x_i$, $i = 0, 1, \ldots, L-1$.

Solutions. Problem 4.1 is the problem of multiplication of the Cauchy (generalized Hilbert) matrix $[h_{ij}]$, $h_{ij} = 1/(x_i - w_j)$, by the vector $\mathbf{a} = [a_0, \ldots, a_{K-1}]^T$. Such a problem can also be viewed as an extension (from polynomials to rational functions) of Problem 1.2.2 ($POL \cdot EVAL$) of multipoint polynomial evaluation. The straightforward algorithm requires $L(3K-1)$ ops in order to compute $R(x_i)$ for all i. The methods of chapters 1 and 2 lead to faster solutions but also to the problems of numerical stability. Suppose now that the values $R(x_i)$ are sought within a fixed positive absolute error bound ϵ, denote that $\alpha = \max_k |a_k/w_k|$, and let $|x_i/w_k| < q < 1$ for a constant q and for all i and k. Then $J(2K + 2L - 1)$ ops suffice as long as

$$J \geq \log(\frac{\alpha K}{(1-q)\epsilon}) / \log(\frac{1}{q}).$$

This means a substantial improvement of the bound $L(3K-1)$ if, say, $q = 0.5$ and $\log(\frac{\alpha K}{\epsilon}) = O(\log K)$.

To deduce the above cost estimate, substitute the expressions

$$\frac{1}{x - w_k} = -\frac{1}{w_k} \sum_{j=0}^{\infty} (\frac{x}{w_k})^j, \quad k = 0, 1, \ldots, K-1,$$

and represent the sum of the partial fractions $R(x)$ as follows:

$$R(x) = \sum_{j=0}^{\infty} A_j x^j \tag{4.15}$$

where

$$A_j = -\sum_{k=0}^{K-1} a_k / w_k^{j+1}. \tag{4.16}$$

The power series (4.15) converges for sufficiently small $|x|$. Approximate $R(x)$ by the J-term finite power series and estimate the errors of the resulting approximation to $R(x_i)$ in terms of $|x_i/w_k|$ and α. Under our assumptions about the values J, $|x_i/w_k|$ and $|\alpha_k/w_k|$, we have

$$E_J(x_i) = |R(x_i) - \sum_{j=0}^{J-1} A_j x_i^j| \le \alpha K q^J/(1-q) \le \epsilon.$$

Now, it remains to compute first the coefficients $A_0, \dots, A_{J-1}$ of (4.16), by using $J(2K-1)$ ops, and then the values $\sum_{j=0}^{J-1} A_j x_i^j$ of (4.15) for $i = 0, 1, \dots, L-1$, by using $2JL$ ops. ∎

If $|w_k/x_i| < q < 1$, for a constant q and for all k and i , we will arrive at a similar algorithm by substituting $\frac{1}{x-w_k} = \frac{1}{x}\sum_{j=0}^{\infty}(\frac{w_k}{x})^j$.

5. Iterative Approximation Algorithms for Linear Systems. High Precision Solution of Linear Systems Computed by Using a Low Precision.

In this section we will recall some well known iterative algorithms for solving linear systems of equations; we will use them in the next sections; some of the algorithms will remind us of the algebraic iteration techniques of section 3. In sufficiently many iteration steps, the approximation within any fixed error bound can be computed (compare Remark 3.2).

We will first recall a *residual correction* iterative algorithm for the linear system $A\mathbf{x} = \mathbf{b}$, which performs with a lower precision and outputs a high precision solution ([A], [GL], [W65]) and which has some similarity to Algorithm 3.2. Let $\tilde{A}^{-1}$ denote a computed or readily computable approximation to A^{-1}. [In the classical version of this algorithm, $\widetilde{A}^{-1}$ is implicitly represented by approximations to the matrices L, D and M of (2.2.4).] Then set $\mathbf{x}_0 = \mathbf{e}_0 = \mathbf{0}$ and successively compute

$$\mathbf{r}_{i+1} = \mathbf{b} - A\mathbf{x}_i, \tag{5.1}$$

$$\mathbf{e}_{i+1} = \tilde{A}^{-1}\mathbf{r}_{i+1}, \tag{5.2}$$

$$\mathbf{x}_{i+1} = \mathbf{x}_i + \mathbf{e}_{i+1}$$

for $i = 0, 1, \dots$. It is easy to prove that $\mathbf{x} - \mathbf{x}_i = (I - \widetilde{A}^{-1}A)^i(\mathbf{x} - \mathbf{x}_0)$, $i = 0, 1, \dots$, where $\mathbf{x} = A^{-1}\mathbf{b}$, so that $\mathbf{x}_i$ converges to $\mathbf{x}$ as $i \to \infty$ if $\|I - \widetilde{A}^{-1}A\| < 1$ (for a matrix norm such that $\|B^s\| \le \|B\|^s$ for any matrix B and any natural s) and if the computations are performed with the infinite precision. Furthermore, according to the well-known heuristic ([GL], p. 127), even if we compute $\mathbf{e}_{i+1}$ with the d-bit precision by using (5.2) and if we compute $\mathbf{r}_{i+1}$ with the $(2d)$-bit precision by using (5.1) where $d > g = \log \text{cond}_\infty(A)$, then $||\mathbf{x}_i - \mathbf{x}||_\infty \leq c(i,d)||\mathbf{x}_0 - \mathbf{x}||_\infty$ for $\mathbf{x} = A^{-1}\mathbf{b}$ and for $c(i,d) = \max\{2^{-d}, 2^{-i(d-g)}\}$. Thus, for $k < d$, k correct bits of each component of the solution vector can be computed in the order of $k/(d-g)$ iteration steps performed with the d-bit and $(2d)$-bit precision. Actually, the algorithm can be modified so that the single d-bit precision would suffice throughout. Furthermore, in [P91], [P92d], [P93d] the residual correction algorithm has been modified so as to ensure that the low precision of $O(\log(in))$ bits suffices throughout each loop i of the computation provided that the matrix A is filled with integers $(A)_{u,v}$ such that either $(A)_{u,v} = 0$ or $2^s \leq |(A)_{u,v}| < 2^{s+h}$ for two fixed integers s and h and for all u and v (compare exercise 24).

The residual correction algorithm is a special case of a large class of iterative approximation algorithms for linear systems of equations. Such algorithms are particularly effective for large sparse linear systems, for which much less storage space is usually needed in the iterative algorithms than in the direct methods, such as Gaussian elimination. Every iteration of such iterative algorithms essentially amounts to one or few multiplications of the input matrix or its submatrices by vectors, so that about $F(A)$ to $2F(A)$ words of storage space suffice with using special data structures where $F(A)$ is the number of nonzero entries of A. The cited results of [P91], [P92d], [P93d] can be extended to many such algorithms.

To see some examples of iterative schemes, represent $A = [a_{ij}]$ as $L + D + U$ where $D = \text{diag}(a_{00}, \ldots, a_{n-1,n-1})$, L and U^T are proper lower triangular matrices, and as $A = N - P$ where N is a nonsingular and readily invertible matrix. Then choose an initial approximation $\mathbf{x}_0$ to the solution and recursively compute updated approximation vectors $\mathbf{x}_{s+1}$ such that

$$N\mathbf{x}_{s+1} = \mathbf{b} + P\mathbf{x}_s \tag{5.3}$$

for $s = 0, 1, \ldots$ [compare (3.3) for $pL = P$]. In particular, for $N = D$, $P = -(L+U)$, we arrive at Jacobi iteration, $D\mathbf{x}_{s+1} = \mathbf{b} - (L+U)\mathbf{x}_s$, and for $N = D + L$, $P = -U$, at Gauss-Seidel iteration, $(D+L)\mathbf{x}_{s+1} = \mathbf{b} - U\mathbf{x}_s$. These are two major iterative algorithms for linear systems, used as two basic examples of various (and usually more advanced) practical methods (see [GL], [Var], [You], [FGN] and [P87a]).

Such algorithms are numerically stable as soon as they converge. Their convergence rate depends on the properties of the matrix A. Here is a basic result:

Fact 5.1. *The iteration (5.3) converges to the solution* $\mathbf{x} = A^{-1}\mathbf{b}$ *to the linear system* $A\mathbf{x} = \mathbf{b}$ *for any initial vector* $\mathbf{x}_0$ *and for any vector* $\mathbf{b}$ *if and only if the spectral radius* $\rho = \rho(N^{-1}P) < 1$ *(see Definition 2.1.7). Furthermore,* $||\mathbf{x} - \mathbf{x}_s|| \leq c\rho^s||\mathbf{x} - \mathbf{x}_0||$ *for all* s *and for any fixed vector norm where the constant* c *depends on the vector norm and does not depend on* s.

Proof. (5.3) implies that

$$\mathbf{x}_{s+1} = N^{-1}\mathbf{b} + N^{-1}P\mathbf{x}_s = (I + N^{-1}P + (N^{-1}P)^2 + \cdots + (N^{-1}P)^s)N^{-1}\mathbf{b} + (N^{-1}P)^{s+1}\mathbf{x}_0.$$

Therefore, $\rho < 1$ if and only if (for any pair of vectors $\mathbf{b}$ and $\mathbf{x}_0$) simultaneously the sum in the brackets converges to $(I - N^{-1}P)^{-1}N^{-1}\mathbf{b}$ as $s \to \infty$, and the remaining term, $(N^{-1}P)^{s+1}\mathbf{x}_0$, converges to $\mathbf{0}$, that is, if and only if

$$\lim_{s\to\infty} \mathbf{x}_{s+1} = (I - N^{-1}P)^{-1}N^{-1}\mathbf{b} = [N(I - N^{-1}P)]^{-1}\mathbf{b} = (N - P)^{-1}\mathbf{b} = A^{-1}\mathbf{b}.$$

We also have that $(N-P)\mathbf{x} = \mathbf{b}$, $\mathbf{x} = N^{-1}P\mathbf{x}+N^{-1}\mathbf{b}$, $\mathbf{x}-\mathbf{x}_{s+1} = N^{-1}P(\mathbf{x}-\mathbf{x}_s) = (N^{-1}P)^{s+1}(\mathbf{x}-\mathbf{x}_0)$. From the relation $\mathbf{x}-\mathbf{x}_s = (N^{-1}P)^s(\mathbf{x}-\mathbf{x}_0)$, we obtain that $||\mathbf{x} - \mathbf{x}_s|| \leq ||(N^{-1}P)^s||\ ||\mathbf{x} - \mathbf{x}_0||$. The inequality $||\mathbf{x} - \mathbf{x}_s|| \leq c\rho^s||\mathbf{x} - \mathbf{x}_0||$ follows since $\lim_{s\to\infty}(||(N^{-1}P)^s||)^{1/s} = \rho$ (see [BCM], p. 239). ∎

Some iterative methods (of conjugate gradient and Lanczos classes) give the exact solution $\mathbf{x} = A^{-1}\mathbf{b}$ in at most n iteration steps provided that the computations are performed with the infinite precision (for a nonsingular $n\times n$ matrix A). Each step essentially amounts to one or two multiplications of A by a vector, which makes the approach highly attractive for sparse and/or structured linear systems, for which such a multiplication has a lower complexity.

Convergence rates of these and several other methods, as well as their numerical stability, can be improved if the linear system is h.p.d. or at least Hermitian, and this motivates application of the two symmetrization techniques of chapter 2, that is, the transition from $A\mathbf{x} = \mathbf{b}$ to the systems $A^HA\mathbf{x} = A^H\mathbf{b}$ (h.p.d.) or $\begin{pmatrix} O & A^H \\ A & O \end{pmatrix}\begin{pmatrix} \mathbf{x} \\ \mathbf{c} \end{pmatrix} = \begin{pmatrix} A^H\mathbf{c} \\ \mathbf{b} \end{pmatrix}$ (Hermitian) for any fixed vector $\mathbf{c}$. Note that $\text{cond}_2(A^HA) = (\text{cond}_2(A))^2$, $\text{cond}_g\begin{pmatrix} O & A^H \\ A & O \end{pmatrix} = \text{cond}_g(A)$ for $g = 1, 2, \infty$.

Clearly, the iteration (5.3) is a numerical version of the algebraic iteration (3.3). We will show more similarity among the algebraic p-adic algorithms and their numerical counterparts by presenting the numerical version of Newton's iteration of Algorithm 3.1 [compare also (2.7.1)]. Actually, Newton's iteration is one of the most general and most important numerical algorithms, but we will only apply it to the matrix equation $I - AX^{-1} = O$, in which case it takes the form

$$X_{k+1} = X_k(2I - AX_k) = (2I - X_kA)X_k, k = 0, 1, \dots . \tag{5.4}$$

In this form, we have already used it in Algorithms 2.7.1 and 2.11.1 (compare also [B-I], [B-IC], [PS] and [PR89]). This iteration converges to A^+ as $k \to \infty$, provided that

$$X_0 = \alpha A^H, \ \alpha = \frac{2}{\sigma_1^2 + \sigma_r^2}, \tag{5.5}$$

and that $\sigma_1 \geq \sigma_2 \geq \cdots \geq \sigma_r > 0$ denote all the r singular values of a matrix A of rank r.

Indeed, let $\rho_j^{(i)}$ denote a singular value of X_iA. Then since $I - AX_{k+1} = (I - AX_k)^2$, $I - X_{k+1}A = (I - X_kA)^2$, it can be easily deduced that $1 - \rho_j^{(k+1)} = (\rho_j^{(k)} - 1)^2$, for all j and k. Applying these equations together with the relations

$$\frac{2}{1 + (\text{cond}_2(A))^2} = \rho_r^{(0)} \leq \rho_j^{(0)} \leq \rho_1^{(0)} = \frac{2}{1 + 1/(\text{cond}_2(A))^2},$$

which hold for all j and for $\text{cond}_2(A) = \sigma_1/\sigma_r$, we arrive at the following result:

Proposition 5.1. *Under (5.5), $k = O(s \log \text{cond}_2(A))$ iteration steps (5.4) suffice in order to ensure that $||(A^+ - X_k)A||_2 = \max_j |\rho_j^{(k)} - 1| < 2^{-2^s}$ for $j = 1, 2, \dots, r$, as $s \to \infty$.*

Since every step (5.4) essentially reduces to two matrix multiplications, Proposition 5.1 gives us an estimate for the complexity of approximating A^+ in terms of the norm of the output error and of $\text{cond}_2(A)$. The estimate for the complexity is quite small unless $\text{cond}_2(A)$ is very large. We refer the reader to [PS] on the more specific complexity estimates for the iteration (5.4) and for its improved versions that ensure a substantial convergence acceleration on the proof of the strong numerical stability of such computations, even for a rank deficient matrix A, and on their extension to computing the numerical generalized inverses, $A^+(\epsilon)$, of A. The method, however, does not generally preserve sparsity and/or structure of A and X_0 in the transition to X_k for larger k.

Remark 5.1. The asymptotic estimate of Proposition 5.1 still holds for the alternative choices of $\alpha = 1/||A^H A||_1$ or $\alpha = 1/(||A^H||_\infty ||A||_1)$ in (5.5) (see [B-I], [B-IC] and [PR89]). These values of α are readily available for a given matrix A.

6. Data Compression and Decreasing the Storage Space for the Discrete Solution of PDE's: Compact Multigrid.

Next, we will present a solution of a linear system of equations arising from the N-point discretization of a linear *partial differential equation* (*PDE*). This is an important practical problem, and we will decrease the estimated bit-cost of its solution by means of computing with a low precision (compare Remarks 1.1 and 9.2 on the advantages of computing with a lower precision). The efficacy of this approach is due to some special techniques for representation and compression of data (rather than to any innovation in the classical arithmetic scheme of multigrid algorithms); these techniques are not captured under the arithmetic (RAM or circuit) models. Similar comments apply to the approach of our section 9.

We will first decrease the storage space. The straightforward representation of the solution within the discretization error requires $O(N \log N)$ bits, but we will follow [PR90], [PR92], [PR93] and will arrive at a compact representation with $O(N)$ bits, which, in the case of the constant coefficient linear PDE's, we will compute in section 7 by using a low precision and $O(N)$ bit-operations (see section 3 of the next chapter on the parallel cost of this computation).

Our data compression techniques will be somewhat similar to the residual correction technique of the previous section, but we will incorporate them in a variant of the multigrid scheme, customarily used for the solution of PDE's (see more details in the next section). Following [PR90], [PR92], [PR93], we will call this algorithm *compact multigrid.*

Discretization of a PDE

Let us next specify these results starting with some auxiliary facts and definitions. We will study linear PDE's on the unit d-dimensional cube, discretized over a family of d-dimensional lattices $L_0, L_1, \ldots, L_k$, where each internal point of L_j lies at the distance $h_j = 2^{-j}$ from its $2d$ nearest neighbors. There are exactly $|L_j| = N_j = 2^{dj}$ points in L_j for $j = 0, 1, \ldots, k$, and the overall number of points equals $N_k = N = 2^{dk}$, where $k = (\log N)/d$, provided that the boundary points whose coordinates only differ by 0 or 1 from each other are identified as a single point. (On the actual discretization grids for PDE's, all the boundary points are distinct, so that the j-th grid contains slightly more than N_j points.)

Let $u(\mathbf{x})$, a function in the d-dimensional vector of variables $\mathbf{x}$, represent the solution to a PDE, and let $u_j(\mathbf{x})$ for a fixed $\mathbf{x} \in L_j$ denote the respective component of the N_j-dimensional vector $\mathbf{u}_j$ that represents the solution to the linear system,

$$D_j \mathbf{u}_j = \mathbf{b}_j, \tag{6.1}$$

of the difference equations generated by a discretization of the PDE over the lattice L_j. Then $\Delta_j(\mathbf{x}) = u(\mathbf{x}) - u_j(\mathbf{x})$ for $\mathbf{x} \in L_j$ denotes the discretization error function on L_j for $j = 1, \ldots, k$.

Discretization of the PDE over L_j is defined by replacing all the derivatives in the PDE by finite differences according to some customary formulae, such as

$$\begin{aligned}
\frac{\partial u(x,t)}{\partial x} &= \frac{1}{2h}(u(x+h,t) - u(x-h,t)) + O(h^2),\\
\frac{\partial^2 u(x,t)}{\partial x^2} &= \frac{1}{h^2}(u(x+h,t) - 2u(x,t) + u(x-h,t)) + O(h^2),\\
\frac{\partial^2 u(x,t)}{\partial x \partial t} &= \frac{1}{hk}(u(x+h,t+k) - u(x+h,t-k) - u(x-h,t+k) +\\
&\quad u(x-h,t-k)) + O((h+k)^2).
\end{aligned}$$

These expressions rely on Taylor's expansion of $u(x,t)$ (see [A] or [CdB]), and the terms represented with the "O" notation above have coefficients proportional to some (higher order) derivatives. These terms are deleted when we replace the partial derivatives by finite differences. Here we suppose that the unknown function $u(x,t)$ is smooth enough to support the above relations.

The resulting bounds [of the form $O(h^\alpha)$ for positive constants α] on the errors of the approximation to the partial derivatives are translated into the bounds $O(h_j^\beta)\text{cond}_2(D_j)$, for positive constants β, on the errors of the approximation to $u(\mathbf{x})$ by $u_j(\mathbf{x})$. These bounds turn into the bounds (6.4), (6.5) for a large class of PDE's and of their discretizations such that

$$\text{cond}_2(D_j) = O(1/h_j^\nu), \quad \nu < \beta. \tag{6.2}$$

It is also easy to observe that the above discretization of a linear PDE gives us matrices D_j with $O(1)$ nonzero coefficients per row as $j \to \infty$.

Let $u_0(\mathbf{x}) = 0$ for $\mathbf{x} \in L_0$ and let $\hat{u}_{j-1}(\mathbf{x})$ for $j = 1, 2, \ldots, k$ denote the prolongation of $u_{j-1}(\mathbf{y})$ from L_{j-1} to L_j, obtained by means of the interpolation (averaging) of the values of $u_{j-1}(\mathbf{y})$ at an appropriate array of points of L_{j-1} lying near $\mathbf{x}$. Specifically, we assume that $\hat{u}_{j-1}(\mathbf{x}) = u_{j-1}(\mathbf{x})$ if $\mathbf{x} \in L_{j-1}$, and $\hat{u}_{j-1}(\mathbf{x})$ is the average of $u_{j-1}(\mathbf{y})$ over all $\mathbf{y}$ such that $\mathbf{y} \in L_j$ and, say, $|\mathbf{y} - \mathbf{x}| = h_j$ if $\mathbf{x} \in L_j - L_{j-1}$. Then

$$u_j(\mathbf{x}) = \hat{u}_{j-1}(\mathbf{x}) + e_j(\mathbf{x}), \quad \mathbf{x} \in L_j,\ j = 1, \ldots, k, \tag{6.3}$$

where $e_j(\mathbf{x})$ denotes the interpolation error on L_j.

Weak pseudo regularity assumption

We will further assume that the discretization and interpolation errors satisfy the two following bounds, which we will call the assumptions of *the weak pseudo regularity* of the PDE:

$$|\Delta_j(\mathbf{x})| \le 2^{c-\alpha j}, \tag{6.4}$$

$$|e_j(\mathbf{x})| \le 2^{c-\alpha j} \tag{6.5}$$

for all $\mathbf{x} \in L_j$, $j = 1, \ldots, k$, and for fixed $c \ge 0$ and $\alpha > 0$. Below, we will deduce (6.5) from (6.4). In turn, the assumption (6.4) follows from the assumption (6.2) and holds for a very large class of PDE's (see, for instance, [Am], p. 29; [LP] and [Ri]) including the well-posed linear PDE's, as well as many nonlinear PDE's. Such an assumption is routinely made in the analysis of the multigrid methods (e.g., [Br84], [Br77], [Br72], [BDu]); in particular, the auxiliary grid problems are said to be "solved to the level of truncation" defined by (6.4) (see [McC87], p. 26).

As a part of the weak pseudo regularity assumption, let us further assume that $u(\mathbf{x})$ has been scaled so that $|u_j(\mathbf{x})| \le 1$ for $\mathbf{x} \in L_j$ and for all j and that every $e_j(\mathbf{x})$ is represented with (that is, rounded off to) α bits.

Let us next show that (6.4) for any fixed pair of α and j implies (6.5) for some fixed pair of α and j. First rewrite (6.4) as follows:

$$|\Delta_j(\mathbf{x}_j)| \le a h_j^{\gamma}, \quad \mathbf{x}_j \in L_j, \tag{6.6}$$

where a and γ are positive constants and h_j is the length of a side of the mesh of L_j, so that $h_{j-1} = 2h_j$ for the lattices L_j that we have chosen. Let $\mathbf{x}_{j-1} \in L_{j-1} \subset L_j$, $\mathbf{x}_j \in L_j$, $|\mathbf{x}_j - \mathbf{x}_{j-1}| = g h_j$ and $|e_j(\mathbf{x}_j)| \le |u_j(\mathbf{x}_j) - u_{j-1}(\mathbf{x}_{j-1})|$.

Then deduce from (6.6) that (for some fixed positive g and θ, $0 \leq \theta \leq 1$) we have

$$|e_j(\mathbf{x}_j)| = |u_j(\mathbf{x}_j) - u_{j-1}(\mathbf{x}_{j-1})| \leq |u(\mathbf{x}_j) - u_j(\mathbf{x}_j)| + |u(\mathbf{x}_{j-1}) - u_{j-1}(\mathbf{x}_{j-1})| + |u(\mathbf{x}_j) - u(\mathbf{x}_{j-1})| \leq ah_j^\gamma + ah_{j-1}^\gamma + g|u'(\mathbf{x}_j + \theta h_j)|h_j = a^* h_j^\gamma,$$

that is,

$$|e_j(\mathbf{x}_j)| \leq a^* h_j^\gamma \tag{6.7}$$

where

$$a^* = a(1+(h_{j-1}/h_j)^\gamma) + g|u'(\mathbf{x}_j+\theta h_j)|h_j^{1-\gamma} \leq (1+2^\gamma)a + ag \max|u'(\mathbf{x})|h_j^{1-\gamma}.$$

The bounds (6.6) and (6.7) turn into the bounds (6.4) and (6.5) for $c = \log a^*$ (or for $c = \log a$), $\alpha = -\gamma(\log h_j)/j$, and $\mathbf{x} = \mathbf{x}_j$. If $\mathbf{x}_j \in L_j$, then we just replace $\mathbf{x}_{j-1}$ by $\mathbf{x}_j$ above and cancel the term $|u(\mathbf{x}_j) - u(\mathbf{x}_{j-1})|$.

Remark 6.1. We may replace the bounds (6.4), (6.5) and $|u_j(\mathbf{x})| \leq 1$ by the bounds

$$||\,\mathbf{\Delta}_j|| \leq 2^{c-\alpha j}||\mathbf{u}_j||,$$

$$||\mathbf{e}_j|| \leq 2^{c-\alpha j}||\mathbf{u}_j||,$$

for a fixed vector norm, provided that $\mathbf{\Delta}_j, \mathbf{e}_j$ and $\mathbf{u}_j$ are considered N_j-dimensional vectors with the components $\Delta_j(\mathbf{x})$, $e_j(\mathbf{x})$ and $u_j(\mathbf{x})$ for $\mathbf{x} \in L_j$, respectively. This modification would not change our resulting estimates for the cost of the Compact Multigrid.

(Strong) pseudo regularity assumption

Studying the recovery of the output values from their compressed representation, we will assume *pseudo regularity,* which means the weak pseudo regularity together with the assumption that the prolongation from L_{j-1} to L_j only requires $O(N_j)$ bit-operations (which is clearly the case for the interpolation by averaging).

In the next section, we will assume *strong pseudo regularity,* that is, in addition to the pseudo regularity, we will assume that

1) a fixed iterative algorithm [such as (5.3) or a multigrid algorithm of the next section] for linear systems with matrices D_j uses $O(1)$ multiplications of submatrices of D_j by vectors for every j, in order to decrease, by the factor independent of j and N, the norm of the error of the

approximation to the solution $u_j(\mathbf{x})$ of the system (6.1) (*the linear order of convergence assumption*); as well as our other assumptions, this assumption routinely holds for a large class of PDE's encountered in scientific study and engineering;

2) the entries of the matrices D_j for all j, as well as the components of $\mathbf{b}_j$, are integers having magnitudes $O(1)$ as $j \to \infty$ or turn into such integers after their truncation and after scaling the system (6.1); such a *bounded coefficients assumption,* as well as the next *assumption of sparsity,* holds for the piecewise constant coefficient PDE's;
3) every row of the matrix D_j has $O(1)$ nonzero entries (*sparsity* assumption).

For convenience, we will assume the fixed point binary representation, but shifting to the floating point representation would not cause any substantial increase of these estimates.

Compression of the output data

Now assume the weak pseudo regularity relations and compress approximations to all the N values of $u_k(\mathbf{x})$ on L_k within absolute errors of at most $2^{c-\alpha k}$, so as to decrease the storage space required.

For the straightforward fixed point binary representation of these values of $u_k(\mathbf{x})$, we generally need $N\lceil \alpha k - c\rceil$ bits.

As an alternative, let us store $u_k(\mathbf{x})$ on L_k in the compressed form by recursively approximating, within $2^{c-\alpha j-\alpha}$, the fixed point binary values $e_j(\mathbf{x})$ for $\mathbf{x} \in L_j$, $j = 1, \ldots, k$. The storage space of $2^d\lceil \alpha - c\rceil + \alpha(N_2 + N_3 + \cdots + N_k) < 2^d\lceil \alpha - c\rceil + 2\alpha N = O(N)$ bits suffices for this compressed information, which means saving roughly the factor of $k/2 = 0.5 \log N$ bits against the straightforward representation.

Recovery of the solution values from the compressed data.

Next, assume the pseudo regularity and recover $u_k(\mathbf{x})$ on L_k from the compressed information given by $e_j(\mathbf{x})$ on L_j for $j = 1, \ldots, k$. Start with $u_0(\mathbf{x}) = 0$ for $\mathbf{x} \in L_0$ and recursively, for $j = 1, \ldots, k$, compute the values

a) $\hat{u}_{j-1}(\mathbf{x})$ on L_j, by prolongation of $u_{j-1}(\mathbf{x})$ from L_{j-1} to L_j, and then
b) $u_j(\mathbf{x})$ on L_j, by applying the equations (6.3).

Perform both Stages a) and b) with the precision $2^{c-\alpha j-\alpha}$. The Stage b) amounts to appending α bits of $e_j(\mathbf{x})$ to the available string of bits of the fixed point binary representation of $\hat{u}_{j-1}(\mathbf{x})$ for each $\mathbf{x} \in L_j$. The Stage a) amounts to scanning the values of $u_{j-1}(\mathbf{x})$ on L_{j-1} and to the summation of few β-bit binary numbers [where, say, $\beta = O(\alpha)$]. These binary numbers are defined by the β *least significant* bits in the representation of $u_{j-1}(\mathbf{x})$ for appropriate $\mathbf{x}$ from L_{j-1}. Since $\sum_j N_j = O(N)$, the computational complexity estimates for Stages a) and b) stay within the desired bound $O(N \log N)$.

In many practical applications, the compressed solutions need not be decompressed. For example, for the time dependent PDE's, the most customary solution methods compute the solution at a discrete sequence of, say, T time steps. In each time step, a PDE is approximately solved by using an N-point discretization of the PDE fixed at such a time value and by using the approximate solution (obtained in the compressed form at the previous time step) as an initial approximation to the current solution. Thus, the solutions at these time steps need not be decompressed, except for the solution at the final time step. Our total bit-cost in this case, including decompression of the final solution and T calls for compact multigrid, would be $O(N(T + \log N))$ bit-operations [requiring $O(T \log N + \log^2 N)$ time and using $N/\log N$ bit-serial processors]. Here, we need the linear order of convergence assumption; if it holds initially, we shall preserve it by using sufficiently small time steps.

7. Computing the Compressed Solution by Means of the Compact Multigrid Algorithms.

In this section, we will assume the strong pseudo regularity of the PDE. The time cost of computing the compressed data structure is dominated by the time required to obtain the solution vectors $\mathbf{e}_j$ for the linear systems of equations over L_j, for $j = 1, \ldots, k$:

$$D_j \mathbf{e}_j = \mathbf{r}_j. \tag{7.1}$$

Here

$$\mathbf{r}_j = \mathbf{b}_j - D_j \hat{\mathbf{u}}_{j-1}, \tag{7.2}$$

the matrices D_j and the vectors $\mathbf{b}_j$ are from the linear systems (6.1), and the vectors $\mathbf{e}_j$ and $\hat{\mathbf{u}}_{j-1}$ have the components $e_j(\mathbf{x})$ and $\hat{u}_{j-1}(\mathbf{x})$ corresponding to the points $\mathbf{x} \in L_j$ and defined by (6.1) and (6.3).

We will follow the routine of the multigrid methods (compare [McC87], [FMcB88], [FMcB91]) and will evaluate the vectors $\mathbf{e}_j$ recursively for $j = 1, \ldots, k$ by applying the so-called V-*cycle multigrid scheme* and arriving at the desired Compact Multigrid algorithm, in which we will exploit the compressed representation of the solution. Initially, we will let $u_0(\mathbf{x}) = 0$ for

$\mathbf{x} \in L_0$, and at stage j, we will successively compute the following vectors (for all $\mathbf{x} \in L_j$):

a) $\hat{u}_{j-1}(\mathbf{x})$ [by prolongation of $u_{j-1}(\mathbf{x})$ from L_{j-1} to L_j],

b) $r_j(\mathbf{x})$ [by using the equation (7.2)],

c) $e_j(\mathbf{x})$ [by solving the linear system (7.1) "to the level of truncation"],

d) $u_j(\mathbf{x})$ in the compressed form [by using the equations (6.3), as in section 6].

We will then restrict $u_{j+1}(\mathbf{x})$ to $u_j(\mathbf{x})$ for $j = k-1, k-2, \ldots, 1$. For simplicity, let this stage contain no smoothing iterations, unlike some customary variants of the multigrid scheme. Then we will recursively repeat such a loop, customarily called V-cycle.

Part 1) of the strong pseudo regularity assumption means that for all j, the errors of the approximations to $u_j(\mathbf{x})$ decrease by a constant factor independent of j and N when Stages a)–d) are repeated once, with only $O(1)$ iteration steps used at Stage c), for solving linear systems (7.1) for every fixed j. Such convergence results have been proved for the customary multigrid algorithms applied to a large class of PDE's (see [BDu], [Ha77], [HT], [Ha80], [Ha85], [McC86], [McCT], [FMcB87], [DD]).

Let us estimate the time cost of these computations, dominated by the time needed for solving the linear systems (7.1).

The size $|L_j| = 2^{dj}$ of the linear system (7.1) increases by 2^d times as j grows by 1. Even if we assume that the solution time for the system (7.1) is linear in $|L_j|$, all the k such systems are solved almost as fast as the single system (6.1) for $j = k$, giving us the uncompressed output values $u_k(\mathbf{x})$ for $\mathbf{x} \in L_k$. Namely, the solution time in the transition from the single system to all the k systems grows by at most $1/(1-2^{-d})$ times in terms of ops involved. The bit-operation count is even more favorable to the solution of the systems (7.1) for all j, as opposed to the single system (6.1) for $j = k$, because the output values $e_j(\mathbf{x})$, satisfying the systems (7.1), are sought with the lower precision of α bits.

Furthermore, we solve the linear systems (7.1) by iterative methods where each step is essentially reduced to a constant number (say, one or two) multiplications of a matrix D_j or its submatrices by vectors. Due to the linear convergence assumption that we made, a constant number of iterations suffices at each step j in order to compute the $\alpha = O(1)$ desired bits of $e_j(\mathbf{x})$.

The computational cost of multiplication of D_j by a vector is $O(N_j)$ ops for a sparse discretization matrix D_j [having $O(1)$ nonzero entries in each row]. The next two propositions summarize our estimates:

Proposition 7.1. *$O(N)$ ops suffice to compute the vectors $\mathbf{e}_j$ for all j, that is, to compute the smooth compressed solution to a strongly pseudo regular PDE discretized over the lattice L_k.*

Furthermore, we only need $O(1)$ bits in order to represent $e_j(\mathbf{x})$ for every $\mathbf{x} \in L_j$ and every j. Since D_j has only $O(1)$ nonzero entries per row and since these entries are integers having magnitudes $O(1)$ [due to part 2) of the strong pseudo regularity assumption], it suffices to use $O(1)$ bits to represent $r_j(\mathbf{x})$. [These $O(1)$ bits do not generally define the entire associated component of $\mathbf{b}_j$, since $|r_j(\mathbf{x})|$ may be much less than $||\mathbf{b}_j||_\infty$]. Thus, we will perform all the arithmetic operations with $O(1)$-bit operands and will arrive at Proposition 7.2.

Proposition 7.2. *$O(N)$ bit-operations and $O(N)$ storage space under the Boolean model of computation suffice in order to compute the compressed solution to a strongly pseudo regular PDE by using the Compact Multigrid algorithm.*

Extensions of the results.

The above results can be immediately extended to the case of more general sequences of the sets $S_0, S_1, \ldots, S_k$ of the discretization of the PDE's, provided that each set S_j consists of $c_j\sigma^j$ points where $0 < c < c_j < c^*$, $\sigma > 1$, c, c^* and σ are constants (this includes the grids with step sizes that may vary depending on the direction of the steps), and that the pseudo regularity assumptions are respectively extended to the case of the sets S_j. We also need to assume a constant degree bound $2d$ for all the discretization points, that is, each of them is supposed to have at most $2d$ neighbors: this will imply that each equation of the associated linear algebraic system has at most $2d + 1$ nonzero coefficients. Moreover, the presented approach can be further extended to some nonlinear PDE's, as long as our assumptions [such as (6.4) and (6.5)] hold and as long as dealing with nonlinear systems of difference equations that replace the linear systems (6.1) (obtained by the discretization of PDE's) remains relatively inexpensive. On further extensions, see [PR93].

8. Approximation with an (Arbitrarily) High Precision.

As we have already pointed out in section 1, if we apply a solution algorithm by using a floating point arithmetic, then we shall arrive at an approximate solution to the problem, due to the roundoff of the arithmetic operations. So, instead of devising "exact" algorithms, i.e., algorithms that would output the correct solution if performed with infinite precision arithmetic, we may consider "approximate" algorithms, which, even when performed with the infinite precision, only give us an approximation of the function that we have to compute. In this way, by weakening our requirements to the output, we hope to decrease the computational cost.

The algorithms of section 4 for multipoint polynomial evaluation and for Problem 4.1 ($C(G \cdot H) \cdot VECTOR$) have been devised along this line.

In this section we describe the class of *any precision approximation algorithms* (abbreviated as *APA-algorithms* and due to [BCLR], [B80], [BCLR81], [BLR]), which includes the algorithms of [B84], [Sc82a], [PLS] for polynomial division (see section 4 of our chapter 6), algorithms of [B84a] for polynomial evaluation (see below), and several well-known algorithms for matrix multiplication (see chapter 5 and [BCLR], [BCLR81], [B84a], [P84], [P84a], [P84b]).

Here is a simple example: suppose that we evaluate the pair of bilinear forms $x_1y_1 + x_2y_2$ and x_1y_2 and wish to save multiplications, except for the cost-free multiplications by the powers of 2, which amount to shifts of the radix point of the binary values. In this case, we may first compute x_1y_2 and then approximate $x_1y_1 + x_2y_2$ by computing

$$(x_1 + x_2 2^{-G})(y_1 + y_2 2^G) - 2^G(x_1y_2).$$

This is an APA-algorithm, for we may make the error term $2^{-G}x_2y_1$ arbitrarily small by choosing a large natural G, and we only need to count two multiplications (since multiplications by powers of 2 are assumed to be cost-free), whereas the exact evaluation would have required three.

Next, we will recall two APA-algorithms due to [B84a]; the first algorithm approximates the value of an n-th degree polynomial at a single point, and the second approximates the matrix-by-vector product:

i) For simplicity, suppose that n is an even number and observe the identity

$$\begin{aligned} p(x) = &(\cdots((p_n x^2 + p_{n-1}x + p_{n-2})x^2 + p_{n-3}x + p_{n-4})x^2 + \cdots \\ &+ p_2)x^2 + p_1 x + p_0. \end{aligned}$$

Rewrite it as the matrix -by-vector product,

$$\begin{pmatrix} p_n & p_{n-1} & p_{n-2} \\ z_1 & p_{n-3} & p_{n-4} \\ z_2 & p_{n-5} & p_{n-6} \\ \vdots & \vdots & \vdots \\ z_{n/2-1} & p_1 & p_0 \end{pmatrix} \begin{pmatrix} x^2 \\ x \\ 1 \end{pmatrix} = \begin{pmatrix} z_1 \\ z_2 \\ z_3 \\ \vdots \\ z_{n/2} \end{pmatrix}$$

where $z_{n/2} = p(x)$. The following algorithm computes the auxiliary values s and c and approximations z_i^* to z_i for $i = 1, \ldots, n/2$ (for given input values x, $p_0, \ldots, p_n$).

Algorithm 8.1.
Input: natural number n, the coefficients $p_0, \ldots, p_n$ of the polynomial $p(x) = \sum_{i=0}^{n} p_i x^i$, and a complex number x.
Output: approximation to $p(x)$.
Computation: successively compute the values

$$s = x^2, \quad c = sx,$$
$$z_1^* = (\epsilon p_n + x)(p_{n-1} + \epsilon^{-1}s) + p_{n-2} - \epsilon^{-1}c,$$
$$z_{i+1}^* = (\epsilon z_i^* + x)(p_{n-2i-1} + \epsilon^{-1}s) + p_{n-2i-2} - \epsilon^{-1}c, \; i = 1, \ldots, \frac{n}{2} - 1.$$

Here, ϵ is a nonzero parameter. We let $\epsilon = 2^{-N}$ for a large natural N. Then ϵ converges to 0, and $z_{n/2}^*$ converges to $z_{n/2}$ as N grows to ∞.

The algorithm involves $2n$ additions and $n/2 + 2$ multiplications, not counting $n + 1$ multiplications by ϵ and by ϵ^{-1} (which amount to the shifts of the radix point since ϵ is an integer power of 2), to be compared with n additions and n multiplications required for the exact evaluation of $p(x)$.

ii) To devise an APA-algorithm from [B84a] for an $m \times 2$ matrix-by-vector multiplication, that is, for computing $x_{i0}y_{00} + x_{i1}y_{10}$, $i = 1, \ldots, m$, follow the same line as in Algorithm 8.1. That is, first note that for a nonzero ϵ,

$$x_{i0}y_{00} + x_{i1}y_{10} = (\epsilon^{-1}y_{10} + x_{i0})(y_{00} + \epsilon x_{i1}) - \epsilon^{-1}s_0(Y) - \epsilon q_i(X),$$
$$q_i(X) = x_{i0}x_{i1}, \quad i = 0, 1, \ldots, m-1, \quad s_0(Y) = y_{10}y_{00},$$

then choose $\epsilon = 2^{-g}$, where g is a large natural constant. Now, successively evaluate

$$\epsilon^{-1}y_{10}, \epsilon^{-1}s_0(Y), \; \epsilon^{-1}y_{10} + x_{i0}, \; y_{00} + \epsilon x_{i1}, \; (\epsilon^{-1}y_{10} + x_{i0})(y_{00} + \epsilon x_{i1}),$$

and, finally, $(\epsilon^{-1}y_{10}+x_{i0})(y_{00}+\epsilon x_{i1})-\epsilon^{-1}s_0(Y)$ for all i, which gives us $x_{i0}y_{00}+x_{i1}y_{10}$ within the small term $\epsilon q_i(X)$. Note that the refinement by rounding off gives us the correct output values if all the input entries are integers and if $\epsilon < 0.5|q_i(X)|$ for all i.

The algorithm involves $3m$ operations $\pm$, $m+1$ shifts of the radix point of binary numbers (multiplications by powers of 2), and $m+1$ other multiplications. This can be an advantage against $2m$ multiplications and m additions of the straightforward algorithm. The algorithm takes similar advantages in the case of matrix-by-vector multiplication with an input of a larger size.

Clearly, the rounding errors generated by an APA-algorithm tend to infinity as $\epsilon \to 0$. This means that a higher precision computation should be used with APA-algorithms. In spite of this fact, APA-algorithms have been used as a basis for designing the most successful known algorithms for some computational problems yielding the record upper bounds on their arithmetic and Boolean (bit) complexity (see exercises 20, 21, and for more details and examples, see [B80], [BCLR81], [B82], [P84], [P84b]). This is due to the possibility of saving ops by using APA-algorithms and to the two following observations: a) the precision needed in the computation grows slowly [logarithmically the input size (see exercise 21)] under the recursive application of the *tensor product* (see section 11 and exercise 49 of chapter 2 for its definition), which enables us to dramatically accentuate any decrease of the arithmetic cost; b) the techniques of interpolation of [B80] (also see [Cod85]) can be applied to reduce the order of the approximation error.

9. Data Compression and Acceleration of Computations via Binary Segmentation.

In this section, we will apply some techniques of binary segmentation to decrease the bit-operation cost of some polynomial and matrix computations (compare exercises 1 and 26). We will consider the variants of the binary segmentation that require a higher precision arithmetic (the cost of which is higher), but already a single higher precision operation suffices, so that the overall Boolean (bit-operation) cost of the solution can be smaller than in other known algorithms (see Remark 9.2 on another variant of binary segmentation). Furthermore, in many cases, the machine precision greatly exceeds the precision needed for the actual computation, as in the case, for example, of many combinatorial and graph computations (compare Appendix A to chapter 2), at the initial stages of p-adic lifting in Algorithms

3.1 and 3.2, in the residual correction algorithm (5.1), (5.2) (particularly, in its refinement presented in [P91], [P92d], [P93d]), and in the Compact Multigrid iteration. In such cases, we may perform the binary segmentation within the machine precision and arrive at a practical and sometimes dramatic improvement of the computations. Unlike the techniques shown in the preceding sections, binary segmentation cannot be described under the model of computation of the arithmetic straight-line programs, since the multiplications of long integers, as well as binary segmentation, is not performed through a sequence of ops starting with the integer input data. Nevertheless, such a noncanonical approach has all potentials to compete with both the classical algorithms and the FFT based fast polynomial arithmetic (compare Remark 9.2) and has good practical promise for some fundamental computations with vectors and matrices (see Example 9.3).

Example 9.1, *interpolation via binary segmentation.* Let $p(x) = \sum_{i=0}^{n} p_i x^i$ be a polynomial with nonnegative integer coefficients, which are all strictly less than 2^g for a positive g. Let $h \geq g$ be an integer. Then the binary values $p_0, \ldots, p_n$ can immediately be recovered as the appropriate segments of the binary value $p(2^h) = \sum_{i=0}^{n} p_i 2^{hi}$, that is, in this case we only need a single interpolation point $x = 2^h$, rather than $n+1$ points used in chapter 1 for Problem 1.2.3 ($POL \cdot INTERP$). If we only need to recover a single coefficient p_q or only the coefficients $p_0, \ldots, p_q$ for $q < n$, we may first reduce $p(x)$ modulo x^{q+1} or we may compute just $p(2^h) \bmod 2^{h(q+1)}$.

The simple algorithm of Example 9.1 is immediately extended to the case where $p_0, \ldots, p_n$ are integers lying strictly between -2^g and 2^g ([Kn], p. 191). Choose an integer $h \geq g+1$ and compute $p(2^h) + \sum_{i=0}^{n} 2^{hi+h-1} = \sum_{i=0}^{n} (p_i + 2^{h-1}) 2^{hi}$; then recover the nonnegative integers $p_i + 2^{h-1} < 2^h$ and, finally, p_i for all i. Surely, the method also works where the coefficients $p_0, \ldots, p_n$ are *Gaussian integers,* that is, complex numbers with integer real and imaginary parts. Also, the entire approach can be extended to ternary and, more generally, p-adic segmentation (compare exercise 26).

Remark 9.1. Binary segmentation relies on correlation between integers and polynomials with integer coefficients, already commented on in chapter 1.

Example 9.2, *convolution of vectors via binary segmentation* ([FP]). Let m and n be two natural numbers, U and V be two positive constants, $\mathbf{u}$ and $\mathbf{v}$ be the coefficient vectors of $u(x)$ and $v(x)$, where $u(x)$ and $v(x)$ denote two polynomials of degrees $m-1$ and $n-1$, respectively, whose coefficients are integers in the range from $-U$ to U and from $-V$ to V, respectively. Then the coefficients of the polynomial $w(x) = u(x)v(x)$ range from $-W$ to W where $W = \min\{m, n\} UV$, so that Example 9.1 suggests that we may recover the convolution of **u** and **v**, that is, the coefficients of the polynomial $u(x)v(x)$, if we compute the binary value $u(2^h)v(2^h)$ for an integer $h \geq 1 + \log(W+1)$. If $u(x)$ and $v(x)$ have real coefficients, we can represent them in the binary form, truncate their fractions, scale the results in order to make them integers, and apply the above algorithm in order to approximate the coefficients of $w(x)$.

Example 9.3, *computing the inner and outer products of two vectors* (section 40 of [P84b] and [P92]). Under the assumptions of Example 9.2, we may compute the inner product $\mathbf{u}^T\mathbf{v} = \sum_{j=0}^{n-1} u_j v_j$ (if $m = n$) and the outer product $[w_{ij}] = \mathbf{u}\mathbf{v}^T = [u_i v_j]$ of two vectors $\mathbf{u} = [u_i]$ and $\mathbf{v} = [v_j]$ by means of binary segmentation of the two integers $p(2^h)$ and $q(2^h)$. Here, $p(2^h) = u(2^h)2^{(n-1)h}v(2^{-h}) = (\sum_{i=0}^{n-1} 2^{ih}u_i)(\sum_{j=0}^{n-1} 2^{(n-1-j)h}v_j) = \sum_{k=0}^{2n-2} z_k 2^{kh}$, $z_{n-1} = \mathbf{u}^T\mathbf{v}$, and in this case h is an integer, $h \geq 1 + \log(W+1)$, $W = nUV$, whereas $q(2^h) = (\sum_{i=0}^{m-1} 2^{inh}u_i)(\sum_{j=0}^{n-1} 2^{jh}v_j) = \sum_{i,j} u_i v_j 2^{(in+j)h}$, $(\mathbf{u}\mathbf{v}^T)_{ij} = u_i u_j$, and in this case h is an integer, $h \geq 1 + \log(W+1)$, $W = UV$. Thus, each of the inner and outer products is computed by means of a single multiplication of two integers and binary segmentation of the product, which promises to be useful in numerous matrix algorithms.

For demonstration, let the vectors **u** and **v** have dimension 14 and have components 0, -1 and 1 and assume the computer precision of 80 bits. Then the straightforward evaluation of the convolution, inner product and outer product of **u** and **v** requires 365, 27 and 196 ops, respectively. With binary segmentation we only need single integer multiplication for the inner product, two such multiplications for convolution and 5 multiplications for the outer product, since we may use a sufficiently high precision. Specifically, even if the computer precision were, say, 69 bits, then we still may multiply $u(2^h)$ by $2^{(n-1)h}v(2^{-h})$, for $n = 14$, $h = 5$, and add $\sum_{i=0}^{26} 2^{5i+4}$ to the product, chopping the resulting sum to 69 bits, either the most or the least significant. The subsequent segmentation of the last, respectively, the first four bits, would in both cases output the inner product of **u** and **v**. To compute the convolution of **u** and **v**, we would have needed two multiplications modulo 2^{80}, that is, the multiplications of $u(2^h)$ by $v(2^h)$ and of $2^{mh}u(2^{-h})$ by $2^{nh}v(2^{-h})$ where $m = n = 14$, $h = 5$. To compute the outer product of **u** and **v**, we would need to obtain a 392-bit output integer. With 80-bit precision, this integer could be replaced by 5 integers, each of them represented with at most 80 bits, and each of them output by a single multiplication.

Remark 9.2. The terminology "interpolation via binary segmentation" is due to [BP86a]; the first application of this method (to convolution) is due to [FP] (see also [P80]). In [P84b], section 40, the approach has been extended to several further computations with matrices and vectors (this has also been an extension of the class of APA-algorithms). Schönhage in [Sc82a] used the algorithm of Example 9.2 and extended it to DFT [by means of reducing DFT to polynomial multiplication as in the solution of Problem 1.2.5 $(GEN \cdot DFT)$] and to polynomial division (see chapter 6). Further extension to the shift of the variable x is immediate [see Problem 1.2.6 $(VAR \cdot SHIFT)$, whose solution in section 2 of chapter 1 can be complemented with scaling to bound the absolute values of the operands and to improve numerical stability]. The estimates of [Sc82a] show that the binary segmentation saves a factor of $\log n$ against the Boolean cost bounds based on the algorithms for Problems 1.2.2–1.2.6 $(POL \cdot EVAL)$, $(POL \cdot INTERP)$, $(SEV \cdot POL \cdot MULT)$, $(GEN \cdot DFT)$, $(VAR \cdot SHIFT)$ and 1.3.1 $(POL \cdot DIVIDE)$. In chapter 6, we will follow [BP86a] and will apply binary segmentation to improve the previously known algorithms for polynomial division in the case of polynomials with integer coefficients. On heuristic applications of binary segmentation to practical computation of polynomial gcd, see [CGG], [DP]; on some theory of such applications, see [Sc88]. Summarizing, all these results promise to make the techniques of binary segmentation *practically useful for some fundamental operations with matrices and vectors* and *practically competitive with both classical and FFT based algorithms for polynomial computations*, in the cases where the input values are represented by sufficiently short integers (note that the computations in finite fields having small positive characteristics can be reduced to this case, and compare exercises 1 and 14).

Remark 9.3. Computation with a lower precision can be performed faster on specialized computers, such as MASPAR, and Example 9.3 enables the users to take similar advantages in computations on any digital computer.

Exercises to Chapter 3.

1. Estimate the asymptotic bit-cost of multiplication of a pair of $(K-1)$-st degree polynomials with integer coefficients ranging from 0 to $p-1$ where $K=2^k$, k is integer. Do this for a) $p=2^{O(K)}$, b) $p=O(K)$ and c) $p=O(1)$ as $K\to\infty$. In each case, first apply the straightforward algorithm, then the algorithm based on three FFT's on the complex $2K$-th roots of 1. Then apply an algorithm based on three FFT's on $2K$ points over the ring $\mathbf{Z}_M$ of integers modulo M, first for $M=K^{K/2}+1$ and then for a prime $M=hK+1$, $h=O(K)$. Then apply the algorithm of [Nus] (see the proof of Theorem 1.7.1). Finally, apply an alternative algorithm based on Example 9.2. In all cases estimate the bit-operation cost and the bit-memory cost. Similarly, compare various approaches to polynomial division and computing polynomial gcd where the input coefficients are bounded integers (use the results of chapter 6).
2. Consider the problem of computing $\{x_i\}_{i=1,\dots,n}$, where
$$x_{i+1}=\frac{10}{3}x_i-x_{i-1},$$
$$x_1=1,\ x_2=\frac{1}{3}.$$
Prove that this problem is ill-conditioned (according to the definitions of section 1) by showing that $x_i=(\frac{1}{3})^{i-1}$, whereas the solution $\{\tilde{x}_i\}_{i=1,n}$, obtained by means of the same recursion but with the perturbed initial values $x_1=1$, $x_2=\frac{1}{3}+\epsilon$, is
$$\tilde{x}_i=\frac{3}{8}\epsilon 3^{i-1}+(1-\frac{3}{8}\epsilon)(\frac{1}{3})^{i-1}.$$
3. Consider the problem of solving the linear system $A\mathbf{x}=\mathbf{b}$ where the matrix A is such that $a_{ii}=\alpha$, $i=1,\dots,n$, $a_{i,i+1}=a_{i+1,i}=-1$, $i=1,\dots,n-1$, $a_{ij}=0$ elsewhere, $\alpha\geq 3$. Prove that
   a) the problem is well-conditioned,
   b) Gaussian elimination is numerically stable,
   c) Gaussian elimination applied to the system obtained by moving the first equation to the last position is numerically unstable.
4. For a tridiagonal matrix T, show numerical instability of the representation of T^{-1} based on the solution of exercise 13 of chapter 2.
5. Express the polynomials $R_i(x)$ of the pseudo remainder sequence for $u(x)$ and $v(x)$ as subresultants and use Fact 2.2 in order to bound the magnitudes of the coefficients of these polynomials (*without* their scaling).

6. (Communicated by L. Lovász and G. Szekeres.) Let B be an $m \times n$ matrix where $m \geq n$. Prove that $\det(B^T B)$ equals the sum of the squares of all the $n \times n$ subdeterminants of B ([Bour], [Fic]). Use this result to estimate the average value of $\det A$ for an $n \times n$ matrix A filled with the values -1 and 1, assuming a random uniform distribution of -1 and 1 in each column of A. Compare this value with the bounds of Fact 2.2.
7. Extend Fact 2.2 to the case of matrix polynomials.
8. Let A be an $n \times n$ matrix filled with univariate or multivariate polynomials of bounded degrees. Apply Gaussian elimination with no pivoting as a rational (infinite precision) process in order to compute $\det A$. Examine the growth of the coefficients of the auxiliary polynomials arising in this process. Compare the bounds on these coefficients with the bounds of exercise 7.
9. Restate exercise 8 for matrices filled with integers rather than with polynomials. Estimate from above the magnitudes of the auxiliary integers arising in the process of Gaussian elimination applied to such a matrix and compare the resulting estimates with Fact 2.2 and with the estimates of [Edm].
10. Given natural N and k and k pairs of integers a_i, b_i, $|a_i| \leq N$, $|b_i| \leq N$, $\gcd(a_i, b_i) = 1$, $i = 1, \ldots, k$, choose a real interval $\mathcal{J}$ and k random integers $c_1, \ldots, c_k$ in this interval and try to estimate the probability that $q = \operatorname{lcm}(b_1, \ldots, b_k)$ where $\sum_i c_i a_i / b_i = p/q$ and where p and q are two mutually prime integers.
11. Let $q = 4$ in Corollary 2.1 and $||A||_2 = n^4$. Try to choose N and r so as to minimize $n^r ||A||_2^{1/N}$ subject to the inequalities assumed in Corollary 2.1.
12. Given two positive integers R and S, both being powers of 2, and two polynomials of degrees at most $K - 1$, such that $K = RS$,

$$u(x) = \sum_{g=0}^{R-1} u_g(x) x^{gS} , \qquad v(x) = \sum_{g=0}^{R-1} v_g(x) x^{gS}$$

where

$$\deg u_g(x) < S , \quad \deg v_g(x) < S \text{ for all } g .$$

Estimate that $12K \log(2K) + 6K \log(2R) - 3S \log(2S) + O(K)$ ops suffice in the following algorithm for Problem 1.2.4a ($POL \cdot MULT$):

1. Evaluate $u_g(x)$, $v_g(x)$, for $g = 0, \ldots, R-1$, on the $2S$-th roots of 1, by means of $2R$ DFT's, each on $2S$ points.

2. Set $u_j(x) = v_j(x) = 0$ if $j \geq R$. Evaluate the $2R-1$ polynomials $w_h(x) = \sum_{g=0}^{h} u_g(x)v_{h-g}(x)$, for $h = 0, 1, \ldots, 2R-2$, on the $2S$-th roots of 1 [by computing $2S$ convolutions of $2S$ pairs of R-dimensional vectors; that is, one convolution,

$$W_i(y) = U_i(y)\, V_i(y) \ ,$$

is computed for each $2S$-th root of 1, ω^i, $i = 0, \ldots, 2S-1$, where $U_i(y) = \sum_{g=0}^{R-1} u_{i,g}y^g$, $V_i(y) = \sum_{g=0}^{R-1} v_{i,g}y^g$, $y = x^S$, $u_{i,g} = u_g(\omega^i)$, $v_{i,g} = v_g(\omega^i)$, ω is a primitive $2S$-th root of 1; computing these $2S$ convolutions amounts to performing $6S$ DFT's, each on $2R$ points, and $4RS$ additional multiplications]. (Apply the positive and negative wrapped convolution to perform this stage by using slightly fewer ops.)

3. Compute the coefficients of the polynomials $w_h(x)$ for $h = 0, \ldots, 2R-2$ by performing $2R-1$ inverse DFT's, each on $2S$ points.

4. Compute and output the coefficients of the polynomial

$$w(x) = \sum_{h=0}^{2R-2} w_h(x)x^{hS}$$

by performing $2(R-1)(S-1)$ additions.

13. Give your asymptotic bounds on the number of bit-operations needed for computing $\mathbf{x} = A^{-1}\mathbf{b} \bmod p^k$ by means of each of Algorithms 3.1 and 3.2 where A is an $n \times n$ integer matrix, $\det A \neq 0 \bmod p$, $n \to \infty$, $k \to \infty$, $p = O(1)$. Express the answers as functions in k and n.

14. Apply the techniques of section 3 in order to estimate the bit-operation complexity of all the computational problems studied in chapters 1 and 2 assuming that all the input values are integers between $-H$ and H for a fixed H (for the estimates involving polynomial reciprocal, use the upper bound on the magnitude of the entries of the inverse of a triangular Toeplitz matrix from chapter 6 or [BP86a]). Wherever possible, also obtain alternative estimates by using binary segmentation along the line of section 9. Examine, in particular, multipoint polynomial evaluation, using the comments of Remark 4.2.

15. Consider the system $A\mathbf{x} = \mathbf{b}$, where $A = (a_{ij})$ is an $n \times n$ matrix such that $a_{ii} = a_{in} = 1$, $i = 1, \ldots, n$, $a_{ij} = -1$ for $i > j$, $a_{ij} = 0$ elsewhere. Prove that

a) solving this system by means of Gaussian elimination does not require divisions, except for one at the first step of the back substitution stage where the divisor is 2^{n-1};

b) the precision needed for the computation is $n-1$ bits;

c) $||\mathbf{x}||_\infty \le 4||\mathbf{b}||_\infty$.

Observe the following consequences of these results. If the solution is an integer vector, it can be computed by applying Gaussian elimination modulo N, provided that $N > 8||\mathbf{b}||_\infty$ and that 2^{n-1} has its inverse modulo N. For instance, let $n = 1024$, $\mathbf{b} = [b_i]$, and $b_i = n - i + 2$, $i = 1, \ldots, n-1$, $b_n = 1$. In this case the solution is an integer vector, and $||\mathbf{x}||_\infty \le 4n + 4$. For $N = 8n + 9$, we may invert 2^{n-1} modulo N. Therefore, computing with 14 bits is sufficient in order to arrive at the correct output, whereas Gaussian elimination performed over the integers requires the precision of $n - 1 = 1023$ bits.

16. Show that the residual correction algorithm (5.1) may diverge if computations in *both* (5.1) and (5.2) are performed with the d-bit precision, $d = \lceil \log \mathrm{cond}_2(A) \rceil + 1$. Moreover, consider the modification of this algorithm where (5.1) is replaced by the following vector equation: $\mathbf{r}_{i+1} = \mathbf{r}_i - A\mathbf{e}_i$. Show numerical instability of this modification. Try to modify the algorithm (5.1), (5.2) so as to yield convergence where all the computations are with the d-bit precision, for the d defined above.

17. a) Estimate the number of the Jacobi iteration steps required in order to decrease by 10^3 times the 1-norm of the initial error provided that the entries of the $n \times n$ input matrix $A = [a_{ij}]$ satisfy the bounds $a_{jj} \ge 3\sum |a_{ij}|$ for all j, where the sum is in i from 1 to $j-1$ and from $j+1$ to n.

    b) Under the latter assumption, let $X_0 = \mathrm{diag}(1/a_{11}, \ldots, 1/a_{nn})$ and estimate the 1-norm of the error of the approximation to A^{-1} by the matrix X_{10} obtained in 10 Newton's iteration steps (5.4).

    c) Investigate the numerical stability of both algorithms [of a) and b)] according to the definitions of section 1.

    Hint: use the error bound (1.3) and Fact 5.1 in a) and the estimate of Proposition 5.1 in b).

18. Interpret the iterative algorithms (5.3) as the residual correction algorithms. Assume the computation precision of $d = \lceil \log \mathrm{cond}_1(A) \rceil$ $+1$ bits and estimate the bit-operation cost of Jacobi iteration of the exercise 17a).

19. Let $A \in \mathbf{F}_{n,n}$ and $\mathcal{P} : \mathbf{F}_{n,n} \to \mathbf{F}_{n,n}$ be an operator (projector) such that $\mathcal{P}(B + C) = \mathcal{P}(B) + \mathcal{P}(C)$, $||\mathcal{P}(B)|| \le ||B||$, for any $B, C \in \mathbf{F}_{n,n}$ [for motivation, see an example of such a projector in part b) below].

Consider the following modification of Newton's iteration (5.4):

$$X_0 \in \mathbf{F}_{n,n},$$
$$Y_{k+1} = X_k(2I - AX_k),$$
$$X_{k+1} = \mathcal{P}(Y_{k+1}).$$

a) Study the convergence of this method. Let $F_k = A^{-1} - X_k$, $E = \mathcal{P}(A^{-1}) - A^{-1}$, and $E_k = \mathcal{P}(A^{-1}) - X_k$. Prove that

$$||F_{k+1}|| \le ||E|| + ||F_k||^2||A||,$$
$$||E_{k+1}|| \le (||E_k|| + ||E||)^2||A||.$$

Therefore,

$$||E_{k+1}|| \le 4||A||\ ||E_k||^2 \text{ if } ||E|| \le ||E_k||,$$

that is, $||E_{k+1}|| \le 4||E_k||^2$ as long as $||E_k|| \ge ||E||$.

b) Now suppose that A is a 9-diagonal band matrix such that $|a_{ii}| > 4\sum_{j=0,\ j\neq i}^{n-1} |a_{ij}|$, $i = 0, \dots n-1$, and that $\mathcal{P}$ is the projection on the 9 central diagonals, i.e., $(\mathcal{P}(X))_{ij} = x_{ij}$ for $|i-j| \le 4$, $(\mathcal{P}(X))_{ij} = 0$ otherwise. Apply the above iteration in order to approximate the central diagonals of A^{-1}. Determine the cost per iteration, determine the highest accuracy that can be obtained in the result, and estimate the number of iteration steps sufficient to reach it. Then use the output of this modified Newton's iteration as an initial approximation X_0 for Newton's iteration (5.4). Compare the overall cost of this computation by the resulting algorithm with the overall cost of Newton's iteration (5.4) applied in order to approximate the same diagonals with the same precision.

20. At least 3 multiplications are required in order to compute the following functions (bilinear forms):

$$f_1(\mathbf{x},\mathbf{y}) = x_1y_1 + x_2y_2 = \mathbf{x}^T \begin{pmatrix} 1 & 0 \\ 0 & 1 \end{pmatrix} \mathbf{y},$$
$$f_2(\mathbf{x},\mathbf{y}) = x_1y_2 = \mathbf{x}^T \begin{pmatrix} 0 & 1 \\ 0 & 0 \end{pmatrix} \mathbf{y}.$$

These functions however, can be approximated with any precision by performing just 2 multiplications, provided that multiplications by the powers of 2 (radix point shifts) are considedred cost-free. Indeed, choose $\epsilon = 2^{-g}$ and compute either the values

$$\tilde{f}_1(\mathbf{x},\mathbf{y}) = f_1(\mathbf{x},\mathbf{y}) + \epsilon x_2y_1 = (x_1 + \epsilon x_2)(y_1 + \epsilon^{-1}y_2) - \epsilon^{-1}x_1y_2,$$
$$\tilde{f}_2(\mathbf{x},\mathbf{y}) = f_2(\mathbf{x},\mathbf{y}) = x_1y_2 \qquad (E.1)$$

or the values

$$\begin{aligned}\hat{f}_1(\mathbf{x},\mathbf{y}) &= f_1(\mathbf{x},\mathbf{y}) = a+b,\\ \hat{f}_2(\mathbf{x},\mathbf{y}) &= f_2(\mathbf{x},\mathbf{y}) + \epsilon^2 x_2 y_1 = \epsilon(a-b),\\ a &= (x_1+\epsilon x_2)(y_1+\epsilon^{-1}y_2)/2,\\ b &= (x_1-\epsilon x_2)(y_1-\epsilon^{-1}y_2)/2.\end{aligned} \tag{E.2}$$

Suppose that the d-bit floating point arithmetic (with the precision $u = 2^{-d}$) is used. Prove that the rounding errors and the output approximation errors of the computation of this pair of bilinear forms are $O(\epsilon^{-1}u)$, $O(\epsilon)$, respectively, if (E.1) is used, and $O(\epsilon^{-1}u)$, $O(\epsilon^2)$, respectively, if (E.2) is used. Prove that by choosing $g = d/2$ and $g = d/3$ in the cases (E.1) and (E.2), respectively, we arrive at the total output approximation errors $O(2^{-d/2})$ and $O(2^{-2d/3})$, respectively. More generally, if $-\beta$ is the smallest negative exponent of ϵ that appears in some computation, then the rounding error is $O(u\epsilon^{-\beta})$. If the output approximation error is $O(\epsilon^{\alpha})$ and if $\epsilon = u^{1/(\alpha+\beta)}$, we have both rounding and output errors bounded by $O(u^{\alpha/(\alpha+\beta)})$, that is, $O(d/(1+\beta/\alpha))$-bit precision computation is sufficient to obtain $O(d)$ correct bits in the output.

21. Suppose that the $m \times n$ bilinear forms $f_i(\mathbf{x},\mathbf{y}) = \mathbf{x}^T A_i \mathbf{y}$, $i = 1,\dots,k$, can be approximated with any precision by an APA-algorithm that uses t multiplications but does not use their commutative property and has the output approximation error $O(\epsilon^{\alpha})$. Moreover, assume that $-\beta$ is the smallest negative exponent of ϵ that appears in the computation. Let $\otimes$ denote the *tensor product* (the *Kronecker product*), which we have already defined in exercise 49 of chapter 2 as follows: $A \otimes B = [a_{ij}B]$ is the block matrix having blocks $a_{ij}B$. Consider the $K = k^p$ bilinear forms

$$f_{i_1,\dots,i_p}(\mathbf{X},\mathbf{Y}) = \mathbf{X}^T\left(A_{i_1} \otimes A_{i_2} \otimes \cdots \otimes A_{i_p}\right)\mathbf{Y}$$

where $\mathbf{X},\mathbf{Y}$ are two vectors consisting of $M = m^p$, $N = n^p$ components, respectively. Prove that the coefficients of the above bilinear forms can be approximated within the error bound $O(\epsilon^{\alpha})$ by means of an APA-algorithm where t^p multiplications are involved and where the smallest negative exponent of ϵ involved in the computation is $-\beta p$. It follows that $O(d/(1+c\log K))$-bit precision computation is sufficient to provide d correct bits in each nonvanishing output value, where $c = (\beta/\alpha)\log k$. Apply this result to the bilinear forms $f_1(\mathbf{x},\mathbf{y})$, $f_2(\mathbf{x},\mathbf{y})$ and to the algorithms based on the relations (E.1), (E.2) of exercise 20.

22. Apply the interpolation via binary segmentation in order to improve the presented solution of Problem 1.4.5 ($POL \cdot COMP$) and the second presented solution of Problem 1.6.3 ($POL \cdot DECOMP$).
23. Show two monic polynomials of degree 3 with the coefficients lying between -10 and 10 whose quotient (in their division with remainder) has a coefficient greater than 10^{20}. May this occur if the two polynomials have integer coefficients lying between -10 and 10?
Hint: See the error analysis in chapter 6 or in [BP86a].
24. ([P91], [P92d]). For two integers, g and h, $g < h$, the set of numbers v such that

$$|v| = j2^g, \quad 0 \le j < 2^{h-g}, \quad j \text{ is integer},$$

is denoted $P(g,h)$ and is said to be *the precision interval* (g,h), and we also define $P(-,h) = \{v, |v| < 2^h\}$, the *precision ray* $(-\infty, h)$. If $v \in P(g,h)$, $v \ge 0$, and $w = t + v + u2^h$ for two nonnegative integers u and t such that $t < 2^g$, then the operator $S(g,h)$ that transforms w into v is called the (g,h)-*binary segmentation operator*, so that $v = S(g,h)w$; we will also set $S(-,h)w = t + v$, $S(-,h)(-w) = -t - v$, $-v = S(g,h)(-w)$. The following simple observations enable us to extend the binary segmentation from sums, differences and products to the operands of summation, subtraction and multiplication. (Note that this application of binary segmentation is quite different from ones covered in section 9.)

1. For positive integers k, $K = 2^k$, $i_1, \dots, i_K$ and for two integers g and h, $g < h$, we have:

$$|S(g,h)\sum_{m=1}^{K} i_m - S(g,h)\sum_{m=1}^{K} S(g-k,h)i_m| < 2^g.$$

2. Given three integers h, i and j such that $i > 0$, $j > 0$, $|i - j| < 2^h$, let us denote

$$d = d(i,j,h) = S(-,h+1)i - S(-,h+1)j.$$

Then

$$i - j = d \quad \text{if} \quad |d| < 2^h,$$
$$i - j = -(2^{h+1} - |d|)\text{sign } d \text{ , otherwise.}$$

3. Given six integers a, b, g, h, i and j, $a < b, g < h$, $j \in P(a,b)$, we have:

$$|S(g,h)(ij) - S(g,h)((S(g-b,h-a)i)j)| < 2^g.$$

Apply these properties above to decrease the precision of the operands in the computation of the inner product of two vectors, $p = \sum_i u_i v_i$, where it is known that $|p| < 2^h$, even though $|u_i v_i|$ may exceed 2^h for some i.

Then consider the residual correction algorithm of section 5. Assume that the entries of the input matrix of the coefficients lie in the fixed precision interval $P(a,b)$, for two fixed constants a and b. Furthermore, assume that for two fixed integers h and g, and for $b \leq -\log \|I - \tilde{A}^{-1}A\|_\infty$, we have

$$\log \|\mathbf{r}_i\|_\infty \leq h + \log i - bi\ , \quad \log \|\mathbf{e}_i\|_\infty \leq g + \log i - bi\ .$$

Use the above properties to perform the residual correction algorithm with all operands segmented to a lower precision of $O(\log(in))$ bits on the i-th iteration loop, by means of removing *both* most and least significant bits in their representation. Extend these results to perform Jacobi's and Gauss-Seidel's iteration of section 5 with a lower precision (by removing both most and least significant bits of the operands).

25. Try to use the techniques of exercise 24 to generalize the compact multigrid algorithms of sections 6 and 7.
26. Extend the approach of section 9 to ternary segmentation. In particular, estimate how many ternary digits are needed for the operations with the two vectors $\mathbf{u}$ and $\mathbf{v}$ of dimension 14 with the components 0, -1 and 1 considered at the end of section 9.
27. Apply the approach of Example 9.3 to computation of the shortest distances in a graph by means of the algorithm of Appendix A to chapter 2.
28 (*Numerical stabilization by means of equilibration*, inspired by a discussion with E. Kaltofen). Consider the solution of Problem 2.2.8 ($L{\cdot}SQUARES$) based on the matrix equation (2.2.9). Assume that the coefficients c_i of the characteristic polynomial $c_A(\lambda) = \det(\lambda I - A) = \sum_{i=0}^{n} c_i \lambda^i$ have magnitudes of the order $\|A^{n-i}\|$ for an appropriate matrix norm, $i = 0, 1, \ldots, n$. Typically, $\|A^+\mathbf{b}\|$ is much smaller than $\max_i\{\|A^i\|\,|c_i|\}$, and this causes numerical instability of the evaluation of $A^+\mathbf{b}$ based on (2.2.9), since in this case an output of a smaller norm is computed as a linear combination of large auxiliary values. Similar problems arise if we rely on a modification of (2.2.9) where the minimum polynomial $m_A(\lambda)$ replaces $c_A(\lambda)$. Investigate the numerical stability of the same process after the preliminary normalization of A by scaling, by means of premultiplication and/or postmultiplication of A by

diagonal matrices or, even simpler, by the constants $\|A\|^{-1}$ for appropriate matrix norms. Try to extend the same idea of *equilibration by preconditioning* to approximating the matrix eigenvalues via the evaluation of the coefficients of its characteristic polynomial (in the latter case, however, see [W65] on the additional difficulties caused by the ill-conditioning of approximating polynomial zeros).

29 ([P92e]). Estimate the bit-complexity of the application of the Chinese remainder theorem to recovering the value $a \bmod p$ from the values $a \bmod p_i$, $i = 1, \dots, n$, where $p_1, \dots, p_n$ are n distinct primes, $p = p_1 p_2 \cdots p_n$, $\log p_i = O(n)$, for all i. Employ the algorithm of [Kam87], which simplifies integer division where the divisor is small relative to the dividend.

30. Extend Proposition 4.1 to the case of the inverse DFT. Then perform a rounding error analysis of the algorithm for polynomial multiplication based on the technique of evaluation/interpolation on the Fourier set.

31. Consider the function $f(x, y) = x^2 - y^2$. Prove that the problem of computing $f(x, y)$ for $(x, y) \in \Omega \subset \mathbf{R}^2$ is ill-conditioned in terms of the relative error, if Ω intersects one of the two lines $y = x$ and $y = -x$. Consider the two solution algorithms based on the following relations:

$$f(x, y) = (x * x) - (y * y),$$
$$f(x, y) = (x - y) * (x + y).$$

Prove that the first algorithm is numerically unstable, whereas the second algorithm is stable. Moreover, for a natural d, prove that the algorithmic relative error of the second algorithm, performed with the d-bit precision, is bounded from above by $3u$ for any input data, where $u = 2^{-d}$.

32 (compare Appendix A). Let $f(\mathbf{x})$ be a rational function in real variables $\mathbf{x} = (x_1, \dots, x_n)$. Consider the set of perturbed input values $\tilde{x}_i = x_i(1+\epsilon_i)$, $i = 1, \dots, n$. Prove that if $f(\mathbf{x})$ is defined on the line segment connecting two points, $\mathbf{x} = (x_1, \dots, x_n)$, $\tilde{\mathbf{x}} = (\tilde{x}_1, \dots, \tilde{x}_n)$, and does not vanish in $\mathbf{x}$, then, for the inherent relative error $\epsilon_{inh} = (f(\tilde{\mathbf{x}}) - f(\mathbf{x}))/f(\mathbf{x})$, the following relation holds:

$$\epsilon_{inh} = \sum_{i=1}^{n} c_i \epsilon_i + O(\sum_{i=1}^{N} |\epsilon_i|^2),$$
$$c_i = \frac{x_i f'_{x_i}(\mathbf{x})}{f(\mathbf{x})}, \quad i = 1, \dots, n,$$

where $f'_{x_i}(x) = \partial f(x)/\partial x_i$. Prove that, for the function $f(\mathbf{x}) = \sum_{i=1}^{n} x_i$, the inherent error is at most $\max|\epsilon_i|$, if $\mathbf{x}$ has all nonnegative (or nonpositive) components. Moreover, by using the relations $fl(a \circ b) = (a \circ b)(1+\epsilon)$, $|\epsilon| < u$, $u = 2^{-d}$, where $fl(\cdot)$ denotes the computation performed with the d-bit floating point arithmetic, prove that the algorithmic error of the computation by the formula $f(\mathbf{x}) = (\cdots(((x_1 + x_2) + x_3) + \cdots + x_{n-1}) + x_n$ has modulus less than $u(n-1) + O(u^2)$. Prove that the algorithmic error generated by the fan-in method has modulus less than $u \log n + O(u^2)$.

33. Represent a rational number a/b (a, b integers) with the pair (a, b) so that the following composition rules apply: $(a, b) * (c, e) = (ac, be)$, $(a, b)/(c, e) = (ae, bc)$, $(a, b) \pm (c, e) = (ae \pm bc, be)$. Let $d = \lceil \max\{\log|a|, \log|b|\}\rceil$ denote the number of bits associated with the ratio a/b. By using Hadamard's inequality, evaluate an upper bound on the number of bits sufficient to represent the solution $\mathbf{x}$ to the linear system $A\mathbf{x} = \mathbf{b}$, where the $n \times n$ matrix A and the vector $\mathbf{b}$ have rational entries, each represented with at most d bits. Prove that $O(n^2 d)$ bits are sufficient to represent any components of $\mathbf{x}$; moreover, prove that $O(n \log n + nd)$ bits suffice if A has integer entries.

34. Let $u(x)$, $v(x)$ be monic polynomials of degrees n and m, respectively, $m < n$, having integer coefficients, each represented with at most d bits.
    a) Prove that the matrices $B(u, v)$ and $H(u, v)$ defined in section 9 of chapter 2, have integer entries, each represented with $O(d + \log n)$ and $O(d + n \log n)$ bits, respectively.
    b) Perform the analysis of the precision of the computation by Algorithm 2.11.1, applied to compute the coefficients of the characteristic polynomial of $B(u, v)$ by using rational arithmetic. For multiplication of polynomials, apply the evaluation/interpolation on the Fourier set and use exercise 30.
    c) Perform the analysis of the precision of computation needed by Algorithm 2.11.2 for the solution of a linear system $B(u, v)\mathbf{x} = \mathbf{b}$.
    d) Use the results of parts a), b) and c) to estimate the precision of the computation of $\gcd(u(x), v(x))$ by means of Algorithm 2.9.1.

35 ([DGR]). The approximation algorithm of section 4 for multiplication of Cauchy (generalized Hilbert) matrix by a vector relies on Taylor's expansion of the function $1/(x - x_j)$.
    a) Find appropriate techniques for approximating this function by linear combinations of the ratios $T_n(x)/(x - t_k)$, $k = 1, \ldots, n$,

where $T_n(x)$ is the Chebyshev polynomial of degree n and $t_k = -\cos((2k-1)\pi/(2n))$ is one of its zeros (a Chebyshev node). (Note that 2 evaluations of square roots of positive numbers and $2\log n + c$ arithmetic operations for a fixed constant c suffice to compute $T_n(a)$ for any fixed real a.) Apply such an approximation to improve the approximation algorithm of section 4 for Problem 4.1 ($C(G \cdot H) \cdot VECTOR$).

b) Extend the above techniques to solve a *generalized real interpolation problem* by relying on the Lagrange interpolation formula (1.4.1): for an n-th degree polynomial given by its values on a set of $n+1$ real points $x_0, \ldots, x_n$, approximate its values on another set of real points $y_0, \ldots, y_n$.

In both cases obtain the solution by using $O(nb + n\log n)$ ops where 2^{-b} is the output error bound.

36. Let a_1, b_1, a_2, b_2 be floating point numbers, $w_1 = a_1 + ib_1$, $w_2 = a_2 + ib_2$. By using the relations $fl(a \circ b) = (a \circ b)(1+\epsilon)$, $|\epsilon| < u = 2^{-d}$, where $fl(\cdot)$ denotes the computation performed with the d-bit floating point arithmetic and $\circ$ denotes any arithmetic operation, prove that $\mathrm{fl}(w_1 + w_2) = (w_1 + w_2)(1+\phi)$, $|\phi| < u$, $\mathrm{fl}(w_1 w_2) = (w_1 w_2)(1+\psi)$, $|\psi| < 2\sqrt{2}u(1+u)$, provided that no overflow and underflow conditions occur and that multiplication of complex numbers is performed by means of the customary algorithm requiring 4 real multiplications and 2 real additions. Derive an analogous result for division.

37 ([Bre76], see also [Alt], [BB], [LO]). For a complex x, for $s > 1$, and for $f(x)$ being one of the elementary functions $\log x$, $\exp(x)$, $\sin(x)$, $cos(x)$ and $\tan(x)$ or a reverse of such a function, approximate $f(x)$ within the error 2^{-s} by using $O(\mu(s)\log s)$ bit-operations provided that $\mu(s)$ bit-operations suffice for multiplication of two integers modulo 2^s [compare formula (1.2.4) and (1.2.5)].

Appendix A. Floating Point Numbers and Error Analysis.

In this appendix we recall some basic results and techniques for operating with floating point numbers and for error analysis. For more details we refer the reader to [A], [CdB], [W63], [Ster] and [BBCM].

We start with a theorem on the unique representation of a nonzero real number in a given base β.

Theorem A1. *Let β be an integer, $\beta \geq 2$. For any real $x \neq 0$, there exist a unique integer p and a unique sequence of integers $\{d_i\}_{i=1,2,\ldots}$, such that $x = \text{sgn}(x)\beta^p \sum_{i=1}^{+\infty} d_i\beta^{-i}$, where $d_1 \neq 0$, $0 \leq d_i < \beta$, for all i, $d_h \neq \beta - 1$ for an infinite sequence of natural h, $\text{sgn}(x) = 1$ if $x > 0$, $\text{sgn}(x) = -1$ if $x < 0$.*

The number β is the *base* of the representation, p is the *exponent*, $d_1, d_2, \ldots$ are the *digits*, and $\sum_{i=1}^{+\infty} d_i\beta^{-i}$ is the *fractional part (fraction, mantissa)* of the representation of x. Setting the condition $d_1 \neq 0$ is called *normalization*, and, together with the assumption that $d_h \neq \beta - 1$ for infinitely many natural h, this condition guarantees the uniqueness of the representation.

With a quadruple (β, t, m, M) of four positive integers (where $\beta \geq 2$), we associate the set of *floating point* numbers as follows

$$\mathcal{F}(\beta, t, m, M) = \{0\} \cup \{x \in \mathbf{R} : \quad x = \text{sgn}(x)\beta^p \sum_{i=1}^{t} d_i\beta^{-i},$$
$$0 \leq d_i < \beta,\ i = 1, 2, \ldots, t,\ d_1 \neq 0,\ -m \leq p \leq M\}.$$

Let $x \in \mathbf{R}$ be nonzero and have the representation given in Theorem A.1. If $p > M$ or $p < -m$, then we say that x is not representable in $\mathcal{F}(\beta, t, m, M)$; moreover, we say that *overflow* occurs if $p > M$, *underflow* occurs if $p < -m$. Any real x, such that $\beta^{-m} \leq |x| \leq \beta^M$, can be approximated by the floating point number $\text{trunc}(x) = \text{sgn}(x)\beta^p \sum_{i=1}^{t} d_i\beta^{-i}$, called the *truncated value* of x and lying in $\mathcal{F}(\beta, t, m, M)$. We also define $\text{round}(x)$, the *rounded value* of x, as

$$\text{round}(x) = \begin{cases} \text{trunc(x)} & \text{if } d_{t+1} < \beta/2, \\ \text{trunc}(x) + \beta^{p-t} & \text{if } d_{t+1} \geq \beta/2. \end{cases}$$

Hereafter we will use the common notation $\tilde{x}$ for both $\text{round}(x)$ and $\text{trunc}(x)$. Measuring the representation error by means of the *relative error* $(\tilde{x} - x)/x$, we may easily prove that

$$|(\text{trunc}(x) - x)/x| < \beta^{1-t}, \ |(\text{trunc}(x) - x)/\text{trunc}(x)| < \beta^{1-t},$$
$$|(\text{round}(x) - x)/x| < \frac{1}{2}\beta^{1-t}, \ |(\text{round}(x) - x)/\text{round}(x)| < \frac{1}{2}\beta^{1-t},$$

provided that x is representable in $F(\beta, t, m, M)$ for a fixed quadruple (β, t, m, M) and that the rounding process does not lead to overflow. In other words, for a real floating point number x, *the relative error of its representation in $F(\beta, t, m, M)$ is bounded by a constant independent of x*, provided that overflow and underflow do not occur. This fact can also be simply expressed by the following formulae:

$$\begin{aligned} \operatorname{trunc}(x) &= x(1+\epsilon_1) = x/(1+\eta_1), \\ \operatorname{round}(x) &= x(1+\epsilon_2) = x/(1+\eta_2), \\ |\epsilon_1|, |\eta_1| &< \beta^{1-t}, |\epsilon_2|, |\eta_2| < \frac{1}{2}\beta^{1-t}. \end{aligned} \tag{A.1}$$

The quantities $u = \beta^{1-t}$, in the case of truncation, and $u = \frac{1}{2}\beta^{1-t}$, in the case of rounding, called the *machine precision*, define the tight upper bounds on the modulus of the relative *representation error* $(\tilde{x} - x)/x$.

If we are required to evaluate a rational function $f(\mathbf{x})$ for $\mathbf{x} = (x_1, \ldots, x_n)$, and if we represent the input data $(x_1, \ldots, x_n)$ by means of machine numbers $(\tilde{x}_1, \ldots, \tilde{x}_n) = \tilde{\mathbf{x}}$, then we actually compute the function value at a different point. The relative error

$$\epsilon_{inh}(\mathbf{x}) = \frac{f(\tilde{\mathbf{x}}) - f(\mathbf{x})}{f(\mathbf{x})}$$

will be called the *inherent error*.

Let $\epsilon_i = (\tilde{x}_i - x_i)/x_i$ denote the representation errors, and suppose that the function $f(\mathbf{x})$ is defined on the segment line connecting the points $\mathbf{x}$ and $\tilde{\mathbf{x}}$. Then we may relate the inherent and the representation errors by using Taylor's differential formula:

$$\begin{aligned} \epsilon_{inh}(\mathbf{x}) &= \sum_{i=1}^{n} c_i \epsilon_i, \\ c_i &= x_i \frac{f'_{x_i}(\mathbf{z})}{f(\mathbf{z})}, \\ \mathbf{z} &= \theta\mathbf{x} + (1-\theta)\tilde{\mathbf{x}}, \ 0 < \theta < 1, \end{aligned} \tag{A.2}$$

where $f'_{x_i}(\mathbf{z})$ denotes the partial derivative of the function $f(\mathbf{x})$ with respect to x_i. The numbers c_i, called the *amplification factors*, measure the sensitivity of the function $f(\mathbf{x})$ to small perturbations of the input values and express the numerical conditioning of the function.

The reader can easily find simple examples that the set $\mathcal{F}(\beta, t, m, M)$ is not closed under arithmetic operations. On the other hand, we may define floating point operations by rounding or truncating the corresponding real arithmetic operations (ops), under which $\mathcal{F}(\beta, t, m, M)$ is closed. The arithmetic operations defined in this way do not satisfy some formal properties of the corresponding arithmetic operations over the real numbers (such as associativity and distributivity), but due to (A.1), they always satisfy the inequality

$$(a \ \widetilde{\text{op}}\ b) = (a \text{ op } b)(1+\epsilon), \quad |\epsilon| < u, \tag{A.3}$$

where $a, b \in \mathcal{F}(\beta, t, m, M)$, the value a op b is representable in $\mathcal{F}(\beta, t, m, M)$, "$\widetilde{\text{op}}$" denotes the operation obtained by rounding or truncating the result of the corresponding arithmetic operation ("op"), u is the machine precision, and ϵ is the *roundoff* error.

A set of aritmetic operations implemented in $\mathcal{F}(\beta, t, m, M)$ in such a way that (A.3) holds is called *floating point arithmetic.*

In the computation of a rational function $f(\mathbf{x})$ at a point $\tilde{\mathbf{x}}$, where $\tilde{\mathbf{x}} = (\tilde{x}_1, \dots, \tilde{x}_n)$ is an n-tuple of machine numbers, the actually computed value, affected by the roundoff error generated in each arithmetic operation, is different from $f(\tilde{\mathbf{x}})$; let us denote it $\phi(\tilde{\mathbf{x}})$. The *algorithmic error,*

$$\epsilon_{alg}(\tilde{\mathbf{x}}) = \frac{\phi(\tilde{\mathbf{x}}) - f(\tilde{\mathbf{x}})}{f(\tilde{\mathbf{x}})},$$

is related to the inherent error and to the *total* error,

$$\epsilon_{tot}(\mathbf{x}) = \frac{\phi(\tilde{\mathbf{x}}) - f(\mathbf{x})}{f(\mathbf{x})},$$

by means of the following equation:

$$\epsilon_{tot}(\mathbf{x}) = \epsilon_{inh}(\mathbf{x}) + \epsilon_{alg}(\tilde{\mathbf{x}}) + \epsilon_{inh}(\mathbf{x})\epsilon_{alg}(\tilde{\mathbf{x}}).$$

Equation (A.3) is a useful tool for estimating an upper bound on the absolute value of the algorithmic error; this is the objective of the *forward error analysis.*

A different effective way to estimating the algorithmic error is the *backward error analysis.* It consists in determining upper bounds on the relative perturbations $\delta_i = \frac{|x_i - y_i|}{|x_i|}$, $i = 1, \dots, n$, where the n-tuple $\mathbf{y} = (y_1, \dots, y_n)$ of *real* numbers is such that $f(\mathbf{y}) = \phi(\tilde{\mathbf{x}})$.

Once such bounds have been determined, it is easy to arrive at a bound on the algorithmic error by means of the relations (A.2) where ϵ_i is replaced by δ_i, $\tilde{\mathbf{x}}$ is replaced by $\mathbf{y}$, and $\mathbf{x}$ by $\tilde{\mathbf{x}}$, thus reducing the analysis of the algorithmic error to the analysis of the inherent error. In fact, the algorithmic error can be viewed as the inherent error due to the perturbation of the input data given by δ_i. If the perturbations δ_i have the same order of magnitude as does the machine precision, the algorithm is customarily considered numerically stable.

The forward and the backward error analyses are usually performed under the worst case assumption, that is, it is assumed that the roundoff error generated in each floating point operation has its maximum value u. Of course, this can be far above the average case error, and, indeed, the worst case assumption generally leads to overly pessimistic upper bounds. A more realistic analysis is the *statistical error analysis* where each roundoff error is viewed as a random variable having a known probability distribution. The objective of this analysis is to bound the probability that the modulus of the algorithmic error is lower than a given value.

Error analysis can be alternatively performed by implementing *interval arithmetic* (see [AlH]), where each real number x is replaced by a pair of floating point numbers (a, b) defining a sufficiently narrow interval such that $x \in [a, b]$. The operation on intervals are implemented in such a way that if $x \in [a, b]$, $y \in [c, d]$, then, $x \; \widetilde{\mathrm{op}} \; y \in [p, q]$ where $[p, q] = [a, b] \; \mathrm{op} \; [c, d]$.

Although the statistical error analysis and interval arithmetic give more valuable results than the worst case analysis, they require more intensive analysis and additional information about the errors, which is not always available, so in this book we restrict ourselves with the (forward and backward) worst case error analysis.

4

Parallel Polynomial and Matrix Computations

Summary.

In this chapter we will survey the techniques of the design of algorithms that effectively solve the computational problems of the previous chapters on parallel computers, reaching a substantial (and frequently dramatic) acceleration of the known sequential algorithms. In some cases this requires us to revise the sequential case approach or even to change it completely. We systematically review the state of the art in this area, in particular, the computational complexity estimates, the algorithms supporting them, and the major techniques for the design of effective parallel algorithms. We include several recent research results and techniques, particularly in sections 4 and 6 and in the appendices. The results of Appendices B and C on computing a maximal independent subset of a vector set and the techniques of the design of parallel algorithms demonstrated in Appendix A may be of interest for the designers of combinatorial and graph algorithms.

1. Introduction. Basic Concepts, Results and Definitions.

We will start this chapter with a list of its contents. In the first section we recall some basic concepts and results on parallel algorithms and then briefly comment on some practical aspects of parallel computing. Those readers who wish to focus more on the algorithm design and less on formal models and implementation, may initially skip this section. In section 2 we review the parallel arithmetic complexity of the algorithms of chapters 1 and 2 for computations with polynomials and structured matrices. In section 3 we treat both Boolean and arithmetic parallel complexity of matrix multiplication. Then, in the same section and in sections 4 and 5, we study other computations with general matrices. In sections 2–5 we estimate the arithmetic complexity of parallel computations over the fields of characteristic 0 and, whenever needed, assume that the field supports FFT. In Remark 2.1, section 6 and Appendix C, we shift to the case of any field of constants and review how this affects our results in the entire area of computations with polynomials, including the computation of gcd's and the entries of the Euclidean scheme, and with structured and general matrices. In sections 7 and 8, we apply Newton's iteration to improve some parallel computations with well-conditioned (both general and structured) matrices. In section 9, we revisit some basic matrix computations and estimate their Boolean and arithmetic cost, in the case of (general and structured) integer input matrices. In the arithmetic case, we estimate the precision of computations too. In Appendix A we demonstrate the general techniques of stream contraction, recursive restarting, and supereffective slowdown of parallel computation, used for designing effective parallel algorithms. In Appendices B and C we present recent parallel algorithms for Problem 2.2.10a ($SUBMATRIX$) and for the equivalent problem of computing a maximal linearly independent subset of a vector set, which has many applications to combinatorial and graph computations. We present some recent techniques and results in sections 4–6, 9 and Appendices A, B and C.

PRAM models.

Our high level description of the parallel algorithms is computer independent, and to estimate the complexity of our parallel computations, we will assume the *Parallel Random Access Machine (PRAM)* models, which are further subdivided into Boolean, arithmetic and algebraic PRAM models and on which we refer the reader to section 1 of chapter 1 and Remarks 1.1 and 1.2 of this section. We will measure the cost of parallel computations by the *parallel time* and *processor bounds*; their product will represent the *potential work* of the algorithm. The parallel time and processor bounds will be defined within constant factors (see Definition 1.1). For our computational problems with matrices and polynomials, the same asymptotic bounds (within constant factors) hold for the circuit complexity of these problems (see Remark 4.1 on the exceptional cases and see Appendix A to chapter 1 on Boolean and arithmetic circuits), provided that the circuit depth, size and the ratio of the size and the depth correspond to the parallel time, the potential work and the number of processors under the PRAM model, respectively.

Table 1.1 enables us to extend the parallel arithmetic complexity estimates obtained under the arithmetic PRAM models or under the arithmetic circuit model to the parallel Boolean complexity estimates, obtained under the Boolean PRAM models or under the Boolean circuit model, respectively; that is, we will simply replace each arithmetic operation performed with the

h-bit precision by several Boolean operations or by a Boolean circuit whose size and depth are bounded according to Table 1.1 (compare (1.2.4), (3.1.1), [Of62], [Of65], [Sa] and [RT]), which corresponds to the estimates for the potential work and parallel time, respectively, required in order to perform such an arithmetic operation under the Boolean PRAM model.

Operation	Size	Depth
Addition/Subtraction	$O(h)$	$O(\log h)$
Multiplication	$O(h\log h\log\log h)$	$O(\log h)$
Division	$O(h\log h\log\log h)$	$O(\log h\log\log h)$

Table 1.1. The size and depth of Boolean circuits performing arithmetic operations.

We refer the reader to [EG88], [KR], [Q94], [J], [Le92], [CLR], [LLMPW], [Parb], [AHMP], [ARa], [Bor77], [BCP], [BGH], [CR], [E], [G86], [Ran], [Ruz], [Val], [Val90] for further abundant information on PRAM algorithms and various models of parallel computations.

The B-principle (a variant of Brent's scheduling principle) and its applications.

As this is customary and realistic for matrix and polynomial computations ([Bre74], [KR] and [EG88]), we will adopt the following simple *B-principle*, which can be viewed as a variant of Brent's well-known scheduling principle for parallel computing: a single processor may simulate s processors in $O(s)$ time. Therefore, by slowing down parallel computations by $O(s)$ times, one may decrease the number of processors from Ps to P, as long as P and s are positive integers. Thus, we will define our upper bounds t (on parallel time) and p (on the number of processors) up to within such scaling factors s, so that our parallel algorithms can be adjusted to the available number of processors. In particular, any parallel algorithm running in time t on p processors can be simulated in sequential time $O(tp)$ by a single processor. Thus the bound on the potential work $w = tp$ may at best reach the record sequential time bound (within a constant factor), in the optimum case (see Definition 1.2).

Here is another example of various useful applications of this principle:

Example 1.1. We may apply the fan-in method to compute the sum $s = \sum_{i=0}^{n-1} a_{i,0}$ (where $n = 2^k$, k integer) by using $k = \log n$ steps and $n/2$

processors as follows:

$$a_{i,h} = a_{2i,h-1} + a_{2i+1,h-1}, \ i = 0, 1, \ldots, 2^{k-h} - 1, \ h = 1, \ldots, k, \ s = a_{0,k}.$$

The potential work equals $kn/2$ and exceeds the sequential complexity of $n-1$ ops by more than $k/2$ times, but we may slow down the first $\lceil \log k \rceil$ steps so as to use only $\lceil n/k \rceil$ processors at these steps. In this way, the h-th step is slowed down by the factor of $O(k/2^h)$, $h = 1, \ldots, \lceil \log k \rceil$. The subsequent $k - \lceil \log k \rceil$ steps can be performed with no slowdown and again with $\lceil n/k \rceil$ processors. Then the potential work will decrease to $O(n)$, that is, to the level within a constant factor from the sequential time of this algorithm, whereas the parallel time will stay at the level of $O(k)$ steps.

The following simple result generalizes the above technique (in the spirit of [Bre74]) in order to effectively reorganize a parallel algorithm so as to increase its processor efficiency:

Proposition 1.1. *Let the computation consist of S stages, such that p_s processors concurrently perform t_s time-steps at stage s, $s = 1, \ldots, S$. Then, under the B-principle, this computation can be performed by using $O(t)$ time steps on $p = \lceil \sum_{s=1}^{S} t_s p_s / t \rceil$ processors where $t = \sum_{s=1}^{S} t_s$ (versus the original t steps and $\max_s p_s$ processors).*

Proof. Let S_+ be the set of all natural s such that $p_s > p$, and $S_- = S - S_+$. Let us slow down the computation at stages $s \in S_+$ by $O(p_s/p)$ times, so as to have at most p processors working at each stage. Then the overall computational time will grow to $t_- + t_+$ where

$$t_- = \sum_{s \in S_-} t_s \leq t,$$

$$t_+ = O(\sum_{s \in S_+} t_s(p_s/p)) = O(t).$$

■

In particular, if $t_s = 1$ for all s, then $t = S$, and Proposition 1.1 implies the time bound $O(t) = O(S)$, and the processor bound $p = \lceil \sum_{s=1}^{S} p_s/t \rceil = \lceil \sum_{s=1}^{S} p_s/S \rceil$. If, in addition, $p_s \leq \theta^{s-1} p_1$ for a constant θ, $0 < \theta < 1$, then $p < \lceil p_1/((1-\theta)S) \rceil$. In Example 1.1, $\theta = 1/2$, $S = \log n$, $p_1 = n/2$, and we obtain that $p \leq \lceil n/\log n \rceil$ processors support the time bound $O(\log n)$.

Parallel complexity, NC, processor efficiency and some open problems.

Definition 1.1. We will write $O_A(t,p)$ and $O_B(t,p)$ to denote the parallel complexity estimates where the upper estimates $O(t)$ and $O(p)$ simultaneously hold for both asymptotic parallel time and the number of processors, over the arithmetic and the Boolean PRAM models of parallel computing, respectively, and then we will assume that the problem can also be solved in $O(st)$ (arithmetic or Boolean) parallel time using p/s (arithmetic or bit-serial) processors for any pair of natural s and p/s. We will write $O_A(T) = O_A(T,1)$ and $O_B(T) = O_B(T,1)$ to denote the asymptotic bounds on the arithmetic and Boolean potential work of a parallel algorithm, respectively, so that, according to the B-principle, $O_A(t,p)$ and $O_B(t,p)$ also mean $O_A(tp)$ and $O_B(tp)$, respectively, but not vice versa in general.

We will define the input size of a computational problem as usual, that is, by the number of input parameters under the arithmetic model and by the number of bits defining the input under the Boolean model (to represent the size of rectangular input matrices, under the arithmetic model, we will use the pairs of numbers).

The next definition of NC (the abbreviation for Nick's class, named after Nicholas Pippenger) is theoretically important for the classification of how the solution of computational problems may resist parallel acceleration even under a very weak restriction on the number of processors:

Definition 1.2. A computational problem with input size N *is in* NC^k, for a nonnegative integer k, under the arithmetic or Boolean models of parallel computation, if this problem can be solved by using $O((\log N)^k)$ parallel arithmetic or Boolean steps, respectively, and simultaneously a polynomial number, $N^{O(1)}$, of processors. NC is defined as the union of NC^k for all nonnegative k. An algorithm supporting the bounds $O((\log N)^k, N^{O(1)})$ or $O_B((\log N)^k,\ N^{O(1)})$ for a specified constant k is called an NC^k*-algorithm.* An NC-algorithm is any NC^k-algorithm where k is not required to be specified. An NC-algorithm is *processor efficient* (or *work efficient*) if its potential work w is within a polylogarithmic factor from the record sequential time bound T for the same problem, and is *work optimum* if $w = O(T)$. If the computation involves randomization, then NC^k, NC, NC^k-algorithms and NC-algorithms are called RNC^k, RNC, RNC^k-algorithms and RNC-algorithms, respectively.

Together with the B-principle, this definition implies that a computational problem is not in NC if it is not in P, that is, if it does not have a sequential algorithm running in $N^{O(1)}$ time, that is, in time polynomial in N.

There are several important computational problems in P for which we do not know if they are in NC or not. (See [GHR], [GRy], [KR], in addition to our next comments, and compare [Mul94] on a recent progress on parallel complexity of the linear programming computations.)

Some of these problems are known to be P-complete: any other problem in P can be reduced in NC to such a P-complete problem, so that any NC-algorithm for such a problem can be extended to an NC-algorithm for any other problem in P. The list of P-complete algebraic and optimization problems includes: solving a system of linear inequalities $A\mathbf{x} \geq \mathbf{b}$ over the rationals, solving a system of linear equations $A\mathbf{x} = \mathbf{b}$ over the rationals subject to the constraint $\mathbf{x} > \mathbf{0}$, the linear programming problem, Gaussian elimination with partial pivoting (where the output is the set of the signs of the pivot elements), and computing the iterated mod function over the integers. The iterated mod function, $\text{Mod}(m_0, m_1, \ldots, m_k)$, is recursively defined by the equations: $\text{Mod}(m_0, m_1, \ldots, m_{i+1}) = \text{Mod}(m_0, m_1, \ldots, m_i)$ modulo m_{i+1}, $i = 0, \ldots, k-1$. There are two versions of this problem, depending on whether $m_0, \ldots, m_k$ are integers or polynomials. In the former case, the problem (as we said) is P-complete; in the latter case, it is in NC ([KRu]): an interesting fact to remember!

On the other hand, there are several computational problems, for which both questions about their membership in NC and about their P-completeness remain open. The list of such problems currently includes: computing the gcd of two integers, computing all the remainders in the extended Euclidean scheme for two integers (this problem is abbreviated as EUGCD), testing if their gcd is 1, computing (modulo a prime) the inverse of an integer and its integer power, and the integer linear programming problem in k variables, for a fixed k (k-ILP).

Obtaining NC-algorithms for these problems or proving their P-completeness would be a breakthrough (to see the importance of this topic for our study, recall, for instance, the role of the gcd of integers for Algorithm 3.3.3); furthermore, it would be interesting even to establish new nontrivial NC-reductions and/or NC-equivalence among these problems. The latest progress in this area is the proof of NC-equivalence of EUGCD and the optimization version of 2-ILP (see [SPL-K], also compare [G84] [De], [L-K,P]).

For all the fundamental computations with matrices and polynomials presented in chapters 1 and 2, however, NC-solution algorithms are well known. Moreover, the following remarkable general result and its extension in [Ka88] reduce the design of NC-algorithms for polynomial and matrix computations to the design of sequential algorithms running in $N^{O(1)}$ ops, and the latter task is usually straightforward for matrix and polynomial computations:

Theorem 1.1 ([VSBR], also see [MRK]). *If a set of multivariate polynomials of degrees $N^{O(1)}$, N being the input size, can be evaluated in $N^{O(1)}$ ops, then this set can be evaluated at the cost $O_A(\log^2 N, N^{O(1)})$.*

[Ka88] extended this result to rational functions.

The remaining open problems in this area are thus narrowed to:

a) studying the computational problems that are not reduced to the evaluation of multivariate rational functions;

b) acceleration of computations in NC^2 to yield the time bounds $o(\log^2 N)$ [say, $O(\log N)$] using $N^{O(1)}$ processors;

c) improving processor (work) efficiency of the known NC-algorithms that are not work optimum.

For an example of a challenge from class a), we recall the interesting problem of modular polynomial exponentiation, that is, of computing (the coefficients of) the polynomial $p(x)^d \bmod m(x)$ for two given polynomials $p(x)$, $m(x)$ and a natural d. No NC-algorithms are known for this problem (see problem IX in section 4 of [KR]), even under the extension of our arithmetic model where computing an integer exponent of a given scalar is a unit cost operation. (Note that, for $m(x) = x$ and for $p(x) = p$ being a fixed scalar, the polynomial exponentiation turns into scalar exponentiation, that is, into the problem of computing p^d, which requires at least $\log d$ parallel time under our original arithmetic model of computing, [BM], p.127.)

For task b), we recall the general techniques of *stream contraction*, which we will exploit in the Appendix and in chapter 6 (also compare [PR93], [HPR], [BP]), and the Markov chain-random walk techniques, which are the basis of the algorithms supporting Theorem 4.2 (see exercises 5 and 6 and chapter 6, for examples of other techniques).

In the following sections, we will focus on *the problem of improving the work and processor efficiency of the known NC but processor inefficient algorithms for several matrix and polynomial computations.* For some computational problems, the ratio of the smallest known upper bounds on the potential work w of the parallel NC-algorithms to the smallest known upper bounds on the sequential time T for solving the same computational tasks exceed n^c, where c is a positive constant and n is the input size. In such cases removing or decreasing the factor n^c is a great theoretical challenge, which is also a practical problem: the users are more interested in keeping the ratio T/w close to 1 than in having a polylogarithmic time bound. The researchers sometimes feel in the opposite way, and this feeling can be partly justified due to the power of the general techniques of *supereffective slowdown* and *recursive restarting* (see the next subsection).

Our objectives and some practical considerations. Supereffective slowdown of parallel computations.

Our main subject in this chapter will be the design of processor (work) optimum or processor (work) efficient NC-algorithms for computations with polynomials and with general and structured matrices, assuming that the entire parallel cost of the solution is dominated by its arithmetic computation part. This simplifying assumption has been introduced for the sake of the technical integrity of our presentation, even though it ignores some important practical aspects of parallel computation, such as processor communication and synchronization, data fetching, and printing the output of the solution.

Most of the parallel computers perform arithmetic operations (ops) much faster than they fetch data, so that it is practically important to increase the number of ops per data fetch by choosing appropriate algorithms. In particular, this objective has been achieved in computational linear algebra by means of systematically replacing matrix-by-vector multiplications by matrix multiplications (compare the editions of [GL] in 1983 and 1989, with dramatically increased attention to matrix multiplication in 1989). Indeed, applying the straightforward algorithms, we need $n(n+1)$ data fetches and $2n^2 - n$ ops for computing $A\mathbf{v}$ and $2n^2$ fetches and $2n^3 - n^2$ ops for computing AB for a pair of $n \times n$ matrices A and B and an n-dimensional vector $\mathbf{v}$. The latter computation implies the ratio $\dfrac{2n^3 - n^2}{2n^2}$, which is, of course, much more favorable than the ratio $\dfrac{2n^2 - n}{n^2 + n}$, implied by the former computation.

Another customary way of decreasing nonarithmetic time is by decreasing the number of processors, which is going to be our major objective when we select or design NC-algorithms. Actually, several of our NC-algorithms for polynomial and structured matrix computations are processor efficient and, moreover, only require a number of processors, which is linear in the input size. In all other cases, this number is at most quadratic, including such problems as computing the polynomial gcd and solving a general triangular or a Toeplitz-like linear system of equations. For the three latter problems, all the known processo-efficient parallel algorithms are not in NC, except for the computation over the rationals (see section 9). We may, however, dramatically accelerate them (although not to the polylog time level) with no or a relatively small sacrifice in their processor efficiency, by means of the techniques of *supereffective slowdown of parallel computations.* Such a possibility has already been shown in Example 1.1, where an appropriate slowdown [by $O(1)$ times] of a parallel algorithm enabled us to improve its processor efficiency by more than a constant factor; that is, the effect was much greater than in the usual application of the B-principle (compare Proposition 1.1). For a large class of computations, a similar effect is obtained by means of a distinct special technique, called *recursive restarting* and demonstrated in Appendix A, in sections 8 and 9, in exercises 8b), c), d), e) and in [BP], [P88], [P89b], [P92c], [PP] and [PP,a]. In fact, the technique of recursive restarting enables us to utilize some fast but processor-inefficient algorithms as auxiliary blocks of the design of fast and processor-efficient parallel algorithms (see Appendix A). In Appendix A we will also show how the (already cited) *stream contraction* techniques (which exploit pipelining of recursive algorithms and which are used in [PR91], [BP] and [HPR]) can be combined with the recursive restarting and the supereffective slowdown techniques in order to improve the processor efficiency of some important computations, with no sacrifice in their running time.

Besides studying the structured matrix case, we will also estimate the parallel complexity of the NC-computations with general matrices; this has theoretical applications to estimating parallel complexity of combinatorial computations (see [GP85;88], [Lo], [MVV], and [NSV]). The following citation is from an article surveying the known parallel algorithms for combinatorial problems: "While the motivation of these algorithms may come from different branches of mathematics, they eventually boil down to linear algebra, most often to matrix inversion" ([Lo], p. 394).

Practical limitations on NC-computations with matrices and some recent progress.

In computations with $n \times n$ general (dense and unstructured) matrices for large n, except for Problem 2.2.1 ($M \cdot VECTOR$), we usually need more than n^2 processors to support the polylog time bounds. In this case, the cost of processor communication and synchronization (which we do not include in our estimates) makes the overall parallel time at best linear in n under the models of the mesh-connected processor array on the plane and of multiprocessors [Q], which correspond to the computers currently performing most of the practical computations with large general matrices. The practical value of the NC-algorithms was further limited in this case, since, until 1991, in such computations (except for matrix-by-matrix and matrix-by-vector multiplications), all the known NC-algorithms required a substantially greater potential work than the best algorithms having linear (in n) parallel time (compare [GPS], [Or] and [OV]) and, in addition, were prone to numerical stability problems. Both of these defects can be avoided by means of the transition to computations over finite fields and p-adic lifting (see section 3 of chapter 3 and section 9 of this chapter); these tools are customary for algebraic and symbolic computing, but are not accepted in the practice of numerical matrix computations (compare chapter 3). Processor efficiency can be alternatively repaired by means of the supereffective slowdown and recursive restarting techniques (compare exercise 8), which promise to be valuable in practice of numerical matrix computations. Furthermore, in sections 4–6, we will present some recent randomized algorithms for operations with general matrices. These algorithms are both in (randomized) NC and processor efficient, so that, with the incorporation of p-adic lifting, they are of potential practical value for algebraic and symbolic computing (compare [Ka93]). On the other hand, in our study parallel matrix multiplication (being both fast and processor efficient) serves as the major block of parallel algorithms for matrix computations. This agrees well with the rapidly increasing popularity of parallel matrix multiplication in numerical computational practice.

Further comments on PRAM models.

We will conclude this section with some further comments on the PRAM models.

Remark 1.1 (compare [EG88], [KR], [Q94], [J], [Le92]). The PRAM models allow p processors to simultaneously access p distinct locations of their shared global memory. There exist three customary modifications depending on whether several processors may or may not also access the same memory location for writing and/or reading simultaneously. They are called EREW, CREW and CRCW PRAM, where C, E, R and W stand for concurrent, exclusive, read and write, respectively. This list is in the order of increasing the computational power. The most powerful CRCW PRAM models are further classified according to how they resolve the conflicts where more than one processor attempts to write into the same memory location, most powerful being the priority CRCW PRAM model, under which the priority for writing into the memory location is given to a specified candidate processor, typically, the candidate processor with the lowest index. The computational power of all these PRAM models varies little, however; that is, all these machine models may simulate each other by using the same parallel time (within the factor $\log N$) and the same number of processors, N being the input size. The factor $\log N$ can make the actual difference already for the parallel time bound for the numerical summation of N elements of Boolean algebra: this bound is of the order $\log N$ on the EREW PRAM but is $O(1)$ on a randomized CRCW PRAM [Rei]. In our study, we will only deal with numerical summation, and the record arithmetic time bound for the summation of N numbers is $O(\log N)$ under any PRAM model (compare Example 1.1 and Fact 2.1). The readers may, however, prefer to assume the CREW PRAM model, rather than EREW PRAM model, and to use the concurrent read mode for fast broadcasting the matrix entries to appropriately chosen processors, which is convenient for the PRAM implementation of the matrix algorithms of this chapter.

Remark 1.2. The assumption about the simultaneous access to the shared memory, used in the PRAM models, has not been supported by the current computer hardware so far, but it is customarily accepted in the literature on parallel algorithms as a means of describing them in a way independent of their implementation on the specific computers. Many of the parallel algorithms that we present assuming PRAM models are, however, practical; they can be easily adjusted to the implementation on the specific parallel computers, and (as we have already pointed out) our high-level description of our algorithms is machine independent. We actually need PRAM models just to specify the complexity estimates. On the other hand, some experts "emphasize the view that the PRAM models are not just theoretical constructs, but are potentially the basis of parallel programming languages and machine designs" ([Val]). To confirm this point of view by the current progress in the development of the high speed computer technology, [Val] cites some impressive recent results on the simulation of PRAM's on realistic models of parallel computation; for instance, CRCW PRAM has been simulated on such models with the increase of the time and processor bounds by at most logarithmic factors; moreover, the processor bound may be kept unchanged if a randomized simulation is allowed ([AHMP], [Ran], [Val90]).

2. Parallel Complexity of Computations with Polynomials and Structured Matrices.

In Table 2.1 we list the current record upper estimates for the arithmetic parallel complexity of some polynomial computational problems from chapter 1, thus extending the earlier taxonomy of [Cook85], and in Table 2.2 we list the similar parallel cost estimates for the computations by some classical algorithms. At the end of this section, we will comment on the parallel computations with structured matrices and, in Remark 2.2, on the practical significance of parallel complexity estimates for computations with polynomials. Next, we will briefly review some estimates for the parallel arithmetic complexity, which can be extended to the Boolean complexity estimates for the same computational problems by using Table 1.1 and the following fact of general interest (compare Example 1.1):

Fact 2.1 (see [Of62], [Of65], [Kr]). *Given s natural numbers less than 2^h, their sum can be computed at the arithmetic cost $O_A(\log s, s/\log s)$ and at the Boolean cost $O_B(\log (sh), sh/ \log (sh))$.*

In Table 2.2 we will state the estimates for the parallel arithmetic complexity of computations over the fields of characteristic 0 supporting DFT (the reader may assume the complex field), and we will comment on the extension to an arbitrary field in Remark 2.1 and in section 6. We will start with the following simple auxiliary algorithm, which has numerous extensions and applications to parallel computations (see [Of62], [EG88], [KR], [LLMPW], [Reif93] and [Q]):

Algorithm 2.1, *parallel prefix computation or parallel cascade computation.*

Input: natural h, $n = 2^h$, and the components $a_1, \ldots, a_n$ of a vector $\mathbf{a} = [a_i]$.

Output: $s_k(\mathbf{a}) = \sum_{i=1}^k a_i$ for $k = 1, \ldots, n$.

Computation:

1. Compute $b_i = a_{2i-1} + a_{2i}$, $i = 1, \ldots, n/2$.
2. Apply Algorithm 2.1 to compute and output $s_k(\mathbf{b}) = \sum_{i=1}^k b_i$, $k = 1, \ldots, n/2$; set $s_0(\mathbf{b}) = 0$.
3. Compute $s_{2k-1}(\mathbf{a}) = s_{k-1}(\mathbf{b}) + a_{2k-1}$; $s_{2k}(\mathbf{a}) = s_k(\mathbf{b})$ for $k = 1, \ldots, n/2$.

The algorithm can be implemented at the cost $O_A(\log n, n/\log n)$ and can be immediately extended (at the same asymptotic cost) to the computation of $p_k(\mathbf{a}) = \prod_{i=1}^k a_i$, for $k = 1, \ldots, n$, for given $a_1, \ldots, a_n$; to the computation of $x_k(\mathbf{a}, \mathbf{b}) = a_k x_{k-1} + b_k$, for $k = 1, \ldots, n$, for given $a_1, \ldots, a_n$, $b_1, \ldots, b_n$, $x_0 = 0$ (see exercise 4 or [Q], chapter 7); and to several combinatorial computations (see [EG88], [KR]).

Let us list some other estimates for parallel complexity of arithmetic computations.

The problem of computing the coefficients $q_i = p_i a^i$ of the polynomial

$$q(y) = p(x) = \sum_i p_i x^i = \sum_i p_i a^i y^i \tag{2.1}$$

[where $y = x/a$ and where we are given a and the coefficients p_i of $p(x)$] has parallel complexity $O(\log n, n/\log n)$. (To compute all the powers a^i, apply an extension of Algorithm 2.1.) The complexity of the evaluation of the polynomial $p(x)$ of (2.1) at a single point x is $O_A(\log n, n/\log n)$ (see [BM], section 6.2, for quite tight lower and upper time bounds), and the complexity of the forward and inverse DFT's on $O(n)$ points is $O_A(\log n, n)$, with small overhead constants (see [Pease]).

The latter parallel complexity bound is fundamental: it is immediately extended to the same asymptotic complexity bound for multiplication of two polynomials, that is, for the convolution of two vectors [Problem 1.2.4a ($POL \cdot MULT$) for $s = 2$] and for the shift of the variable x [Problem 1.2.6 ($VAR \cdot SHIFT$)]; it is also easily extended to the bound $O_A(\log D, sD)$ on the complexity of the convolution of s vectors of dimensions $d_j + 1$, $j = 1, \ldots, s$, such that $D = 1 + \sum_{j=1}^{s} d_j$ [Problem 1.2.4 ($SEV \cdot POL \cdot MULT$)], and to the bound $O_A(\log^2 n, n/\log n)$ on the complexity of computing the quotient and the remainder of the division of two polynomials of degrees at most n [Problem 1.3.1 ($POL \cdot DIVIDE$)]. The latter bound has been obtained by using the Sieveking-Kung algorithm and the B-principle. In chapter 6 we will improve this bound to $O_A(\log n, n)$, provided that we compute the quotient and the remainder approximately but with an arbitrarily high output precision by means of Any Precision Approximation (APA) algorithms (see this definition in chapter 3). We also reach the bound $O_A(h \log n, (n/h)(1+2^{-h}\log^{(h)} n))$, where we exactly compute the output and where h is any value such that $1 \leq h \leq \log^* n$. Here, $\log^{(0)} n = n$, $\log^{(h)} n = \log \log^{(h-1)} n$, $h = 1, \ldots, \log^* n$, $\log^* n = \max\{h : \log^{(h)} n \geq 0\}$ ([BP]); throughout this section and, in particular, in Table 2.1, we will let $h = \log^* n$, which implies the bound $O_A(\log n \log^* n, n/\log^* n)$.

Now assume that the fast parallel polynomial multiplication and division algorithms have been incorporated into the fast algorithms of chapter 1 for Problems 1.2.2 ($POL \cdot EVAL$) and 1.2.3 ($POL \cdot INTERP$) of polynomial interpolation and multipoint evaluation. Then we arrive at the parallel complexity bounds $O_A(\log^2 n \log^* n,\ m/\log^* n)$ (for the exact m-point evaluation), $O_A(\log^2 n, m)$ (for any precision m-point approximation, APA) and similarly, $O_A(\log^2 n \log^* n, n/\log^* n)$ and $O_A(\log^2 n, n)$ (for interpolation). These bounds are extended to the estimates $O_A(\log^2 m \log^* m, m^2/\log^* m)$ and (in the APA case) $O_A(\log^2 m, m^2)$, for the complexity of the composition of two polynomials of degrees at most m [Problem 1.4.5 ($POL \cdot COMP$)]. To derive the parallel complexity bounds of Table 2.1 for Problems 1.4.6 ($SER{\cdot}COMP$) and 1.6.1 ($SER{\cdot}REVERSE$) of the composition and the reversion modulo z^{n+1} of power series, proceed as in chapter 1, but evaluate modulo z^{n+1} the required powers of $t_1(z)^i$, $i = 2, \ldots, g$, in two stages (see Problem 1.4.6 for the definition of g): first precompute all the powers $(t_1(z))^J \bmod z^{n+1}$, for $J = 2^j$, $j = 1, 2, \ldots, \lceil \log g \rceil$, by means of repeated squaring; then compute each remaining power of $t_1(z)$ modulo z^{n+1} as the product of at most $\lceil \log g \rceil$ precomputed powers, by using the algorithm for Problem 1.2.4 ($SEV \cdot POL \cdot MULT$) of polynomial multiplication. Table 2.1 also contains the estimates for approximating the zeros of an n-th degree polynomial (see chapter 7), which can be extended to the solution of Problems 2.3.1 ($EIGENVALUES$) and 2.3.3 (SVD) (see volume 2).

All the problems listed above are in NC under both arithmetic and Boolean circuit models. Euclidean algorithm — presented for computing gcd's, partial fraction decomposition, Chinese remainder computation, rational interpolation, and so on — generally requires up to n recursive steps, and this does not place the computations in NC. The parallel arithmetic time for computing the gcd and all other entries of the extended Euclidean scheme for two polynomials, however, can be decreased to $O(\log^3 n)$ by using $O(n^2/\log^2 n)$ processors; this exploits a reduction of these problems to computations with structured matrices (see chapter 2 and our comments later on in this section).

Let us briefly review the estimates for the parallel complexity of such computations, relying on our algorithms of chapter 2. We first recall the bound $O_A(\log n, rn)$ on the complexity of multiplication of an $n \times n$ Toeplitz-like or Hankel-like matrix A by a vector, provided that A is given with its F-generator of length r, F being a displacement operator of (2.11.1)-(2.11.4), (2.11.18), (2.11.19). [Indeed, this computation can be reduced to $2r$ polynomial multiplications or to $2r$ multiplications of circulant matrices by vectors, and further, to $(4r+2)$ FFT's on $O(n)$ points; we refer the reader to [GO], [GO1], [Bo], [DiFZ], [BDiF] on some further improvements]. The converse problem of computing a solution [Problem 2.2.3 ($LIN{\cdot}SOLVE$)] or a least-squares solution [Problem 2.2.8 ($L \cdot SQUARES$)] to a linear system of n equations with n unknowns and with structured coefficient matrix A, as well as computing the coefficients of $c_A(\lambda) = \det(\lambda I - A)$ [Problem 2.2.6 ($CHAR \cdot POL$)], can be solved at the cost $O_A(\log^2 n, n^2/\log n)$ if we apply Algorithms 2.11.1 and 2.11.2 and perform polynomial multiplication and division by using the cited fast parallel algorithms and their straightforward extension to the case of bivariate polynomial multiplication.

The solution can be applied or immediately extended in order to solve (at the same asymptotic parallel computational cost) Problems 2.2.4 ($DETERMINANT$) of computing $\det A$, 2.2.3a ($LIN \cdot SOLVE1$) of computing a solution to a linear system if it is consistent, and 2.2.10 ($M \cdot RANK$) of computing the matrix rank, provided that the input matrix A is the sum of a Toeplitz-like matrix and a Hankel-like matrix (see, in particular, Algorithms 2.11.1, 2.11.2 and Remark 2.11.1). If A is an $n \times n$ nonsingular Toeplitz-like matrix given with its F-generator of length r (and with its first column if $F = F^+$ or if $F = F^\pm$, or with its last column if $F = F^-$), then at the same asymptotic computational cost we may compute the $\tilde{F}$-generator of length at most r for A^{-1} (and the first column of A^{-1} if $F = F^+$ or if $F = F^\pm$, or the last column of A^{-1} if $F = F^-$), where $\tilde{F} = F_{(Z^T,Z)}$ if $F = F_{(Z,Z^T)}$, $\tilde{F} = F_{(Z,Z^T)}$ if $F = F_{(Z^T,Z)}$ [see (2.11.1), (2.11.2)], and $F = \tilde{F}$ for any operator F of (2.11.18), (2.11.19), (2.11.23) such that $F(A^{-1}) = -A^{-1}F(A)A^{-1}$ [compare (2.11.21)]. It is sufficient to consider the latter case (where $F = \tilde{F}$), due to Fact 2.11.6, and to apply the following simple modification of Algorithm 2.11.2.

Algorithm 2.2.

Input: natural n and r; an F-generator U, V of length at most r for the $n \times n$ nonsingular matrix A [where F is any operator of (2.11.18), (2.11.19), (2.11.23), so that $F(A) = UV^T$]; and one of the two vectors $A\mathbf{e}^{(0)}$ if $F = F^+$ or $F = F^\pm$, or $A\mathbf{e}^{(n-1)}$ if $F = F^-$.

Output: the first (respectively, the last) column of A^{-1} and an F-generator G, H of length at most r for A^{-1}, so that $F(A^{-1}) = GH^T$.

Computation:

1. Apply Algorithm 2.11.1 in order to compute the coefficients c_i, $i = 0, \ldots, n-1$, of the characteristic polynomial of A, an F-generator modulo λ^{n+1} and the first (or last) column of $\sum_{i=0}^n \lambda^i A^i$.
2. Compute $G = -A^{-1}U$, $H^T = V^T A^{-1}$ in the following way: first compute G_i and H_i such that $\sum_{i=0}^n \lambda^i A^i U = \sum_{i=0}^n G_i \lambda^i$, $\sum_{i=0}^n \lambda^i V^T A^i = \sum_{i=0}^n H_i^T \lambda^i$, then apply the Cayley-Hamilton theorem and set $G =$

$(1/c_n)\sum_{i=0}^{n-1} c_i G_i$, $H = -(1/c_n)\sum_{i=0}^{n-1} c_i H_i$. Compute the first (or last) column of A^{-1} as $-(1/c_n)\sum_{i=0}^{n-1} c_i \mathbf{a}_i$, where $\mathbf{a}_i$ is the first (respectively, last) column of A^i.

The computations at Stage 2 are based on Theorems 2.11.3 and 2.11.4 and Corollary 12.1, and the cost of performing Algorithm 2.2 is $O_A(\log^2 n, n^2r^2/\log n)$, which turns into $O_A(\log^2 n, n^2/\log n)$, if $r = O(1)$.

Our parallel algorithms for Toeplitz-like computations and the complexity estimate $O_A(\log^2 n, n^2/\log n)$ are then easily extended to the evaluation of the (g, h) Padé approximation, with $g + h = O(n)$ (due to Algorithm 2.5.1); of the minimum span for a linear recurrence sequence [see Problem 1.5.3 ($RECUR \cdot SPAN$) and its solutions in sections 5 of chapters 1 and 2]; and of the gcd and the lcm of two polynomials of degree n (due to Algorithms 1.5.3 and 1.5.4 or, alternatively, to Algorithm 2.9.1). We refer the reader to exercise 4d on a modification of the latter algorithms that supports the bounds $O_A(n^a \log^2 n,\ n^{2-2a}/\log n)$ on the complexity of Toeplitz and Toeplitz-like solver and related computational problems; these bounds are valid for all a, $0 < a \leq 1$.

The block LU factorization of a Bezout matrix can be computed by means of Algorithm 2.10.1 at the cost $O_A(\log^3 n, n^2/\log^2 n)$ (see Proposition 5.2). The same technique can be applied to a Hankel matrix. As a consequence, the cost of the computation of all the polynomials generated by the Euclidean scheme is $O(\log^3 n, n^2/\log^2 n)$.

In spite of this progress, we still have $tp \geq nT/\log n$, with p, t, T denoting the current record bounds (within constant factors) on the number of processors, parallel and sequential solution time, respectively (compare Definition 1.1), required in the $O(\log^2 n)$ parallel time algorithms that compute the displacement generator of the inverses of Toeplitz and Toeplitz-like matrices, as well as in their extensions listed above. The design of processor-efficient NC-algorithms for the same computational problems remains a major theoretical challenge in this research area. In sections 8 and 9, we describe some partial progress in this direction: in section 8 we will follow [P92c] to show how a well-conditioned Toeplitz-like linear system can be solved numerically in NC by using $O(n)$ processors, and in section 9 we will extend this result to the processor-efficient exact solution of this problem in RNC over the rationals (and, therefore, to the processor-efficient exact solution in RNC of some important related problems, such as the computation of the polynomial gcd over the rationals). The processor-efficient NC-algorithms

	Sequential Arithmetic Time		Parallel Arithmetic Complexity	
Problem	Lower Bounds	Upper Bounds	Time	Processors
evaluation at a point (no preconditioning)	n (±) n (*,+)	n (±) n (*,+)	log n	n/log n
evaluation at a point (with preconditioning)	n(±) (n+1)/2 (*,+)	n+2 (±) (n+2)/2 (*,+)	log n	n/log n
evaluation and interpolation at n-th roots of 1 (FFT)	c n	n log n	log n	n
evaluation at m points. Interpolation at m=n points	m log n	$m \log^2 n$	$\log^2 n$	m (APA output)
			$\log^2 n \log^* n$	$m/\log^* n$
multiplication (degrees m and n, N=m+n)	c N	N log N	log N	N
division (degrees m and n, k=m-n)	c k	k log k	log k	k (APA output)
			$\log k \log^* k$	$k/\log^* k$
gcd computation (degrees m and n, N=m+n)	c N	$N \log^2 N$	N log N	log N
			$\log^2 N$	$N^2/\log N$
computing a zero of a polynomial with error $< 2^{-b}$	n+log b	n log b log n	$\log^2 n \log b$	n
computing all the zeros with error $< 2^{-b}$	n+log b	$n^2 \log b \log n$	n log n log b	n
			$\log^3(nb)$	$(bn)^{O(1)}$
composition of two polynomials of degree n	$c n^2$	$n^2 \log^2 n$	$\log^2 n$	n^2 (APA output)
			$\log^2 n \log^* n$	$n^2/\log^* n$
composition and reversion modulo z^n	c n	$(n \log n)^{3/2}$	$\log^2 n$	$(n^3/\log n)^{1/2}$ (APA output)
			$\log^2 n \log^* n$	$(n^2/\log n)^{1/2}/\log^* n$
decomposition of a polynomial of degree N	c N	$N \log^2 N$	$\log^2 N \log^* N$	$N/\log^* N$
			$\log^2 N$	N (APA output)

Table 2.1. Arithmetic Complexity of Polynomial Computations (compare [P92a]), (c denotes constants not less than 1, upper bounds are within constant factors, except for the evaluation at a single point).

of section 9 compute the exact solution of linear systems $A\mathbf{x} = \mathbf{b}$ of n equations over the rationals, which requires to use the precision of computations of the order $n \log \|A\|_2$. This is acceptable for symbolic and algebraic computing but not for practical numerical computations where n is large.

Due to the results of chapter 2, the parallel complexity bounds for both polynomial evaluation and interpolation on n points are immediately extended to the evaluation of the products $A\mathbf{v}$ and $A^{-1}\mathbf{v}$ for an $n \times n$ Vandermonde or Cauchy (generalized Hilbert) matrix A. Furthermore, we may apply the approach of chapter 2 and reduce such computations, as well as the evaluation of the rank, the determinant and the inverse (if the determinant is not 0) of an $n \times n$ Vandermonde-like or Cauchy (Hilbert)-like matrix A to the similar computations with Toeplitz-like matrices (we leave the details to the reader). By using our algorithms for computations with structured matrices, we also arrive at the cost bounds $O_A(\log^3 n, n^2/\log^2 n)$ for computing all the polynomials generated by the extended Euclidean scheme for two polynomials of degree n (due to Algorithm 2.10.1). The sequential cost is of the order n^2 (lower and upper bounds), so our NC-solution is processor-efficient in this case.

	upper bound	time	processors
multiplication (for degrees m and n)	mn	k=log min(m,n)	mn/k
division (for degrees m and n, n≤m)	(m-n)n	m-n	n
gcd computation (for degrees m and n) (via Euclid's algorithm)	$m \log^2 m$	m log m	log m

Table 2.2. Arithmetic Cost of Performing the Straightforward Polynomial Algorithms (estimated up to within small constant factors).

Remark 2.1. The estimates of this section for the parallel complexity of computations with structured matrices and polynomials hold over any field of constants $\mathbf{F}$ that has sufficiently many elements and supports DFT. In addition, in the case of computations involving the transition from the power sums of all the zeros to the coefficients of the n-th degree polynomials, we need to assume (in order to arrive at the complexity estimates of this section) that $\mathbf{F}$ supports division by $n!$. Over an arbitrary field or ring $\mathbf{F}$, the known complexity bounds change, as we have studied in section 7 of chapter 1 and in Remark 2.3.2 and will study in section 6 of this chapter. For convenience, here is a list of the major changes of the estimated cost of the basic operations, which define the changes in the cost of other computations of this section:

a) an extra factor of $\lceil \log n / \log p \rceil$ appears in the time-complexity estimates for matrix computations involving the transition from the power sums of all the zeros of the n-th degree characteristic polynomial of a matrix to its coefficients over a field of characteristic p where $1 < p < n$ (see section 6);

b) an extra factor log log n appears in the processor bound for polynomial multiplication (Theorem 1.7.1) unless **F** supports performing DFT on a set of $O(n)$ points;

c) polynomial division has complexity $O_A(\log n \ \log\log n, n)$ over any ring with unity, **F** ([RT] and [BP]); this accommodates the influence of part b) too. If **F** supports generalized DFT (Problem 1.2.5) (that is, if an element x of **F** is available such that **F** contains its reciprocal x^{-1}, and if $1, x, x^2, \ldots, x^{N-1}$ are distinct for N of the order n), then the better bounds of [BP], cited earlier, apply [with the correction of the processor bound according to part b)]; **F** contains an element x with the above properties as long as there are at least N distinct invertible elements in **F**;

d) the algorithms that involve N distinct elements can always be performed in an extension **E** of **F**, with **E** being the field of univariate polynomials over **F** modulo an irreducible polynomial of degree $d = \lceil \log N / \log |\mathbf{F}| \rceil$, where $|\mathbf{F}|$ denotes the cardinality of **F**; the resulting change of the parallel complexity estimates amounts to extra factors $O(\log d)$ for the time bound and $O(d \log\log d)$ for the processor bound (see Theorem 1.7.1); furthermore, in the case of computations entirely reducible to operations with polynomials (and this includes many computations with structured matrices), such a change can be limited to the extra factor $O(d)$ in processor bounds since the operations with k-variate polynomials in **E** can be reduced to operations with $(k+1)$-variate polynomials in **F** having degree d in the newly added variable. In particular, in the randomized algorithms of this chapter for $n \times n$ matrix computations, we only need to choose random elements in $n^{O(1)}$-element sets. Then the extensions of degrees $d = O(\log n)$ always suffice. Moreover, $d = O(1)$ if, say, $|\mathbf{F}| > n$, which, in particular, is the case where the characteristic of **F** is zero or exceeds n.

Remark 2.2. Comparison of the complexity estimates of Tables 2.1 and 2.2 shows substantial superiority of the FFT based polynomial arithmetic over the classical one, in the case of parallel computing. Moreover, this conclusion does not change if we take into account the overhead constants omitted in both Tables 2.1 and 2.2 (these constants are small in both cases). This consideration shall give the decisive edge to the practical choice of the FFT based parallel algorithms, even if the user does not have enough motivation to apply them as sequential algorithms.

3. Parallel Complexity of Matrix Multiplication. Parallel Compact Multigrid.

The simple parallelization of the straightforward algorithm for $m \times n$ by $n \times p$ matrix multiplication supports the upper bound $O_A(\log\,(mnp), mnp/\log\,n)$ on the parallel complexity of this computational problem, which, in particular, implies the bound $O_A(\log\,n, n^2/\log\,n)$ on the complexity of multiplication of an $n \times n$ matrix by a vector.

If, however, the input matrix is sparse, say, if it has size $n \times n$ and has at most $s = O(n^\theta)$ nonzero entries per row where θ is a constant, $\theta < 1$, then its multiplication by a vector costs $O_A(\log\,s, ns/\log\,s)$. In particular, in the Compact Multigrid algorithm of section 7 of chapter 3, we dealt with $N \times N$ matrices having only $s = O(1)$ nonzero entries per row. We may exploit this observation and extend Propositions 3.7.1 and 3.7.2 as follows:

Proposition 3.1. *The smooth compressed solution to a strongly pseudo regular PDE discretized over an N-point lattice can be computed at the cost $O_B(\log\,N, N/\log\,N)$.*

Proof. The algorithm supporting Propositions 3.7.1 and 3.7.2 has stages $j = 1, \ldots, k$, $k = (\log\,N)/d$; at stage j we use $O(1)$ time and $N_j = 2^{dj}$ bit-serial processors. This leads to a parallel Compact Multigrid algorithm that surely supports the bound $O_B(\log\,N, N)$. Moreover, we may improve the processor bound by using the B-principle along the line of Example 1.1 and of the proof of Proposition 1.1. Indeed, in the parallel compact multigrid algorithm, the first $(\log\,N)/d - \log\log\,N$ stages only require $\lceil N/\log\,N \rceil$ bit-serial processors, and at each of the last $\log\log\,N$ stages, that is, the j-th stages, for $j = (\log\,N)/d - \log\log\,N + 1, \ldots, (\log\,N)/d$, we will slow down the computation so as to use time $O(2^{dj}(\log\,N)/N)$ and only $\lceil N/\log\,N \rceil$ bit-serial processors. The overall time of the resulting parallel algorithm remains $O(\log\,N)$. ∎

Let us now turn to parallel computations with general $n \times n$ matrices. Hereafter, in this chapter, $\bar{\omega}$ will denote any fixed value exceeding the exponent ω of matrix multiplication, $2 \le \omega < 2.376$ [see Problem 2.2.2 $(M \cdot MULTIPLY)$].

Proposition 3.2 ([P87]). *The product UV of $n \times n$ matrices U and V can be computed at the arithmetic cost $O_A(\log n, n^{\bar{\omega}})$ and, if the matrices U and V are filled with integers, then also at the Boolean cost $O_B(\log (dn), n^{\bar{\omega}}\mu(d)/\log d)$ where $\mu(d)$ is defined by (3.1.1), and*

$$d = O(\log (n||U||||V||)). \tag{3.1}$$

We will next recall some properties of bilinear algorithms, which we will use in the proof of Proposition 3.2 and in chapter 5.

Definition 3.1, *bilinear algorithms for matrix multiplication.* Given a pair of $m \times n$ and $n \times p$ matrices $X = [x_{ij}]$, $Y = [y_{jk}]$, compute XY in the following order: First evaluate the linear forms in the x-variables and in the y-variables,

$$L_q = \sum_{i,j} f_{ijq}x_{ij}, \quad L'_q = \sum_{j,k} f^*_{jkq}y_{jk}, \tag{3.2}$$

then the products $P_q = L_q L'_q$ for $q = 0, 1, \ldots, M-1$, and finally the entries $\sum_j x_{ij}y_{jk}$ of XY, as the linear combinations

$$\sum_j x_{ij}y_{jk} = \sum_{q=0}^{M-1} f^{**}_{kiq}L_q L'_q, \tag{3.3}$$

where f_{ijq}, f^*_{jkq} and f^{**}_{kiq} are constants such that (3.2) and (3.3) are the identities in the indeterminates x_{ij}, y_{jk}, for $i = 0, 1, \ldots, m-1$; $j = 0, 1, \ldots, n-1$; $k = 0, 1, \ldots, p-1$. M, the total number of all multiplications of L_q by L'_q, is called the *rank of the algorithm*, and the multiplications of L_q by L'_q are called the *bilinear steps* of the algorithm.

Proof of Proposition 3.2. We may assume that $n = s^h$ for a fixed but sufficiently large s and for $h \to \infty$, and that there exists a basis bilinear algorithm of rank $M \le s^{\bar{\omega}}$ for computing the product XY of a pair of $s \times s$ matrices, $X = [x_{ij}]$ and $Y = [y_{jk}]$ (see Theorem 5.3.1 in chapter 5; or [BM], sect. 2.5; or [P84b]). Furthermore, it is possible to assume, without loss of generality, that the values f_{ijq}, f^*_{jkq} and f^{**}_{kiq} in (3.2), (3.3) are rational for all i, j, k and q (see [P84b], Corollary 5.6, p. 26).

The basic bilinear algorithm (3.2), (3.3) can be applied where all the x_{ij}, y_{jk} and, consequently, all the L_q and L'_q denote $s \times s$ matrices for a natural s; then we will recursively apply this basic bilinear algorithm with successive substitutions of $s^g \times s^g$ matrices for x_{ij} and y_{jk} (for all i, j and k), where one needs to multiply a pair of $n \times n$ matrices with $n = s^{g+1}$, $g =$

$h-1, h-2, \ldots, 1$. This defines a recursive bilinear algorithm for computing UV, and we have that

$$t_{A,g+1} \le t_{A,g} + 2\log s + \log M + 4,$$
$$p_{A,g+1} \le \max\{2s^{2g+2},\ Ms^{2g},\ p_{A,g}M\},$$

where $t_{A,l}$ and $p_{A,l}$ denote the arithmetic parallel time and the number of processors used in the above recursive bilinear algorithm for $s^l \times s^l$ matrix multiplication. Since $M \le s^{\bar{\omega}}$, this immediately implies the complexity bound $O_A(\log n, n^{\bar{\omega}})$ for the $n \times n$ matrix multiplication.

Next, let us scale (3.2) and (3.3) so as to turn all the rational constants f_{ijq}, f^*_{jkq} and f^{**}_{kjq} into integers, that is, let us evaluate $Q\sum_j x_{ij}y_{jk}$ (rather than $\sum_j x_{ij}y_{jk}$) for all i and k where Q is a common multiple (say, the product) of all the denominators of f_{ijq}, f^*_{jkq} and f^{**}_{kiq}, for all i, j, k and q. In this case, recursive application of the basis bilinear algorithm will output the product of two given $s^h \times s^h$ matrices and of the scalar multiplier Q^h; the subsequent division of the output by this multiplier (one arithmetic step, n^2 processors) will give us the desired matrix product.

We will not need to change the integer constants Qf_{ijq}, Qf^*_{jkq} and Qf^{**}_{kiq} as $h \to \infty$, so that each multiplication by such a constant can be replaced by $O(1)$ additions/subtractions. Then all the arithmetic steps of the recursive algorithm turn into additions/subtractions, except for the single bilinear multiplication step (which multiplies L_q by L'_q) and except for the single step of the division by Q^h. Now, Proposition 3.1 immediately follows from the estimates of Table 1.1 and Fact 2.1, except that we need to show that the d-bit precision of computation will suffice for d of (3.1). The latter fact follows from (3.2) and (3.3), because the moduli of all the input values, other than the constants f_{ijq}, f^*_{jkq} and f^{**}_{kiq}, cannot exceed $||U||$ and $||V||$ and because our scaling may increase the magnitudes of the values involved in the recursive bilinear algorithm by at most the factor $|Q|^h < C^h$ for a constant C, which means adding at most $\log(|Q|^h) = O(h) = O(\log n)$ bits to the precision of the computation. The complexity of the divisions by Q^h is small enough, because Q is a positive integer constant. ∎

Remark 3.1. The straightforward application of the estimates of Table 1.1 and of Fact 2.1 to the arithmetic time bound $O_A(\log n)$ only implies the inferior Boolean time bound $O_B(\log d \log n)$.

Remark 3.2. Multiplications of L_q by L'_q can be performed concurrently for all q, in a single parallel multiplication step. Similarly, multiplications by all the coefficients f_{ijq} and f^*_{jkq} in (3.2) can be performed concurrently, in a single parallel multiplication step [which may also be replaced by $O(1)$ addition/subtraction steps, for the fixed integers f_{ijq} and f^*_{jkq}], and the same can be said about f^{**}_{kiq} in (3.3). Thus, if $n = s^g$, then in all the g calls to the basis algorithm, $2g+1$ parallel multiplication steps suffice, even if we do not replace multiplications by f_{ijq}, f^*_{jkq}, f^{**}_{kiq} by additions/subtractions.

Remark 3.3. If $M = s^\omega$ for some fixed basis bilinear algorithm for the $s \times s$ matrix multiplication, then $n^{\bar{\omega}}$ can be replaced by $n^\omega/\log n$ in the statement of Proposition 3.2 (compare exercise 7).

Hereafter let $P(n)$ denote any processor bound such that the bounds $O_A(\log n, P(n))$ hold for the parallel complexity of $n \times n$ matrix multiplication. Clearly,

$$n^2/P(n) = O(\log n) \ , \tag{3.4}$$

whereas Proposition 3.2, applied for any $\bar{\omega} > \omega$, in particular, for $\bar{\omega} = 2.376$, implies that we may set

$$P(n) = O(n^{\bar{\omega}}) \ . \tag{3.5}$$

Propositions 2.2.1 and 3.2 together imply

Corollary 3.1. *For an $n \times n$ matrix A and an n-dimensional vector* $\mathbf{v}$, *the arithmetic complexity of computing the Krylov matrix $K(A, \mathbf{v}, m)$ [see Problem 2.2.1a (KRYLOV)] is $O_A(\log^2 n, P(n))$, for $m = O(n)$, $P(n) < n^{2.376}$.*

4. Parallel Algorithms for the Inverse and the Characteristic Polynomial of a Matrix (with Applications).

In this and next sections, we will review the best NC-algorithms (see Definition 1.2) available for the general matrix computations, as well as the techniques for designing these algorithms. We believe that some of these techniques are themselves interesting, and the resulting asymptotic complexity estimates (which have been extended to many combinatorial computations too) are of theoretical importance; furthermore, some randomized NC-algorithms for the large scale computations with unstructured matrices seem to be of some potential practical value in the area of algebraic computation with infinite precision.

We will utilize some techniques developed in chapter 2. In particular, we will cite the following extension of Theorem 2.1.3:

Theorem 4.1 ([KaSi]). *Given a family of arithmetic circuits of size s and depth d that compute a multivariate rational function R, it is possible* *to compute all the first partial derivatives of R on arithmetic circuits having depth $O(d)$ and size at most $4s$.*

Remark 4.1. We will only need Theorem 4.1 in order to extend parallel algorithms for Problem 2.2.4 (*DETERMINANT*) to the solution of Problem 2.2.5 (*INVERT*). Due to Theorem 4.1, such extensions do not change the asymptotic bounds on the depth and size of the associated arithmetic circuits. Under the arithmetic PRAM models, the techniques of [KaSi] give us similar extensions, with the asymptotic time bounds unchanged in the transition from Problem 2.2.4 (*DETERMINANT*) to Problem 2.2.5 (*INVERT*) and with the processor bound increasing by the factor $O(\log^{\alpha} n)$, for $\alpha \leq 2$, in the case of an $n \times n$ input matrix. We will accommodate the latter change in (4.6). The reader may alternatively restate all our estimates by using the circuit models and setting the above α to equal 0 (see the first paragraph of the "PRAM models" subsection of section 1).

Let us now begin our review. For the inversion of a triangular matrix A, we may first apply the factorization shown in Remark 2.2.1 and then apply Proposition 1.1 in order to invert A (over any field of constants) at the cost

$$O_A(\log^2 n, P(n)/\log n). \tag{4.1}$$

This processor-efficient NC-algorithm is actually due to Heller, 1973 (see [BM], page 146). Recall, however, that the triangular linear system $A\mathbf{x} = \mathbf{b}$ can be solved in $O(n^2)$ ops, so that devising a processor-efficient NC-algorithm for such a system remains an important open problem (compare exercise 8b).

We may invert a general matrix A by means of computing the balanced recursive factorization (2.2.1)-(2.2.3), but this requires n parallel arithmetic steps, since computing the Schur complement S of (2.2.3) must precede the next recursive step. Similar problems arise for a parallel version of the LSP factorization algorithm of chapter 2. We arrive at NC-algorithms by applying Proposition 2.2.1 and the reductions that support Proposition 2.2.2.

Proposition 4.1. *There exist NC-algorithms for Problems 2.2.3a ($LIN \cdot SOLVE1$), 2.2.3–2.2.6 ($LIN \cdot SOLVE$, $DETERMINANT$, $INVERT$, $CHAR \cdot POL$), 2.2.8 ($L \cdot SQUARES$), 2.2.10 ($M \cdot RANK$) and 2.2.12 ($GEN \cdot INVERSE$) over any field of constants of characteristic 0, over which these problems are defined. Furthermore, for Problems 2.2.3–2.2.6, the algorithms also work over any field of characteristic greater than* *n. In all cases, for the above computational problems, the algorithms yield the cost bounds*

$$O_A(\log^2 n, nP(n)) . \tag{4.2}$$

Proof ([Cs]). Due to Remark 2.6.1 and since we work over the fields allowing division by $n!$, Problem 2.2.6 ($CHAR \cdot POL$) of computing the coefficients $c_0, \dots, c_{n-1}$ of the characteristic polynomial of A, $\lambda^n + \sum_{i=0}^{n-1} c_i \lambda^i$, reduces to computing the traces of the first n powers of A [which is in turn reduced to Problem 2.2.2a ($M \cdot POWERS$)]; the cost of this reduction is $O_A(\log^2 n, n/\log n)$. This is surely within the bound (4.2), which bounds the cost of the solution of Problem 2.2.2a ($M \cdot POWERS$). Knowing $c_0, \dots, c_{n-1}$, we immediately obtain $\det(A) = (-1)^n c_0$ and also arrive at A^+ by applying (2.2.9), which takes the form $A^{-1} = -\sum_{i=1}^{n}(c_i/c_0)A^{i-1}$ if A is nonsingular (compare Theorem 2.1.1). The reductions of computational problems shown in section 2 of chapter 2 lead to the extension of the estimates (4.2) to other problems. (In particular, we use the immediate extension from $L \cdot SQUARES$ to $LIN \cdot SOLVE1$.) ∎

[Datta] extends (4.2) to Problem 2.2.7 ($LU \cdot FACTORS$) for an h.p.d. input matrix (over the fields of characteristic 0 or greater than n) and to Problem 2.2.9 ($QR \cdot FACTORS$) over the subfields of the complex field (see Proposition 5.2).

Proposition 4.2 ([PSa], [GP89]). *It is possible to improve the bounds (4.2) for Problems 2.2.3–2.2.6 ($LIN \cdot SOLVE$, $DETERMINANT$, $INVERT$, $CHAR \cdot POL$), 2.2.8 ($L \cdot SQUARES$) to the bounds*

$$O_A(\log^2 n, \tilde{P}(n)) , \tag{4.3}$$

provided that $\tilde{P}(n) = \max\{P(n^{1.25}, n, n^{1.25}), P(n^{0.5}, n^2, n^{0.5})\}$ and that the bound $O_A(\log n, P(m,n,p))$ holds for the parallel complexity of $m \times n$ by $n \times p$ matrix multiplication. Furthermore, one may set $\tilde{P}(n) \le n^\beta$ for a positive β such that $\beta < \omega + 0.5 < 2.851$.

Proposition 4.2 is obtained by combining Theorem 4.1 with the algorithm of [GP89].

The algorithms supporting Propositions 4.1 and 4.2 are in NC but are not processor-efficient. In sections 7 and 8, we will improve them to processor-efficient NC-algorithms for Problems 2.2.3-2.2.6, assuming that the input matrix is well-conditioned or using Boolean PRAM and randomization of Las Vegas type. Furthermore, we have

Proposition 4.3 ([P92b], [KP]). *There exists a (Las Vegas) randomized processor-efficient algorithm that, within the cost bounds*

$$O_A(\log^2 n, P^*(n)) , \tag{4.4}$$

$$P^*(n) = \max\{P(n), n^2 \log\log n/\log n\} \tag{4.5}$$

[compare (3.4), (3.5)], solves Problems 2.2.3 ($LIN \cdot SOLVE$), 2.2.4 ($DETERMINANT$) and 2.2.6a ($MIN \cdot POL$) over the fields of constants of characteristic 0, and also solves Problems 2.2.3 ($LIN \cdot SOLVE$) and 2.2.4 ($DETERMINANT$) over the fields of characteristic greater than n.

Proof. To devise the desired algorithm for $LIN \cdot SOLVE$ and $DETERMINANT$, we first apply the algorithm of the proof of Proposition 2.2.2 that supports the randomized reduction

$$(RECUR \cdot SPAN, M \cdot VECTOR, KRYLOV) \succeq MIN \cdot POL$$

of Problem 2.2.6a ($MIN \cdot POL$) of computing the minimum polynomial of a matrix A to Problems 1.5.3 ($RECUR \cdot SPAN$) of computing the minimum span of a linear recurrence sequence, 2.2.1a ($KRYLOV$), and 2.2.1 ($M \cdot VECTOR$).

The latter two problems can surely be solved within the cost bounds (4.4). It remains to deduce (4.4) for Problem 1.5.3 ($RECUR \cdot SPAN$). In chapter 1, its solution was reduced to Problem 1.5.2b ($PADÉ$), which Algorithm 2.5.1 further reduced to the evaluation of the rank r of an $n \times n$ auxiliary Toeplitz matrix T and to solving a nonsingular Toeplitz linear system of r equations.

The latter step can be performed within the cost bounds (4.4) by means of Algorithm 2.11.2, which also computes the rank r over the fields having characteristic 0. ∎

For the purpose of solving Problems 2.2.3 ($LIN \cdot SOLVE$) and 2.2.4 ($DETERMINANT$) for a nonsingular input matrix A, we may follow [KP] and, *over any field*, avoid the computation of the rank r of the auxiliary matrix T by just ensuring (with a high probability) that this rank equals n or, equivalently, that $\det T \neq 0$ (and the nonsingularity of T can be verified by means of Algorithm 2.11.2).

Indeed, from section 5 of chapter 2, we know that the matrix T is nonsingular if $m_A(\lambda) = c_A(\lambda)$ for the input matrix A (see Algorithm 2.5.1), and with a high probability we may ensure the latter equation by applying

Procedure 1.7.1 and Regularizations 2.13.1–2.13.5 unless A is singular. For instance, we may apply Regularizations 2.13.3 and 2.13.4 and, instead of working with the input matrix A, solve Problems 2.2.3 ($LIN \cdot SOLVE$) and 2.2.4 ($DETERMINANT$) for the matrix $\widetilde{A} = AHY$, where $Y = \mathrm{diag}(y_1, \ldots, y_n)$, $H = [h_{i,j}]$, $h_{i,j} = h_{i \pm j}$, H is an $n \times n$ Hankel or Toeplitz matrix, and the entries y_g and h_k are random elements chosen independently in a sufficiently large subset of the field of constants $\mathbf{F}$ or of its extension [it suffices to use subsets of cardinality $n^{O(1)}$ (see Remark 2.1d)].

Then, on one hand, $m_{\widetilde{A}}(\lambda) = c_{\widetilde{A}}(\lambda)$ with a high probability, unless $\det A = 0$, and on the other hand, the solutions to Problems 2.2.3 ($LIN \cdot SOLVE$) and 2.2.4 ($DETERMINANT$) for the input A can be immediately recovered from the solutions for the input $\widetilde{A}$. Finally, if Algorithm 2.11.2 detects that $\det \widetilde{T} = 0$ and hence $m_{\widetilde{A}}(\lambda) \neq c_{\widetilde{A}}(\lambda)$, we conclude that $\det \widetilde{A} = \det A = 0$ with a high probability, which completes the randomized solutions of Problems 2.2.3 and 2.2.4. Thus, since randomization was already needed for the reduction from Problem 1.5.3 ($RECUR \cdot SPAN$) anyway, we may, with no loss of generality, assume that

$$m_A(\lambda) = c_A(\lambda)$$

and avoid computing the rank of the auxiliary matrix.

Theorem 4.1 and Remark 4.1 enable us to extend the result for $\det A$ to computing A^{-1}. (See a distinct approach in Remark C.3 of Appendix C).

Corollary 4.1. *Problem 2.2.5 ($INVERT$) can be solved at the cost*

$$O_A \left(\log^2 n,\ P^*(n) \log^\alpha n \right), \qquad \alpha \le 2. \tag{4.6}$$

Remark 4.2. The above algorithm only involves divisions by $2, 3 \ldots, n$ (at the stage of the solution of Newton's identities), by the determinants of the Toeplitz matrix of the coefficients of the auxiliary linear system (in the solution of $RECUR \cdot SPAN$) and, for Problems 2.2.3 ($LIN \cdot SOLVE$) and 2.2.5 ($INVERT$), by $\det A$.

Similarly to the proof of Proposition 4.1, we now use the techniques of section 2 of chapter 2 to extend the randomized complexity bounds (4.4) and (4.6) to several other problems of matrix computations. In particular, we extend (2.2.9) to the computation of $A^+\mathbf{b}$ provided that A, $\mathbf{b}$ and $m_A(\lambda)$ are available, and we also solve Problem 2.2.10 ($M \cdot RANK$): we compute $\mathrm{rank}\, A = \mathrm{rank}\, \widetilde{A}$ from the coefficients of $c_{\widetilde{A}}(\lambda)$ for the matrix $\widetilde{A}$ defined above. In both cases we arrive at (4.4) (over the fields of characteristic 0 or greater than n). Then we extend this computation to randomized algorithms supporting (4.4) for Problems 2.2.10b ($M \cdot PRECNDTN$) and 2.2.3a ($LIN \cdot SOLVE1$) and supporting (4.6) for Problem 2.2.3b ($NULL \cdot SPACE$). Finally, for Problems 2.2.7 ($LU \cdot FACTORS$) and 2.2.9 ($QR \cdot FACTORS$) of triangular and orthogonal factorization, we arrive at processor-efficient and (Las Vegas) randomized solutions in NC^3 (see the next section) by using the above matrix inversion algorithm in order to invert the northwestern blocks [denoted B in (2.2.1)-(2.2.3)] in the recursive evaluation of the recursive factorization (2.2.1)-(2.2.3).

We also recall the output-sensitive approach of section 13 of chapter 2 to the solution of Problem 2.2.10 ($M \cdot RANK$) and immediately extend (2.13.6), (2.13.7) to the following randomized and output-sensitive parallel complexity estimate (over the fields of characteristic 0 or greater than n):

$$O_A(\log n + \log {}^3 r, (n^2 \log n + (\log {}^3 r) P^*(r))/(\log n + \log {}^3 r)), \qquad (4.7)$$

for an $n \times n$ input matrix of rank r. The reader may further extend this to the output-sensitive estimates for Problems 2.2.3a, 2.2.3b and 2.2.10b.

We conclude this section by recalling the following result of [CF] where Problem 2.2.3 ($LIN \cdot SOLVE$), for a well-conditioned input matrix of coefficients, is approximatively solved by a Monte Carlo algorithm in $O(\log n)$ parallel arithmetic steps with a polynomial number of processors:

Theorem 4.2. *Let A be an $n \times n$ matrix such that $||A||_\infty \le 1 - c/n^k$, $c > 0$, let $\mathbf{b}$ be a vector of dimension n, and let δ, $\epsilon < 1$ be positive constants. Then, an approximate solution $\hat{\mathbf{x}}$ of the system $\mathbf{x} = A\mathbf{x} + \mathbf{b}$ satisfying the inequality*

$$\text{probability}(||\mathbf{x} - \hat{\mathbf{x}}||_\infty / ||\mathbf{x}||_\infty < \epsilon) > 1 - \delta$$

can be computed at the cost $O_A(\log n, n^{3k+3}\epsilon^{-1}(\log \epsilon^{-1})/\delta)$ by means of a Monte Carlo randomized algorithm. Moreover, we may relax the condition $||A||_\infty \le 1 - c/n^k$, still keeping the same time bound with a polynomial number of processors, if $\log(1/\epsilon) = O(\log n)$ and if $\log \operatorname{cond}(I - A) = O(\log n)$.

Remark 4.3. The processor bound of Theorem 4.2 is exponential in n if we require approximation with a precision less than 2^{-n} (say), but such a bound may still help us to spare some unreasonable attempts of proving lower bounds such as $\log {}^2 n$ on the randomized parallel time for approximating the solution to a linear system.

5. Extensions of Parallel Matrix Inversion.

We apply the matrix inversion algorithms of the previous section in order to invert all the northwestern blocks B in the recursive factorization (2.2.1)–(2.2.3) and arrive at the following results:

Proposition 5.1 ([P87]). *Let there exist the LU factorization of a fixed $n \times n$ matrix A. Then Problem 2.2.7 ($LU \cdot FACTORS$) for this matrix A can be solved over the fields of characteristic 0 or greater than n, both deterministically, at the cost*

$$O_A(\log^3 n,\ \tilde{P}(n)/\log n)\ , \tag{5.1}$$

for $\tilde{P}(n)$ of (4.3), $\tilde{P}(n) \leq n^\beta$, $\beta < \omega + 0.5$, $\beta < 2.851$, and by a randomized (Las Vegas) algorithm, at the cost

$$O_A(\log^3 n, P^*(n)\log^{\alpha-1} n)\ , \tag{5.2}$$

where $P^*(n)$ is defined by (4.5), $P^*(n) < n^{2.376}$, and $\alpha \leq 2$ is defined in Remark 4.1.

Proof. Recall from exercise 4 of chapter 2 that there exists the (balanced) recursive factorization (2.2.1) if and only if there exists the LU factorization of A, and, furthermore, that there exists an LU factorization of a matrix A with nonsingular factors L and U if and only if A is strongly nonsingular. Compute the recursive factorization (2.2.1) and recover an LDM-factorization of A. Deduce (5.1) and (5.2) by combining the bounds (4.3) and (4.6) on the complexity of matrix inversion and Proposition 3.2 and by using the B-principle, as in the proof of Proposition 1.1. ∎

We refer the reader to Procedure 1.7.1 and Regularizations 2.13.1–2.13.3 on randomized transition from a nonsingular matrix A to a strongly nonsingular $V = BAC$.

Proposition 5.2. *For $m = O(n)$ and for an $m \times n$ matrix A (of full rank), Problem 2.2.9 ($QR \cdot FACTORS$) of orthogonal factorization of A can be solved at the deterministic cost bounded by (4.2) and (5.1) as well as at the (Las Vegas) probabilistic cost bounded by (5.2) (in both cases increased by the cost of $n \times m$ by $m \times n$ matrix multiplication for larger m).*

Proof. The bound (4.2) is due to [Datta]. Other bounds of Proposition 5.2 follow from (2.2.6), (2.2.7), Fact 2.2.1 and Proposition 5.1, since $A^H A$ (or AA^H if $m < n$) is a Hermitian positive definite (h.p.d.) matrix so that there exists its recursive factorization of the form (2.2.1). ∎

Corollary 5.1 ([P87]). *For an $n \times n$ matrix A and a vector $\mathbf{v}$ of dimension n, let the Krylov matrix $K(A, \mathbf{v})$ be nonsingular. Then Problem 2.3.2 ($H \cdot REDUCE$) of Hessenberg reduction for such a matrix A can be solved at the deterministic cost bounded by (4.2) and (5.1) as well as at the (Las Vegas) probabilistic cost (5.2).*

Corollary 5.1 is immediately implied by Corollary 3.1 and Proposition 5.2 and by the following auxiliary result:

Fact 5.1. *If the matrix $K(A, \mathbf{v})$ is nonsingular, $K(A, \mathbf{v}) = QR$, $Q^H Q = I$, and the matrix R is upper triangular, then $Q^H AQ = H$, where the matrix H is in the upper Hessenberg form.*

Proof [compare the relations (2.3.7); [Par], p. 253, or [GL], p. 368]. Let $Q = [\mathbf{q}_1, \mathbf{q}_2, \ldots, \mathbf{q}_n]$. Recall that Q is orthogonal, R is upper triangular, and $Q = K(A, \mathbf{v})R^{-1}$. It follows that the vector $\mathbf{q}_i$ is orthogonal to any linear combination of the first $i-1$ columns of $K(A, \mathbf{v})R^{-1}$ and, consequently, to the first $i-2$ columns of $AK(A, \mathbf{v})R^{-1} = AQ$, for $i = 3, 4, \ldots, n$. Therefore, $\mathbf{q}_i^H A\mathbf{q}_j = 0$ if $i > j+1$. ∎

Our solution of Problems 2.3.2a ($T \cdot REDUCE$) and 2.3.2b ($T \cdot REDUCE1$) given by Algorithm 2.10.2 is more involved and leads us to improved asymptotic cost bounds.

Proposition 5.3. *Algorithm 2.10.1 for computing the block triangular LDL^T factorization of the matrix $JB(u, v)J$ can be carried out at the cost bounded by $O_A(\log^3 n, n^2/\log^2 n)$, provided that $B(u, v)$ is the $n \times n$ Bezout matrix defined by the two polynomials $u(x)$ and $v(x)$ and that J is the reversion matrix of Definition 2.4.1. The same complexity bound holds for the block triangular factorization of an $n \times n$ Bezout or Hankel matrix.*

Proof. The algorithm is carried out in $O(\log n)$ steps. At the general recursive step i, we first compute the rank of a Hankel-like matrix. This is performed at Stage 1 of Algorithm 2.10.1 at the cost $O_A(\log^2 n, n^2/\log n)$, by means of Algorithm 2.11.1 (see Remark 2.11.1) applied to the Hankel-like matrix $JB(u, v)J$. Stage 2 consists of

a) computing the inverse of the $k \times k$ Hankel-like matrix Y_{m_i};
b) computing the $(n-k) \times k$ Hankel-like matrix $GY_{m_i}^{-1}$ and the $(n-k) \times (n-k)$ Hankel-like matrix S_{m_i};
c) computing the polynomials $r_i(x)$ and $r_{i+1}(x)$ associated with the matrix S_{m_i};
d) computing the polynomials $u^{(i)}(x)$ and $v^{(i)}(x)$ such that $H(u^{(i)}(x), v^{(i)}(x)) = (H(u, v))_{m_i}$.

The cost of computing F-generators of the Hankel-like matrices in parts a) and b) is dominated by the cost of solving a Hankel linear system, i.e., $O_A(\log^2 n,\ n^2/\log n)$. Parts c) and d) are reduced to solving a triangular Toeplitz linear system (compare Remark 2.9.1) and to solving a Hankel linear system [compare Proposition 2.10.5 and (2.9.6a)].

The cost of performing Stage 4 is $O_A(\log n, n^2)$. Stage 3 recursively applies the same computations to submatrices of smaller sizes. Thus, the overall cost of computing the matrices L and D is $O_A(\log^3 n, n^2/\log^2 n)$, due to the application of the B-principle (similarly to the proof of Proposition 1.1). ∎

Due to the correlation between the Euclidean scheme and the block LDL^T factorization of the matrix $JB(u,v)J$, analyzed in section 10 of chapter 2, we have the following result:

Corollary 5.2. *The coefficients of the polynomials generated by the Euclidean scheme applied to two polynomials $u(x)$, $v(x)$ of degrees n, m, respectively, $n > m$, can be computed at the cost $O_A(\log^3 n, n^2/\log n)$.*

Proposition 5.4. *Randomized Algorithm 2.10.2, for the reduction of an $n \times n$ matrix to its block tridiagonal form, can be carried out at the cost*

$$O_A(\log^3 n, P^*(n)/\log n) \text{ or } O_A(\log^2 n, n^3/\log n).$$

Proof. The cost of the Krylov sequence computation at Stage 2 is $O_A(\log^2 n,\ P^*(n))$. The subsequent computation of the scalars $\mathbf{p}^H A^i \mathbf{q}$, $i = 0, \ldots, 2n$, at Stage 3 can be performed at the cost $O_A(\log n, n^2)$. The complexity of Stage 4 is dominated by the application of Algorithm 2.10.1 at the cost $O_A(\log^3 n, n^2/\log^2 n)$ or $O_A(\log^2 n, n^3/\log n)$. At the Stage 5, the product $H^{(1)}L^{-T}$ can be computed at the cost $O_A(\log n, n^2)$, since $H^{(1)}$ is a Hankel matrix. Moreover, the entries along the three main diagonals of $L^{-1}H^{(1)}L^{-T}$ are computed by performing $n \times n$ matrix multiplications at the cost $O_A(\log n, P(n))$. Summarizing and using the B-principle, we obtain the estimates of Proposition 5.4. ∎

As we have already pointed out in chapter 3, all the above reductions of Problems 2.3.2 ($H \cdot REDUCE$), 2.3.2a ($T \cdot REDUCE$) and 2.3.2b ($T \cdot REDUCE1$) can be considered as auxiliary rational steps of generally nonrational computation of the eigendecomposition (2.1.1) and the SVD of (2.1.2), (2.1.3) for a matrix A. Let us specify the other rational steps in the approximate evaluation of the SVD and complement them by a nonrational step performed by means of the algorithm of [BP91] (see chapter 8).

Algorithm 5.1, for Problem 2.3.3 (SVD).

Input: natural m, n and the entries a_{ij} of an $m \times n$ matrix $A = [a_{ij}]$.

Output: approximations to the entries of the matrices U, Σ and V defining $A = U\Sigma V^H$, the SVD of A.

Computation: choose a random vector $\mathbf{v}$ and compute the matrix $B = A^H A$ and then the Krylov matrix $K = K(B, \mathbf{v})$. Then compute the orthogonal factorization of K to test if the matrix K is nonsingular.

a) If so, reduce B to the tridiagonal form (apply Corollary 5.1 and Fact 5.1 and note that the Hessenberg reduction keeps B Hermitian and therefore makes it tridiagonal). Then apply an algorithm of [BP91] (which we will recall in volume 2) and approximate at first the eigendecomposition of B, $B = A^H A = V\Sigma^2 V^H$, $V^H V = I$, and then the SVD of A, $A = U\Sigma V^H$, $U^H U = I$, $U = AV\Sigma^{-1}$.

b) If $K = K(B, \mathbf{v})$ is singular, apply a little less efficient algorithm: first compute the characteristic polynomial $\det(\lambda I - B)$, then approximate its zeros, which are the eigenvalues of B (by means of the same algorithm of [BP91]), and finally approximate the eigendecomposition of B (given the eigenvalues of B) and the SVD of A. ■

Instead of the characteristic polynomial of B, we could have computed the minimum polynomial of B. Its zeros would have given us all the eigenvalues of B but, generally, not their multiplicity, whose computation would have been an additional problem.

6. Extension to Computations over Any Field of Constants.

The parallel arithmetic complexity estimates of section 3 surely hold over any field of constants. Let us next review the extension of the results of sections 4 and 5 to the case of any field, and then we will similarly extend the results of section 2 (also see Appendix C on further, more recent progress).

In Remarks 2.3.2, 2.5.4 and 2.6.1, we showed some difficulties of such an extension, and now we will show how to overcome these difficulties with no great increase of the complexity estimates.

We will not need to consider Problems 2.2.1 ($M \cdot VECTOR$), 2.2.1a ($KRYLOV$), 2.2.2 ($M \cdot MULTIPLY$) and 2.2.2a ($M \cdot POWERS$) (whose solution was obtained over any field), as well as Problems 2.2.8 ($L \cdot SQUARES$), 2.2.9 ($QR \cdot FACTORS$) and 2.2.12 ($GEN \cdot INVERSE$) (whose solution has no value over the fields of positive characteristics, as we observed in Remark 2.3.2). Thus, we will next consider the remaining problems of section 2 of chapter 2 and of chapter 1, partitioning our presentation into two main parts, where we cover computations with general matrices and with dense structured matrices (also applied to computations with polynomials), respectively, and we will further partition each part.

Computations with general matrices. $LIN \cdot SOLVE$ and $DETERMINANT$.

The algorithms of [GP89] and of our section 4 support the estimates (4.3) (deterministic) for Problems 2.2.3–2.2.6 ($LIN \cdot SOLVE$, $DETERMINANT$, $INVERT$, $CHAR \cdot POL$) and (4.4) (randomized) for Problems 2.2.3–2.2.5 and 2.2.6a ($MIN \cdot POL$) over the fields of characteristics 0 or greater than n (allowing division by $n!$), but they do involve divisions by $2, 3, \ldots, n$ at the stage of the transition from the power sums s_k of the eigenvalues of the input matrix A (equal to the traces of the powers of A) to the characteristic polynomial of A. Therefore, they do not apply, say, over the rings of integers modulo m, for $m \leq n$.

Two alternative deterministic algorithms, one due to Berkowitz [Ber] and based on the earlier algorithm by Samuelson ([FF]) and another due to Chistov [Chi], support the following result:

Proposition 6.1. *The deterministic estimate (4.2) of Proposition 4.1 for the complexity of Problems 2.2.3–2.2.6 holds over any field of constants.*

Proof. Let us next describe *Chistov's algorithm* and estimate its parallel cost. Let $(A)_{i,j}$ denote the entry (i,j) of an $n \times n$ matrix A, A_k the $k \times k$ leading principal submatrix of A, I_k the $k \times k$ identity matrix, and $I = I_n$. Then Fact 3.2.1 implies that $1/\det A = \prod_{k=1}^{n}(A_k^{-1})_{k,k}$ and, similarly, $1/\det(I - \lambda A) = \prod_{k=1}^{n}((I_k - \lambda A_k)^{-1})_{k,k}$. On the other hand, for any $k \times k$ matrix B, we have: $(I_k - \lambda B)^{-1} = \sum_{i=0}^{\infty}(\lambda B)^i = \prod_{h=0}^{\infty}(I_k + (\lambda B)^{2^h})$. Therefore,

$$s(\lambda) = 1/\det(I - \lambda A) =$$

$$\prod_{k=1}^{n}((I_k - \lambda A_k)^{-1})_{k,k} = \prod_{k=1}^{n} \prod_{h=0}^{\lceil \log n \rceil} (I_k + (\lambda A_k)^{2^h})_{k,k} \bmod \lambda^{n+1}.$$

Thus, we may evaluate the powers $(A_k)^{2^h}$, $h = 0, 1, \ldots, \lceil \log n \rceil$; $k = 1, \ldots, n$, at the overall cost $O_A(\log^2 n, n\,P(n))$ of (4.2), then compute the product modulo λ^{n+1} of all the binomials $(I_k + (\lambda A_k)^{2^h})_{k,k}$ for all k and h, and finally recover the coefficients of $\det(I - \lambda A) = s(\lambda)^{-1} \bmod \lambda^{n+1}$. The $n\lceil \log n \rceil$ multiplications of polynomials modulo λ^{n+1} cost $O_A(\log^2 n, n^3/\log n)$, even if we use the straightforward algorithm for polynomial multiplication; this also bounds the cost of the inversion modulo λ^{n+1} of the power series $s(\lambda)$. Thus, we arrive at the overall complexity bound $O_A(\log^2 n, nP(n))$ for computing the characteristic polynomial of A (Problem 2.2.6) and, consequently, for the solution of Problems 2.2.3–2.2.5 [compare the equation (2.2.9)]. Let us next show how to compute (remaining within the same cost bounds) the coefficients of the (reverse) characteristic polynomials of A_k for $k = 1, \ldots, n$:

$$z_k(\lambda) = \det(I_k - \lambda A_k) = \sum_{i=0}^{k} c_{i,k} \lambda^{k-i} = 1/\prod_{j=1}^{k}((I_j - \lambda A_j)^{-1})_{jj}, \qquad (6.1)$$

where $c_{0,k} = (-1)^k \det A_k$, $\deg z_k(\lambda) \leq k$, for all k, and where we invert modulo λ^{n+1} all the matrices $I_j - \lambda A_j$, $j = 1, \ldots, n$, rather than just $I - \lambda A$. Formally, we will proceed as follows:

Algorithm 6.1.

Input: a natural n and the entries of an $n \times n$ matrix $A = [a_{ij}]$.

Output: the coefficients $c_{i,k}$, $i = 0, \ldots, k-1$, of the characteristic polynomials of the $k \times k$ leading principal submatrices A_k, for $k = 1, 2, \ldots, n$,

$$c_k(\lambda) = \det(\lambda I_k - A_k) = \sum_{i=0}^{k} c_{i,k} \lambda^i, \quad c_{k,k} = 1, \ c_{0,k} = (-1)^k \det A_k.$$

Computation:

1. Repeatedly square the matrices A_h to compute the scalars $(A_h^{2^g})_{h,h}$, $g = 1, 2, \ldots, \lceil \log h \rceil$, and then the coefficients of the polynomials

$$b_h(\lambda) = ((I_h - \lambda A_h)^{-1})_{h,h} \bmod \lambda^{n+1} =$$
$$\prod_{g=0}^{\lceil \log h \rceil} \left(I_h + (\lambda A_h)^{2^g}\right)_{h,h} \bmod \lambda^{n+1}, \quad h = 1, \ldots, n.$$

2. Compute modulo λ^{k+1} the products,

$$p_k(\lambda) = \prod_{j=1}^{k} b_j(\lambda) \bmod \lambda^{k+1}, \quad k = 1, \ldots, n,$$

by applying the parallel prefix computation algorithm (see Algorithm 2.1).

3. Invert $p_k(\lambda)$ modulo λ^{k+1}: for every k, $k = 1, \ldots, n$, set $z_{0,k}(\lambda) = 1$, apply $h_k = \lceil \log(k+1) \rceil$ steps,

$$z_{g+1,k}(\lambda) = z_{g,k}(\lambda)(2 - p_k(\lambda) z_{g,k}(\lambda)) \bmod \lambda^{2^{g+1}}, \quad g = 0, \ldots, h_k - 1,$$

and output $z_k(\lambda) = z_{h_k,k}(\lambda) = 1/p_k(\lambda) \bmod \lambda^{k+1}$.

To prove the correctness of Algorithm 6.1, observe that its Stage 3 is Newton's iteration for polynomial division, and deduce by induction on k that for all k and g, we have

$$1 - z_{g,k}(\lambda)p_k(\lambda) = 0 \bmod \lambda^{2^g},$$

since surely $p_k(\lambda) = 1 \bmod \lambda$ for all k. Correctness of the algorithm now follows from (6.1). ■

At Stage 1 of Algorithm 6.1, we may also compute (within the same asymptotic cost bound) the vectors $A_k^{-1}\mathbf{v}_k$ for any fixed set of k vectors $\mathbf{v}_k$ of dimension k and for $k = 1, \ldots, n$ [see (2.2.9)], *which solves Problem 2.2.3 (LIN · SOLVE) for all* A_k. Furthermore, we may apply Theorem 4.1 and Remark 4.1 in order to extend the solution of Problem 2.2.6 ($CHAR \cdot POL$) for all A_k to the solution of Problem 2.2.5 ($INVERT$) for all A_k at the cost $O_A(\log^2 n,\ P(n)\, n \log^{\alpha} n)$, $\alpha \leq 2$.

Let us next improve the complexity bound (4.2) by using randomization and the technique of recursive parallel triangulation (due to [KP,a]).

Proposition 6.2 ([KP,a]). *The randomized processor efficient NC-algorithm of section 4 for Problems 2.2.3 (LIN · SOLVE) and 2.2.4 (DETERMINANT) can be extended to solve these problems over any field* **F** *of constants having characteristic* p *and a sufficiently large cardinality (compare Remark 2.1d) at a randomized cost bounded by*

$$O_A(\gamma(n,p)\log^2 n\ ,\ P^*(n)\,/\,\gamma(n,p)\,), \tag{6.2}$$

where $P^*(n)$ *is defined by (4.5),* $\gamma(n,p) = 1$ *if* $p = 0$, $\gamma(n,p) = \lceil \log n / \log p \rceil$ *otherwise.*

Proof. We only need to complement our algorithms of section 4 by an algorithm for an auxiliary nonsingular Toeplitz-like linear system $H\mathbf{y} = \mathbf{v}$ of n equations, which should run at the cost bounded by (6.2). Furthermore, with no loss of generality, we may assume that

$$m_H(\lambda) = c_H(\lambda) = \sum_{i=0}^{n} c_i \lambda^i\ , \quad c_n = 1\ .$$

Indeed, either of the Regularizations 2.13.1 and 2.13.5 replaces the original auxiliary Toeplitz system $T\mathbf{x} = \mathbf{b}$, arising in the proof of Theorem 4.3, by the new Toeplitz-like system of the form $H\mathbf{y} = \mathbf{v}$ (or, equivalently,

$KTW\mathbf{y} = K\mathbf{b}$, where $H = KTW$, $\mathbf{v} = K\mathbf{b}$, $\mathbf{x} = W\mathbf{y}$); specifically, the system $T\mathbf{x} = \mathbf{b}$ is replaced either by the linear system

$$HC(I + Z^T)\mathbf{y} = \mathbf{b}$$

[where $\mathbf{x} = C(I+Z^T)\mathbf{y}$, $K = I$, $W = C(I+Z^T)$ and C is a matrix given with its random $F_{(Z,Z^T)}$-generator of length 2 if we apply Regularization 2.13.1] or by the linear system

$$UHL\mathbf{y} = U\mathbf{b}$$

[where $\mathbf{x} = L\mathbf{y}$, $K = U$, $W = L$, and U^T and L are random lower triangular Toeplitz matrices, except that $(U)_{0,0} = 1$, if we apply Regularization 2.13.5]. We have (with a high probability) $m_{KTW}(\lambda) = c_{KTW}(\lambda)$ in both cases. The resulting asymptotic complexity estimates are the same and are deduced similarly in both of the above approaches, but the latter approach (based on Regularization 13.5) promises to be more effective in practice; in particular, it involves fewer random values.

Now, assuming that $m_H(\lambda) = c_H(\lambda)$, we may surely compute the values $s_k = \text{trace}(H^k)$, $k = 0, 1, \ldots, n$, at the cost bounded by (6.2).

For the transition from the values s_k to the coefficients c_i, we will use Newton's identities (1.4.8) with p_i replaced by c_i for $i = 0, 1, \ldots, n$, that is,

$$\sum_{i=1}^{k-1} c_{n-i}s_{k-i} + kc_{n-k} = -s_k \,, \qquad k = 1, \ldots, n \,.$$

We will describe the algorithm in the case of fields $\mathbf{F}$ having characteristic 2, since the case of any characteristic p, for $1 < p \leq n$, is treated similarly. Replace the coefficients k by 0 if k is even, by 1, otherwise, and represent the system of Newton's identities as the vector equation,

$$V\mathbf{c} = -\mathbf{s} \,,$$

$$V = \begin{bmatrix} 1 & & & & & \\ s_1 & 0 & & & & O \\ s_2 & s_1 & 1 & & & \\ s_3 & s_2 & s_1 & 0 & & \\ s_4 & s_3 & s_2 & s_1 & 1 & \\ \cdot & \cdot & \cdot & \cdot & \cdot & \cdot \end{bmatrix},$$

$$\mathbf{c} = [c_{n-1}, \ldots, c_0]^T \,, \qquad \mathbf{s} = [s_1, \ldots, s_n]^T \,.$$

This is a singular linear system over $\mathbf{F}$.

We now remove all the even numbered equations and the respective entries of V and $\mathbf{s}$ and reorder the variables so that the system turns into the following one:

$$[B, C]\begin{bmatrix}\mathbf{c}_1\\ \mathbf{c}_0\end{bmatrix} = -\mathbf{s}_1\ , \tag{6.3}$$

$$B = \begin{bmatrix} 1 & & & \\ s_2 & 1 & & O \\ s_4 & s_2 & 1 & \\ s_6 & s_4 & s_2 & 1 \\ \cdot & \cdot & \cdot & \cdot \end{bmatrix}, \quad C = \begin{bmatrix} 0 & & & \\ s_1 & 0 & & O \\ s_3 & s_1 & 0 & \\ s_5 & s_3 & s_1 & 0 \\ \cdot & \cdot & \cdot & \cdot \end{bmatrix},$$

$$\mathbf{c}_1 = [c_{n-1}, c_{n-3}, \ldots]^T\ , \quad \mathbf{c}_0 = [c_{n-2}, c_{n-4}, \ldots]^T\ , \quad \mathbf{s}_1 = [s_1, s_3, \ldots]^T\ .$$

Next let us complement this system (of $\lceil n/2 \rceil$ equations) by $\lfloor n/2 \rfloor$ new equations,

$$[E, G]\begin{bmatrix}\mathbf{c}_1\\ \mathbf{c}_0\end{bmatrix} = \mathbf{q}\ , \tag{6.4}$$

so as to arrive at a nonsingular linear system

$$\begin{bmatrix} B & C \\ E & G \end{bmatrix}\begin{bmatrix}\mathbf{c}_1\\ \mathbf{c}_0\end{bmatrix} = \begin{bmatrix}-\mathbf{s}_1\\ \mathbf{q}\end{bmatrix}. \tag{6.5}$$

Exercise 27 of chapter 1 shows that generally we cannot achieve this by using the system (1.4.9) over finite fields, and we will look elsewhere.

We will proceed as in the proof of (4.4) for Problem 2.2.6a ($MIN{\cdot}POL$) in section 4, that is, we will first compute the Krylov matrix $K(H, \mathbf{v}, 2n-1)$ and then the linear recurrence sequence, represented by the vector $\mathbf{t}^T = \mathbf{u}^T K(H, \mathbf{v}, 2n-1)$, for two random vectors $\mathbf{u}$ and $\mathbf{v}$. This will give us a nonsingular Toeplitz linear system of n equations, $T^*\mathbf{c} = \tilde{\mathbf{t}}$, where $(\tilde{\mathbf{t}})_i = -\mathbf{u}^T H^{n+i}\mathbf{v}$, $i = 0, \ldots, n-1$, and $\mathbf{c}$ denotes the coefficient vector of the polynomial $m_H(\lambda) = c_H(\lambda)$.

Reorder the equations and variables of this system according to the permutation

$$2i + 1 \to i\ , \quad 2i \to \lceil n/2 \rceil + i\ , \quad i = 0, 1, \ldots, \lceil n/2 \rceil - 1$$

[which actually has already been applied to the columns of the matrix V and resulted in (6.3)], and thus arrive at the system

$$\begin{bmatrix} T_{00} & T_{01} \\ T_{10} & T_{11} \end{bmatrix}\begin{bmatrix}\mathbf{c}_1\\ \mathbf{c}_0\end{bmatrix} = \begin{bmatrix}\mathbf{t}_1\\ \mathbf{t}_0\end{bmatrix}$$

where T_{ij} are Toeplitz matrices for $i, j = 0, 1$. Then premultiply both sides by a random $\lfloor n/2 \rfloor \times n$ matrix $\widehat{T}$. Similarly to the proof of Lemma 2.13.1, we verify that in this way we arrive at the desired system (6.4) such that the system (6.5) [obtained by combining the systems (6.3) and (6.4)] is nonsingular (with a high probability), even if we let $\widehat{T}$ be the Toeplitz matrix with zeros below its diagonal, ones on the diagonal, and random $n-1$ values above the diagonal.

Now apply the factorization (2.2.2) to the coefficient matrix of (6.5). In this way, the solution of (6.5) is reduced essentially to computing the matrices B^{-1} and $S = G - EB^{-1}C$ and to solving a linear system with the coefficient matrix S, which is a half-size Problem 2.2.3 ($LIN \cdot SOLVE$).

The complexity of computing B^{-1} and S is $O_A(\log^2 n, P(n)/\log n)$ since B is a triangular matrix. At the cost bounded by (4.4), we have reduced the original Problems 2.2.3 ($LIN \cdot SOLVE$) and 2.2.4 ($DETERMINANT$) to half-size Problem 2.2.3 ($LIN \cdot SOLVE$). Recursively applying this reduction at most $\lceil \log n \rceil$ times (we call this process *recursive triangulation*), we solve Problems 2.2.3 ($LIN \cdot SOLVE$) and 2.2.4 ($DETERMINANT$) over any field of characteristic 2, at the overall cost $O_A(\log^3 n, P^*(n)/\log n)$, which gives us (6.2) for $p = 2$. (As we will see later on, the recursive triangulation, excluding the initial stage of the reduction to a Toeplitz linear system, can be performed in time $O(\log^3 n)$ on fewer processors, due to the structure of the matrices B, C, T_0, T_1, $\widehat{T}$.)

Over a field $\mathbf{F}$ of constants having a positive characteristic p, a similar algorithm requires $\lceil \log n/\log p \rceil$ recursive steps (since exactly $\lfloor n/p \rfloor$ diagonal entries of the matrix V of the system of Newton's identities vanish over $\mathbf{F}$), and we arrive at the bounds (6.2) for any p. ∎

Computations with General Matrices. Extensions from $DETERMINANT$.

We apply Theorem 4.1 and Remark 4.1 to extend the randomized solution of $DETERMINANT$ to $INVERT$ (see an alternative in Remark C.3 of Appendix C) so as to reach the randomized complexity bounds

$$O_A(\gamma(n,p)\log^2 n,\ P^*(n)\log^\alpha n/\gamma(n,p))$$

where $\alpha \le 2$ is defined in Remark 4.1, $P^*(n)$ and $\gamma(n,p)$ are defined in (6.2).

As in the previous section, we may extend the matrix inversion algorithm to the solution of Problem 2.2.7 ($LU \cdot FACTORS$). In this case, we yield the randomized complexity bounds

$$O_A\left(\gamma(n,p)\log^3 n, P^*(n)(\log^{\alpha-1} n) \,/\, \gamma(n,p)\right) .$$

To obtain further extensions of our solutions of Problems 2.2.3–2.2.5, we need to solve Problem 2.2.10 ($M \cdot RANK$).

The deterministic solution suggested in chapter 2 (and based on the LSP factorization) solves Problems 2.2.10 ($M \cdot RANK$), 2.2.10a ($SUBMATRIX$) and 2.2.10b ($M \cdot PRECNDTN$) at the computational cost $O_A(h, (m+n)M(h)/h^2)$, $h = \min\{m, n\}$ (over any field).

Let us next recall a deterministic solution in NC. To compute rank A for an $n \times n$ symmetric matrix A over an arbitrary field of constants, follow [Mul], compute $c_{YA}(\lambda) = \det(\lambda I - YA)$, where $Y = \text{diag}\,(1, y, y^2, \ldots, y^{n-1})$, y is an auxiliary parameter, and obtain that rank $A = n - k$, where $k = k(YA)$ is the highest power of λ dividing $c_{YA}(\lambda)$. To relax the symmetry assumption for A, recall that rank $\begin{pmatrix} O & A \\ A^T & O \end{pmatrix} = 2$ rank A.

Remark 6.1. Later on, we will need to extend the equation rank $A = n - k$ to the case where $k = k(YA)$ is replaced by $k = k(YAY)$. The proof of the correctness for such an extension is implicit in [Mul].

The cost of this solution is bounded by $O_A(\log^2 n, n^{\omega+3} \log\log n)$, $\omega < 2.376$, since the output coefficients of $c_{YA}(\lambda) = \det(\lambda I - YA)$ are polynomials in y of degrees up to $\sum_{i=1}^{n-1} i = n(n-1)/2$, and the ops involved in computing $c_{YA}(\lambda)$ are actually ops over such polynomials in y. The known extensions to the deterministic NC-solutions of Problems 2.2.10a ($SUBMATRIX$) and, furthermore, 2.2.3a ($LIN \cdot SOLVE1$), 2.2.3b ($NULL \cdot SPACE$) and 2.2.10b ($M \cdot PRECNDTN$) (over any field) require to increase by n times the processor bound of the solution of Problem 2.2.10 ($M \cdot RANK$).

Next we will apply randomization to decrease the latter estimates for the complexity of computing the matrix rank. This randomization can always be turned into one of Las Vegas type; that is, we may certify the solution by solving Problems 2.2.5 ($INVERT$) and 2.2.2 ($M \cdot MULTIPLY$). Indeed, for an $n \times n$ matrix A of rank r, we may assume without loss of generality that $A = \begin{pmatrix} B & C \\ D & E \end{pmatrix}$, where B is a nonsingular $r \times r$ matrix (otherwise, we may apply Regularization 2.13.2 based on Lemma 2.13.1 and then verify nonsingularity of B by computing $\det B$). Then for $F = \begin{pmatrix} I_r & -B^{-1}C \\ O & I_{n-r} \end{pmatrix}$, we have $AF = \begin{pmatrix} B & O \\ D & O \end{pmatrix}$, and by computing AF, we verify if $r = \text{rank}\, A$ has been computed correctly [compare one of the solutions of Problem 2.2.3b ($NULL \cdot SPACE$) in chapter 2].

Our results on randomized solution of $M \cdot RANK$ are summarized in the next proposition (see a further improvement in Appendix C):

Proposition 6.3 ([KP,a]). *Problem 2.2.10 ($M \cdot RANK$) can be solved over a field* $\mathbf{F}$ *having characteristic* p *and a sufficiently large cardinality, at randomized cost bounded by each of the two following expressions:*

a) $O_A(t_{n,r}, (P(n)\log^2 n + (r\log r)^2 \log\log r)/t_{n,r})$,
$t_{n,r} = \gamma(n,p)\log^2 n + \gamma(r,p)\log^3 r$;

b) $O_A(\gamma(n,p)\log^2 n, n^3 \log\log n/(\gamma(n,p)\log n))$, (6.6a)

provided that r *is the degree of the minimum polynomial of the input matrix (so that* $r \le n$*),* $P(n)$ *is defined in section 3 [see (3.4), (3.5)] and* $\gamma(s,p)$ *in (6.1).*

c) *Furthermore, if* $p = 0$ *or* $p > n$*, then Problem 2.2.10 ($M \cdot RANK$) can be solved at the randomized cost bounded by (4.4).*

Remark 6.2. The expressions for processor bounds of parts a) and b) are equal to $P(n)$ within polylogarithmic factors so that they are $O(n^{\bar\omega})$ for $\bar\omega < 2.376$ [see (3.5)].

Next we will deduce Proposition 6.3. We will first reduce Problem 2.2.10 ($M \cdot RANK$) to $MIN \cdot POL$, by following [KS91], [KP,a].

Lemma 6.1 ([KS91]). *Let* $A = \begin{bmatrix} B & C \\ E & G \end{bmatrix}$ *be an* $n \times n$ *matrix of rank* $r < n$*; let* B *be its* $r \times r$ *nonsingular submatrix, and let* $H = H(B,C,E) = B + B^{-1}CE$ *be nonsingular. Then a)* $c_A(\lambda) = c_H(\lambda)\lambda^{n-r}$, $m_A(\lambda) = \lambda m_H(\lambda)$, $c_H(0)m_H(0) \ne 0$. *If, furthermore,* $c_H(\lambda) = m_H(\lambda)$*, then b)* $m_A(\lambda) = \lambda c_H(\lambda)$ *and* $r = \deg c_H(\lambda) = \deg m_A(\lambda) - 1$.

Proof. $c_H(0)m_H(0) \ne 0$ since H is nonsingular. Applying a similarity transformation, define the matrix

$$\begin{bmatrix} I & B^{-1}C \\ O & I \end{bmatrix} \begin{bmatrix} B & C \\ E & G \end{bmatrix} \begin{bmatrix} I & -B^{-1}C \\ O & I \end{bmatrix} = \begin{bmatrix} H & O \\ E & O \end{bmatrix}.$$

Recall that the similarity transformations of A change neither $c_A(\lambda)$ nor $m_A(\lambda)$ and easily verify the part a) of the lemma. The part b) immediately follows. ∎

Seeking $r = \text{rank } W$ for a given $n \times n$ matrix W, we may first try to find an $n \times n$ matrix A such that rank $A = r =$ rank W and all the assumptions of Lemma 6.1 hold for A, and then compute and output $r = \deg m_A(\lambda) - 1$.

To satisfy with a high probability the assumptions of Lemma 6.1, that is, the nonsingularity of the $r \times r$ matrices B and H [from part a) of the lemma] and the equation $c_H(\lambda) = m_H(\lambda)$ of part b), we will follow [KS91] and first apply Regularization 2.13.2 and Procedure 1.7.1. This transforms W into $V = UWL$ for two lower triangular Toeplitz matrices L and U^T defined in Lemma 2.13.1 and such that, with a high probability, rank $V = r =$ rank W, and the $r \times r$ leading principal block $\tilde{B}$ of V is strongly nonsingular (by Lemma 2.13.1).

Now set $A = VX$ for a random diagonal matrix $X = \text{diag}(x_0, \dots, x_{n-1})$ $= \text{diag}(X_0, X_1)$, such that $X_0 = \text{diag}(x_0, \dots, x_{r-1})$ and $X_1 = \text{diag}(x_r, \dots,$ $x_{n-1})$ (see the last subsection of this section on some alternative approaches). Then X is nonsingular with a high probability, and therefore rank $W =$ rank $V =$ rank $A =$ rank $\tilde{B} =$ $\text{rank}(\tilde{B}X_0)$, so the matrix $B = \tilde{B}X_0$ is nonsingular.

Denote that

$$V = \begin{bmatrix} \tilde{B} & \tilde{C} \\ \tilde{E} & \tilde{G} \end{bmatrix},$$

$$H = H(\tilde{B}X_0, \tilde{C}X_1, \tilde{E}X_0) = X_0^{-1}\tilde{B}^{-1}\tilde{C}X_1\tilde{E}X_0 + \tilde{B}X_0,$$

$$H = \widetilde{H}X_0,$$

$\widetilde{H} = \tilde{B} + X_0^{-1}\tilde{B}^{-1}\tilde{C}X_1\tilde{E}$. The $k \times k$ leading principal minor (subdeterminant) of H is either a polynomial in the entries of X_1 and has total degree at most k, or else if $\tilde{B}^{-1}\tilde{C}X_1\tilde{E}$ vanishes, this minor is a polynomial in the entries of X_0 of degree at most k. Therefore, with a high probability, $\widetilde{H}$ and H are strongly nonsingular matrices (as follows from Lemma 1.5.1), so the assumptions of part a) of Lemma 6.1 hold. Moreover, strong nonsingularity of $\widetilde{H}$ enables us to apply Corollary 2.13.1 and to deduce that $m_H(\lambda) = c_H(\lambda)$, which is the remaining assumption of part b) of Lemma 6.1. Due to the above arguments, we may apply Lemma 6.1 and probabilistically reduce the computation of rank A to the evaluation of $m_A(\lambda)$ [Problem 2.2.6a $(MIN \cdot POL)$].

The randomized algorithm of section 4 evaluates $m_A(\lambda)$ over any field, and its cost is clearly within the bounds (6.2) if we exclude the stage of computing the rank r of the auxiliary Toeplitz matrix T, $r =$ deg $m_A(\lambda)$. It remains to estimate the complexity of the latter stage.

To compute $r =$ rank T, we may apply Algorithm 2.8.2 (binary search) or its output sensitive modification [compare (2.13.6), (2.13.7) and (4.7)] to the Toeplitz-like matrix $V = UTL$, where U^T and L are random unit lower triangular Toeplitz matrices. (By Lemma 2.13.1, the $r \times r$ leading principal submatrix of V is strongly nonsingular with a high probability.)

In each of $\lceil \log n \rceil$ or each of $2\lfloor \log n \rfloor + 1$ steps of such a search, we shall compute the determinant of a leading principal submatrix of V, which is a Toeplitz-like matrix. Later on [see (6.9)], we will show how to compute the determinant of an $r \times r$ Toeplitz-like matrix at the randomized cost $O_A(\gamma(r,p)\log^2 r,\ r^2 \log\log r\,/\,(\gamma(r,p)\log r))$, $r \le n$, which for the entire search (involving the evaluation of at most $2\lceil \log r \rceil + 1$ such determinants) implies the estimate

$$O_A\left(\gamma(r,p)\log^3 r,\ r^2\log\log r\,/\,(\gamma(r,p)\log r)\right) . \qquad (6.6b)$$

Combining this estimate with (6.2) and using the B-principle, we obtain the bound of part a) of Proposition 6.3.

To obtain the bound of part b), we compute $r = \operatorname{rank} T$ by applying Algorithm 6.1 to the evaluation of the determinants of all the n leading principal submatrices of T, with Stage 1 of this algorithm reduced to application of Algorithm 2.11.1.

Let us next deduce part c) of Proposition 6.3, that is, decrease the above bounds on the parallel cost in the case where the field of constants has characteristic $p = 0$ or $p > n$ and therefore allows division by $n!$.

To achieve this, we will apply a modification of the approach of [Mul]. We first fix a random value y and arrive at the Hankel matrix $H^* = YJTY$, where $Y = \operatorname{diag}(1, y, y^2, \ldots, y^{n-1})$ and J is the reversion matrix. Next, we compute the characteristic polynomial $c_{H^*}(\lambda)$ of the Hankel matrix H^* (which is a simple computation in the field allowing division by $n!$), then recall Remark 6.1 (for $A = JT$), read rank H^* from the coefficients of $c_{H^*}(\lambda)$, and output rank $T =$ rank H^*.

The cost of this randomized computation of the rank of a Toeplitz matrix T is dominated by the cost of the evaluation of the coefficient vector of $c_{H^*}(\lambda)$. Due to Algorithm 2.11.1 or, alternatively, Remark 2.11.4, this cost is bounded by

$$O_A(\log^2 n, n^2 \log\log n/\log n)$$

over the fields of characteristics $p = 0$ and $p > n$, which implies the desired bound (4.4) of part c) on the complexity of computing the rank of an $n \times n$ matrix over such fields. ∎

Now, Regularization 2.13.2 and Lemma 2.13.1 enable us to extend the randomized solution of Problem 2.2.10 ($M \cdot RANK$) to the randomized solution of Problems 2.2.10b ($M \cdot PRECNDTN$) and 2.2.3a ($LIN \cdot SOLVE1$)

preserving the asymptotic complexity estimates of Proposition 6.3 for Problem 2.10 ($M \cdot RANK$). Then, as in section 2 of chapter 2, we reduce Problem 2.2.3b ($NULL \cdot SPACE$) to $M \cdot RANK$ and $INVERT$. In Appendix B, we will recall a result from [E91] and its further refinement from [ChR], which extend the processor efficient NC-algorithms for $M \cdot RANK$ and $INVERT$ to Problem 2.2.10a ($SUBMATRIX$).

Deterministic Computations with Dense Structured Matrices and Polynomials.

We will next extend our algorithms for computations with polynomials and structured matrices to the case of any field. We will actually study Toeplitz, Toeplitz-like and Toeplitz-like+Hankel-like computations; their extension to computations with polynomials and other dense structured matrices can be performed over any field with little change against the case of the complex field (see section 7 of chapter 1 and sections 11 and 12 of chapter 2).

To reflect the dependence of the estimated complexity of our computations on the chosen field (or ring) of constants $\mathbf{F}$, we will use the following functions:

$$\phi_{\mathbf{F}}(s) = \begin{cases} s & \text{if the field } \mathbf{F} \text{ supports FFT,} \\ s \log \log s & \text{otherwise,} \end{cases}$$

$$\psi_{\mathbf{F}}(s) = \begin{cases} 1 & \text{if the field } \mathbf{F} \text{ allows division by } n!, \\ s & \text{otherwise.} \end{cases}$$

Let us next show some extensions (to any field) of our deterministic algorithms for Problems 2.2.3–2.2.6 of the computation of the solution to a linear system, the inverse, the determinant and the characteristic polynomial of a matrix, assuming an $n \times n$ Toeplitz, Toeplitz-like and Toeplitz-like+Hankel-like input matrix A. These extensions will enable us to deduce the deterministic complexity bounds

$$O_A(\log^2 n, n\phi_{\mathbf{F}}(n)\psi_{\mathbf{F}}(n)/\log n) \tag{6.7}$$

for these problems.

(6.7) is supported by Algorithms 2.11.1 and 2.11.2 in the case where the field $\mathbf{F}$ allows division by $n!$. Otherwise, we will rely on Chistov's (rather than Csanky's) Algorithm 6.1 for the deterministic computation over any field $\mathbf{F}$ of the coefficients of the characteristic polynomials of the $n \times n$ input matrix A and of all its $k \times k$ leading principal submatrices A_k, $k = 1, \dots, n$, $A = A_n$, and we will only need to change Stage 1 of Algorithm 6.1, so as to evaluate the polynomials $b_h(\lambda) \bmod \lambda^{n+1}$, for all h, by means of n times applying Algorithm 2.11.1.

At Stage 1 of Algorithm 6.1, we may also compute the vectors $A_k^{-1}\mathbf{v}_k$ for any fixed set of vectors $\mathbf{v}_k$ of dimension k, $k = 1, \dots, n$ (see Algorithm 2.11.2), which gives us the solution of the linear systems $A_k\mathbf{x}_k = \mathbf{v}_k$ and similarly (displacement generators for) the inverse matrices A_k^{-1}.

The cost estimates for application of Algorithms 2.11.1 and 2.11.2 to Toeplitz-like+Hankel-like input matrices A are extended to Algorithm 6.1, except that the processor bounds increase by n times, and we arrive at the complexity estimates (6.7) for $\psi_{\mathbf{F}}(n) = n$.

Remark 6.3. For an $n \times n$ Toeplitz-like+Hankel-like input matrix A, Algorithm 6.1 computes $\det A_h$, for $h = 1, \dots, n$ and over any field $\mathbf{F}$, at the cost deterministically bounded by (6.7) for $\psi_{\mathbf{F}}(n) = n$, and this gives us $r = \operatorname{rank} A$ in the cases where it is known that the $r \times r$ leading principal submatrix of A is nonsingular.

Furthermore, we may apply the techniques of sections 11 and 12 of chapter 2 and immediately extend our results for Toeplitz-like+Hankel-like computations to computations with other structured matrices and with polynomials over arbitrary fields.

Over the fields of characteristic $p = 0$, we easily extend (6.7) to Problems 2.2.8 ($L{\cdot}SQUARES$), 2.2.10 ($M{\cdot}RANK$) and 2.2.3a ($LIN{\cdot}SOLVE1$) with Toeplitz-like+Hankel-like input matrices.

For polynomial computations, we observe that Algorithm 2.5.1 for Padé approximation runs at the cost bounded by (6.7) over any field $\mathbf{F}$ if at its Stages 1 and 2 we apply Algorithm 6.1 (observe that the matrix TJ in Algorithm 2.5.1 satisfies the condition stated in Remark 6.3). The same bounds are then immediately extended to the complexity of computing the gcd and the lcm of two polynomials of degrees at most n and to computing the span of the minimum length for a linear recurrence (for the gcd, we may alternatively use Algorithm 2.9.1, performing its Stage 1 by means of Algorithm 6.1, and then Remark 6.3 still applies).

By combining Algorithm 2.10.1 and our results on parallel computations with structured matrices, we arrive at the cost bounds $O_A(\log^3 n, n\phi_{\mathbf{F}}(n)\psi_{\mathbf{F}}(n)/\log^2 n)$ for deterministic evaluation, over any field $\mathbf{F}$, of all the polynomials of the extended Euclidean scheme for two input polynomials of degrees at most n. Note that $\psi_{\mathbf{F}}(n) = 1$ and that this NC-algorithm is processor-efficient if $\mathbf{F}$ supports division by $n!$.

Randomized Computations with Dense Structured Matrices. $LIN \cdot SOLVE$ and $DETERMINANT$.

Finally, we will follow [P93a] in order to improve some of the above deterministic complexity estimates by using *randomized algorithms* for *Toeplitz-like+Hankel-like* computations.

In this subsection, we will study the computations over any field $\mathbf{F}$; in particular, we will allow characteristics from 2 to n. We will start this study with revisiting our randomized solution of Problems 2.2.3 ($LIN \cdot SOLVE$) and 2.2.4 ($DETERMINANT$) and now assuming that the input matrix A is a Toeplitz-like+Hankel-like matrix. In this case, we will proceed similarly to the case of a general matrix A: we reduce the original problem to solving Toeplitz-like+Hankel-like linear systems, whose size recursively decreases by the factor p.

We recall that the first step of the recursive triangulation essentially reduces to the evaluation of:

a) $\text{trace}(A^i)$, $i = 0, 1, \ldots, 2n - 1$ [which defines the matrices B and C of (6.3)];

b) the Toeplitz matrix, T_1, defined by the vector $\mathbf{u}^T K$, $K = K(A, \mathbf{v}, 2n-1)$ (for two random vectors $\mathbf{u}$ and $\mathbf{v}$), and the matrix $\widehat{T}T_1$, for a random rectangular Toeplitz matrix $\widehat{T}$ with zeros below its diagonal [which defines the matrices E and G of (6.4)];

c) the matrix $S = G - EB^{-1}C$.

The second step is similar but with the $\lfloor n/p \rfloor \times \lfloor n/p \rfloor$ matrix S replacing the $n \times n$ matrix A, and we also proceed similarly at the further steps of the recursive triangulation, with the p-fold size reduction of the input matrix in each step.

In the case of a Toeplitz-like+Hankel-like input matrix A, the complexity of the two Stages a) and b) of the first step of the recursive triangulation is bounded by

$$O_A(\log^2 n, n\phi_{\mathbf{F}}(n)/\log n)\ . \tag{6.8}$$

Moreover, the same cost bound applies to the evaluation of $\text{trace}(S^i)$, for all i, and of $K(S, \mathbf{v}, 2n - 1)$, that is, to Stages a) and b) where the input matrix A is replaced by the matrix S. Indeed, the matrix S has size $\lfloor n/p \rfloor \times \lfloor n/p \rfloor$ and will be obtained with its displacement generator of length $O(p)$. Therefore, the cost bound (6.8) for the above computations with S follows by combining Algorithms 2.11.1, 2.11.2 and Corollary 2.12.1. Each next step of the recursive triangulation decreases (by at least p times) the size of the auxiliary linear system and preserves the bound $O(p)$ on the length of the displacement generator available for its coefficient matrix. Therefore, (6.8) bounds the cost of performing the Stages a) and b) of every step of recursive triangulation, and we will extend this bound to Stage c) too. Moreover, due to the p-fold size decrease of the input to every recursive step of the triangulation, as soon as we prove the extension to the bound (6.8) on the cost of performing Stage c), we may apply Proposition 1.1 and arrive at

Proposition 6.4 ([P93a]). *Problem 2.2.3 ($LIN \cdot SOLVE$) and 2.2.4 ($DETERMINANT$) can be solved over any field of constants* $\mathbf{F}$ *having characteristic* p *and sufficiently many elements (compare Remark 2.1d) at the randomized cost bounded by*

$$O_A(\gamma(n,p)\log^2 n, n\phi_{\mathbf{F}}(n)/(\gamma(n,p)\log n)), \tag{6.9}$$

provided that the input matrix is an $n \times n$ *Toeplitz-like+Hankel-like matrix;* $\gamma(n,p) = 1$ *if* $p = 0$, $\gamma(n,p) = \lceil \log n / \log p \rceil$ *otherwise [as in (6.2)].* [Note that application of (6.9) over the fields of constants allowing division by $n!$ covers the bounds (6.2).]

To prove Proposition 6.4, it remains to compute the displacement generator of length $O(p)$ for S, at the cost bounded by (6.8). Towards this goal, we will next revisit the evaluation of the matrices B^{-1}, $B^{-1}C$ and $S = G - EB^{-1}C$ of (6.3)–(6.5), which we have earlier described in some detail in the case of $p = 2$ and will extend this evaluation to any $p \geq 2$.

Let V denote the coefficient matrix of the original triangular singular system of n Newton's identities, let U denote the $n \times n$ coefficient matrix of the nonsingular Toeplitz linear system generated by the vector $\mathbf{t}^T = \mathbf{u}^T K(A, 2n-1, \mathbf{v})$, and let $\widehat{T}$ denote the $(n/p) \times n$ Toeplitz compression matrix. Let us denote that $n = kp$ and assume that k is an integer. Otherwise, we may just shift from A to the block matrix $\text{diag}(A, I)$ of size $N \times N$, $N = \lceil k \rceil p$. Note that V has exactly $k = n/p$ zeros on its diagonal.

Define the permutation π of the set $\{0, 1, \ldots, n-1\}$, enumerating the rows and columns of U and V, such that

$$\pi(ip + j) \to jk + i\,, \qquad i = 0, \ldots, k-1;\; j = 0, \ldots, p-1\,.$$

Apply this permutation to both rows and columns of each matrix U and V and denote the resulting matrices $\pi(U)$ and $\pi(V)$, respectively. Observe that both of these matrices turn into $p \times p$ block Toeplitz matrices with

$k \times k$ Toeplitz blocks. Furthermore, the blocks of $\pi(V)$ are lower triangular matrices. Represent

$$\pi(V) = \begin{bmatrix} B & C \\ B_1 & C_1 \end{bmatrix} , \qquad \pi(U) = [U_0, U_1] ,$$

where $B \in \mathbf{F}_{n-k,n-k}$, $C \in \mathbf{F}_{n-k,k}$, $U_0 \in \mathbf{F}_{n,n-k}$, $U_1 \in \mathbf{F}_{n,k}$, and observe that B is a nonsingular matrix, U_0 and U_1 are matrices of full rank. Denote $\widehat{T}U_0 = E$, $\widehat{T}U_1 = G$. Now, the first step of the recursive triangulation applied to the original Toeplitz-like linear system has been reduced to the transition from the linear system (6.5) to the system

$$S\mathbf{c}_0 = \mathbf{q} - EB^{-1}\mathbf{s}_1 , \qquad S = G - EB^{-1}C .$$

The computational complexity of this transition is dominated by the complexity of computing the matrices U, V [already shown to be bounded by (6.8)], and by the complexity of computing the matrices B^{-1}, $B^{-1}C$ and S, to be estimated next. (We do not include the computational cost of generating the random Toeplitz matrix $\widehat{T}$.)

a) *Evaluation of* B^{-1}. Since B is a nonsingular $(p-1) \times (p-1)$ block Toeplitz matrix, we only need to find a block generator of $F(B^{-1})$ for $F = F_{(Z^T,Z)}$ (say). We apply Algorithm 2.7.1, which essentially consists of $O(\log p)$ multiplications of bivariate block polynomials; each of them has degrees $p-1$ and 2^j in its two variables, respectively, $j = 0, 1, \ldots, \lceil \log (p-1) \rceil$, and outputs the block traces of B^h and the block vectors $B^h\mathbf{w}$, $h = 0, 1, \ldots, 2p-3$, where $\mathbf{w}$ is a random block vector. Since B is a $(p-1)\times(p-1)$ block matrix and $\mathbf{F}$ has characteristic p, we may apply the solution of Problem 1.4.8 ($I \cdot POWER \cdot SUMS$) from chapter 1 and compute the block characteristic polynomial of B at the cost of computing modulo x^{2^j} (successively for each j) $O(1)$ products of block polynomials in x, where j ranges from 0 to $\lceil \log p \rceil$.

Finally, we apply the matrix equation (2.2.9), which we scale by dividing it by the block determinant of B, and output a block generator of length 2 for B^{-1}.

Since each block of B is a triangular $k \times k$ Toeplitz matrix, its inversion (if it is nonsingular) is reduced to Problem 1.3.4 ($POL \cdot RECIPR$) and gives us the inverse of the block at the cost $O_A(\log k \ \log \log k, k)$ (compare Remark 2.1). Multiplication of two blocks can be reduced to multiplication of two univariate polynomials of degrees at most k, since, generally, multiplication of s-variate block polynomials in $x_0, \ldots, x_{s-1}$, reduced modulo

$x_i^{d(i)}$, for $i = 0, \ldots, s-1$, amounts to multiplication of $(s+1)$-variate polynomials reduced modulo $x_i^{d(i)}$, for $i = 0, \ldots, s$ and for $d(s) = O(k)$. We recall Theorem 1.7.1 and (1.8.3) to bound the cost of such an operation by $O_A(\log N, \phi_{\mathbf{F}}(N))$, for $N = \prod_{i=0}^{s}(2d(i)+1)$. [We only need to apply this bound for $s \leq 2$ and for $N = O(p^2k) = O(pn)$.] Summarizing all the above estimates, we thus compute a displacement block generator of length at most 2 for B^{-1} at the cost bounded by $O_A(\log N \ \log p, \phi_{\mathbf{F}}(N)/\log p) = O_A(\log n \ \log p, \phi_{\mathbf{F}}(np)/\log p)$ [which is surely within the bound (6.8)], and we observe that the $k \times k$ blocks of B^{-1} are lower triangular Toeplitz matrices, since so are the blocks of B (compare exercise 20 of chapter 2).

b) *Computing the block vector* $B^{-1}C$. This is reduced to four block multiplications of $(p-1)\times(p-1)$ triangular Toeplitz block matrices with $k \times k$ Toeplitz blocks (that define a block generator for B^{-1}) by the block vector C. Since the blocks of B^{-1}, C, and therefore $B^{-1}C$ are $k \times k$ triangular Toeplitz matrices, obtain the bound $O_A(\log n, \phi_{\mathbf{F}}(n))$ on the computational cost of performing this stage [which is surely within the bound (6.8)].

c) *Computing* $U_0(B^{-1}C)$. Both multiplicands are Toeplitz block matrices of sizes $p \times (p-1)$ and $(p-1) \times 1$, respectively, with $k \times k$ Toeplitz blocks. The multiplication is reduced to k multiplications of bivariate polynomials of degrees $O(p)$ and $O(k)$ in the two variables, so the overall cost of this stage is bounded by $O_A(\log n, k\phi_{\mathbf{F}}(n))$ and therefore, surely, by (6.8), as clearly, also is the cost of the subsequent computation of $U_1 - U_0(B^{-1}C)$.

d) *Finally,* the latter $p \times 1$ block matrix is premultiplied by a $1 \times p$ block matrix $\widehat{T}$, with $k \times k$ Toeplitz or Toeplitz-like blocks. We output a displacement generator of length $O(p)$ for the $k \times k$ product, S. The cost of this stage, $O_A(\log n, k\phi_{\mathbf{F}}(n))$, is again clearly dominated by (6.8).

This completes the proof of the bound (6.8) on the cost of the auxiliary evaluation of a displacement generator of length $O(p)$ for S, and now, for solving the original computational problems, the overall cost bound (6.9) follows. ∎

Extensions of Toeplitz-like $LIN \cdot SOLVE$. Computations with Dense Structured Matrices and Polynomials.

The randomized bound (6.9) is immediately extended, via Algorithms 2.11.1 and 2.11.2, to the evaluation (over any field) of a short displacement generator for the inverse of a Toeplitz-like matrix and then of all the entries of this inverse.

As we did with the bound (6.2) in the case of general input matrices, we may now extend the bounds (6.6a), (6.6b) (over any field of constants) and (6.9) (over any field of constants supporting division by $n!$) to various other randomized computations with Toeplitz-like+Hankel-like input matrices, in particular, to solving Problems 2.2.3a ($LIN \cdot SOLVE1$), 2.2.3b ($NULL \cdot SPACE$), 2.2.6a ($MIN \cdot POL$), 2.2.10 ($M \cdot RANK$) and 2.2.10b ($M \cdot PRECNDTN$) (with Toeplitz-like+Hankel-like input matrices).

The extension of the bounds (6.6a) and (6.6b) on the complexity of $M \cdot RANK$ from the case of Toeplitz to Toeplitz-like+Hankel-like input matrix is straightforward. (We just need to apply first Regularization 2.13.2 and then either binary search or Algorithm 6.1.) As an exercise, the reader may apply the output sensitive approach of [ChR] and express the complexity of $M \cdot RANK$ for an $n \times n$ Toeplitz-like+Hankel-like input matrix T in terms of n and r=rank T [compare (2.13.6), (2.13.7)].

The improvement to the cost bound (6.9) [or (6.2), (4.4)] (over the fields allowing division by $n!$) is also straightforward in the case of a Toeplitz input matrix T, but not in the case of a Toeplitz-like+Hankel-like input matrix T, where the matrix $YJTY$ for $Y = \operatorname{diag}(1, y, y^2, \ldots, y^{n-1})$ may have no Hankel-like structure anymore.

We will next follow [P93a] to show the reduction of Problem 2.2.10 ($M \cdot RANK$) from the case of a Toeplitz-like+Hankel-like input matrix T to the case of a Toeplitz input matrix, which will enable us to extend the complexity bounds (6.9) to the case of computing the rank of such a matrix T over fields allowing division by $n!$. We will apply our previous algorithm (designed for the reduction from the case of the general $n \times n$ input matrix to the Toeplitz case via the reduction of $M \cdot RANK$ to $MIN \cdot POL$) until we arrive at the auxiliary $n \times n$ matrix

$$V = UTL = \begin{bmatrix} B(V) & C(V) \\ E(V) & G(V) \end{bmatrix} ,$$

such that $B(V)$ is a strongly nonsingular $r \times r$ matrix, and

$$\text{rank } V = \text{ rank } T = \text{ rank } B(V) = r \ .$$

Then, at the next step, we need to ensure the assumptions of Lemma 6.1 for the matrix $A = MVN$ by choosing appropriate matrices M and N. Furthermore, we now assume that V is a Toeplitz-like+Hankel-like matrix, and we shall choose M and N being Toeplitz-like+Hankel-like matrices too. Our previous choice, $M = I$, $N = \operatorname{diag}(x_0, \ldots, x_{n-1})$, where $x_0, \ldots, x_{n-1}$ are indeterminates, violated the latter requirement so that we need a distinct approach. Therefore, we now set

$$N = S(I + Z) \ ,$$

letting M and S be generic Toeplitz-like+Hankel-like matrices, associated with an appropriate operator F. We choose an operator F such that rank $F(M) \leq 3d+11$, rank $F(S) \leq 2d+6$, and each of the matrices M and S is represented by its F-generator, defined according to Theorem 2.11.2, with the generic vectors $\mathbf{g}_i$, $\mathbf{h}_i$ in the expansions of the matrices $F(M)$ and $F(S)$, that is, we assume that these matrices are generated by vectors filled with indeterminates. Furthermore, here we specify that $F = F_{(Z,Z^T)}$ and that rank $F(V) = d = O(1)$.

We will verify the assumptions of Lemma 6.1 for $A = MVN$ and for generic Toeplitz-like matrices M and S; thus, by replacing the indeterminates by random values, according to Procedure 1.7.1, we will satisfy these assumptions with a high probability.

Specifically, we need to show that (for generic Toeplitz-like+Hankel-like matrices M and S) we have rank $A =$ rank V, the matrices $B = B(A)$ and $H = B + B^{-1}C(A)E(A)$ are nonsingular, where $A = \begin{bmatrix} B(A) & C(A) \\ E(A) & G(A) \end{bmatrix}$, $B = B(A) = A_{(r,r)}$ is an $r \times r$ matrix. We immediately deduce these relations from the similar properties of the matrix V and its submatrices. We, however, also need to show that $m_H(\lambda) = c_H(\lambda)$. To prove the latter equation, we need the following simple corollary of Theorem 2.11.2:

Lemma 6.2. *Let*

$$W = \begin{bmatrix} W_{00} & W_{01} \\ W_{10} & W_{11} \end{bmatrix}, \; W(0) = \begin{bmatrix} W_{00} & O \\ W_{10} & O \end{bmatrix}, \; W(1) = \begin{bmatrix} O & W_{01} \\ O & W_{11} \end{bmatrix},$$

and let $W(i,j)$ be obtained from W by substituting $W_{gh} = O$ unless $g = i$, $h = j$, for $i,j = 0,1$, so that $W(0,0) = \mathrm{diag}(W_{00}, O)$, $W(0,1) = \mathrm{diag}(W_{0,1}, O)J$, $W(1,0) = \mathrm{diag}(O, W_{10})J$, $W(1,1) = \mathrm{diag}(O, W_{11})$, $W(j) = W(0,j) + W(1,j)$, $j = 0,1$. Let $F = F_{(Z,Z^T)}$, $d = \mathrm{rank}\, F(W)$, $d(j) = \mathrm{rank}\, F(W(j))$, $d(i,j) = \mathrm{rank}\, F(W(i,j))$, $d_{ij} = \mathrm{rank}\, F(W_{ij})$, $i,j = 0,1$. Then $d_{ij} \leq d+i+j$, $d(i,j) \leq d+2$, $d(j) \leq d+1$, $d(i,j) \leq d_{i,j} + 2 - i - j$; $i,j = 0,1$.

Now, we will verify that $m_H(\lambda) = c_H(\lambda)$ by applying the assignment techniques (similarly to how we handled Regularization 2.13.1). Specifically, we will show that some assignment of values for the indeterminates turns the matrix $A = MVN = MVS(I+Z)$ into $\mathrm{diag}((I_r+Z_r), O)$, where Z_r denotes the $r \times r$ down-shift matrix of Definition 2.4.1 [and then $m_H(\lambda) = c_H(\lambda)$ for these values of the indeterminates and, therefore, also generically]. Let $S = \begin{bmatrix} I & -B^{-1}(V)C(V) \\ O & I \end{bmatrix}$ where $V = \begin{bmatrix} B(V) & C(V) \\ E(V) & G(V) \end{bmatrix}$ and observe that in this case, $VS = \begin{bmatrix} B(V) & O \\ E(V) & O \end{bmatrix}$. Then set $K = \operatorname{diag}(B^{-1}(V), O)$, $L = \begin{bmatrix} I & O \\ -E(V)B^{-1}(V) & I \end{bmatrix}$, $M = KL$, and verify that $MVS = \operatorname{diag}(I_r, O)$, so that

$$A = MVN = MVS(I + Z) = \operatorname{diag}(I_r + Z_r,\, O)\ .$$

It remains to verify that the above choice of S and M is consistent with the bounds $2d+6$ and $3d+11$ on rank $F(M)$ and rank $F(N)$, respectively [provided that rank $F(V) = d$]. Let us denote $\bar{B} = B(V)$, $\bar{C} = C(V)$, $\bar{E} = E(V)$, rank $F(W) = d(W)$, for any matrix W. Let us now apply Lemma 6.2 and deduce that $d(\bar{B}) \le d$, $d(\bar{C}) \le d+1$, $d(\bar{E}) \le d+1$. Apply Theorem 2.11.1 and Facts 2.11.6 and 2.12.1 and deduce that $d(\bar{B}^{-1}) \le d+2$, $d(\bar{B}^{-1}\bar{C}) \le 2d+4$, $d(\bar{E}\bar{B}^{-1}) \le 2d+4$. Now from Lemma 6.2 and from the simple observation that rank $F(W+I) \le 1+$ rank $F(W)$ (for any matrix W), deduce that $d(S) \le 2d+6$, $d(K) \le d+4$, $d(L) \le 2d+6$, and from Corollary 2.12.1, deduce that $d(M) \le 3d+11$.

This completes verification of the assumptions of Lemma 6.1.

Now, the evaluation of rank T has been probabilistically reduced to the evaluation of the degree of the minimum polynomial $m_A(\lambda)$, and this in turn probabilistically reduces to the evaluation of the rank of an $n \times n$ Toeplitz matrix.

In this way we complete the extension of the bounds (6.9) to the bounds on the complexity of the computation of rank A for any Toeplitz-like+Hankel-like matrix A (over any field allowing division by $n!$).

As in section 12 of chapter 2, we may also extend the complexity bounds (6.6a), (6.6b) and (6.9) to the case of computations with other dense structured matrices, and by combining the algorithms of chapter 2 with these randomized bounds, we may extend them to computing the gcd and the lcm of two polynomials, the (m,n) Padé approximation for $m = O(n)$, and the minimum span of a linear recurrence sequence. We may also combine (6.9) with Algorithm 2.10.1 or with the algorithm of [P89], page 1478, cited at the end of section 8 of chapter 2, to obtain the estimates

$$O_A\left(\gamma(n,p)\log^3 n,\ n\phi_{\mathbf{F}}(n)\,/\,(\gamma(n,p)\log^2 n)\right)$$

for the randomized complexity of the evaluation of all the entries of the extended Euclidean scheme for two polynomials over any field. According to the latter estimates, we have a processor-efficient randomized solution in this case.

7. Numerical Inversion of Well-Conditioned General Matrices.

Let us reexamine Problems 2.2.5 ($INVERT$) and 2.2.12 ($GEN \cdot INVERSE$), assuming that the $n \times n$ input matrix A is *asymptotically well-conditioned*, that is,

$$\log \text{cond}_2(A) = O(\log n), \tag{7.1}$$

and that we need to *solve the problem numerically,* by computing a matrix B such that

$$||A^+ - B||_2 \leq \epsilon ||A^+||_2, \tag{7.2}$$

for a fixed tolerance ϵ to the relative error norm of the output approximation.

A processor-efficient NC-algorithm for this task is given by Newton's iteration (3.5.4). Proposition 3.5.1 immediately implies

Fact 7.1. *Under (7.1), for $\epsilon \geq \exp(-n^g)$ and for a fixed constant g, the solutions of Problems 2.2.5 ($INVERT$) and 2.2.12 ($GEN \cdot INVERSE$) can be approximated, within the output error bound (7.2), at the cost* $O_A(\log^2 n, P(n))$.

In [P85] and [P87] this result has been extended in order to devise processor-efficient NC-algorithms for the factorization Problems 2.2.7 ($LU \cdot FACTORS$) and 2.2.9 ($QR \cdot FACTORS$) provided that the input matrix A is well-conditioned and [this is only needed for Problem 2.2.7 ($LU \cdot FACTORS$)] is also Hermitian positive definite. The proof of these extensions is similar to the proofs of Propositions 5.1 and 5.2.

From the practical point of view, Newton's iteration is particularly attractive for parallel computing, since its main building block is matrix multiplication (highly effective for parallel implementation). Besides, numerical stability can be ensured even when the input matrix is rank deficient ([PS]).

8. Numerical Inversion of Well-Conditioned Toeplitz-like+Hankel-like Matrices.

Several algorithms that we recalled in chapter 2 invert an $n \times n$ Toeplitz-like matrix by using $O(n \log^2 n)$ ops, but all of them run no faster than in linear parallel time. The NC-algorithms of sections 2, 4, 5 and 6 can be applied, but their potential work bound exceeds $n \log^2 n$ too much.

Processor-efficient NC-algorithms are available, however, for the numerical inversion of $n \times n$ matrices that simultaneously are Toeplitz-like and satisfy the relation (7.1) (being asymptotically well-conditioned), (see [P88], [P89b], [P92c], [P93e]).

Specifically, an $n \times n$ matrix A given with its F-generator of length at most r, where r is a constant independent of n and F is any displacement operator of (2.11.1)-(2.11.2), can be numerically inverted [with tolerance $\epsilon = 2^{-b}$ to the output errors] at the cost

$$O_A((\log b)\log^2 n, n \log n \ \log\log n), \tag{8.1}$$

provided that A is an asymptotically well-conditioned matrix (see [P92c]). The algorithm supporting these bounds relies on Newton's iteration (3.5.4) and can be extended to the numerical generalized (pseudo) inversion of well-conditioned rectangular Toeplitz-like matrices having full rank.

The overhead constant hidden in the "O" estimates of (8.1) is large in the general case but becomes small in the important special case where a good initial approximation to A^{-1} is readily available, and, moreover, the estimated complexity bound decreases to $O_A(\log^2 n, n)$ in this case. (Such a special case is typical in the practically important signal processing computations where the coefficients of a Toeplitz or Toeplitz-like linear system change slowly and the system must be solved in real time, that is, as fast as possible. In another class of applications to signal processing, the input Toeplitz matrix is close to a readily invertible band Toeplitz matrix.) We will next recall the algorithm of [P88], [P89b], [P92c], [P93e], for this special case.

By virtue of Proposition 3.5.1, Newton's iteration (3.5.4), $X_{k+1} = X_k(2I - AX_k)$, converges rapidly to A^{-1} if $\|X_0A - I\|_2$ is substantially less than 1, since $\|X_{i+1}A - I\|_2 = \|X_iA - I\|_2^2$, for all i. Every k-th step essentially amounts to two matrix multiplications and has a low computational cost as long as the matrix X_k is available with its short F-generator, for F of (2.11.1), (2.11.2). This follows from Corollary 2.12.1, which also implies that

$$r(F(X_{k+1})) \leq 2r(F(X_k)) + r(F(A)) + 3.$$

Here and hereafter, $r(F(Y))$ denotes the length of an F-generator available for a matrix Y. [We may slightly improve the latter bound if we replace F by one of the operators (2.11.18), (2.11.19).] Therefore, $r(F(X_k)) = O(\log n)$ if $r(F(X_0)) = O(1)$ and if, say, $k = \log\log n$, but such a rank may reach the value n in $O(\log n)$ steps, and this means a high cost of the subsequent Newton steps.

To exclude such a growth of the cost, we will periodically apply [in every $O(\log\log n)$ or even in every $O(1)$ steps] *restarting* procedure, based on the algorithm of section 11 of chapter 2 whose outline follows Theorem 2.11.2. This algorithm computes F-generators of lengths $r(F(X(X_k)))$ for some sufficiently close approximations $X(X_k)$ to X_k (and thus to A^{-1}) such that

$$r(F(X(X_k))) \le r = \text{rank }(F(A^{-1})) \le \text{rank }(\tilde{F}(A^{-1})) + 2 = \text{rank }(F(A)) + 2$$

where $\tilde{F}$ and F are the dual operators of (2.11.1)-(2.11.2) (see Theorem 2.11.1 and Corollary 2.11.1). Then we will restart Newton's iteration (3.5.4) with X_k replaced by $X(X_k)$, that is, we will write

$$X_{k+1} = X(X_k)(2I - AX(X_k)) \tag{8.2}$$

and will recursively repeat this process so as to bound the lengths of the F-generators for all the computed approximations X_k and $X(X_k)$ to A^{-1} and, consequently, to bound the computational cost of both performing each step (3.5.4) and computing $X(X_k)$ for a given X_k. Furthermore, due to Lemma 2.11.1,

$$||X(X_k) - A^{-1}||_2 \le c(n)||X_k - A^{-1}||_2,$$

$$c(n) = 1 + 2n(r(F(X_k)) - r(F(X(X_k)))).$$

Therefore, for $\kappa = \text{cond}_2(A)$, we have:

$$\rho(X(X_k)) \le c(n)\kappa\rho(X_k).$$

Here and hereafter, $\rho(W) = \rho_A(W) = ||I - WA||_2$ for any matrix W. The approximation to A^{-1} deteriorates in the transition from X_k to $X(X_k)$ but is again improved in the transition to X_{k+1}, as long as $X(X_k)$ remains close to A^{-1}.

Specifically, since $\rho(X_{h+1}) \le \rho^2(X(X_h))$ for all h, it is sufficient to ensure that

$$(c(n)\kappa)^2\rho(X_k) < 1/2, \quad \text{for } \rho(X_k) = \rho_A(X_k) \text{ and some } k = O(\log\log n), \tag{8.3}$$

in order to preserve a sufficiently rapid convergence of Newton's iteration (3.5.4), in the transition to (8.2). Then we may apply the iteration (3.5.4), except that in every k steps (3.5.4), we perform a single step (1.2) and thus ensure squaring the value $\rho(X_i)$ in every $k+1$ steps. To reach the bound

$$\rho(X_s) \le 2^{-b}, \tag{8.4}$$

for a fixed positive b, we need $s = O((k+1)\log b)$ iteration steps [for k of (8.3)]. If (8.3) holds for a fixed $k = O(1)$, then $O(\log b)$ steps suffice. If, in addition, the input matrix A is a Toeplitz-like matrix, then we may perform each restarting procedure at the cost $O(\log n, n/\log n)$ and reach the bound (8.4) at the overall cost

$$O_A((\log b)(\log n), n). \tag{8.5}$$

A similar argument based on using the operator $F = F^{\pm}$, of Example 2.11.4, yields the following extension of these bounds:

Proposition 8.1. *Let $n \times n$ matrices A and X_0 be such that (8.3) holds for $k = O(\log\log n)$. Let these two matrices be given with their F-generators of lengths $O(1)$ for F being one of the displacements operators F^+, F^-, $F^{\pm}$, F_+ and F_-, of section 11 of chapter 2. Let b be a positive constant. Then an F-generator of length $O(1)$ for a matrix X_s satisfying the inequalities $\rho(X_s) \le 2^{-b}$, $\|X_s - A^{-1}\|_2 \le 2^{-b}\|A^{-1}\|$ can be computed at the cost $O_A((\log b)(\log n)k, n)$.*

If we are only given a short displacement generator of a Toeplitz-like+Hankel-like $n \times n$ matrix A, which is asymptotically well-conditioned (but if we are not given any initial approximation to A^{-1}), then we may approximate A^{-1} at a slightly higher cost, bounded by (8.1), ([P92c]). The idea is to obtain the desired initial approximation X_0 by combining the two techniques a) of an artificial (variable) diagonal (of [P85], [P87]) and b) of homotopy. We will next outline this algorithm (not used in our book) referring the reader to [P92c] and [P93e] for the details.

Without loss of generality, assume that A is an h.p.d. matrix; otherwise, recall that $(A^HA)^{-1}A = A^{-1}$, $\text{cond}_2(A^HA) = (\text{cond}_2(A))^2$, and rank $(F(A^HA)) \le 2$ rank $(F(A)) + 1$ (see Corollary 2.12.1).

Now observe that $I - A_0X = -A/b$, for $A_0 = A + bI$, $X_0 = I/b$, and that $||A_0^{-1}||_2 \le (1/b)||(I + A/b)^{-1}||_2 \le 2/b$ if $b \ge 2||A||_2$. By choosing a larger value for b, ensure the inequality (8.3), for A replaced by A_0 and for $X_0 = I/b$. Then compute an approximation to A_0^{-1} within relative error norm σ_0 in $O(\log\log(1/\sigma_0))$ Newton's steps. Choose σ_0 such that (8.3) would hold for $X_k = \tilde{A}_0^{-1}$ (the output approximation to A_0^{-1}), for a fixed constant $p < 1$, and for A replaced by $A_1 = A + pbI$. Let the matrix $\tilde{A}_0^{-1}$ serve as an initial approximation to A_1^{-1} in Newton's iteration (3.5.4) for $A = A_1$. This iteration outputs an approximation $\tilde{A}_1^{-1}$ to A_1^{-1}, within a relative error norm σ_1. σ_1 is chosen so as to insure (8.3) for $X_k = \tilde{A}_1^{-1}$ and for A replaced by $A_2 = A + p^2bI$. Then recursively repeat this process of approximating

$A_j^{-1} = (A + p^j bI)^{-1}$ by $\tilde{A}_j^{-1}$ for $j = 0, 1, \dots, m$, until you arrive at a matrix $\tilde{A}_m^{-1}$ being sufficiently close to A^{-1}. Actually, $||A_j A_{j-1}^{-1} - I||_2 \leq 1 - p$ for all j, and we require that $||\tilde{A}_j^{-1} - A_j^{-1}||_2 \leq \delta ||A_j^{-1}||_2$, $\delta = p(1-p)/(2(p+\kappa^+))$, where we set $\kappa^+ = n||\tilde{A}_j^{-1}||\ ||A_j||/(1 - \alpha(j)) \geq \kappa$, $\alpha(j) = n||I - \tilde{A}_j^{-1} A_j||_1 \geq ||I - \tilde{A}_j^{-1} A_j||_2$. With this tolerance value δ for the output error norms, relatively few Newton's steps (3.5.4), starting with $X_0 = \tilde{A}_{j-1}^{-1}$ and ending with $X_s = \tilde{A}_j^{-1}$, shall ensure that $||A_j \tilde{A}_j^{-1} - I||_2 \leq q = 1 - p(1+p)/2$ for every j. Choosing $m \geq \ln\left(\frac{n\kappa^+}{(1-p)^2}\right) / \ln\left(\frac{1}{p}\right)$, we also ensure that $||AA_m^{-1} - I||_2 \leq 1 - p$, $||A\tilde{A}_m^{-1} - I||_2 \leq q$, and it remains to apply Newton's iteration (3.5.4) to A with $X_0 = \tilde{A}_m^{-1}$.

To combine these homotopy techniques with the recursive restarting procedure, we may choose, say, $q = 1/32$, $p = 0.97902$ and $m = 47.14 \ln($ $2269.7 n\kappa^+)$. Then, for every fixed j, we may start with $X_0 = \tilde{A}_{j-1}^{-1}$ and quickly arrive at (8.3) with A replaced by A_j. At this point, X_k is sufficiently close to A_j^{-1} so that we may replace X_k by $X(X_k)$ and restart Newton's iteration, periodically replacing the computed matrices by their approximations of bounded F-ranks. To define an initial approximation X_0 in the Newton iteration (3.5.4) for inverting A_{j+1}, for each $j \leq m$, we use the approximation to A_j^{-1}, within $\delta \leq p(1-p)/(2(p+\kappa))$, computed at the previous recursive step.

9. Parallel Boolean Complexity of Matrix Computations. Parallel Arithmetic Complexity in the Case of Integer Input Matrices. Some Further Extensions.

In this section we will first follow [P85] and [P87] in order to devise algorithms supporting the record estimates for the parallel Boolean complexity of Problems 2.2.4–2.2.6 ($DETERMINANT$, $INVERT$, $CHAR \cdot POL$), where all the input values are integers or are made integers by scaling the input. (By using the techniques of section 2 of chapter 2, the reader may easily extend these results to yield parallel solution of $LIN \cdot SOLVE$, $M \cdot RANK$, $LU \cdot FACTORS$ and so on.) Then we will follow [P93c] and will combine these algorithms with Proposition 8.1 to reach the record bounds on the parallel Boolean complexity for the same problems and for $M \cdot RANK$ in the Toeplitz-like+Hankel-like case. Finally, we will list some further extensions.

All the Las Vegas randomized algorithms of this section are in RNC and are processor-efficient under the Boolean computational model. Furthermore, they require the precision of computation, which is on the optimum level of the output precision (within a small constant factor). This precision is too high for the large scale numerical computation by these algorithms, which is the usual phenomenon when one computes the exact solution to a linear system of equations over the rationals; the algorithms, however, may be practically interesting for symbolic computing. Some of our algorithms give RNC and processor-efficient solutions to $LIN \cdot SOLVE$, $DETERMINANT$, $M \cdot RANK$ and, consequently, to several related computational problems, in both cases of general and Toeplitz-like+Hankel-like input matrices, under both Boolean and arithmetic PRAM models (and, again, these algorithms require the asymptotically optimum computational precision).

We will represent the output matrix A^{-1} of Problem 2.2.5 ($INVERT$) by the pair (adj A, $\det A$), so as to make the output values integers lying between $-N$ and N, $N = (1 + ||A||)^n$ (see Fact 3.2.2). Until the end of this section, $||.||$ will denote the 2-norm $||.||_2$.

Our first Boolean complexity estimates will rely on the algorithm of [GP89] for Problems 2.2.4–2.2.6 of computing the determinant, the inverse and the characteristic polynomial of a matrix, respectively. In this algorithm, we only need to use divisions by $2, 3, \ldots, n$, additions, subtractions and multiplications, and we arrive at the arithmetic cost bound (4.3). Thus we choose a prime $p > n$, choose an integer h such that $p^h > 2N$, perform all the computations modulo p^h (including the divisions by 2,3,..., n, whose results are integers in the range from 0 to $p^h - 1$), and recover the integer output from its value modulo p^h. We arrive at

Proposition 9.1 ([P85], [P87]). *Problems 2.2.4–2.2.6 ($DETERMINANT$, $INVERT$ and $CHAR \cdot POL$) can be deterministically solved at the Boolean cost*

$$O_B(\log n \ \log (\tilde{d}n), \widetilde{P}(n)\, \mu(\tilde{d})), \tag{9.1}$$

where $\widetilde{P}(n)$ is from (4.3), $\mu(h)$ is defined in (3.1.1), and $\tilde{d} = O(n \log (1 + ||A||))$.

Next, we will improve the bound (9.1) by using p-adic lifting (Algorithm 3.3.1). We will need to use the single random integer parameter p in order to avoid singularities.

Let $\mathbf{v}$ be a vector, A be a matrix, both filled with integers, and p be a prime, such that

$$\det A \neq 0 \bmod p, \tag{9.2}$$

$$\det K(A, \mathrm{v}) \neq 0 \bmod p, \tag{9.3}$$

and $p > n$. (With these restrictions on p, divisions by 0 mod p will be avoided in our computations.) Apply Corollary 3.1 and the algorithm of [GP89] to compute modulo p the matrices A^{-1} and $K^{-1}(A, \mathbf{v})$. This implies the bound (9.1), with $\tilde{d}$ replaced by $\log p$. Then apply Newton-Hensel's Algorithm 3.3.1 in order to compute $A^{-1} \bmod p^H$ and $K^{-1}(A, \mathbf{v}) \bmod p^H$, $H = 2^h$, and choose h such that $p^H > 2(1 + ||A||)^n$. Then compute the vector $\mathbf{c}(A) = K^{-1}(A, \mathbf{v})A^n\mathbf{v} \bmod p^H$. [By virtue of the Cayley-Hamilton Theorem 2.1.1, the vector $\mathbf{c}$ of the coefficients of the characteristic polynomial of A satisfies the vector equation $K(A, \mathbf{v})\mathbf{c} = A^n\mathbf{v}$ for any fixed vector $\mathbf{v}$, so $\mathbf{c} = \mathbf{c}(A) \bmod p$.] The entries of the vector $\mathbf{c}(A)$ and of the matrix adj $A = A^{-1} \det A$ are integers, whose absolute values are less than $(1 + ||A||)^n$. Thus, we immediately recover adj A and $\mathbf{c}$ from their values modulo p^H. This solves Problems 2.2.4–2.2.6 at a relatively high arithmetic cost but at a low Boolean cost:

Theorem 9.1 ([P85], [P87]). *Let A be a matrix, $\mathbf{v}$ a vector (both filled with integers) and p a prime such that $p > n$ and (9.3) holds. Then Problems 2.2.4 (DETERMINANT) and 2.2.6 (CHAR · POL), of computing $\det A$ and the coefficients of $c_A(\lambda) = \det(\lambda I - A)$, can be solved at the parallel Boolean cost bounded by*

$$O_B(\log n \, \log (dn), P(n)\, \mu(d)), \tag{9.4}$$

for $P(n)$ of (3.5) and (4.1), $\mu(h)$ defined by (3.1.1), and $d = O(n \log (1 + ||A||))$. If, in addition, (9.2) holds, then the bound (9.4) also holds for the parallel Boolean complexity of Problem 2.2.5 (INVERT) of computing A^{-1}.

Let us try to relax the assumptions (9.2) and (9.3). If [under (9.3)] we solve Problem 2.2.4 (*DETERMINANT*) of computing $\det A$, we may immediately check if (9.2) holds. If $\det A = 0 \bmod p$, then we may choose and test new primes p (sequentially or in parallel), until we ensure (9.2). If $\det A \neq 0$, (9.2) must hold for at least one of n distinct primes $p \geq ||A||$, since $|\det A| \leq ||A||^n$, and similarly if we replace A by $K(A, \mathbf{v})$ (see Fact 3.2.2). Corollary 3.2.1 enables us to estimate the probability that (9.2) and (9.3) hold for a random choice of p. According to these estimates, such a probability converges to 1 as $n \to \infty$ (so that the algorithm supporting Theorem 9.1 is a randomized and processor-efficient NC-algorithm for Problems 2.2.4–2.2.6) provided that

$$\det A \neq 0, \tag{9.5}$$

$$\det K(A, \mathbf{v}) \neq 0, \tag{9.6}$$

and p is a random prime in the interval of (3.2.3) for $f(n)$ of Corollary 3.2.1. By solving Problem 2.2.4 ($DETERMINANT$), we may test the assumption (9.5), which is only needed to support the estimate (9.4) for matrix inversion, so this assumption means no loss of generality.

For a random choice of a vector $\mathbf{v} = (v_0, \dots, v_{n-1})^T$, (9.6) holds with a high probability provided that $m_A(\lambda)$, the minimum polynomial of A, has degree n or, equivalently, that

$$c_A(\lambda) = \det(\lambda I - A) = m_A(\lambda). \tag{9.7}$$

Indeed, (9.7) implies that the matrices $I, A, \dots, A^{n-1}$ are linearly independent; therefore, $\det K(A, \mathbf{v})$ is not equal to 0 for some $\mathbf{v}$. It remains to observe that $\det K(A, \mathbf{v})$ is a polynomial in $v_0, \dots, v_{n-1}$ of the overall degree at most n and to apply Corollary 1.5.1.

Finally, for Problems 2.2.4 ($DETERMINANT$) and 2.2.5 ($INVERT$), we may probabilistically ensure (9.7) by applying, say, Regularization 2.13.5; the shift from A to a new input matrix changes the overall computational cost bounds by at most a constant factor. Summarizing, we arrive at the following results:

Theorem 9.2 ([P85], [P87]).

a) The randomized (Las Vegas) complexity bounds (9.4) for Problem 2.2.4 ($DETERMINANT$) and 2.2.5 ($INVERT$) of computing $\det A$ *and the inverse A^{-1} hold with a probability converging to 1 as $n \to \infty$ for an $n \times n$ input matrix A with integer entries.*

b) The same is true for Problem 2.2.6 ($CHAR \cdot POL$) of computing all the coefficients of $\det(\lambda I - A)$ *provided that (9.7) holds.*

Combining the randomized algorithm for Problem 2.2.6a ($MIN \cdot POL$) of section 4 with Newton-Hensel's lifting Algorithm 3.3.1, the reader may arrive at the same favorable probabilistic parallel Boolean complexity estimates for Problem 2.2.6a ($MIN \cdot POL$). This is also an alternative way of proving part a) of Theorem 9.2; moreover, this way leads to smaller overhead constants and seems to be more attractive for practical application, except that it needs an order of n random parameters, versus a single parameter p.

In [P85] and [P87], a slightly different approach, using the artificial diagonal techniques, also gave the randomized bound

$$O_A((\log n)\log (n \log \|A\|),\ P(n)). \tag{9.8}$$

We will next recall a simplified version of this approach, also used in [P85] and [P87], Corollary 5.7, and leading to the computation at the cost bounded by $O_A((\log^2 n)\log(n\log\|A\|), P(n)/\log n)$ with the optimum asymptotic precision of $O_A(n\log(n\log\|A\|))$ bits, for an $n\times n$ integer input matrix A (compare also with [P93c]). At the end of this section, we will observe that the same algorithm, combined with our algorithms of the previous section, gives us an RNC and processor-efficient solution of Toeplitz-like+Hankel-like linear system over the rationals, which can be immediately extended to an RNC and processor-efficient algorithm for polynomial gcd over the rationals. In this version of the approach of [P85] and [P87], we first apply the technique of artificial variable diagonal, that is, we invert the auxiliary diagonally dominant matrix

$$A(p) = (A \bmod p) + 32pn^3 I, \tag{9.9}$$

for p being a random prime in the interval (3.2.3), with $f(n)$ of Corollary 3.2.1, for $b=1$. We set $X_0 = I/(32pn^3)$ so that $\|I - X_0A(p)\|_\infty < 1/(30n^2)$, and rapidly approximate $A^{-1}(p)$ via Newton's iteration (3.5.4).

From (2.2.1) and (2.2.3) we deduce that

$$\det A = \det B \ \det S \tag{9.10}$$

for $A = \begin{pmatrix} B & C \\ E & G \end{pmatrix}$, $S = G - EB^{-1}C$. With no loss of generality, we assume that $n = 2^k$, k is an integer, $k \geq 10$ (say), B, C, E, G are $(n/2)\times(n/2)$ matrices, and we recursively apply (9.10), with A replaced by $A(p)$, to reduce *DETERMINANT* for $A(p)$ to a series of the computations of the products and inverses of matrices of sizes recursively decreasing by factor 2. In k recursive steps, we will decompose $\det A(p)$ into a product of the determinants of 1×1 matrices. We will call these 1×1 matrices *associated with* $\det A$.

In the analysis of this algorithm, we will need some auxiliary estimates.

Lemma 9.1. *Let B, C, E, G be $m\times m$ matrices, B nonsingular, $S = G - EB^{-1}C$, $A = (a_{i,j}) = \begin{pmatrix} B & C \\ E & G \end{pmatrix}$. Let $\|.\|$ denote $\|.\|_h$ for $h=1$ or for $h=\infty$, let a, c be two positive scalars, $c > a$, and let $\|A - cI_{2m}\| \leq a$. Then $\|S - cI_m\| \leq ac/(c-a) + a^2c$.*

Proof. Denote that $W = I_m - B/c$ so that $(B/c)^{-1} = cB^{-1} = \sum_{i=0}^{\infty} W^i$. Deduce from the assumption of Lemma 9.1 that $\|C\| \leq a$, $\|E\| \leq a$, $\|W\| \leq$

a/c, $\|B^{-1} - I_m/c\| \leq 1/(c-a)$, $\|G - cI_m\| \leq a$. Therefore, $\|S - cI_m\| \leq \|G - cI_m\| + \|E\|\|B^{-1}\|\|C\| \leq a\left(1 + \frac{a}{c-a} + \frac{a}{c}\right) = ac/(c-a) + (a^2/c)$. ∎

Corollary 9.1. *All the n matrices associated with* $\det A$ *have norms less than* $pn + 32pn^3$.

Proof. We will first deduce the strict bound $32pn^3 + 2pn$, by applying Lemma 9.1, at first to $A = A(p)$, and then, recursively, to the half size leading principal submatrix and to its Schur complement in the input matrix A of the previous application of Lemma 9.1. In all these applications, $c = 32pn^3$, whereas the value a grows very slowly, in each application of Lemma 9.1. In the first recursive step, we may set $a = a_0 = pn$. Then Lemma 9.1 gives us $a = a_1 = pn\left(1 + (16n^2)^{-1} + (32n^2(32n^2-1))^{-1}\right)$ for the next step, and, according to a simple inspection, $a = a_k < 2pn$ in k steps; we also note that the norms of all matrices associated with $\det A(p)$ are less than $c + a_k$. To decrease all the bounds on $a = a_i$, $i = 1, 2, \ldots, k$, by the factor 2 and thus to arrive at the desired bound of Corollary 9.1, we modify our first step. Instead of the application of Lemma 9.1 to $A = A(p)$, we repeat the proof of this lemma, setting $B = B(p)$, $C = C(p)$, $E = E(p)$ and $G = G(p)$. ∎

It suffices to approximate all the matrices associated with $\det A(p)$ up to within the relative output error bound $E/(4n)$ (say) in order to ensure approximating $\det A(p)$ within the relative output error bound $E/2$ and the absolute error bound $|\det A(p)|E/2$. Due to Corollary 9.1, we have $|\det A(p)| < (32pn^3 + pn)^n$. Therefore, by setting $E = (32pn^3 + pn)^{-n}$, we ensure that $|\det A(p)|E/2 < 1/2$, and this enables us to recover the exact value of $\det A(p)$ by means of rounding off to the nearest integer.

It remains to approximate the single entry of each matrix associated with $\det A(p)$ so as to stay within the relative error bound $E/(4n)$. To achieve this, we perform all the computations with a sufficiently high precision and apply sufficiently many steps of Newton's iteration for the inversion of the auxiliary matrices, where the matrices $I_s/(32pn^3)$, for appropriate s, serve as the initial approximations X_0. Routine error analysis (compare Appendix A of chapter 3), based on the bounds (3.1.2), (3.1.3) and Proposition 3.5.1 (where, in our case of inverting strongly diagonally dominant matrices, we may set $\text{cond}_2 A < 2$), suffices to show that already $O(\log(n \log p))$ steps of each Newton's iteration, performed with the precision of $O(n \log(n \log p))$ bits, guarantee the desired error bound $E/(4n)$ for $E = (32pn^3 + pn)^{-n}$. In this way, at the cost bounded according to (9.8), we compute $\det A(p)$ within an absolute error less than $1/2$. Rounding off to the nearest integers gives us the exact values at first of $\det A(p)$ and then of adj $A(p)=A^{-1}(p)\det A(p)$, (compare Remark 3.3.1). Recall (9.9), obtain that $A(p)$ mod $p=A$ mod p, and compute $(\det A)$ mod $p=(\det A(p))$ mod p, (adj A) mod $p=$(adj $A(p)$) mod p, and, if (9.2) holds, then also

$$A^{-1} \bmod p = (((\text{adj } A) \bmod p)/((\det A(p)) \bmod p)) \bmod p.$$

Apply similar techniques and, by means of the same recursive process and within the same asymptotic cost bounds, compute modulo p the matrices B^{-1} and S, as well as other matrices involved in the recursive algorithm for computing $\det A$. This algorithm, based on (9.10), also computes modulo p the determinants of B and S, as well as other factors modulo p of $\det A$ mod p, defined via recursive application of (9.10).

Next choose $h = \lfloor \log (2\|A\|^n)/\log p \rfloor + 1$ such that

$$p^H > 2\|A\|^n \geq \max\{|\det A|, \|\text{adj } A\|\} \tag{9.11}$$

for $H = 2^h$ (compare Fact 3.2.2). Apply h steps of Newton-Hensel's Algorithm 3.1 to compute A^{-1} mod p^H, B^{-1} mod p^H, S mod p^H. Extend this process to the evaluation modulo p^H of all the matrices involved in the factorization of $\det A$, according to the recursive application of (9.10). (Here we need strong nonsingularity of A, which can be ensured via symmetrization.) In particular, in this way, we compute the n matrices of size 1×1 whose entries, being multiplied together modulo p^H, will give us $\det A$ mod p^H. Finally, we recall (9.11) and apply the simple modular rounding-off technique (see section 3 of chapter 3) to recover the integer output values $\det A$ and adj A from $(\det A)$ mod p^H and (adj A) mod $p^H=$ $(A^{-1}$ mod $p^H)((\det A)$ mod $p^H)$ mod p^H.

We verify easily that the overall arithmetic cost of these computations is bounded by (9.8), as promised, and the optimum asymptotic precision of $O(n \log (n \log \|A\|))$ bits suffices. Moreover, we also recursively apply this algorithm in the binary search process to compute the maximum size of the nonsingular leading principal submatrix of the matrix A^TA, over the rationals, which gives us rank(A^TA) = rank A. This implies randomized solution of Problem 2.2.10 ($M \cdot RANK$), at the asymptotic cost

$$O_A(\log^2 n \ \log (n \log \|A\|), P(n)), \tag{9.12}$$

and the asymptotic bound on the precision of this computation remains at the optimum level, of $O(n \log (n \log \|A\|))$ bits.

The reader may apply the output sensitive modification and/or the concurrent divide-and-conquer modification of the binary search algorithm for matrix rank [compare (2.13.16) and Theorem C.3 in Appendix C] and estimate the resulting impact of these modifications on the complexity estimates (9.12).

Finally, if A is a Toeplitz-like+Hankel-like matrix, then the same algorithms, combined with the techniques of recursive restarting of the previous section, immediately enable us to obtain solutions, over the rationals, for all the computational problems cited above. Indeed, all operations with matrices are now reduced to the operations with their short displacement generators, and each $k \times k$ matrix-by-vector multiplication is now performed at the cost $O_A(\log k, k)$. In all cases, except for $M \cdot RANK$, the randomized overall asymptotic arithmetic cost of computing the solution is bounded by $O_A(\log^2 n \log (n \log \|A\|), n)$, and, moreover, the asymptotically optimum precision of $O(n \log (n \log (\|A\|))$ bits suffices in these computations. For $M \cdot RANK$, the arithmetic time bound grows by a logarithmic factor due to the application of the binary search process of Algorithm 2.8.2 or its output sentive and/or concurrent divide-and-conquer modifications.

Note that, in our case, we have the assumption (8.3) held already for $k = 0$ [for the input matrix A, this is due to (9.9) and, for the auxiliary matrices defined by (9.10), to exercise 4d of chapter 2]. Due to (8.3), effectiveness of recursive restarting of (8.2) is guaranteed in this case. Recursive restarting is also needed at the stage of application of Newton-Hensel's Algorithm 3.3.1. At this stage, we may again apply the algorithm that we outlined after Theorem 2.11.2, but we may also use any other solution of Problem 2.2.11b ($G \cdot COMPRESS$).

Due to the results of section 8 (or alternatively 9 and 10) of chapter 2, the latter solutions of Problems 2.2.3 ($LIN \cdot SOLVE$) and 2.2.10 ($M \cdot RANK$) in the Toeplitz-like (or even Toeplitz) case, together with the asymptotic estimates for the parallel arithmetic cost and the bit-precision of the computations, are immediately extended to yield RNC and processor-efficient algorithms for Problems 1.5.1 ($POL \cdot GCD$), 1.5.1a ($POL \cdot LCM$), 1.5.2b ($PAD\acute{E}$) and 1.5.3 ($RECUR \cdot SPAN$) over the rationals.

Exercises to Chapter 4.

1. a) Recall the algorithms of chapter 1 based on FFT, substitute the parallel cost of FFT and verify the respective estimates of Table 1.2.

   b) Substitute the bounds of Table 1.1 and of Fact 2.1 and extend the estimates of Table 1.2 to the asymptotic Boolean complexity estimates, assuming all the computations (including FFT) performed modulo a fixed (large) prime.

   c) Summarize (in the form of a digraph) the main reduction techniques that transform various problems of this chapter to each other over the complex field **F** of constants and over **F** being an arbitrary field.

2. Supply the overhead constants to the asymptotic estimates of this chapter, assuming the upper bound $(t,p) \le (3\log n, 2n)$ for FFT on n points.

3. Prove the arithmetic bound of Fact 2.1 (by using the B-principle).

4. Extend Algorithm 2.1 to the evaluation of the partial products $\prod_{i=1}^{k} a_i$ for $k = 1,\dots,n$ and for the input $a_1,\dots,a_n$ and to the evaluation of the terms of the linear recurrence $x_k = a_k x_{k-1} + b_k$, for $k = 1,\dots,n$ and for the input $x_0 = 0, a_1,\dots,a_n, b_1,\dots,b_n$.

5. Let $T(x)$ be a polynomial, $T(0) = 1$, $K = 2^k$, k be a positive integer. Assume that the field of constants supports DFT on K^2 points.

   a) Exploit the identities $1/T(x) = 1/(1-(1-T(x))) = \sum_{i=0}^{\infty}(1-T(x))^i = \prod_{j=0}^{\infty}(1+(1-T(x))^J)$, $J = 2^j$, in order to compute the polynomial $v_K(x) = (1/T(x)) \bmod x^K$ by using $O(\log K)$ steps with K^2 processors. [Apply the evaluation-interpolation techniques in order to compute the coefficients of $(1-T(x))^i$, $i = 0,1,\dots,K-1$.]

   b) Arrive at the same estimates by first recursively computing the values $w_H(x) = \prod_{j=0}^{h}(1+(1-T(x))^J)$, $J = 2^j$, $H = 2^h$, $h = 0,1,\dots,\log K$, for x taking the values of all the K^2-th roots of 1, and then by recovering the coefficients of $w_K(x)$ by means of interpolation.

   c) Compute the coefficients of the K-th power of a polynomial of degree $K-1$ at the cost $O_A(\log K, K^2)$.

6 ([Pr]). Let $V = (x_i^j)$ be a nonsingular $n \times n$ Vandermonde matrix. Prove that $(V^T)^{-1} = LU$, where $U = L_0(1)^T D_0^{-1} \cdots L_{n-1}(1)^T D_{n-1}^{-1}$,

$L = L_{n-1}(x_n) \dots L_0(x_1)$,

$$L_k(\alpha) = \begin{pmatrix} I_k & O \\ O & B \end{pmatrix}, \quad B = \begin{pmatrix} 1 & & & O \\ -\alpha & 1 & & \\ & \ddots & \ddots & \\ O & & -\alpha & 1 \end{pmatrix}, \quad D_k = \begin{pmatrix} I_{k+1} & O \\ O & D \end{pmatrix},$$

$D = \text{diag}\,(x_{k+2} - x_1, \dots, x_n - x_{n-k-1})$, (compare [GL]). Apply the above relations to prove that $(L)_{ij} = \sigma_{i-j}(-x_1, \dots, -x_{i-1})$ for $j \leq i$, $(L)_{ij} = 0$, otherwise [where $\sigma_s(a_1, \dots, a_t)$ is the s-th degree elementary symmetric function of $a_1, \dots, a_t$], and $(U)_{ij} = PREFIX_j(\alpha_{i1}, \dots, \alpha_{in})$ $= \prod_{s=1}^{j} \alpha_{is}$ for $j \geq i$, $\alpha_{ij}(x_i - x_j) = 1$ for $i \neq j$, $\alpha_{ii} = 1$; $(U)_{ij} = 0$ for $j < i$. Conclude that the entries of L and U can be computed at the cost of $O_A(\log n, n^2 \log\log n)$ over any field $\mathbf{F}$ of constants, in which the matrix V is nonsingular, or at the cost $O_A(\log n, n^2)$ if, in addition, $\mathbf{F}$ supports FFT. Extend this result to polynomial interpolation.

7. a) Prove that a bilinear algorithm for $s \times s$ matrix multiplication using r bilinear multiplications implies the bound of $O(n^{3\log r/\log s})$ ops on the asymptotic complexity of $n \times n$ matrix multiplication; then extend this bound to $O_A(\log n, n^{3\log r/\log s}/\log n)$ provided that $3\log r/\log s > 2$.

   b) Try to prove the bound $O_A(\log n, n^{\omega})$ on the complexity of $n \times n$ matrix multiplication (open problem).

8. a) Supply the details of the proof of (4.1) for the inversion of an $n \times n$ triangular matrix A by means of the recursive factorization (2.2.1)-(2.2.3). Then deduce the bound $O_A(\log^2 n, P(n))$, which is only slightly inferior to (4.1), by relying on the matrix equations

$$A^{-1} = (I - (I - A))^{-1} = \sum_{i=0}^{n-1} (I - A)^i = \prod_{j=0}^{h} (I + (I - A)^{2^j})$$

   provided that $h = \lceil \log n \rceil - 1$ and that $I - A$ is a *proper triangular matrix* with its diagonal filled with zeros. Extend this result to an arbitrary nonsingular triangular matrix A by means of its scaling.

   b) ([PP], [PP,a], [P,a]). Deduce from (4.1) the bounds $O_A(n^{\alpha}\log^2 n, n^{2-\alpha}/\log^2 n)$, $\alpha = 1 - 1/(\bar{\omega} - 1)$, for solving a nonsingular triangular linear system of n equations, for any $\bar{\omega}$ satisfying $2 \leq \omega \leq \bar{\omega} \leq 3$. Try to obtain a further improvement. *Hint:* Compute an unbalanced factorization of Remark 2.2.1 and apply it recursively. Alternatively, apply block Gaussian elimination, choosing $k \times k$ diagonal blocks for an appropriate k, and concurrently inverting all these n/k blocks at the overall cost $O_A(\log^2 k, (n/k)P(k)/\log k)$, [compare (4.1)].

c) ([PP], [PP,a]). Rely on $k \times k$ matrix multiplication at the cost $O_A(\log k, k^3/\log k)$ and on $k \times k$ matrix inversion at the cost $O_A(\log^2 k, k^4/\log k)$ over any field **F** of constants (compare sections 3 and 6) and deduce the deterministic bounds $O_A(n^{1/3}\log^2 n, n^{8/3}/\log n)$ for the inversion of an $n \times n$ strongly nonsingular matrix over any field of constants. Extend these bounds to $O_A(n^{1/5}\log^2 n, n^{4/5}\log^2 n)$ by relying on the estimates $O_A(\log k, k^3/\log n)$ and $O_A(\log^2 k, k^{7/2}/\log n)$ for $k \times k$ matrix multiplication and inversion, respectively. *Hint:* Compute an unbalanced recursive factorization (2.2.2).

d) ([PP], [PP,a]). Over the field of complex constants, extend the approach of 8c) to solving a nonsingular Toeplitz-like linear system of n equations at the cost $O_A(n^{1/2}\log^2 n, n/\log n)$ and, more generally, at the cost $O_A(n^a\log^2 n, n^{2-2a}/\log n)$ for any a, $0 \le a < 1$. Then extend these bounds to computing the gcd of two polynomials of degree n. Can we devise a work optimum NC-algorithm for computing all the entries of the extended Euclidean scheme for two polynomials along this line?

e) ([P93a]). Algorithm 2.8.2 for the binary search can be interpreted as an example of supereffective slowdown: the objective of the binary search is to replace n concurrent calls to a subroutine by $\log n$ successive calls to it. Thus, if each call has the complexity bounds $O_A(t, p)$, then the entire solution has the complexity bounds $O_A(t, np)$ without binary search and $O_A(t\log n, p)$ with it. As an exercise, replace the binary search by the k-ary search that supports the complexity bounds $O_A(t\log n/\log k, (k-1)p)$ for any $k \ge 2$. Specify these bounds for $k = n^d$, $0 < d \le 1$, and $k = 2^{b(\log n)^c}$, $0 \le c \le 1 \le b$.

9 ([PSa]). Reduce the evaluation of trace(A^k), for $k = 1, 2, \ldots, n$ and for an $n \times n$ matrix A, to rectangular matrix multiplication in order to arrive at the estimate $O_A(\log^2 n, n^{\omega+0.5})$ for the complexity of this problem and, therefore, of Problem 2.2.6 ($CHAR \cdot POL$). Extend this complexity bound to Problems 2.2.3 ($LIN \cdot SOLVE$) (use Cayley-Hamilton Theorem 2.1.1 and Proposition 2.2.1) and 2.2.5 ($INVERT$) (apply Theorem 4.1 or, alternatively, rectangular matrix multiplication).

10. Examine [GP89], prove (4.3), and specify β in $\widetilde{P}(n) = O(n^\beta)$.

11. Estimate the probability of a failure of the randomized algorithms of this chapter as a function of the cardinality of the set of available distinct elements supporting the randomization. For the field of integers modulo a prime p, show how the parallel complexity estimates change in the transition to the extension fields, that is, express these estimates in terms of p and the input size I. Specify these estimates for $p = 2$, $p = 5$ and $p = n/2$ where the input is an $n \times n$ matrix, n is even (compare [KP,a]).
12 ([E91]). Reduce Problems 2.2.7b ($PLU \cdot FACTORS$) and 2.2.9a ($QRP \cdot FACTORS$) to Problems 2.2.5 ($INVERT$) and 2.2.10 ($M \cdot RANK$) and show the resulting randomized NC-complexity bounds.
13. Modify Problems 2.2.9 ($QR \cdot FACTORS$) and 2.2.9a ($QRP \cdot FACTORS$) so as to preserve the orthogonality of the factors Q but to relax the normalization assumption, that is, require that $Q^H Q$ be a nonsingular diagonal matrix but not necessarily the identity matrix I. Compute such factorizations over any field of constants and supply the parallel complexity estimates.
14. Over the field of characteristic p, $0 < p \leq n$, try to find a deterministic construction that would complement the system (1.4.8) of Newton's identities to obtain a nonsingular linear system of equations defining the vector of the coefficients of the characteristic polynomial of an $n \times n$ matrix (compare Appendix C).
15. Estimate the parallel computational complexity of
    a) Problem 2.2.7 ($LU \cdot FACTORS$) with a strongly nonsingular Toeplitz-like input matrix,
    b) Problems 2.2.6 ($CHAR \cdot POL$) and 2.2.6a ($MIN \cdot POL$) for Vandermonde and Vandermonde-like matrices.
16. Supply the details of the proof of the bound (9.1).
17 a) ([P87]). Estimate the parallel Boolean cost of the computation by the following randomized and processor-efficient NC-algorithm for Problems 2.2.3 ($LIN \cdot SOLVE$), 2.2.4 ($DETERMINANT$) and 2.2.6 ($CHAR \cdot POL$) (the estimates are only slightly inferior to ones of Theorem 9.1):

    Input: an $n \times n$ matrix A [and for Problem 2.2.3 ($LIN \cdot SOLVE$), also an n-dimensional vector **b**] filled with integers.

    Output: solutions to Problems 2.2.3 ($LIN \cdot SOLVE$), 2.2.4 ($DETERMINANT$) and 2.2.6 ($CHAR \cdot POL$).

    Computation:

0 (Initialization). Choose n random integers that form a vector $\mathbf{v}$ having norm of the order $n^{O(1)}$, choose $2n$ distinct integers $\lambda_1, \lambda_2, \dots, \lambda_{2n}$ (say, let $\lambda_j = j$ or choose λ_j at random), and compute the coefficients of the $2n+1$ polynomials $\prod_{j=1}^{2n}(\lambda - \lambda_j)$, $\prod_{k \neq j}(\lambda - \lambda_k)$, $j = 1, \dots, 2n$. Here and hereafter, $\prod_{k \neq j}$ denotes the product in k ranging from 1 to $2n$ but not taking the value j.

1. Compute the Krylov matrix $K(A, \mathbf{v}, 2n)$.

2. Compute the $2n+1$ vectors

$$\mathbf{b} = \left(\prod_{j=1}^{2n}(A - \lambda_j I)\right)\mathbf{v}, \; \mathbf{x}(j) = \left(\prod_{k \neq j}(A - \lambda_k I)\right)\mathbf{v}, \; j = 1, \dots, 2n,$$

which satisfy the $2n$ following linear systems of equations:

$$(A - \lambda_j I)\mathbf{x}(j) = \mathbf{b}, \quad j = 1, \dots, 2n.$$

3. Compute the coefficients of the polynomials $q(\lambda)$ (of degrees at most $n-1$) and $p(\lambda)$ (monic and of degree at most n) such that $x_1(j) = q(\lambda_j)/p(\lambda_j)$ for $j = 1, \dots, 2n$, where $x_1(j)$ is the first component of the vector $\mathbf{x}(j)$. If $p(\lambda)$ has degree n, then denote:

$$p(\lambda) = \det(\lambda I - A) = \sum_{i=0}^{n} c_i x^i \, ,$$

output $(-1)^n c_0$ for Problem 2.2.4 (*DETERMINANT*), output $c_0, \dots, c_{n-1}$ for Problem 2.2.6 (*CHAR · POL*), and go to Stage 4 for Problem 2.2.3 (*LIN · SOLVE*); otherwise, if $\deg p(\lambda) < n$, repeat the computations for a new choice of a random vector $\mathbf{v}$ and a distinct set of $\lambda_1, \dots, \lambda_{2n}$.

4. Compute the Krylov matrix $K(A, \mathbf{b}, n-1)$, then compute the vector $c_0 A^{-1}\mathbf{b} = s(A)\mathbf{b}$, where $s(\lambda) = (p(\lambda) - c_0)/\lambda$. If $c_0 \neq 0$, obtain $A^{-1}\mathbf{b}$ by means of the division by c_0.

b) Apply the same algorithm for Toeplitz-like input matrices and estimate its arithmetic complexity by choosing a set of scaled Fourier points as $\lambda_1, \dots, \lambda_{2n}$.

18 ([Reif93a]). Use randomization to accelerate the parallel prefix computation (Algorithm 2.1).

19. Let $D = \text{diag}\,(A)$ for an $n \times n$ matrix A. Call a matrix A *asymptotically strongly diagonally dominant* if $||I - D^{-1}A||_\infty < 1 - n^{-c}$ and/or if $||I - AD^{-1}||_1 \leq 1 - n^{-c}$ for a fixed constant c (compare Definition 2.1.13).

Show that if A is strongly diagonally dominant, then AD^{-1} and/or $D^{-1}A$ are well-conditioned matrices and extend the complexity bound (4.1) to numerical inversion of an asymptotically strongly diagonally dominant matrix.

20. Prove the bounds (2.1.5) by using the interlacing property of the eigenvalues of a Hermitian matrix ([GL], p. 411, or [Par], p. 186). Then deduce the bound $O_A(\log^3 n, P(n))$ on the complexity of numerical solution of Problems 2.2.7 ($LU \cdot FACTORS$) (with an h.p.d. and well-conditioned input matrix A) and 2.2.9 ($QR \cdot FACTORS$) (with a well-conditioned input matrix A).
21. Supply the details of the algorithm of section 8, applied in the case of a well-conditioned Toeplitz-like+Hankel-like input matrix. Try to modify this algorithm by replacing Newton's iteration by the steepest descent algorithms (compare [CdB], [LV]).
22. Rely on Algorithm 6.2 in order to estimate the parallel randomized arithmetic complexity of Problem 2.2.10a ($SUBMATRIX$) in the case of an $n \times n$ Toeplitz-like input matrix.
23. Based on the randomized solutions of Problems 2.2.4, 2.2.5 and 2.2.10, devise an NC and work-efficient algorithm that reduces an $n \times n$ matrix to the Frobenius normal form ([Gies]) and to the Smith normal form ([Ja]), (open problem) and estimate the parallel arithmetic cost of these reductions. Extend the algorithms and their cost estimates to the solution of Problems 2.2.6 ($CHAR \cdot POL$). Do this exercise separately for a general input matrix and for a Toeplitz-like input matrix.

24 ([P94a]). Solve a Toeplitz-like system of equations at the cost $O_A(\log n, n^2)$ by combining Algorithm 2.11.2 with Example 3.9.2 (provided that all the input values are bounded integers). Apply scaling to extend the result to the case of inputs represented by the ratios of bounded integers. Estimate the Boolean cost of this computation. Extend the approach to various matrix computations by using Example 3.9.3.

25. Decrease the number of indeterminates in Lemma 2.13.1 by turning the entries of U into the $(n+1)$-st powers of the respective entries of L. Examine how this affects the number of random parameters and the probability estimates in the randomized algorithms that rely on Lemma 2.13.1.

Appendix A. An Example of Application of Stream Contraction, Recursive Restarting and Supereffective Slowdown to Parallel Polynomial Computations.

In this appendix we will follow [BP] and will show a supereffective slowdown of fast parallel evaluation modulo z^N of the d-th root of a polynomial $p(z)$ such that $p(0) = p^{1/d}(0) = 1$. This will give us a chance to demonstrate some general techniques of designing effective parallel algorithms. To simplify the presentation, we assume that $d = 2$.

Let $w_k(z)$ denote $p^{1/2}(z) \bmod z^k$, $k = 1, 2, 3, \ldots$. Precompute $p^{-1}(z) \bmod z^n$, for a fixed n, assume that $w_k(z)$ is known, for a fixed k, $n > k \geq 1$ (in particular, $w_1(z) = 1$ is known), set $v_0(x) = w_k(z)$, apply Newton's iteration to the equation

$$w^{-2}(z)\, p(z) - 1 = 0 \ ,$$

and recursively compute

$$v_{i+1}(z) = 0.5 v_i(z)(3 - (p^{-1}(z) \bmod z^n) v_i^2(z)) \ , \quad i = 0, 1, \ldots \ . \tag{A.1}$$

Then it can be shown that

$$v_{i+1}(z) - p^{1/2}(z) = -(v_i(z) - p^{1/2}(z))^2 (p^{-1/2}(z) + 0.5 w_i(z)\, p^{-1}(z)) \ ,$$

and therefore,

$$v_j(z) = w_{kJ}(z) = p^{1/2}(z) \bmod z^{kJ} \ , \quad J = 2^j \ , \quad j = 0, 1, \ldots \ .$$

The latter relations still hold if we replace $v_j(z)$ by $v_j(z) \bmod z^{2kJ}$ and $p^{-1}(z)$ by $p^{-1}(z) \bmod z^{2kJ}$, so that Newton's step (A.1) for $i = j$ is performed at the cost $O_A(\log\,(kJ), kJ)$. In $\lceil \log\,(n/k) \rceil$ steps we will compute $w_n(z) = p^{1/2}(z) \bmod z^n$ at the overall cost bounded by

$$O_A(\log n \ \lceil \log\,(n/k) \rceil \ , n / \lceil \log\,(n/k) \rceil) \ , \tag{A.2}$$

which turns into

$$O_A(\log^2 n \ , n/\log n) \quad \text{for} \quad k = 1 \ .$$

A distinct application of Newton's iteration (see [BP]) accelerates the computation and leads to the bounds

$$O_A(\log n \ , k(n/k)^a \log n) \ , \tag{A.3}$$

$a = \log 3 = 1.5849\ldots$, on the cost of computing $w_n(z) = p^{1/2}(z) \bmod z^n$ provided that $k \leq n$ and that $w_k(z) = p^{1/2}(z) \bmod z^k$ is available. The algorithm supporting (A.2) is based on the *stream contraction* technique, according to which each iteration of a recursive process should be merged with the preceding iteration, in a pipelined fashion, as this is done in section 9, in our algorithm for solving Toeplitz linear systems, and in [BP] and in chapter 6 for polynomial division (also compare applications of such techniques, in [PR91] and [HPR], to computation of paths in graphs). In our case we contract the stream of Newton's steps (A.1) by dropping the stage of the recovery of the coefficients of $v_j(z)$. Such a stage is usually performed by means of the inverse DFT's in the algorithms for polynomial computations based on Newton's iteration. Now, we reduce $p^{-1}(z)$ modulo $z^{k2^{i+1}}$ in (A.1) but do not reduce $v_i(z)$ modulo any power of z. Then the degree of $v_i(z)$ equals $3^i k \sum_{g=0}^{i} (2/3)^g$, which is less than $3^{i+1}k$, for all i, and instead of the coefficients of $v_i(z)$, we only compute the values of the polynomials $p^{-1}(z) \bmod z^{k2^{i+1}}$ and $v_i(z)$ on the set of N Fourier points (that is, on the set of the N-th roots of 1), for $N = O(n^a)$ and for $i = 0, 1, \ldots$.

More precisely, we first concurrently compute the values $p^{-1}(z) \bmod z^{k2^{i+1}}$ on the latter Fourier set for all i and then recursively apply (A.1) for $i = 0, 1, \ldots$, to compute the values $v_{i+1}(z)$ on this set, by using at most $5N$ ops for each i. In this way we arrive at the estimates (A.3), which show that our parallel computation is fast but not processor-efficient: its potential work is $n^a k^{1-a} \log^2 n$.

By applying the *recursive restarting* technique and the *supereffective slowdown* technique, we will, however, arrive at the fast and processor-efficient algorithm, yielding the bounds

$$O_A(\log N \ \log\log N, N/\log\log N) \tag{A.4}$$

on the complexity of the evaluation of $p^{1/2}(x) \bmod x^N$.

Indeed, first apply the algorithm supporting (A.3), for $k = 1$, $n = N^b$, $b = 0.63$, and output $w_n(z)$ (with $n = N^b$), at the cost $O_A(\log n, n^a \log n) = O_A(\log N, N^{ab} \log N)$, where $ab < 0.999$. Then apply the algorithm again, with $k = N^b$, $n = N^{(2-b)b}$, and output $w_n(z)$, at the cost $O_A(\log N, N^{ab+b-ab^2} \log N)$. Recursively apply the same process and output $w_n(z)$ for $n = n(i) = N^{1-(1-b)^i}$, $k = n(i-1)$, $1 - b = 0.37$, at the cost of the i-th recursive step bounded by

$$O_A(\log N,\ N^{1-(1-b)^{i-1}(1-ab)} \log N)\ , \quad i = 0, 1, \ldots\ . \tag{A.5}$$

For i of the order log log n, this will already give us the order of N coefficients of $p^{1/2}(z)$ at the overall cost $O_A(\log N \log\log N, N\log N)$.

The first version of Newton's iteration [see (A.2)] now enables us to extend this computation and to obtain the remaining part of the first N coefficients of $p^{1/2}(x)$ at the additional cost $O_A(\log N, N)$ [or $O_A(\log N \log\log N, N/\log\log N)$ after a slowdown by log log N times] and at the overall cost $O_A(\log N \log\log N, N\log N)$.

Finally, let us improve these bounds to $O_A(\log N \log\log N, N/\log\log N)$ of (A.4). Indeed, the latter cost bounds hold if we replace N by $\widetilde{N} = N/(\log N \log\log N)$ in the above construction and apply it to compute $p^{1/2}(x) \bmod x^{\widetilde{N}}$. These asymptotic complexity bounds do not increase in the transition from $p^{1/2}(x) \bmod x^{\widetilde{N}}$ to $p^{1/2}(x) \bmod x^N$ by means of the first version of Newton's iteration [see (A.2) for $n = N$, $k = \widetilde{N}$], and we arrive at (A.4).

On the other hand, for any positive ϵ, we may replace N by $\widetilde{N}$ in (A.5) and choose i in (A.5) so as to arrive at the bound

$$O_A(\log \widetilde{N}, \widetilde{N})$$

on the cost of computing $p^{1/2}(z) \bmod z^N$ for $N = \widetilde{N}^{1/(1+\epsilon)}$. This cost bound [alternative to (A.4)] can be rewritten as follows:

$$O_A(\log N, N^{1+\epsilon}) . \tag{A.6}$$

The approach and the resulting estimates are immediately extended to the evaluation of $p^{1/d}(z) \bmod z^N$ for any fixed d and to other algebraic computational problems, for which Newton's iteration only involves polynomials (and no rational functions) in $v_i(x)$. In particular, the bounds (A.4) and (A.6) can be extended to a large class of computational problems specified in the following theorem:

Theorem A.1. *Let P denote a given problem of computing a string $s(N)$ of N scalars. Suppose that for $c > 1$ and for a pair of integers k and m, $1 \le k < m$, two alternative "black box" parallel algorithms, Algorithms A.1 and A.2, are available that on the given substring of the first k scalars of $s(N)$ compute its next $m - k$ scalars, at the cost bounded by $O_A(\log m \log (m/k), m/\log (m/k))$ and $O_A(\log m, k(m/k)^c)$, respectively. Then recursive application of Algorithm A.2, for some appropriate recursive choice of the pairs $k = k_i$, $m = m_i$, $i = 0, 1, \ldots$, such that $k_{i+1} = m_i > k_i$, followed by application of Algorithm A.1, supports the solution of the problem P of size N at the cost bounded by (A.4) and (A.6).*

In section 3 of chapter 6 we will improve (A.4) for a special class of computational problems, exemplified by polynomial division (we have already cited the improved bounds in section 2).

Recursive restarting and the B-principle seem to be known for some time as a customary means of the design of parallel algorithms for combinatorial and graph computations, but recently the power of these techniques has been demonstrated in another important area: in [PP] and [PP,a], supereffective slowdown of some fundamental parallel computations in linear algebra and in path algebras has been achieved by means of recursively stopping and restarting some iterative parallel computation process and by decreasing the size of the input with each restarting. In the latter approach, we apply fast but processor-inefficient algorithms to decrease the input size in each stopping and restarting. The idea is to have processor inefficiency only relative to the decrease of the input size, but not relative to the input size itself. Thus the input size decreases rather slowly but steadily in all the recursive steps, which ultimately suffices for obtaining a considerable decrease of the overall parallel computational cost [compare exercise 8 b), c), d)].

Appendix B. Computing a Maximal Linearly Independent Subset of a Vector Set and Solving the *SUBMATRIX* Problem.

We will first follow [E91] to extend our parallel algorithms for $M\cdot RANK$ and $INVERT$ to the solution of $SUBMATRIX$. (In the equivalent form of computing *a maximal linearly independent subset of a set of n-dimensional vectors,* Problem 2.2.10a ($SUBMATRIX$) is known to have several applications to combinatorial and graph computations.) Then we will describe an improvement of the algorithm of [E91], due to [ChR], and its output-sensitive modification, also due to [ChR]. (We extend the estimates of [E91] and [ChR] to any field of constants by using our results of section 6. In Appendix C we will show further improvements.)

The solution of [E91] for Problem 2.2.10a ($SUBMATRIX$) involves the solution of two auxiliary problems (compare Definition 2.1.8).

Problem B.1 ($V\cdot COMPRESS$), *vector compression. Given an $m\times k$ matrix A of rank r, compute an $r\times k$ matrix B such that $\mathcal{N}(A)=\mathcal{N}(B)$.*

Solution. Note that $r\le m$ and compute [at the cost $O_A(\log m, mk)$] the matrix $B=RA$, where R is a random $r\times m$ Toeplitz matrix. To test the correctness of this randomized solution, one may just check if BQ is a nonsingular matrix, by solving Problem 2.2.4a ($SINGULAR?$) for a random $k\times r$ matrix Q. (It suffices to choose random Toeplitz matrices as R and Q, thus using fewer random values.) ■

Problem B.2 ($V\cdot SPLIT$), *vector split. Given two matrices $A\in \mathbf{F}_{h,m_1}$, $B\in\mathbf{F}_{h,m_2}$, $r=$ rank A, compute two matrices $M\in\mathbf{F}_{h,h}$ such that $\mathcal{R}(A)=\mathcal{N}(M)$ and $E=MB\in\mathbf{F}_{h,m_2}$. Note that $\mathcal{N}(E)=\{\mathbf{v}\colon B\mathbf{v}\in\mathcal{R}(A)\}$.*

Solution. First fix a sufficiently large set S and define a random Toeplitz matrix $T\in\mathbf{F}_{m_1,r}$ with the entries in S. Then compute, at the cost $O_A(\log m_1,\ hm_1)$, the matrix $G=AT$, which has rank r with a high probability. Then compute a matrix $H\in\mathbf{F}_{h,h-r}$, whose columns form a basis for $\mathcal{N}(G^T)$ [this stage solves Problems 2.2.3b ($NULL\cdot SPACE$) for an $r\times h$ matrix G^T of full rank, at the randomized cost $O_A(\gamma(r,p)\log^2 r, P^*(r)\log^\alpha r/\gamma(r,p))$, $\alpha\le 2$, of the inversion of an $r\times r$ matrix increased by $O_A(\log r, P(r,r,h-r))$, the cost of $r\times r$ by $r\times(h-r)$ matrix multiplication]. Finally, set $M=[H,O]^T$ and compute $E=MB$, at the cost $O_A(\log h, P(h,h,m_2))$. ■

Solution of Problem 2.2.10a ($SUBMATRIX$), ([E91]). First compute $r=$ rank A, then a maximal linearly independent subset S of the column set of A (forming an $m \times r$ submatrix B of A), and, finally, a maximal linearly independent subset of the row set of B. It is sufficient to specify the computation of B, since the next, final stage is performed similarly. ■

Algorithm B.1, *computing a maximal linearly independent subset of a set of vectors.*

Input: a matrix $A \in \mathbf{F}_{m,n}$.

Output: lexicographically first maximal linearly independent subset of the set of the columns of A.

Computation:

1. Compute r = rank A and a matrix $C \in \mathbf{F}_{r,n}$ such that $\mathcal{N}(A) = \mathcal{N}(C)$, by solving Problem B.1 of vector compression. Denote $C = C_n = C^{(1)} = [\mathbf{c}_1, \mathbf{c}_2, \ldots, \mathbf{c}_n]$, $C_j = [\mathbf{c}_1, \ldots, \mathbf{c}_j]$, $C^{(j)} = [\mathbf{c}_j, \ldots, \mathbf{c}_n]$, $j = 1, \ldots, n$, so that C_j is the leftmost $r \times j$ submatrix of C, $C^{(j)}$ is the rightmost $r \times (n + 1 - j)$ submatrix of C.
2. Apply Algorithm 2.8.2 (binary search) to compute unique k such that rank $C_k = \lceil r/2 \rceil >$ rank C_{k-1}.
3. Solve Problem B.2 ($V \cdot SPLIT$) with C_k and $C^{(k+1)}$ used as the input matrices A and B, respectively. Let E denote the output matrix, such that $E = MC^{(k+1)} \in \mathbf{F}_{r,n-k}$, for $M \in \mathbf{F}_{r,r}$, $\mathcal{N}(E) = \{\mathbf{v}\colon C^{(k+1)}\mathbf{v} \in \mathcal{R}(C_k)\}$.
4. Recursively apply Stages 1–5 of this algorithm to
   a) the column set of C_k,
   b) the column set of E.

   Do parts a) and b) concurrently. In each case, output the indices of the lexicographically first maximal linearly independent subset of the column set of the input matrix. Let $i_1, \ldots, i_{\lceil r/2 \rceil - 1}$ and $j_1, \ldots, j_{\lfloor r/2 \rfloor}$ enumerate these output subsets in the cases a) and b), respectively.
5. In parallel for $1 \leq g \leq r$, set and output

$$h_g = \begin{cases} i_g & \text{if } \ g \leq \text{rank } C_k, \\ k + j_{g-rank\ C_k} & \text{if } \ g > \text{rank } C_k. \end{cases}$$

The correctness of the algorithm is immediately verified.

Cheriyan and Reif in [ChR] observed that, even without its balancing Stage 2, Algorithm B.1 remains correct if, at Stage 3, k is set equal to $\lfloor n/2 \rfloor$ (say). We will refer to this modification of Algorithm B.1 as to **Algorithm B.1a.**

The overall computational cost of performing Algorithm B.1a is dominated by the estimated cost of the following operations: recursive computation of ranks at Stage 1, the first vector compression, the multiplications of general matrices (M by B), and the matrix inversions involved in the solution of the $NULL \cdot SPACE$ problems (both of the two latter groups of matrix operartions are needed for the solution of the $V \cdot SPLIT$ problems). Algorithm B.1a requires $\lceil \log n \rceil$ recursive calls; the i-th call requires to solve 2^i problems of $INVERT$ and $M \cdot RANK$ of sizes at most $b(i) \times b(i)$, $b(i) = \min\{r, n2^{-i}\}$, for $INVERT$ and $b(i) \times n2^{-i}$ for $M \cdot RANK$, except for the initial computation of rank A, where $A \in \mathbf{F}_{m,n}$.

Applying the estimate of section 6 for the randomized parallel complexity of $INVERT$ and using the B-principle, we bound the overall cost of all the matrix inversions by

$$O_A(\gamma(r,p)(\log n)\log^2 r, (n/r)P^*(r)(\log^\alpha r)/(\gamma(r,p)\log n)) \qquad (B.1)$$

with $P^*(r)$, $\gamma(r,p)$ and α defined according to (4.5), (6.2) and Remark 4.1. This dominates the estimated cost of all the matrix multiplications involved. To bound the overall cost of computing matrix ranks, we apply the estimates of Proposition 6.3; in the first of these estimates we omit the elaboration of the polylogarithmic factors in the processor bounds (compare Remark 6.2). In this way we obtain

Proposition B.1. *Problem 2.2.10a ($SUBMATRIX$) for an $n \times n$ matrix of rank r can be solved over a field of constants $\mathbf{F}$ having characteristic p and cardinality $|\mathbf{F}|$ at the cost bounded by the following expressions:*

a) $O_A(\gamma(n,p)\log^3 n + \gamma(r,p)(\log^3 r)\log n, n^{\bar{\omega}})$,

b) $O_A(t^*, n^3(\log n)(\log\log n)/t^*)$, $t^* = \gamma(n,p)\log^3 n$,

c) $O_A(\log^3 n, (P^*(n) + (\log^{\alpha+2} r)P^*(r)n/r)/\log^2 n)$,

where $\bar{\omega} < 2.376$, and it is assumed that $|\mathbf{F}|$ is sufficiently large (compare Remark 2.1d) and that the bound of part c) applies if $p = 0$ or $p > n$.

Further elaboration of the processor bounds in part a) and the extension of our estimates to the general case where $m \neq n$ are left to the reader. Both of these tasks can be achieved based on Proposition 6.3, and we refer the reader to Appendix C on further improvements.

Next we will show an output-sensitive improvement of the solution of Problem 2.2.10a ($SUBMATRIX$), provided that the rank r of an $n \times n$ input matrix A is much smaller than n (and we refer the reader to Appendix C for a further improvement, by different means).

With no loss of generality, we will consider the solution of the equivalent problem of computing a maximal linearly independent subset S of the row set of A. The algorithm of [BGH] first computes the ranks, $r_k = \text{rank } A^{(k)}$, $k = 1, \ldots, n$, for the submatrices $A^{(k)}$ of A formed by replacing the last $n - k$ rows of A by zero rows. The algorithm includes row k of A into the output set S if and only if $r_k > r_{k-1}$.

In the output-sensitive improvement of this algorithm, due to [ChR], one computes rank $B^{(k)}$, where $B^{(k)} = UA^{(k)}L$, U^T and L are random unit lower triangular matrices. Due to Lemma 2.13.1, the $r_k \times r_k$ leading principal submatrix of $B^{(k)}$ is nonsingular with a high probability. Since $r_k \leq r$, we may, therefore, compute r_k as the rank of $C^{(k)}$, the $r \times r$ leading principal submatrix of $B^{(k)}$. Thus, the computation of a maximal linearly independent subset S is reduced to

1) evaluation of the $r \times r$ matrices $C^{(k)}$ for $k = 1, \ldots, n$ and
2) computing rank $C^{(k)}$ for $k = 1, \ldots, n$.

We will refer to this computation as to **Algorithm B.2.**

By applying the parallel prefix algorithm (Algorithm 2.1) and cleverly exploiting the Toeplitz structure of the matrices U and L, one may compute the matrices $C^{(1)}, \ldots, C^{(n)}$ at the cost $O_A(\log n, n^2)$ (see [ChR], section 3). At Stage 2, we apply Proposition 6.3 and arrive at the following estimates for the complexity of computing a maximal linearly independent set and solving *SUBMATRIX* over a field of constants that has characteristic p and a sufficiently large cardinality (see further improvements in Appendix C):

a) $O_A(t_1, (n^2 \log n \log^2 r + n \log^2 r(P(r) + r^2 \log\log r))/t_1)$, where $t_1 = \log n + \gamma(r,p)\log^3 r$,

b) $O_A(t_2, (n^2 \log n + r^3 n \log\log r \log r)/t_2)$, $t_2 = \log n + \gamma(r,p)\log^2 r$.

c) $O_A(\log n + \log^2 r, (n^2 \log n + nP^*(r)\log^2 r)/(\log n + \log^2 r))$, if $p = 0$ or $p > n$;

Here it is assumed that the rank of the input matrix A is known; otherwise, the complexity estimates should grow accordingly.

Appendix C. Improved Randomized Algorithms for Matrix Rank, Maximal Linearly Independent Subset of a Vector Set and the *SUBMATRIX* Problem.

In this appendix we will follow [P93a]. We will first recall the recent work [Sc93] on estimating the parallel complexity of Problem 2.2.6 (*CHAR·POL*), over any field of constants, which extends the earlier works of [Kak], [Kak,a]. The estimates of [Sc93] are strictly inferior to ones of [Chi] and [Ber], but we will apply some techniques of [Sc93], modify them, and combine them with ones of our chapters 1 (section 7) and 2 (sections 11–13), to arrive at some effective randomized algorithms, improving our results of section 6 on the matrix rank computation over any field of constants. (We will also apply similar techniques to obtain two alternative proofs of Proposition 6.2.)

Then we will extend our new result on $M \cdot RANK$ to improve the algorithms of Appendix B for *SUBMATRIX* and for a maximal linearly independent subset of a vector set. Furthermore, for the two latter computational problems, we will show a substantial additional improvement, based on a distinct *concurrent divide-and-conquer technique.*

In this way we will strictly improve the previous record complexity estimates of [ChR] even over the field of characteristic 0. (The estimates of [ChR], for the general case, where the rank can be of the order n, actually rely on a rearrangement of the algorithm of [E91], utilizing the results of [KP], and on the result on $M \cdot RANK$ from [KP,a].)

From the power sums of the zeros of a polynomial to its coefficients: existence of rational expressions.

Let

$$c(x) = \sum_{i=0}^{n} c_i x^i = \prod_{j=1}^{n} (x - z_j),\ c_n = 1,$$

denote a generic monic polynomial of degree n with its zeros $z_1, \ldots, z_n$. Let

$$s_k = \sum_{j=1}^{n} z_j^k,\ k = 0, 1, \ldots, n,$$

denote the k-th power sums of the zeros of $p(x)$. Let p be a prime, $\mathbf{F}$ be a field of characteristic p, $p \le n$, $n+g = gp+h$, $0 < h < p$, $g = \lfloor (n-1)/(p-1) \rfloor$; $N(n,p)$ be the set of all the n integers from 1 to $n+g$ not divided by p. Let $b(x) = x^n c(1/x) = 1 + c_{n-1}x + \cdots + c_0 x^n = 1 + b_1 x + \cdots + b_n x^n$ denote the reverse polynomial such that $b_i = c_{n-i}, i = 1, 2, \ldots, n$.

Using these definitions, we will now state the following recent extension of the earlier results of [Kak] and [Kak,a]:

Theorem C.1 ([Sc93]). *The coefficients* $c_0, \ldots, c_{n-1}$ *of* $c(x)$ *can be expressed as rational functions in the variables* $\{s_i, i \in N(n,p)\}$.

Proof. We will next outline (with some small modifications) the constructive proof of this theorem given in [Sc93]. The proof relies on the following equations:

$$x\,(\ln b(x))' = xb'(x)/b(x) = -\sum_{i=1}^{\infty} s_i x^i$$

or, equivalently,

$$xb'(x) = -b(x)\sum_{i=1}^{\infty} s_i x^i. \qquad (C.1)$$

We equate the coefficients of the $n+g$ smallest powers of x on both sides of (C.1) and arrive at a linear system of $n+g$ equations in n unknowns $b_1, \ldots, b_n$. This system is actually equivalent to (C.1) and will be solved in two steps. At first introduce p auxiliary polynomials:

$$b_k(x^p) = \sum_{i=0}^{d_k} b_{k+ip} x^{pi}\ ,\ k = 0, 1, \ldots, p-1,$$

such that

$$d_k \geq d_{k+1}\ ,\ k = 0, \ldots, p-2, d_0 = d = \lfloor n/p \rfloor, d_{p-1} \geq d-1; \qquad (C.2)$$

$$b(x) = \sum_{k=0}^{p-1} b_k(x^p)x^k\ ,\ xb'(x) = \sum_{k=1}^{p-1} kx^k b_k(x^p), \qquad (C.3)$$

$$b_0(x^p) = 1 + b_p x^p + b_{2p}x^{2p} + \cdots + b_{dp}x^{dp}.$$

Next define $p-1$ auxiliary analytic functions,

$$a_k(x^p) = b_k(x^p)/b_0(x^p) = \sum_{i=0}^{\infty} a_{k,i}x^{pi}\ ,\ k = 1, \ldots, p-1. \qquad (C.4)$$

Divide both sides of (C.1) by $xb_0(x^p)$, apply (C.3) and (C.4), and express the resulting equations in terms of $a_k(x^p), k = 1, \ldots, p-1$, and $s_i, i = 1, 2, \ldots,$ to obtain that

$$\begin{aligned} &a_1(x^p) + (2x)a_2(x^p) + \cdots + (p-1)x^{p-2}a_{p-1}(x^p) = \\ &-(\sum_{i=1}^{\infty} s_i x^{i-1})(1 + xa_1(x^p) + x^2a_2(x^p) + \cdots + x^{p-1}a_{p-1}(x^p)). \end{aligned} \qquad (C.5)$$

For the notational convenience, we now introduce the auxiliary power series $a_p(x^p) = \sum_{i=0}^{\infty} a_{p,i}x^{ip}$, setting $a_{p,i} = 0$ for all i, so that $a_p(x^p) = 0$. Then we rewrite (C.5) as follows:

$$a_1(x^p) + (2x)a_2(x^p) + \cdots + px^{p-1}a_p(x^p) =$$
$$-(\sum_{i=1}^{\infty} s_i x^{i-1})(1 + xa_1(x^p) + x^2 a_2(x^p) + \cdots + x^p a_p(x^p)). \qquad (C.5a)$$

Equating the coefficients of all the powers of x on both sides of the latter equation, we could have obtained an infinite linear system of equations in the coefficients $a_{k,i}, k = 1, \ldots, p; i = 0, 1, \ldots,$ with the matrix obtained as the sum of an infinite proper lower triangular Toeplitz matrix with the first column $[0, s_1, s_2, \ldots]^T$ and of the infinite diagonal matrix $D_\infty =$ diag $(1, 2, 3, \ldots)$ mod p. We will work with the finite subsystem of the first $n+g$ equations of this system, that is, we will equate the coefficients of the $n + g$ smallest powers of x on both sides of (C.5a) and will obtain a linear system of $n + g$ equations in n actual variables,

$$a_{1,0}, a_{2,0}, \ldots, a_{p-1,0}, a_{1,1}, \ldots, a_{p-1,1}, a_{1,2}, \ldots, a_{h,g},$$

and in g fictitious variables, $a_{p,i}, i = 0, 1, \ldots, g - 1$, which we have actually set equal to 0. Denote that

$$\mathbf{s} = [s_1, \ldots, s_{g+n}]^T,$$
$$\mathbf{a} = [a_{1,0}, a_{2,0}, \ldots, a_{p,0}, a_{1,1}, \ldots, a_{p,1}, a_{1,2}, \ldots, a_{h,g}]^T,$$
$$D_p = \text{diag}(1, 2, \ldots, p-1, 0), D_h = (1, 2, \ldots, h).$$

Note that **s** and **a** are two vectors of dimension $n + g$ and define the two matrices of size $(n + g) \times (n + g)$:

$$D = \text{diag}(D_p, D_p, \ldots, D_p, D_h) = \text{ diag } (1, 2, \ldots, n + g) \bmod p,$$

$L = L(Z\mathbf{s}), L$ being the lower triangular $(n + g) \times (n + g)$ Toeplitz matrix with the first column $Z\mathbf{s} = [0, s_1, \ldots, s_{n+g-1}]^T$. Then our linear system of $n + g$ equations, obtained from (C.5a), can be compactly written as follows:

$$-(D + L)\mathrm{a} = \mathrm{s}.$$

Now, we delete equations $p, 2p, \ldots, gp$, as well as components $p, 2p, \ldots,$ gp of the vector **a** and columns $p, 2p, \ldots, gp$ of the matrix $D + L$, and arrive at a nonsingular triangular system of n linear equations, which enables us to express the remaining components $a_{1,0}, a_{2,0}, \ldots, a_{h,g}$ of **a** as polynomials in $s_1, \ldots, s_{g+n}$ over the field **F**. Let us hereafter refer to this system as (C.6).

To prove Theorem C.1 it remains to show that the coefficients of $b(x)$ have rational expressions via $a_{1,0}, \ldots, a_{h,g}$. In fact, it is sufficient to prove the existence of such expressions for the coefficients of $b_0(x)$. Indeed, the coefficients of $b_k(x^p)$, for $k = 1, \ldots, p-1$, can then be obtained by means of $p-1$ multiplications of pairs of polynomials of degrees at most g, by using the equations

$$b_k(y) = b_0(y) \sum_i a_{k,i} y^i \bmod y^{d_k}, \; k = 1, \ldots, p-1, \tag{C.7}$$

where $d_k = \deg b_k(y) \leq \lfloor n/p \rfloor$, [compare (C.2)]. Note that (C.7) expresses the coefficients of $b_k(y)$ as polynomials in $\{a_{k,i}\}$ and in the coefficients of $b_0(y)$.

To deduce rational expressions for the coefficients $b_p, b_{2p}, \ldots, b_{dp}$ of $b_0(x)$ via $\{a_{k,i}\}$, recall that $b_0 = 1$ and obtain from (C.2) and (C.7) that

$$b_k(y) = b_0(y) \sum_i a_{k,i} y^i \bmod y^d, \; d = \lfloor n/p \rfloor, \; k = 1, \ldots, p-1.$$

Exactly $d = \lfloor n/p \rfloor$ powers of y are missing on the left hand sides of all such equations for $k = 1, \ldots, p-1$, and this defines a linear system, hereafter referred to as (C.8), consisting of $d = \lfloor n/p \rfloor$ equations in the coefficients $b_p, b_{2p}, \ldots, b_{dp}$ of $b_0(x)$. A careful inspection done in [Sc93] shows nonsingularity of this linear system in the case of a generic monic input polynomial $c(x)$, and this completes the proof of Theorem C.1. ■

Transition from the power sums to the coefficients: rational algorithms.

Now suppose that we are given all the required power sums $s_1, s_2, \ldots$ and need to recover the unknown coefficients of $c(x)$. We may follow the line of the above proof of Theorem C.1 and reduce the computation of the coefficients $c_0, c_1, \ldots, c_{n-1}$ of the polynomials $b(x)$ and $c(x) = x^n b(1/x)$ to the following stages:

1) solve the nonsingular triangular linear system (C.6) of n equations, at the cost $O_A(\log^2 n,\ P(n)/\log n)$;
2) solve the linear system (C.8) of d equations to obtain the coefficient of $b_0(x)$;
3) compute $b_k(y)$ as the polynomial products modulo $y^{d_k}, d_k \leq d = \lfloor n/p \rfloor, k = 1, \ldots, p-1$ [compare (C.2) and (C.7)], at the cost $O_A(\log(n/p), n)$.

At Stage 2 we need nonsingularity of the linear system (C.8), which is guaranteed in the case of a generic monic input polynomial $c(x)$ and, therefore, is ensured with a high probability in the case of a random monic $c(x)$ [compare Procedure 1.7.1 and Lemma 1.5.1].

Application to $LIN \cdot SOLVE$ and $DETERMINANT$.

Now suppose that we seek a vector $\mathbf{x} = A^{-1}\mathbf{b}$, giving the solution to a linear system $A\mathbf{x} = \mathbf{b}$ for a nonsingular $n \times n$ matrix A [Problem 2.2.3 ($LIN \cdot SOLVE$)]. We reduce this problem to solving the linear system $AV\mathbf{y} = \mathbf{b}$ for a random $n \times n$ matrix V. Then we in turn reduce the latter problem to the evaluation of the coefficients of $c_{AV}(x)$, the characteristic polynomial of AV. For the latter evaluation, we first compute $s_i = \text{trace}(AV)^i, i = 1, 2, \ldots$, which gives us the power sums of the zeros of $c_{AV}(x)$; then we go through Stages 1–3, for the transition from the power sums of the zeros to the coefficients of the polynomial.

We may apply the assignment technique of section 13 of chapter 2 so that we will only need to show that the associated linear system (C.8) is nonsingular at least for one instance of V. (Then the nonsingularity of (C.8) will follow with a high probability for a random matrix V, due to the results of sections 5 and 7 of chapter 1.) Such an instance is given by $V = A^{-1}C$, C being the companion (Frobenius) matrix associated with some polynomial $c(x)$ for which the system (C.8) is nonsingular (the existence of such a polynomial was shown when we proved Theorem C.1).

In sections 4 and 6 we have reduced the solution of $LIN \cdot SOLVE$ and $DETERMINANT$ for a general $n \times n$ input matrix to $LIN \cdot SOLVE$ for a Toeplitz matrix A. We will now simplify the choice of the random matrix V, assuming that A is a nonsingular Toeplitz or even, more generally, Toeplitz-like matrix, having a fixed F_+-rank $r, r = O(1)$, so that A^{-1} has F_--rank r. (Recall that $r = 2$ for a Toeplitz matrix A , $F_+ = F_{(Z,Z^T)}$, $F_- = F_{(Z^T,Z)}$, according to the notation of Example 2.11.3, and recall Theorem 2.11.1.)

If rank $F_+(A) =$ rank $F_-(A^{-1}) = r$, we may restrict the choice of a random $n \times n$ matrix V to the case of matrices with F_--ranks at most $r+3$, that is, we may choose $V = \sum_{i=1}^{r+3} U_i L_i$, where L_i and U_i^T are random $n \times n$ lower triangular Toeplitz matrices. Indeed, rank $F_-(\phi) \leq 2$ for any Frobenius (companion) matrix ϕ so that rank $F_-(A^{-1}C) \leq r + 3$ (due to Fact 2.12.1) and $V = A^{-1}C$ is a desired instance of the matrix V, for which the linear system (C.8) associated with the matrix AV is nonsingular.

Since rank $F_-(AV) \leq 2r + 4$ (due to Fact 2.12.1) and since $r = O(1)$, we may apply Algorithm 2.11.1 and compute the values

$$s_i = \text{ trace } (AV)^i, i = 1, \ldots, N(n,p),\ N(n,p) < 2n,$$

at the cost $O_A(\log^2 n, n^2 \log\log n/\log n)$ over any field of constants having sufficiently many elements (compare Remark 2.1d). Let us also include the cost of the randomized transition from the general to the Toeplitz case of $LIN \cdot SOLVE$, the cost of Stages 1 and 3 of the transition from the values s_i to the coefficients of the characteristic polynomial $c_{AV}(x)$, and the cost of the subsequent evaluation of $\mathbf{y} = (AV)^{-1}\mathbf{b}$ and $\mathbf{x} = V\mathbf{y} = A^{-1}\mathbf{b}$ for a given vector $\mathbf{b}$. The overall cost is bounded by $O_A(\log^2 n, P^*(n))$, $P^*(n)$ being defined according to (4.5), and at this cost we reduce the $LIN \cdot SOLVE$ and $DETERMINANT$ of the $n \times n$ size to $LIN \cdot SOLVE$ of the size $\lfloor n/p \rfloor \times \lfloor n/p \rfloor$, that is, to the solution of the auxiliary linear system (C.8) at Stage 2 of the transition from s_i to $c_{AV}(x)$. Recursive application of such a size reduction for $LIN \cdot SOLVE$ leads us to an alternative proof of Proposition 6.2.

As a further modification, we may replace Stage 2 of solving the linear system (C.8) by the solution of a (d_1, d_0) Padé approximation problem with the input function $\sum_i a_{k,i} y^i$. For the generic polynomial $c(x)$ and, with a high probability, for a random polynomial $c(x)$, a unique solution to the latter problem, defining the desired polynomial $b_0(x)$ of degree exactly $d = d_0$, can be computed by solving a nonsingular Toeplitz linear system of $d + d_1 + 1$ equations, where $d + d_1 + 1 \leq 2\lfloor n/p \rfloor + 1$. We may recursively apply the above argument to the solution of the latter system and thus obtain yet another alternative proof of Proposition 6.2.

Application to $M \cdot RANK$

For a problem 2.2.10 ($M \cdot RANK$) the above approach enables us to improve the results of Proposition 6.3 and to reach the randomized bound (6.2) for $M \cdot RANK$ over any field $\mathbf{F}$ having sufficiently many elements (compare Remark 2.1d).

We will first recall that the general case solution of $M \cdot RANK$ has already been reduced to the case of an $n \times n$ Toeplitz input matrix T. Let r denote the unknown value of rank T. Define two random $n \times n$ lower triangular Toeplitz matrices U^T and L, an $n \times n$ Frobenius matrix ϕ, for $\phi = F$ of (2.3.1), associated with a random monic polynomial $\sum_{i=0}^n p_i x^i, p_n = 1$, and two random Toeplitz-like matrices, T_{10} and T_6, having their F_--ranks at most 10 and 6, respectively. Observe that the F_--ranks of both matrices ϕ

and UTL are at most 2, apply Proposition 2.12.1 and the equations (2.12.6) and compute an F_--generator of length at most 23 for the matrix

$$T^* = T_{10}UTLT_6J\phi J,$$

where J is the $n \times n$ reversion matrix of Definition 2.4.1. Note that, with a high probability, r=rank T^* and that, due to Lemma 2.13.1, $UTL = \begin{pmatrix} B & C \\ E & G \end{pmatrix}$ for some $r \times r$ nonsingular matrix B. Recall (2.2.1) and deduce that

$$T^* = \begin{pmatrix} I_r & O \\ O & O \end{pmatrix} J\phi J,$$

if we set

$$T_{10} = \begin{pmatrix} I & O \\ -EB^{-1} & I \end{pmatrix}, T_6 = \begin{pmatrix} B & C \\ O & I \end{pmatrix}^{-1} = \begin{pmatrix} B^{-1} & -B^{-1}C \\ O & I \end{pmatrix}.$$

Let us next verify that, under this specialization of the matrices T_{10} and T_6, their F_--ranks are indeed bounded from above by 10 and 6, as it was required. Indeed, by the definition of the operator F_+ and F_-, we have

$$\text{rank } F_+(B) \leq \text{ rank } F_+(UTL),$$

$$\text{rank } F_-(-E) \leq \text{ rank } F_-(UTL) + 1 \leq 3,$$

$$\text{rank } F_-\left(\begin{pmatrix} B & C \\ O & I \end{pmatrix}\right) \leq 2 + \text{ rank } F_-(UTL) \leq 4.$$

Due to Corollary 2.11.1,

$$\text{rank } F_+\left(\begin{pmatrix} B & C \\ O & I \end{pmatrix}\right) \leq 2 + \text{ rank } F_-\left(\begin{pmatrix} B & C \\ O & I \end{pmatrix}\right) \leq 6,$$

$$\text{rank } F_+(ULT) \leq 2 + \text{ rank } F_-(UTL) \leq 4,$$

$$\text{rank } F_-(T_6) \leq 2 + \text{ rank } F_+(T_6) = 2 + \text{ rank } F_+\left(\begin{pmatrix} B & C \\ O & I \end{pmatrix}^{-1}\right).$$

Due to Proposition 2.11.1,

$$\text{rank } F_-(T_6) = \text{ rank } F_-\left(\begin{pmatrix} B & C \\ O & I \end{pmatrix}^{-1}\right) = \text{ rank } F_+\left(\begin{pmatrix} B & C \\ O & I \end{pmatrix}\right) \leq 6,$$

$$\text{rank } F_-(B^{-1}) = \text{ rank } F_+(B) \leq 4.$$

By applying Proposition 2.12.1 and the equation (2.12.6), we obtain that rank $F_-(-EB^{-1}) \leq 8$ and then deduce from the definition of the operator F_- that rank $F_-(T_{10}) \leq 10$, thus completing the proof of our claim.

On the other hand, under the above specializations of the matrices T_{10} and T_6, we have

$$c_{T^*}(x) = \sum_{i=n-r}^{n} p_i x^i,\ p_n = 1,$$

and furthermore, the assumptions of both parts a) and b) of Lemma 6.1 hold for any choice of $p_{n-1}, \ldots, p_{n-r}$, $p_{n-r} \neq 0$, for which the polynomial $\sum_{i=n-r}^{n} p_i x^{i-n+r}$ has no multiple zeros. Consequently, with a high probability, the characteristic polynomial of T^* satisfies the next equation:

$$c_{T^*}(x) = x^n + \sum_{i=n-r}^{n-1} p_i x^i,\ p_{n-r} \neq 0, \qquad (C.9)$$

and all the assumptions of Lemma 6.1 hold, for a random choice of the matrices T_{10} and T_6 having their F_--ranks at most 10 and 6, respectively, and for a random set $\{p_{n-r}, \ldots, p_{n-1}\}$. Therefore, it remains to compute the coefficients of $c_{T^*}(x)$, to apply Lemma 6.1, and to output rank T^*, which, with a high probability, is equal to rank T.

To obtain the coefficients of $c_{T^*}(x)$, we first apply Algorithm 2.11.1 and compute the values $s_i = \text{trace}\,(T^*)^i$, for $i = 1, 2, \ldots, N$, $N < 2n$, at the cost bounded by $O_A(\log^2 n, n^2 \log\log n/\log n)$.

For the transition from the values s_i to the coefficients of $c_{T^*}(x)$, we will apply the algorithm of this appendix, with Stage 2 based on the reduction to the computation of the (d_1, d_0) Padé approximation. We immediately extend the proof of Theorem C.1 and deduce that the coefficients $p_{n-r}, \ldots, p_{n-1}$ of (C.9) are rational functions in $s_1, s_2, \ldots$; moreover, the latter problem of Padé approximation does not degenerate, if these coefficients are generic, and consequently, with a high probability, it does not degenerate if they are random. Therefore, the original problem of $M \cdot RANK$ for an $n \times n$ Toeplitz matrix has been reduced to $LIN \cdot SOLVE$ for $k \times k$ Toeplitz matrix with $k = d_0 + d_1 + 1 \leq 2\lfloor n/p \rfloor + 1$.

Now recall Proposition 6.2 and arrive at the following result [compare (4.5), (6.2) and Remark 2.1d]:

Theorem C.2. *Problem 2.2.10 ($M \cdot RANK$) for an $n \times n$ matrix can be solved over any field* $\mathbf{F}$ *of characteristic p having sufficiently many elements, at a randomized cost bounded, according to (6.2), by*

$$O_A(\gamma(n,p)\log^2 n, P^*(n)/\gamma(n,p)),$$

where P^* is defined by (4.5) and

$$\gamma(n,p) = \begin{cases} \log n/\log p & \text{if } 1 < p \le n \\ 1 & \text{if } p = 0 \text{ or } p > n. \end{cases}$$

Remark C.1. Problem 2.2.10 ($M \cdot RANK$) for an $m \times n$ input matrix A, $m > n$, can be reduced, at the cost $O_A(\log m, mn)$, to the same problem for an $n \times n$ matrix TA, where T is a random $n \times m$ Toeplitz matrix; similarly, for $n \times m$ input matrix $B, m > n$, we may shift to BT^T.

Improved computation of a maximal linearly independent subset of a vector set and improved solution of the $SUBMATRIX$ problem.

Let us apply Theorem C.2 and Remark C.1 to improve the estimates for the computational cost of performing Algorithms B.2 and B.3. We will state the resulting complexity estimates for the $SUBMATRIX$ problem (such estimates also apply to the computation of a maximal linearly independent subset of a vector set, of course), but we will *exclude the complexity of the evaluation of the rank of an $n \times n$ input matrix* [bounded by $O_A(\gamma(n,p)\log^2 n, P^*(n)/\gamma(n,p))$].

Since we now rely on Proposition 6.2 and Theorem C.2, our estimates for the the overall asymptotic cost of all the matrix rank computations involved in the general solution of $SUBMATRIX$ by Algorithm B.1a are dominated by our bounds on the overall asymptotic computational cost of matrix multiplications and inversions involved in the solution of the auxiliary Problems B.1 ($V \cdot COMPRESS$) and B.2 ($V \cdot SPLIT$). [These bounds combine (B.1) with $O_A(\log^2 n, n^2/\log n)$, the latter bound due to the $O(\log n)$ stages of multiplication of Toeplitz-by-general matrices of decreasing sizes.] With this observation, we obtain the following estimates based on application of Algorithms B.1a and B.2:

Proposition C.1. *Over any field of constants having characteristic p and sufficiently many elements, Problem 2.2.10a ($SUBMATRIX$), for an $n \times n$ matrix of a given rank r, can be solved by means of a randomized algorithm at the computational cost bounded by*

$$\begin{aligned} &O_A(t, (n\log n + P^*(r)(\log^{2+\alpha} r)/r)n/t), \\ &t = (\log n + \gamma(r,p)\log^2 r)\log n, \end{aligned} \qquad (C.10)$$

or, alternatively, by

$$O_A(t_1, (nP^*(r)\log^2 r + n^2\log n)/t_1), \qquad (C.11)$$

where $t_1 = \log n + \gamma(r,p) \log^2 r$, *and* $P^*(r), \gamma(r,p)$ and α *are defined according to (4.5), (6.2) and Remark 4.1.*

Finally, we will accelerate Algorithm B.1a by incorporating the *concurrent* (in place of the usual) *divide-and-conquer technique.* The simple idea is to partition the original problem of size q^d into q smaller subproblems of size q^{d-1} and recursively apply a concurrent solution of these subproblems on q groups of processors. (See exercise 24 on accelerating the binary search in this way.)

Here is such a concurrent extension of Algorithm B.1a:

Algorithm C.1, *concurrent divide-and-conquer for SUBMATRIX.*
Input, output are as in Algorithms B.1 and B.2.
Initialization: Sets $s_0 = n$, fix an integer q, $2 \le q \le n$, and compute $s_1 = \lceil s_0/q \rceil$.
Computation:

1. Proceed as at Stage 1 of Algorithms B.1 and B.1a; in addition, let $\mathbf{c}_h = \mathbf{0}$ for $h > s_0$, let $C_{i,j}$ denote the matrix $[\mathbf{c}_i, \ldots, \mathbf{c}_j]$ for any pair (i,j) such that $i \le j$, and set $C_j = C_{1,j}$ for $j \ge 1$.
2. For $i = 1, \ldots, q$, solve Problem B.2 of vector splitting with the matrices C_{is_1} and C_{is_1+1, is_1+s_1} used as the input matrices A and B, respectively; let E_i denote the output matrix such that $E_i = M_i C_{is_1+1, is_1+s_1} \in \mathbf{F}_{r,s_1}$, for some $M_i \in \mathbf{F}_{r,r}$, and
$$\mathcal{N}(E_i) = \{\mathbf{v} : C_{is_1+1, is_1+s_1}\mathbf{v} \in \mathcal{R}(C_{is_1})\}.$$
Denote $E_0 = C_{is_1}$.
3. Do concurrently for $i = 0, 1, \ldots, q$: set $s_0 = s_1$ and recursively apply Stages 1–4 of this algorithm to the column set of the matrices E_i with the last recursive step to be performed, where $s_1 = 1$.
4. Let columns $h_1^{(i)}, \ldots, h_{k(i)}^{(i)}$ form the lexicographically first maximal linearly independent subset of the column set of E_i, for $i = 0, 1, \ldots, q$. Then output the integers $h_1^{(i)} + is_1, h_2^{(i)} + is_1, \ldots, h_{k(i)}^{(i)} + is_1$, for $i = 0, 1, \ldots, q$, indicating the columns of the input matrix A that form the lexicographically first maximal linearly independent subset of the column set of A.

To simplify the resulting asymptotic complexity estimates for the solution based on Algorithm C.1, we will now assume (with no loss of generality) that $r > 1$, $n = q^d$ for an integer $d = \log n / \log q$, so that we have

$$s_0 = q^{d-k+1}, \; s_1 = s_0/q = q^{d-k}$$

at level k of the recursion, for $k = 1, \ldots, d$.

We now estimate that Stage 2 of Algorithm C.1 calls for the solution of Problems B.1 ($V \cdot COMPRESS$) and B.2 ($V \cdot SPLIT$) at most q^k times at level k of the recursion for $k = 1, \ldots, d$. Furthermore, the input size of Problem B.1 is $n \times n$ at level 1 of the recursion and is $r \times n/q^k$ at level k of the recursion, for $k = 2, \ldots, d$. The input size of Problem B.2 (at level k of the recursion) is defined by the triple of parameters (h, m_1, m_2) such that

$$h \leq \min\left\{r, q^{d-k}\right\}, m_2 \leq s_1 = q^{d-k} = n/q^k,\ m_1 \leq is_1 = iq^{d-k},$$

where i takes on the q values $1, \ldots, q$, and exactly q^{k-1} calls for the solution of ($V \cdot SPLIT$) at level k of the recursion (out of a total of q^k calls) correspond to each of the q values i.

Applying the estimates for the complexity of Problems B.1 and B.2, based on our complexity bounds for Problems 2.2.2 ($M \cdot MULTIPLY$), 2.2.5 ($INVERT$), 2.2.3b ($NULL \cdot SPACE$) and 2.2.10 ($M \cdot RANK$), we arrive at the following estimates for the computational cost of performing Algorithm C.1:

$$O_A\left(\gamma(r,p)(\log^2 r)d,\ \frac{n\,P^*(r)\log^\alpha r}{\max\{q,r\}\gamma(r,p)d^*}\right), d^* = \left\lceil \frac{\log r}{\log q} \right\rceil, \qquad (C.12)$$

for all the auxiliary matrix inversions;

$$O_A(d\log n, nqr/d^*) \text{ and } O_A(d\log r, P(r)n/(d^*r))$$

for all the auxiliary matrix multiplications at Stage 2 (for Problem B.2). Furthermore, we have the following bounds on the complexity of the auxiliary matrix multiplications at Stage 1 (for Problem B.1): $O_A(\log n, n^2)$ for the first one, and $O_A(d\log r, rn)$ for all other matrix multiplications, which amounts to (C.12) for α replaced by 0 (compare Theorem C.2). These estimates combined dominate the asymptotic cost of all the auxiliary matrix rank computations, which amounts to (C.12) for α replaced by 0 (compare Thorem C.2). Summarizing, we obtain the overall complexity estimates for Problem 2.2.10a ($SUBMATRIX$) (excluding the cost of the initial computation of the rank of the input matrix).

Theorem C.3. *Algorithm C.1 supports the following estimates for the complexity of randomized solution of Problem 2.2.10a ($SUBMATRIX$), for an $n \times n$ input matrix of a rank r (given as an input value):*

$$O_A\left(t, \left(\frac{(\log^{2+\alpha} r)P^*(r)n}{\max\{q,r\}} + \frac{P(r)nd\log r}{r} + nqrd\log n + n^2\log n\right)/(d^*\hat{t})\right),$$

where

$$\hat{t} = (\gamma(r,p)\log^2 r + \log n)d, \; d = \log n/\log q, \; d^* = \left\lceil \frac{\log r}{\log q} \right\rceil,$$

q is any fixed value from 2 to n; $P(r) \leq P^*(r)$ [see (4.5)], *and $P^*(r), \gamma(r,p)$ and α are defined according to (4.5), (6.2) and Remark 4.1.*

Remark C.2. Consider the choice of $q = n^u$, for a fixed positive constant $u \leq 1$. Then $d = 0$, and, for any characteristic p, the bounds of Theorem C.3 are strictly superior to the ones of (C.11), supported by our bounds of Theorem C.2 on the complexity of $(M \cdot RANK)$ and by Algorithm B.2, which amounts to the algorithm of [CLR] for $p = 0$. Compared to the bounds (C.10), the estimates of Theorem C.3 improve the time bound by the factor $\log q = (\log n)/d$, at the expense of roughly q-fold increase of one of the terms in the expression (C.10) for the processor bound. Setting $q = n^u$, we estimate that $d = O(1)$, $\hat{t} = O(\gamma(r,p)\log^2 r + \log n)$, and, within a factor logarithmic in n, the product of the time and processor bounds of Theorem C.3 is $O((\log^{2+\alpha} r)P^*(r)n/r + (\log n)(n + rn^u)n)$.

Remark C.3. An alternative approach to randomized parallel computations (over any field of constants) begins with reducing the basic Problems 2.2.5 ($INVERT$) and 2.2.10 ($M \cdot RANK$) for an $n \times n$ matrix A to ones for the matrix $B = U^{-1}L_2U_1AL_1U_2L^{-1}$, where L, L_1, L_2, U^T, U_1^T and U_2^T are random $n \times n$ lower triangular matrices, and U, U_1, L and L_1 are unit triangular Toeplitz matrices. Then we observe that we may satisfy the latter matrix equation for any fixed pair of $n \times n$ matrices A and B having the same rank if we were allowed to choose appropriate $n \times n$ lower triangular matrices L, L_1, L_2, U^T, U_1^T and U_2^T (apply Lemma 2.13.1 to matrices UBL and U_1AL_1). It follows that with a high probability Algorithm 2.10.2, applied to the matrix B in the case of random L, L_1, L_2, U and U_1, outputs a symmetric tridiagonal matrix T. The cost bounds of this reduction (see Proposition 5.4) dominates the cost bounds of the subsequent solution, that is, $O_A(\log n, n^2/\log n)$ for $INVERT$ ([PSA]) and $O_A(\log^2 n, n\log\log n/\log n)$ for $M \cdot RANK$, over any field. The bound for $M \cdot RANK$ (for T) is obtained due to immediate reduction to $CHAR \cdot POL$ for T, and the latter problem can be solved at the cost $O_A(\log^2 n, n\log\log n/\log n)$ over any field (see [BP91], appendix C). The latter paper gives a near optimum parallel solution of Problem 3.1 ($EIGENVALUES$) for T, thus completing a randomized parallel solution of $EIGENVALUES$ for any symmetric matrix. The reader may extend the above randomized parallel solution for $INVERT$ and $M \cdot RANK$ to various other matrix computations, including $SUBMATRIX$ and the computation of a maximal linearly independent subset of a vector set.

Bibliography.

ibliogra-
9, ACM

J. Sci.

e Num-

nge Ex-

Parallel
Founda-

lgebras,

d Set of

Integral

litz Ma-

of Ma-

Inverses
5.
th Ann.
, 1983.
te Poly-
nputing,

ms. Ap-

Inform.

5, Com-

J. Com-
sa, Italy

ter Sci-

national
Verona,

25, 37–

Toeplitz
03–116,

a Appl.,

SIAM J.

An Introduction to Numerical Analysis, Wiley, 1978.

ling, Polynomial Algorithms for LP over a Subring of the Al- with Applications to LP with Circulant Matrices, *Proc. 32nd np. on Foundation of Computer Science*, 480–487, IEEE Com- ress, 1991.

ery of Algebraic Numbers from Their p-adic Approximations, *SAM Intern. Symp. on Symb. and Algebr. Comp.*, 112–120, w York, 1989.

E. Collins, S. McCallum, Cylindrical Algebraic Decompositions, *ut.*, **13, 4**, 865-889, 1984.

Gader, A Variant of the Gohberg-Semencul Formula Involving ices, *SIAM J. Matrix Anal. Appl.*, **12**, **3**, 534–541, 1991.

Gader, New Decompositions of the Inverse of a Toeplitz Matrix, *Symp. on Math Theory of Networks and Systems (MTNS),* 89.

il, O. Margalit, On the Exponent of the All Pairs Shortest Path *32nd Ann. IEEE Symp. on Foundation of Computer Science,* Computer Society Press, 1991.

lil, O. Margalit, M. Naor, Witnesses for Boolean Matrix Mul- Shortest Paths, *Proc. 33rd Ann. IEEE Symp. on Foundations* 417–426, IEEE Computer Society Press, 1992.

, L. Greengard, V. Rokhlin, The Fast Multipole Method for le Simulation, *Computer Physics Communications,* **48**, 115–

V.B. Gragg, Implementation and Use of the Generalized Schur *mputational and Combinatorial Methods in Systems Theory,* Holland, Amsterdam (C.I. Byrnes and A. Lindquist Editors)

V.G. Gragg, The Generalized Schur Algorithm for the Superfast eplitz Systems, *Rational Approximation and Its Applications s and Physics, Lecture Notes in Mathematics,* **1237**, 315–330, n (J. Gilewicz, M. Pindor and W. Siemaszko Editors) 1987.

V.G. Gragg, Superfast Solution of Real Positive Definite Toeplitz *J. Matrix Anal. Appl.*, **9**, **1**, 61–76, 1988.

, M.-D.A. Huang, Recognizing Primes in Random Polynomial *th Ann. ACM Symp. on Theory of Computing,* 462-469, ACM rk, 1987.

erup, K. Mehlhorn, F. Preparata, Deterministic Simulation of llel Model on More Realistic Ones, *SIAM J. Comput.*, **16, 5,**

Hopcroft, J.D. Ullman, *The Design and Analysis of Computer* ddison-Wesley, Reading, Massachusetts, 1976.

Herzberger, *Introduction to Interval Computations,* Academic rk, 1983.

n, R. Rubinfeld, M. Sudan, Reconstructing Algebraic Functions ata, *Proc. 33rd Ann. IEEE Symp. on Foundations of Computer* 12, IEEE Computer Society Press, 1992.

lication is the Easiest Nontrivial Arithmetic Function, *Theoret- Science,* **36**, 333–339, 1985.

umerical Methods for Partial Differential Equations, Academic rk, 1977.

on, A Method of Local Corrections for Computing the Velocity a Distribution of Vortex Blobs, *J. Comput. Phys.*, **62**, 111–123,

1986.

[Ang] D. Angluin, Computational Learning Theory: Survey and Selected B phy, *Proc. 24th Ann. ACM Symp. on Theory of Computing*, 351–3 Press, New York, 1992.

[Ap] A.W. Appel, An Efficient Program for Many-body Simulation, *SIA Stat. Comput.*, **6**, 85–103, 1985.

[APR] L.M. Adleman, C. Pomerance, R.S. Rumley, On Distinguishing Pri bers from Composite Numbers, *Annals of Math.*, **117**, 173–206, 198

[AR] B. Alpert, V. Rokhlin, A Fast Algorithm for the Evaluation of Lag pansions, *SIAM J. Sci. Stat. Comput.*, **12**, **1**, 158–179, 1990.

[ARa] Y. Aumann, M. O. Rabin, Clock Construction in Fully Asynchronou Systems and PRAM Simulation, *Proc. 33rd Annual IEEE Symp. on tion of Computer Sci.*, 147–156, IEEE Computer Society Press, 199

[ASt] A. Alder, V. Strassen, On the Algorithmic Complexity of Associative *Theoretical Computer Science,* **15**, 201–211, 1981.

[ASU] A.V. Aho, K. Steiglitz, J.D. Ullman, Evaluating Polynomials at Fix Points, *SIAM J. Comput.,* **4**, 533–539, 1975.

[B-AS85] A. Ben-Artzi, T. Shalom, On Inversion of Block Toeplitz Matrices *Equations and Operator Theory,* **8**, 751–779, 1985.

[B-AS86] A. Ben-Artzi, T. Shalom, On Inversion of Toeplitz and Close to Toe trices, *Linear Algebra Appl.,* **75**, 173–192, 1986.

[B-I] A. Ben-Israel, A Note on Iterative Method for Generalized Inversic trices, *Math. Comp.,* **20**, 439–440, 1966.

[B-IC] A. Ben-Israel, D. Cohen, On Iterative Computation of Generalized and Associated Projections, *SIAM J. Numer. Anal.,* **3**, 410–419, 19

[B-O] M. Ben-Or, Lower Bounds for Algebraic Computation Trees, *Proc. ACM Symp. on Theory of Computing,* 80–86, ACM Press, New Yor

[B-OT] M. Ben-Or, P. Tiwari, A Deterministic Algorithm for Sparse Multivar nomial Interpolation, *Proc. 20th Ann. ACM Symp. on Theory of C* 301–309, ACM Press, New York, 1988.

[B80] D. Bini, Relations between Exact and Approximate Bilinear Algorit plications, *Calcolo,* **17**, 87–97, 1980.

[B82] D. Bini, Reply to the Paper "The Instability of Bini's Algorithm" *Proc. Lett.,* **14**, **3**, 144–145, 1982.

[B83] D. Bini, On a Class of Matrices Related to Toeplitz Matrices, TR-8 *puter Science Dept., SUNYA,* Albany, New York, 1983.

[B84] D. Bini, Parallel Solution of Certain Toeplitz Linear Systems, *SIAM put.,* **13**, **2**, 268–276, 1984. Also T.R. B82-04, *I.E.I. of C.N.R.,* P (April 1982).

[B84a] D. Bini, On Commutativity and Approximation, *Theoretical Com ence,* **28**, 135–150, 1984.

[B88] D. Bini, Complexity of Parallel Polynomial Computations, *Proc. Int Meeting on Parallel Computing, Methods, Algorithms, Applications* Sept. 28–30, 1988, S. Evans and C. Nodari Sutti Editors.

[B88a] D. Bini, Matrix Structures in Parallel Matrix Computations, *Calcol* 51, 1988.

[BA] R.R. Bitmead, B.D.O. Anderson, Asymptotically Fast Solution of and Related Systems of Linear Equations, *Linear Algebra Appl.,* **34**, 1980.

[BaF] W.W. Barret, P.J. Feinsilver, Inverses of Band Matrices, *Linear Alge* **41**, 111–130, 1981.

[Bai] D.H. Bailey, Extra High Speed Matrix Multiplication on the Cray-2, *Sci. Stat. Comput.,* **9**, 603–607, 1988.

[Bai,a] D.H. Bailey, A Portable High Performance Multiprecision Package, RNR Technical Report RNR-90-022, NASA Applied Research Branch, NASA Ames Research Center, Moffett Fields, California, 1992.

[Bai,b] D.H. Bailey, Multiprecision Translation and Execution of Fortran Programs, *ACM Transactions on Mathematical Software,* 1992.

[Bak71] N.S. Bakhvalov, On a Stable Evaluation of Polynomials, *J. of Comp. Math. and Math. Physics* (in Russian), **11, 6**, 1568–74, 1971.

[BB] J.W. Borwein, P.B. Borwein, The Arithmetic-Geometric Mean and Fast Computation of Elementary Functions, *SIAM Rev.,* **26, 3**, 351–366, 1984.

[BBB] R. Bevilacqua, N. Bonanni, E. Bozzo, On Algebras of Toeplitz Plus Hankel Matrices, T.R. 33/93 *Dipartimento di Informatica, Università di Pisa,* Pisa 1993.

[BBCM] R. Bevilacqua, D. Bini, M. Capovani, O. Menchi, *Metodi Numerici,* Zanichelli, Bologna, Italy, 1992.

[BBo] D. Bini, E. Bozzo, Fast Discrete Transform by Means of Eigenpolynomials, *Computers & Mathematics (with Applications),* **26, 9**, 35–52, 1993.

[BC83] D. Bini, M. Capovani, Fast Parallel and Sequential Computations and Spectral Properties Concerning Band Toeplitz Matrices, *Calcolo,* **22**, 176–189, 1983.

[BC83a] D. Bini, M. Capovani, Spectral and Computational Properties of Band Symmetric Toeplitz Matrices, *Linear Algebra Appl.,* **52/53**, 99–126, 1983.

[BC87] D. Bini, M. Capovani, Tensor Rank and Border Rank of Band Toeplitz Matrices, *SIAM J. Comput.,* **16**, 252–258, 1987.

[BCL] B. Buchberger, G.E. Collins, R. Loos (eds.), *Computer Algebra: Symbolic and Algebraic Computations,* Springer, Berlin, 1982.

[BCLR] D. Bini, M. Capovani, G. Lotti, F. Romani, $O(n^{2.7799})$ Complexity for Approximate Matrix Multiplication, *Inform. Proc. Lett.,* **8, 5**, 234–235, 1979.

[BCLR81] D. Bini, M. Capovani, G. Lotti, F. Romani, *Complessità Numerica,* Serie di Informatica, Boringhieri, Torino, 1981.

[BCM] D. Bini, M. Capovani, O. Menchi, *Metodi Numerici per l'Algebra Lineare.* Zanichelli, Bologna, 1988.

[BCP] A. Borodin, S. Cook, N. Pippenger, Parallel Computation for Well-Endowed Rings and Space-Bounded Probabilistic Machines, *Information and Control,* **58, 1–3**, 113–136, 1983.

[BDB90] D. Bini, F. Di Benedetto, A New Preconditioner for the Parallel Solution of Positive Definite Toeplitz Systems, *Proc. 2nd ACM Symp. on Parallel Algorithms and Architectures,* 220–223, ACM Press, New York 1990.

[BDiF] E. Bozzo, C. Di Fiore, On the Use of Certain Matrix Algebras Associated with Discrete Trigonometric Transforms in Matrix Displacement Decomposition, to appear in *SIAM J. Matrix Anal. Appl.*.

[BDu] R. Bank, T. Dupont, An Optimal Order Process for Solving Finite Element Equations, *Math. Comp.,* **36**, 35–51, 1981.

[Be] E.R. Berlekamp, *Algebraic Coding Theory,* McGraw-Hill, New York, 1968.

[BeC] R. Bevilacqua, M. Capovani, Proprietà delle Matrici a Banda ad Elementi ed a Blocchi, *Boll. Un. Mat. Ital.,* **B13**, 844–861, 1976.

[BeCR] R. Bevilacqua, B. Codenotti, F. Romani, Parallel Solution of Block Tridiagonal Linear Systems, *Linear Algebra Appl.,* **104**, 39–57, 1988.

[BeLR] R. Bevilacqua, G. Lotti, F. Romani, Storage Compression of Inverses of Band Matrices, *Computers & Mathematics (with Applications),* **20**, 1–11, 1990.

[Ber] S. Berkowitz, On Computing the Determinant in Small Parallel Time Using a Small Number of Processors, *Inform. Proc. Lett.,* **18, 3**, 147–150, 1984.

[BF] D. Bini, P. Favati, On a Matrix Algebra Related to the Hartley Transform, *SIAM J. Matrix Anal. Appl.,* **14, 2**, 500–507, 1993.

[BG] D. Bini, L. Gemignani, On the Complexity of Polynomial Zeros, *SIAM J. Comput.,* **21**, 781–799, 1992.

[BG90] D. Bini, L. Gemignani, On the Euclidean Scheme for Polynomials Having Interlaced Real Zeros, *Proc. 2nd Ann. ACM Symp. on Parallel Algorithms and Architectures*, 254–258, ACM Press, New York, 1990.

[BG92] D. Bini, L. Gemignani, Fast Parallel Computation of the Remainder Sequence via Bezout and Hankel Matrices, *Quaderno 79, Dipartimento di Matematica, Università di Parma,* Parma, Italy, 1992 (to appear in *SIAM J. Comput.*),

[BGH] A. Borodin, J. von zur Gathen, J. Hopcroft, Fast Parallel Matrix and GCD Computation, *Information and Control,* **52, 3**, 241–256, 1982.

[BGP] D. Bini, L. Gemignani, V. Y. Pan, Improved Parallel Computations with Matrices and Polynomials, *Proc. 18th ICALP, Lecture Notes in Computer Science,* **510**, 520–531, Springer, Berlin, 1991.

[BGY] R.P. Brent, F.G. Gustavson, D.Y.Y. Yun, Fast Solution of Toeplitz Systems of Equations and Computation of Padé Approximations, *J. of Algorithms,* **1**, 259–295, 1980.

[BHu] J. Barnes, P. Hut, A Hierarchical O(N log N) Force-Calculation Algorithm, *Nature,* **324**, 446–449, 1986.

[BKR] M. Ben-Or, D. Kozen, J. H. Reif, The Complexity of Elementary Algebra and Geometry, *J. Computer and System Sciences,* **32**, 251–264, 1986.

[BKu] R.P. Brent, H.T. Kung, Fast Algorithms for Manipulating Formal Power Series, *J. of ACM,* **25, 4**, 581–595, 1978.

[BL] D. Bini, G. Lotti, Stability of Fast Algorithms for Matrix Multiplication, *Numer. Math.,* **36**, 63–72, 1980.

[BLR] D. Bini, G. Lotti, F. Romani, Approximate Solutions for the Bilinear Form Computational Problem, *SIAM J. Comput.,* **9**, 692–697, 1979.

[BLRu] M. Blum, M. Luby, R. Rubinfeld, Self-Testing/Correcting with Applications to Numerical Problems, *Proc. 22nd Ann. ACM Symp. on Theory of Computing,* 73–83, ACM Press, New York, 1990.

[BLY] A. Björner, L. Lovasz, A. C.-C. Yao, Linear Decision Trees: Volume Estimates and Topological Bounds, *Proc. 24th Ann. ACM Symp. on Theory of Computing,* 170–177, ACM Press, New York, 1992.

[BM] A. Borodin, I. Munro, *The Computational Complexity of Algebraic and Numeric Problems,* American Elsevier, New York, 1975.

[Bo] E. Bozzo, Matrix Algebras and Discrete Transforms, Tesi di Dottorato in Informatica, *Dipartimento di Informatica, Università di Pisa,* Pisa, Italy, 1993.

[Bo93] E. Bozzo, Algebras of Higher Dimension for Displacement Decomposition and Computations with Toeplitz Plus Hankel Matrices, *Linear Algebra Appl.*, to appear.

[Bor77] A. Borodin, On Relating Time and Space to Size and Depth, *SIAM J. Comput.,* **6**, 733–744, 1977.

[BoT] A. Borodin, P. Tiwari, On the Decidability of Sparse Univariate Polynomial Interpolation, *Proc. 22nd Ann. ACM Symp. on Theory of Computing,* 535–545, ACM Press, New York, 1990.

[Bour] N. Bourbaki, *Elements of Mathematics (Algebra I),* Springer, Berlin, 1989.

[BP] D. Bini, V. Y. Pan, Improved Parallel Polynomial Division, *Proc. 33rd Ann. IEEE Symp. on Foundation of Computer Science,* 131–136, IEEE Computer Society Press, 1992, and *SIAM J. Comput.*, **22**, **3**, 617–626, 1993.

[BP86] D. Bini, V. Y. Pan, Fast Parallel Algorithms for Polynomial Division over Arbitrary Field of Constants, *Computers & Mathematics (with Applications),* **12A, 11**, 1105–1118, 1986.

[BP86a] D. Bini, V. Y. Pan, Polynomial Division and Its Computational Complexity, *J. Complexity,* **2**, 179–203, 1986.

[BP88] D. Bini, V. Y. Pan, Efficient Algorithms for the Evaluation of the Eigenvalues of (Block) Banded Toeplitz Matrices, *Math. Comp.,* **50**, **182**, 431–448, 1988.

[BP91] D. Bini, V. Y. Pan, Parallel Complexity of Tridiagonal Symmetric Eigenvalue Problem, *Proc. 2nd Ann. ACM-SIAM Symposium on Discrete Algorithms,*

384–393, ACM Press, New York, and SIAM Publications, Philadelphia, Pennsylvania, 1991.

[BP91a] D. Bini, V. Y. Pan, On the Evaluation of the Eigenvalues of a Banded Toeplitz Block Matrix, *J. of Complexity*, **7**, 408–424, 1991.

[BP93] D. Bini, V. Y. Pan, Improved Parallel Computations with Toeplitz-like and Hankel-like Matrices, *Linear Algebra Appl.*, **188**, **189**, 3–29, 1993. [Proceedings version, *Proc. ACM-SIGSAM Int. Symp. on Symb. Alg. Comp. (ISSAC '93)*, (M. Bronstein Editor) 193–200, ACM Press, New York, 1993.]

[Br72] A. Brandt, Multi-level Adaptive Technique (MLAT) for Fast Numerical Solutions to Boundary Value Problems, *Proc. 3rd Int. Conf. Numerical Methods in Fluid Mechanics* (Paris, France) 1972, *Lecture Notes in Physics,* **18**, 82–89, Springer, Berlin, 1972.

[Br77] A. Brandt, Multi-level Adaptive Solutions to Boundary Value Problems, *Mathematics Comp.,* **31**, 333–390, 1977.

[Br84] A. Brandt, Multigrid Techniques: 1984 Guide, with Applications to Fluid Dynamics, *Available as GMD Studien Nr. 85, GMD-AIW* Postfach 1240, D-5205, St. Augustin 1, W. Germany, 1984.

[Bre74] R.P. Brent, The Parallel Evaluation of General Arithmetic Expressions, *J. of ACM,* **21**, **2**, 201–206, 1974.

[Bre76] R.P. Brent, Fast Multiple-Precision Evaluation of Elementary Functions, *J. ACM,* **23**, 242–251, 1976.

[BreT] R.P. Brent, J.F. Traub. On the Complexity of Composition and Generalized Composition of Power Series, *SIAM J. Comput.,* **9**, 54–66, 1980.

[Bru] G. Bruun, z-Transform DFT Filters and FFT's, *IEEE Trans. Acoust. Speech, Signal Processing*, **ASSP-26**, 56–63, 1978.

[Bshouty] N.H. Bshouty, A Lower Bound for Matrix Multiplication, *Proc. 29th Ann. IEEE Symp. on Foundation of Computer Science,* 64–67, IEEE Computer Society Press, 1988.

[BSS] L. Blum, M. Shub, S. Smale, On a Theory of Computations Over the Real Numbers: NP-Completeness, Recursive Functions and Universal Machines, *Bull. of Amer. Math. Soc.,* **21**, 1–46, 1989.

[BSt] W. Baur, V. Strassen, On the Complexity of Partial Derivatives, *Theoretical Computer Science,* **22**, 317–330, 1983.

[BT] W.S. Brown, J.F. Traub, On Euclid's Algorithm and the Theory of Subresultants, *J. of ACM,* **18**, **4**, 505–514, 1971.

[Buch] B. Buchberger, Gröbner Bases: An Algorithmic Method in Polynomial Ideal Theory, N. K. Bose, Editor, *Recent Trends in Multidimensional Systems Theory,* D. Reidel Publ. Company, 1985.

[Bun] J.R. Bunch, Stability of Methods for Solving Toeplitz Systems of Equations, *SIAM J. Sci. Stat. Comput.,* **6**, **2**, 349–364, 1985.

[C80] G. Cybenko, The Numerical Stability of the Levinson-Durbin Algorithm for Toeplitz System of Equations, *SIAM J. Sci. Stat. Comput.,* **1**, 303–310, 1980.

[Can] D. G. Cantor, On Arithmetic Algorithms over Finite Fields, *J. of Combinatorial Theory*, Series **A** **50**, 285–300, 1991.

[CaMe] S. Cabay, R. Meleshko, A Weakly Stable Algorithm for Padé Approximants and the Inversion of Hankel Matrices, *SIAM J. Matrix Anal. Appl.*, **14, 3**, 735–765, 1993.

[CB] G. Cybenko, M. Berry, Hyperbolic Householder Algorithms for Factoring Structured Matrices, *SIAM J. Matrix Anal. Appl.,* **11**, **4**, 499–520, 1990.

[CdB] C.D. Conte, C. de Boor, *Elementary Numerical Analysis: an Algorithmic Approach,* McGraw-Hill, New York, 1980.

[CE] J. Canny, I. Emiris, An Efficient Algorithm for the Sparse Mixed Resultant, *Proc. 10th Intern. Symp. Applied Algebra, Algebraic Algorithms and Error-Correcting Codes,* (G. Cohen, T. Mora, O. Moreno Editors), *Lecture Notes in Computer Science,* **673**, 89–104, Springer, Berlin, 1993.

[CF] B. Codenotti, F. Flandoli, A Monte Carlo Method for the Parallel Solution of Linear Systems, *J. Complexity,* **5**, 107–117, 1989.

[CGG] B. W. Char, K. O. Geddes, G. H. Gonnet, "QCDHUE: Heuristic Polynomial GCD Algorithm based on Integer GCD Computation," *Proc. EUROSAM '84, Lecture Notes in Computer Science,* **174**, 285–296, Springer, New York – Berlin, 1984.

[CGR] J. Carier, L. Greengard, V. Rokhlin, A Fast Adaptive Multipole Algorithm for Particle Simulation, *SIAM J. Sci. Stat. Comput.,* **9**, **4**, 669–686, 1988.

[Ch88] T.F. Chan, An Optimal Circulant Preconditioner for Toeplitz Systems, *SIAM J. Sci. Stat. Comput.,* **9**, 766–771, 1988.

[Ch89] R.H. Chan, The Spectrum of a Family of Circulant Preconditioned Toeplitz Systems, *SIAM J. Numer. Anal.,* **26**, 503–506, 1989.

[Ch89a] R.H. Chan, Circulant Preconditioners For Hermitian Toeplitz Systems, *SIAM J. Matrix Anal. Appl.,* **4**, 542–550, 1989.

[Chan] T.F. Chan, Rank Revealing QR-factorization, *Linear Algebra Appl.,* **88**, 67–82, 1987.

[Chi] A.L. Chistov, Fast Parallel Calculation of the Rank of Matrices over a Field of Arbitrary Characteristics, *Proc. FCT 85, Lecture Notes in Computer Science,* **199**, 63–69, Springer, Berlin, 1985.

[ChR] J. Cheriyan, J. Reif, Parallel and Output Sensitive Algorithms for Combinatorial and Linear Algebra Problems, *Proc. 5th Ann. ACM Symp. on Parallel Algorithms and Architecture,* 50–56, ACM Press, New York, 1993.

[ChS] R.H. Chan, G. Strang, Toeplitz Equations by Conjugate Gradients with Circulant Preconditioner, *SIAM J. Sci. Stat. Comput.,* **10**, 104–119, 1989.

[CK] J. Chun, T. Kailath, Divide-and-Conquer Solution of Least-Squares Problems for Matrices with Displacement Structure, *SIAM J. Matrix Anal. Appl.,* **12**, **1**, 128–145, 1991.

[CKa] D. G. Cantor, E. Kaltofen, On Fast Multiplication of Polynomials over Arbitrary Rings, *Acta Inf.*, **28**, **7**, 697–701, 1991.

[CKa87] D.G. Cantor, E. Kaltofen, Fast Multiplication of Polynomials with Coefficients from an Arbitrary Ring, Tech. Report No. 87–35, *Comp. Sci. Dept., Rensselaer Polyt. Inst.*, Troy, New York, 1987.

[CKL-A] J. Chun, T. Kailath, H. Lev-Ari, Fast Parallel Algorithm for QR-factorization of Structured Matrices, *SIAM J. Sci. Stat. Comput.,* **8**, **6**, 899–913, 1987.

[CKL89] J.F. Canny, E. Kaltofen, Y. Lakshman, Solving Systems of Non-Linear Polynomial Equations Faster, *Proc. ACM-SIGSAM Int. Symp. on Symb. and Alg. Comp. (ISSAC '89),* 121–128, ACM Press, New York, 1989.

[CLR] T.H. Cormen, C.F. Leiserson, R.L. Rivest, *Introduction to Algorithms*, MIT Press, Boston, Massachusetts, and McGraw Hill, New York, 1990.

[Cod85] B. Codenotti, Error Analysis of Some Approximate Algorithms, *Portugaliae Mathematica,* **43**, 113–119, 1985–1986.

[Col] G.E. Collins, Subresultants and Reduced Polynomial Remainder Sequences, *J. of ACM,* **14**, 128–142, 1967.

[Cook85] S.A. Cook, A Taxonomy of Problems with Fast Parallel Algorithms, *Inform. and Control,* **64**, 2–22, 1985.

[CPW] R. E. Cline, R. J. Plemmons, G. Worm, Generalized Inverses of Certain Toeplitz Matrices, *Linear Algebra Appl.*, **8**, 25–33, 1974.

[CR] S.A. Cook, R.A. Reckhow, Time Bounded Random Access Machines, *J. Computer & System Sciences,* **7**, 354–375, 1973.

[Cs] L. Csanky, Fast Parallel Matrix Inversion Algorithm, *SIAM J. Comput.,* **5**, **4**, 618–623, 1976.

[CT] J.W. Cooley, J.W. Tukey, An Algorithm for the Machine Calculation of Complex Fourier Series, *Math Comput.,* **19**, 297–301, 1965.

[CW90] D. Coppersmith, S. Winograd, Matrix Multiplication via Arithmetic Progressions, *J. of Symbolic Computations,* **9**, **3**, 251–280, 1990 (short version in *Proc.*

19th Ann. ACM Symp. on Theory of Computing, 1–6, ACM Press, New York, 1987).

[CZ] D.G. Cantor, H. Zassenhaus, A New Algorithm for Factoring Polynomials Over Finite Fields, *Math. Comp.*, **36**, **154**, 587–592, 1981.

[Da] P. Davis, *Circulant Matrices*, Wiley, New York, 1974.

[Datta] K. Datta, Parallel Complexity and Computations of Cholesky's Decomposition and QR Factorization, *International J. Computer Math.*, **18**, 67–82, 1985.

[DB] G. Dahlquist, Å. Björk, *Numerical Methods*, Prentice-Hall, New Jersey, 1974.

[dBG] C. de Boor, G. H. Golub, The Numerically Stable Reconstruction of a Jacobi Matrix from Spectral Data, *Linear Algebra Appl.*, **21**, 245–260, 1978.

[DD] C.C. Douglas, J. Douglas, Jr., A Unified Convergence Theory for Abstract Multilevel Algorithms, Serial and Parallel, *SIAM J. Numer. Anal.*, **30**, **1**, 136–158, 1993.

[De] X. Deng, On Parallel Complexity of Integer Linear Programming, *Proc. First Ann. ACM Symp. on Parallel Algorithms and Architectures*, 110–116, ACM Press, New York, 1989.

[DG] P. Delsarte, Y. Genin, On the Splitting of Classical Algorithms in Linear Prediction Theory, *IEEE Trans. on Acoustics, Speech, and Signal Processing*, **35**, **5**, 645–653, 1987.

[DGR] A. Dutt, M. Gu, V. Rokhlin, Fast Algorithms for Polynomial Interpolation, Integration and Differentiation, Research Report RR-977, *Computer Science Department, Yale University*, 1993.

[dH] F.R. deHoog, On the Solution of Toeplitz Systems, *Linear Algebra Appl.*, **88/89**, 123–138, 1987.

[Di] J.D. Dixon, Estimating Extremal Eigenvalues and Condition Numbers of Matrices, *SIAM J. Numer. Anal.*, **20**, **4**, 812–814, 1983.

[DiB92] F. Di Benedetto, Analysis of Preconditioning Techniques for Ill-Conditioned Toeplitz matrices, *Proc. ERCIM Workshop on Numerical Linear Algebra*, 5–10, Pisa, Italy, May 1992. (To appear in *SIAM J. Sci. Stat. Comput.*).

[DiB93] F. Di Benedetto, Preconditioning of Block Toeplitz Matrices by Sine Transforms, *T.R. 235, Dipartimento di Matematica, Università di Genova*, Genova, 1993.

[DiBFS] F. Di Benedetto, G. Fiorentino, S. Serra, Conjugate Gradient Preconditioning for Toeplitz matrices, *Computers & Mathematics (with Applications)*, **25**, 35–45, 1993.

[DiFZ] C. Di Fiore, P. Zellini, Matrix Decompositions Using Displacement Rank and Classes of Commutative Matrix Algebras, *Linear Algebra Appl.* (to appear).

[DL] R.A. Demillo, R.J. Lipton, A Probabilistic Remark on Algebraic Program Testing, *Information Process. Letters*, **7**, **4**, 193–195, 1978.

[DP] J. Davenport, J. Padjet, HEUGCD: How Elementary Upper Bounds Generate Cheaper Data, Proc. EUROCAL '85, *Lecture Notes in Computer Science*, **204**, 18–28, Springer, New York – Berlin, 1985.

[DST] J.H. Davenport, Y. Siret, E. Tournier, *Computer Algebra: Systems and Algorithms for Algebraic Computations*, Academic Press, 1988.

[Duh] P. Duhamel, Implementation of "Split-radix", FFT Algorithms for Complex, Real and Real-Symmetric Data, *IEEE Trans. Acoust., Speech, Signal Processing*, **34**, 285–295, 1986.

[E] W. Eberly, Very Fast Parallel Polynomial Arithmetic, *SIAM J. Comput.*, **18**, **5**, 955–976, 1989.

[E91] W. Eberly, Efficient Parallel Independent Subsets and Matrix Factorizations, *Proc. 3rd IEEE Conference on Parallel and Distributed Processing*, 204–211, IEEE Computer Society Press, 1991.

[E92] W. Eberly, On Efficient Band Matrix Arithmetic, *Proc. 33rd Ann. IEEE Symp. on Foundation of Computer Science*, 457–463, IEEE Computer Society Press, 1992.

[EC] I. Emiris, J. Canny, A Practical Method for Sparse Resultant, *Proc. ACM-SIGSAM Int. Symp. on Symb. Alg. Comput. (ISSAC '93)*, (M. Bronstein Editor), 183–192, ACM Press, New York, 1993.

[Edm] J. Edmonds, Systems of Distinct Representative and Linear Algebra, *J. Res. Nat. Bureau of Standards,* **713**, **4**, 241–245, 1967.

[EG88] D. Eppstein, Z. Galil, Parallel Algorithmic Techniques for Combinatorial Computation, *Annual Review of Computer Science,* **3**, 233–283, 1988.

[EL] M. Ewerbring, F.T. Luk, Canonical Correlations and Generalized SVD: Applications and New Algorithms, *J. Comput. Applied Math.,* **27**, 37–52, 1989.

[Eng] H.T. Engström, Polynomial Substitutions, *Amer. J. Math.,* **63**, 249–255, 1941.

[FD] P.A. Fuhrmann, N.B. Datta, On Bezoutians, Vandermonde Matrices, and the Lienard-Chipart Stability Criterion, *Linear Algebra Appl.,* **120**, 23–27, 1989.

[FF] D.K. Faddeev, V.N. Faddeeva, *Computational Methods of Linear Algebra,* W.H. Freeman, San Francisco, 1963.

[FGN] R. W. Freund, G. H. Golub, N. M. Nachtigal, Iterative Solution of Linear Systems, *Acta Numerica,* **1**, 57–100, 1992.

[FH] K.V. Fernando, S.J. Hammarling, A Product Induced Singular Value Decomposition For Two Matrices and Balanced Realisation, *Linear Algebra in Signals, Systems and Control* (B.N. Datta et al. Editors), 128–140, SIAM, Philadelphia, Pennsylvania, 1988.

[FHR] T. Fink, G. Heinig, K. Rost, An Inversion Formula and Fast Algorithms for Cauchy-Vandermonde Matrices, *Linear Algebra Appl.,* **183**, 179–191, 1993.

[Fi] M. Fiedler, Hankel and Loewner Matrices, *Linear Algebra Appl.,* **58**, 75–95, 1984.

[Fi72] C.M. Fiduccia, Polynomial Evaluation via the Division Algorithm: The Fast Fourier Transform Revisited, *Proc. 4th Ann. ACM Symp. on Theory of Computing,* 88–93, ACM Press, New York, 1972.

[Fi87] C.M. Fiduccia, A Rational View of the Fast Fourier Transform, *Proc. 25th Allerton Conf. Commun., Control and Computing,* 1987.

[Fic] F.A. Ficken, *Linear Transformations and Matrices,* Prentice-Hall, Englewood Cliffs, New Jersey, 1984.

[Fich] F. Fich, The Parallel Random Access Machine, in *Synthesis of Parallel Algorithms,* (J.H. Reif Editor), 843–899, Morgan Kaufmann Publishers, San Mateo, California, 1993.

[FMcB87] P.O. Frederickson, O.A. McBryan, Superconvergent Multigrid Methods, *Cornell Theory Center*, May 1987.

[FMcB88] P.O. Frederickson, O.A. McBryan, Parallel Superconvergent Multigrid, *Multigrid Methods: Theory, Applications and Supercomputing* (S. McCormick Editor), *Lecture Notes in Pure and Applied Math.,* **100**, 195–210, M. Decker, Inc., 1988.

[FMcB91] P.O. Frederickson, O.A. McBryan, Normalized Convergence Rates for the PSMG Method, *SIAM J. Sci. Stat. Comput.,* **12**, 1, 221–229, 1991.

[FMKL] B. Friedlander, M. Morf, T. Kailath, L. Ljung, New Inversion Formulas for Matrices Classified in Terms of their Distances from Toeplitz Matrices, *Linear Algebra Appl.,* **27**, 31–60, 1979.

[FP] M.J. Fischer, M.S. Paterson, String Matching and Other Products, *SIAM-AMS Proc.,* **7**, 113–125, 1974.

[FPt] M. Fiedler, V. Ptak, Loewner and Bezout Matrices, *Linear Algebra Appl.,* **101**, 187–220, 1988.

[FS] G. Fiorentino, S. Serra, Multigrid Methods for Toeplitz Matrices, *Calcolo,* **28**, 283–305, 1992.

[G84] J. von zur Gathen, Parallel Algorithms for Algebraic Problems, *SIAM J. Comput.,* **13**, **4**, 802–824, 1984.

[G84a] J. von zur Gathen, Hensel and Newton's Methods in Valuation Rings, *Math. Comp.,* **42**, **166**, 637–661, 1984.

[G86] J. von zur Gathen, Parallel Arithmetic Computations: A Survey, *Proc. Math. Foundation of Computer Science, Lecture Notes in Computer Science,* **233**, 93–112, Springer, Berlin, 1986.

[G86a] J. von zur Gathen, Representations and Parallel Computations for Rational Functions, *SIAM J. Comput.,* **15, 2**, 432–452, 1986.

[G88] J. von zur Gathen, Algebraic Complexity Theory, *Ann. Review in Comp. Sci.,* **3**, 317–347, 1988.

[G90] J. von zur Gathen, Functional Decomposition of Polynomials: the Tame Case, *J. Symbolic Computation,* **9**, 281–299, 1990.

[Gad] P. Gader, Displacement Operator Based Decomposition of Matrices Using Circulant or Other Group Matrices, *Linear Algebra Appl.,* **139**, 111–131, 1990.

[Gast] N. Gastinel, Inversion d'une Matrice Generalisant la Matrice de Hilbert, *Chiffres,* **3**, 149–152, 1960.

[GCL] K. O. Geddes, S. R. Czapor, G. Labahn, *Algorithms for Computer Algebra,* Kluwer Academic Publishers, Norwell, Massachusetts, 1992.

[GG] M. H. Gutknecht, W. Gragg, Stable Look-ahead Versions of the Euclidean and Chebyshev Algorithms, IPS Research Report 94-04, ETH-Zentrum, Zürich, Switzerland, 1994.

[Gem91] L. Gemignani, Computing the Inertia of Bezout and Hankel Matrices, *Calcolo,* **28**, 267–274, 1991.

[Gem92] L. Gemignani, Fast Inversion of Hankel and Toeplitz Matrices, *Information Process. Letters,* **41**, 119–123, 1992.

[Gen] Y. Genin, On a Duality Relation in the Theory of Orthogonal Polynomials and Its Application in Signal Processing, *Manuscript,* 1987. (Abstract in Program of *First Intern. Conf. Industrial and Applied Math.*, ICIAM87, 10, INRIA, 1987.)

[Ger] A. Gerasoulis, A Fast Algorithm for the Multiplication of Generalized Hilbert Matrices with Vectors, *Math. Comp.,* **50, 181**, 179–188, 1987.

[GeS] P. Gemmel, M. Sudan, Highly Resilient Correctors for Polynomials, *Inf. Proc. Letters,* **43**, 169–174, 1992.

[GF] I.C. Gohberg, I.A. Feldman, Convolution Equations and Projection Methods for Their Solutions, *Translations of Math. Monographs,* **41**, Amer. Math. Soc., Providence, Rhode Island, 1974.

[GHR] R. Greenlaw, J. H. Hoover, W. L. Ruzzo, A Compendum of Problems Complete for P, Tech. Report TR 91-11, *Computer Science Department, Univ. of Alberta,* Edmonton, Canada, and TR 91-05-01, *Computer Science and Engineering Dept., Univ. of Washington,* Seattle, Washington, 1991.

[Gies] M. Giesbrecht, Fast Algorithms for Matrix Normal Forms, *Proc. 33rd Ann. IEEE Symp. on Foundation of Computer Science*, 121–130, IEEE Computer Society Press, 1992.

[Gies1] M. Giesbrecht, Nearly Optimal Algorithms for Canonical Matrix Forms, submitted to *SIAM J. Comput.*

[GK] I. Gohberg, N. Ya. Krupnik, A Formula for the Inversion of Finite Toeplitz Matrices, *Mat. Issled.,* **7,2,** 272–283 (in Russian), 1972.

[GKa] J. von zur Gathen, E. Kaltofen, Factoring Sparse Multivariate Polynomials, *J. Computer & Systems Sciences,* **31**, 265–287, 1985.

[GKi] S. Goldwasser, J. Kilian, Almost All Primes Can Be Quickly Verified, *Proc. 18th Ann. ACM Symp. on Theory of Computing,* 316–329, ACM Press, New York, 1986.

[GKK] I. Gohberg, T. Kailath, I. Koltracht, Efficient Solution of Linear Systems of Equations with Recursive Structure, *Linear Algebra Appl.,* **80**, 81–113, 1986.

[GKKL] I. Gohberg, T. Kailath, I. Koltracht, P. Lancaster, Linear Complexity Parallel Algorithms for Linear Systems of Equations with Recursive Structure, *Linear Algebra Appl.,* **88/89**, 271–315, 1987.

[GKL] J. von zur Gathen, D. Kozen, S. Landau, Functional Decomposition of Polynomials, *Proc. 28th Ann. IEEE Symp. on Foundation of Computer Science,* 127–131, IEEE Computer Society Press, 1987.

[GKS] D.Y. Grigoriev, Y. Karpinski, M. Singer, Fast Parallel Algorithms for Sparse Multivariate Polynomial Interpolation over Finite Fields, *SIAM J. Comput.,* **19**, **6**, 1059–1063, 1990.

[GL] G.H. Golub, C.F. Van Loan, *Matrix Computations,* Johns Hopkins Univ. Press, Baltimore, Maryland, 1989.

[GLi] J.A. George, J.W. Liu, *Computer Solution of Large Sparse Positive Definite Systems,* Prentice-Hall, New Jersey, 1981.

[GLRSW] P. Gemmel, R. Lipton, R. Rubinfeld, M. Sudan, A. Wigderson, Self-Testing/ Correcting for Polynomials and for Approximate Functions, *Proc. 23rd Ann. ACM Symp. on Theory of Computing,* 32–42, ACM Press, New York, 1991.

[GM] Z. Galil, O. Margalit, All Pair Shorted Distances for Graphs with Small Integer Length Edges, submitted to *Information and Computation.*

[GO] I. Gohhberg, V. Olshevsky, Circulants, Displacements and Decompositions of Matrices, *Integral Equations and Operator Theory,* **15**, **5**, 730–743, 1992.

[GO1] I. Gohberg, V. Olshevsky, Complexity of Multiplication with Vectors for Structured Matrices, *Linear Algebra Appl,* **202**, 163–192, 1994.

[GO2] I. Gohberg, V. Olshevsky, Fast Algorithms with Preprocessing for Multiplication of Transposed Vandermonde Matrix and Cauchy Matrix by Vectors, manuscript, *School of Math. Sciences, Tel-Aviv University,* Israel, 1992.

[GP85;88] Z. Galil, V. Y. Pan, Improved Processor Bounds for Combinatorial Problems in RNC, *Combinatorica,* **8**, **2**, 189–200, 1988, and *Proc. 26th Ann. IEEE Symp. on Foundation of Computer Science,* 490–495, IEEE Computer Society Press, 1985.

[GP89] Z. Galil, V. Y. Pan, Parallel Evaluation of the Determinant and of the Inverse of a Matrix, *Inf. Proc. Lett.,* **30**, 41–45, 1989.

[GPS] K.A. Gallivan, R.J. Plemmons, A.H. Sameh, Parallel Algorithms for Dense Linear Algebra Computations, *SIAM Rev.,* **32**, **1**, 54–135, 1990.

[Gr] W.B. Gragg, The Padé Table and Its Relation to Certain Algorithms of Numerical Analysis, *SIAM Rev.,* **14**, **1**, 1–62, 1972.

[Gre] R.T. Gregory, *Error-Free Computations: Why It Is Needed And Methods For Doing It,* Krieger, New York, 1980.

[Grev] T.N.E. Greville, Some Applications of the Pseudoinverse of a Matrix, *SIAM Rev.,* **2**, 15–22, 1960.

[GrLi] W. B. Gragg, A. Lindquist, On the Partial Realization Problem, *Linear Algebra Appl.,* **50**, 277–319, 1983.

[GrS] J. Grcar, A.H. Sameh, On Certain Parallel Toeplitz Linear System Solvers, *SIAM J. Sci. Stat. Comput.,* **2**, 238–256, 1982.

[GRy] A. Gibbons, W. Rytter, *Efficient Parallel Algorithms,* Cambridge University Press, 1988.

[GS] W. Gentleman, G. Sande, Fast Fourier Transform for Fun and Profit, *Proc. Fall Joint Comput. Conf.,* **29**, 563–578, 1966.

[GSe] I.C. Gohberg, A.A. Semencul, On the Inversion of Finite Toeplitz Matrices and Their Continuous Analogs, *Mat. Issled.,* **2**, 201–233 (in Russian), 1972.

[GSh92] J. von zur Gathen, V. Shoup, Computing Frobenius Maps and Factoring Polynomials, *24 Ann. ACM Symp. on Theory of Computing,* 97–105, ACM Press, New York, 1992, and *J. of Computational Complexity,* **2**, 187–224, 1992.

[Gut] M. H. Gutknecht, A. Completed Theory of the Unsymmetric Lanczos Process and Related Algorithms, Part I, *SIAM J. Matrix Anal. Appl.,* **13**, **2**, 594–639, 1992.

[GY] F.G. Gustavson, D.Y.Y. Yun, Fast Algorithms for Rational Hermite Approximation and Solution of Toeplitz Systems, *IEEE Trans. Circuits and Systems,* **26**, **9**, 750–755, 1979.

[H70] A.S. Householder, *The Numerical Treatment of a Single Nonlinear Equation,* McGraw-Hill, New York, 1970.

[H70a] A. S. Householder, Bezoutians, Elimination and Localization, *SIAM Rev.,* **12**, 73–78, 1970.

[Ha] O.H. Hald, Inverse Eigenvalue Problems for Jacobi Matrices, *Linear Algebra Appl.,* **14**, 63–85, 1976.

[Ha77] W. Hackbusch, On the Convergence of Multi-grid Iteration Applied to Finite Element Equations, Report 77-8, *Universität zu Köln,* 1977.

[Ha80] W. Hackbusch, Convergence of Multi-grid Iterations Applied to Difference Equations, *Math. Comp.,* **34**, 425–440, 1980.

[Ha85] W. Hackbusch, *Multi-Grid Methods and Applications,* Springer, Berlin, 1985.

[Hal] S.B. Halley, Solution of Band Matrix Equations by Projection-Recurrence, *Linear Algebra Appl.*, **32**, 33–48, 1980.

[Hei] G. Heinig, Beitrage zur spektraltheorie von Operatorbuschen und zur algebraischen Theorei von Toeplitzmatrizen, Diss. B, *TH Karl-Marx-Stadt,* 1979.

[Hen74] P. Henrici, *Applied and Computational Complex Analysis,* Wiley, New York, 1974.

[Hen82] P. Henrici, *Essentials of Numerical Analysis with Pocket Calculator Demonstrations,* Wiley, New York, 1982.

[HF] U. Helmke, P.A. Fuhrmann, Bezoutians, *Linear Algebra Appl.,* **124**, 1039–1097, 1989.

[Hi86] N.J. Higham, Efficient Algorithms for Computing the Condition Number of a Tridiagonal Matrix, *SIAM J. Sci. Stat. Comput.,* **7**, **1**, 150–165, 1986.

[Hi87] N.J. Higham, A Survey of Condition Number Estimation for Triangular Matrices, *SIAM Rev.,* **29**, **4**, 575–595, 1987.

[HJu] G. Heinig, V. Jungnickel, Hankel Matrices Generated by the Markov Parameters, *Linear Algebra Appl.,* **76**, 121–135, 1986.

[HLPW] M.T. Heath, A.J. Laub, C.C. Paige, R.C. Ward, Computing the SVD of a Product of Two Matrices, *SIAM J. Sci. Stat. Comput.*, **7**, 1147–1159, 1986.

[Ho] J. Hong, Proving by Example and Gap Theorems, *Proc. 27th Ann. IEEE Symp. on Foundation of Computer Science,* 107–116, IEEE Computer Society Press, 1986.

[Hoc] H. Hochstadt, On the Construction of a Jacobi Matrix from Spectral Data, *Linear Algebra Appl.,* **28**, 113–115, 1979.

[Hou] H. Hou, The Fast Hartley Transform, *IEEE Trans. on Computers*, **C-36**, 147–156, 1987.

[HPR] Y. Han, V. Y. Pan, J. Reif, Efficient Parallel Algorithms for Computing All Pair Shortest Paths in Directed Graphs, *Proc. 4th Ann. ACM Symp. on Parallel Algorithms and Architectures*, 353–362, ACM Press, New York, 1992.

[HR] G. Heinig, K. Rost, *Algebraic Methods for Toeplitz-like Matrices and Operators, Operator Theory,* **13**, Birkhauser, 1984.

[HT] W. Hackbusch, U. Trottenberg (eds.), *Multigrid Methods*, *Lecture Notes in Mathematics,* **960**, Springer, Berlin, 1982.

[HW79] G.H. Hardy, E.M. Wright, *An Introduction to the Theory of Numbers,* Oxford Univ. Press, Oxford, 1979.

[Ike] Y. Ikebe, On Inverses of Hessenberg Matrices, *Linear Algebra Appl.,* **24**, 93–97, 1979.

[IKo] D. Ierardi, D. Kozen, Parallel Resultant Computation, in *Synthesis of Parallel Algorithms,* (J. H. Reif Editor), 679–720, Morgan Kaufman Publishers, San Mateo, California, 1993.

[IMH] O. H. Ibarra, S. Moran, R. Hui, A Generalization of the Fast LUP Matrix Decomposition Algorithm and Applications, *J. of Algorithms,* **3**, 45–56, 1982.

[Ioh] I. S. Iohvidov, *Hankel and Toeplitz Matrices and Forms, Algebraic Theory,* Birkhäuser, Boston, 1982.

[IR] K. Ireland, M. Rosen, *A Classical Introduction to Modern Number Theory,* Springer, Berlin, 1982.

[J] J. Ja Ja, *An Introduction to Parallel Algorithms,* Addison-Wesley, Massachusetts, 1992.

[J83] J. Ja Ja, Time-Space Tradeoffs for Some Algebraic Problems, *J. ACM,* **30**, 657–667, 1983.

[Ja] N. Jacobson, *Basic Algebra,* W. H. Freeman Publ., San Francisco, California, 1974.

[K-G] W. Keller-Gehrig, Fast Algorithms for Characteristic Polynomial, *Theoretical Computer Science,* **36**, 309–317, 1985.

[K87] T. Kailath, Signal Processing Applications of Some Moment Problems, *Proc. AMS Symp. in Applied Math.,* **37**, 71–100, 1987.

[Ka83] E. Kaltofen, Factorization of Polynomials, in *Computer Algebra. Symbolic and Algebraic Computation* (B. Buchberger, G.E. Collins and R. Loos Editors), 95–113, Springer, Berlin, 1983.

[Ka87] E. Kaltofen, Single-Factor Hensel Lifting and Its Application to the Straight-Line Complexity of Certain Polynomials, *Proc. 19th Ann. ACM Symp. on Theory of Comp.,* 443–452, ACM Press, New York, 1987.

[Ka87a] E. Kaltofen, Computer Algebra Algorithms, *Ann. Review in Comp. Sci.,* **2**, 91–118, 1987.

[Ka88] E. Kaltofen, Greatest Common Divisor of Polynomials Given by Straight-Line Program, *J. of ACM* **35**, **1**, 231–264, 1988.

[Ka89] E. Kaltofen, Processor-Efficient Parallel Computation of Polynomial Greatest Common Division, Preliminary Report, *Dept. of Computer Sci., RPI,* Troy, New York, 1989.

[Ka90] E. Kaltofen, Polynomial Factorization in 1982–1986, in *Computers in Math.* (D.V. Chudnovsky, R.D. Jenks Editors), *Lecture Notes in Pure and Applied Math.,* **125**, 285–309, Marcel Dekker, New York, 1990.

[Ka91] E. Kaltofen, Effective Noether Irreducibility Forms and Applications, *Proc. 23rd Annual ACM Symp. on Theory of Computations,* 54–63, ACM Press, New York, 1991.

[Ka92] E. Kaltofen, Polynomial Factorization in 1987–91, in *Proc. of International Symposium LATIN'92 (Latin American Theoretical Informatics),* (I. Simon Editor), *Lecture Notes in Computer Science,* **583**, 294–313, Springer, Berlin, 1992.

[Ka93] E. Kaltofen, Analysis of Coppersmith's Block Wiedemann Algorithm for the Parallel Solution of Sparse Linear Systems, *Proc. 10th Intern. Symp. Applied Algebra, Algebraic Algorithms and Error-Correcting Codes,* (G. Cohen, T. Mora, O. Moreno Editors), *Lecture Notes in Computer Science,* **673**, 195–212, Springer, Berlin, 1993.

[Kak] S. Kakeya, On Fundamental Systems of Symmetric Functions, *Japan J. of Math.,* **2**, 69–80, 1925.

[Kak,a] S. Kakeya, On Fundamental Systems of Symmetric Functions, *Japan J. of Math.,* **4**, 77–85, 1927.

[Kam84] M. Kaminski, An Algorithm for Polynomial Multiplication Which Does Not Depend on the Ring Constants, *J. Algorithms,* **9**, 137–147, 1989. (Proceedings version: Multiplication of Polynomials over the Ring of Integers, *Proc. 25th Ann. IEEE Symp. on Foundation of Computer Science,* 251–254, IEEE Computer Society Press, 1984.).

[Kam87] M. Kaminski, Linear Time Algorithm for Residue Computation and a Fast Algorithm for Division with a Sparse Divisor, *J. ACM,* **34**, **4**, 968–984, 1987.

[Kam89] M. Kaminski, A Note on Probabilistically Verifying Integer and Polynomial Products, *J. ACM,* **36**, **1**, 142–149, 1989.

[KaSi] E. Kaltofen, M. Singer, Size Efficient Parallel Algebraic Circuits for Partial Derivatives, *Tech. Report 90–32, Computer Science Department, RPI,* Troy,

New York, 1990.

[KaT] A E. Kaltofen, B.M. Trager, Computing with Polynomial Given by Black Boxes for Their Evaluations: Greatest Common Divisors, Factorization, Separation of Numerators and Denominators, *J. Symbolic Computation,* **9**, **3**, 301–320, 1990.

[KC] T. Kailath, J. Chun, Generalized Gohberg-Semencul Formulas for Matrix Inversion, *Operator Theory: Advances and Applications,* **40**, 231–246, 1989.

[KKM] T. Kailath, S.-Y. Kung, M. Morf, Displacement Ranks of Matrices and Linear Equations, *J. Math. Anal. Appl.,* **68**, **2**, 395–407, 1979.

[KKS] E. Kaltofen, M.S. Krishnamoorthy, B.D. Saunders, Fast Parallel Computation of Hermite and Smith Forms of Polynomials Matrices, *SIAM J. Alg. Disc. Meth.*, **8**, **4**, 683–690, 1987.

[KKS90] E. Kaltofen, M.S. Krishnamoorthy, B.D. Saunders, Parallel Algorithms for Matrix Normal Forms, *Linear Algebra Appl.*, **136**, 189–208, 1990.

[Kl] J. Klose, Binary Segmentation for Multivariate Polynomials, submitted to *J. of Complexity.*

[KLW] E. Kaltofen, Y.N. Lakshman, J.M. Wiley, Modular Rational Sparse Multivariate Polynomial Interpolation, *Proc. ACM-SIGSAM Int. Symp. on Symb. and Alg. Comp. (ISSAC '90),* 135–139, ACM Press, New York, 1990.

[Kn] D.E. Knuth, *The Art of Computer Programming: Seminumerical Algorithms,* 2, Addison-Wesley, Reading, Massachusetts, 1981.

[KO] A. Karatsuba, Yu. Ofman, Multiplication of Multidigit Numbers on Automata, *Soviet Physics Dokl.*, **7**, 595–596, 1963.

[KP] E. Kaltofen, V. Y. Pan, Processor Efficient Parallel Solution of Linear Systems over an Abstract Field, *Proc. 3rd Ann. ACM Symp. on Parallel Algorithms and Architectures,* 180–191, ACM Press, New York, 1991.

[KP,a] E. Kaltofen, V. Y. Pan, Processor Efficient Parallel Solution of Linear Systems II. The Positive Characteristic and Singular Cases, *Proc. 33rd Ann. IEEE Symp. on Foundation of Computer Science,* 714–723, IEEE Computer Society Press, 1992.

[KR] R. Karp, V. Ramachandran, A Survey of Parallel Algorithms for Shared Memory Machines, *Handbook for Theoretical Computer Science* (J. van Leeuwen Editor), 869–941, North Holland, Amsterdam, 1990.

[Kr] V.M. Krapchenko, Asymptotic Estimation of Addition Time of a Parallel Adder, *Problemy Kobernetiki,* **19**, 107–122, 1967 (in Russian). (English translation in *Syst. Theory Res.*, **19**, 105–122, 1970.)

[KRo] E. Kaltofen, H. Rolletscheck, Computing Greatest Common Divisors and Factorizations in Quadratic Number Fields, *Math. Comp.,* **53**, 188, 697–720, 1989.

[KRu] H. J. Karloff, W. L. Ruzzo, The Complexity of Iterated Mod Function, *Information and Computation*, **80**, 193–204, 1989.

[KS91] E. Kaltofen, B. D. Saunders, On Wiedmann's Method for Solving Sparse Linear Systems, *Proc. AAECC-5, Lecture Note in Computer Science*, **536**, 29–38, Springer, Berlin, 1991.

[KT] H.T. Kung, J.F. Traub, All Algebraic Functions Can be Computed Fast, *J. of ACM,* **25**, **2**, 245–260, 1978.

[KUW] R.M. Karp, E. Upfal, A. Wigderson, Constructing a Perfect Matching Is in Random NC, *Proc. 17-th Ann. ACM Symp. on Theory of Computing,* 22–32, ACM Press, New York, 1985.

[KVM] T. Kailath, A. Viera, M. Morf, Inverses of Toeplitz Operators, Innovations, and Orthogonal Polynomials, *SIAM Rev.,* **20**, **1**, 106–119, 1978.

[L92] E. Linzer, Can Symmetric Toeplitz Solver be Strongly Stable?, *Proc. ISTCS '92*, (D. Dolev, Z. Galil, M.Rodeh Editors), *Lecture Notes in Computer Science*, **601**, 137–146, Springer, Berlin, 1992.

[L-AK] H. Lev-Ari, T. Kailath, Triangular Factorization of Structured Hermitian Matrices, *Operator Theory: Advances and Applications,* **18**, 301–323, 1986.

[L-K,P] Y. Lin-Kriz, V. Y. Pan, On Parallel Complexity of Integer Linear Programming, GCD and Iterated Mod Function, *Proc. 3rd Ann. ACM-SIAM Symp. on Discrete Algorithms,* 124–137, ACM Press, New York, and SIAM Publications, Philadelphia, Pennsylvania, 1992.

[LDC] G. Labahn, Dong Koo Choi, S. Cabay, The Inverses of Block Hankel and Block Toeplitz Matrices, *SIAM J. Comput.* **19,** 98–123, 1990.

[Le92] H.T. Leighton, *Introduction to Parallel Algorithms and Architectures: Arrays, Trees & Hypercubes,* Morgan Kaufmann Publishers, San Mateo, California, 1992.

[Leo] A. Leonard, Vortex Methods for Flow Simulation, *J. Comput. Phys.,* **37**, 289–335, 1980.

[LF] E. Linzer, E. Feig, Implementation of Efficient FFT Algorithms for Fused Multiply-Add Architecture, *IEEE Trans. on Acoust., Speech, Signal Processing.*, **41**, **1**, 93–107, 1993.

[LF,a] E. Linzer, E. Feig, Modified FFTs for Fused Multiply-Add Architecture, *Math. Comp.*, **60**, **201**, 347–361, 1993.

[Li76] S. Linnainmaa, Taylor Expansion of the Accumulated Rounding Errors, *BIT*, **16**, 146–160, 1976.

[Lin] E. Linzer, On the Stable Inversion of Banded Toeplitz Systems, *Linear Algebra Appl.*, **170**, 1–32, 1992.

[LLL] A.K. Lenstra, H.W. Lenstra, L. Lovász, Factoring Polynomials with Rational Coefficients, *Math. Ann.,* **261**, 515–534, 1982.

[LLMPW] T. Leighton, C.E. Leiserson, B. Maggs, S. Plotkin, J. Wein, Theory of Parallel and VLSI Computation, *Lab. for Computer Science, MIT,* Lecture Notes, 1987–90.

[LN] R. Lidl, H. Niederreiter, *Finite Fields, in Encyclopedia of Math. and Its Applications,* **20**, Addison-Wesley, Reading, Massachusetts, 1983.

[LN86] R. Lidl, H. Niederreiter, *Introduction to Finite Fields and Their Applications,* Cambridge University Press, 1986.

[LO] D.W. Lozier, F.W.J. Olver, Numerical Evaluation of Special Functions, in *Math. of Computation, 1943–1993: A Half-Century of Computational Mathematics* (Walter Gautschi Editor), *Proc. of Symposia in Applied Mathematics,* AMS, Providence, Rhode Island, to appear.

[Lo] L. Lovász, Connectivity Algorithms Using Rubber-bands, *Proc. Sixth Conf. on Foundations of Software Technology and Theoretical Computer Science (New Delhi, India), Lecture Notes in Computer Science,* **241**, 394–412, Springer, Berlin, 1986.

[Loos] R. Loos, Computing Rational Zeros of Integral Polynomials by p-adic Expansion, *SIAM J. Comput.,* **12**, 286–293, 1983.

[LP] L. Lapidus, G.F. Pinder, *Numerical Solution of Partial Differential Equations in Science and Engineering,* Wiley, New York, 1982.

[LPSh] J. Laderman, V. Y. Pan, X.-H. Sha, On Practical Acceleration of Matrix Multiplication, *Linear Algebra Appl.*, **162–164**, 557–558, 1992.

[LRT] R.J. Lipton, D. Rose, R.E. Tarjan, Generalized Nested Dissection, *SIAM J. Numer. Anal.,* **16, 2**, 346–358, 1979.

[LTi] P. Lancaster, M. Tismenetsky, *The Theory of Matrices,* Academic Press, 1985.

[LV] E. Linzer, M. Vetterli, Iterative Toeplitz Solvers with Local Quadratic Convergence, *Computing,* **49**, 339–347, 1993.

[M93] B. Mishra, *Algorithmic Algebra,* Springer, New York, 1993.

[Mar] O. Margalit, Algorithms for the All Pairs Shortest Paths Problems, Ph.D. Thesis, *Computer Science Dept., Tel-Aviv University,* Israel, 1993.

[MC] R.T. Moenck, J.H. Carter, Approximate Algorithms to Derive Exact Solutions to Systems of Linear Equations, *Proc. EUROSAM, Lecture Notes in Computer Science,* **72**, 63–73, Springer, Berlin, 1979.

[McC86] S. McCormick (Editor), Proceeding of the 2nd Copper Mountain Multigrid Conference, *Appl. Math. Comp.,* **19** (special issue), 1–372, 1986.

[McC87] S. McCormick (Editor), *Multigrid Methods,* SIAM, Philadelphia, Pennsylvania, 1987.

[McCT] S. McCormick, U. Trottenberg (Editors), Multigrid Methods, *Appl. Math. Comp.,* **13**, 213–474, 1983.

[McES] R.J. McEliece, J.B. Shearer, A Property of Euclid's Algorithm and an Application to Padé Approximation, *SIAM J. Appl. Math.,* **34, 4**, 611–615, 1978.

[Mi] M. Mignotte, Some Inequalities About Univariate Polynomials, *Proc. 1981 ACM Symp. on Symbolic and Algebraic Computation,* 195–199, ACM Press, New York, 1981.

[Mil] G. L. Miller, Riemann's hypothesis and Tests for Primality, *Journal of Computer and System Sciences,* **13, 3**, 300–317, 1976.

[Mil92] V.S.Miller, Factoring Polynomials via Relations Finding, *Proc.ISTCS'92, Lecture Notes in Computer Science,* **601**, 115–121, Springer, Berlin, 1992.

[MM] M. Marcus, H. Minc, *A Survey of Matrix Theory and Matrix Inequalities,* Allyn and Bacon, Boston, Massachusetts, 1964.

[Mo] R. Moenck, Fast Computation of GCDs, *Proc. 5th ACM Ann. Symp. on Theory of Computing,* 142–171, ACM Press, New York, 1973.

[Morf] M. Morf, Doubling Algorithms for Toeplitz and Related Equations, *Proc. IEEE Internat. Conf. on ASSP,* 954–959, IEEE Computer Society Press, 1980.

[MRK] G.L. Miller, V. Ramachandran, E. Kaltofen, Efficient Parallel Evaluation of Straight-line Code and Arithmetic Circuits, *SIAM J. Comput.,* **17, 4**, 687–695, 1988.

[MST88] Y. Mansour, B. Scheiber, P. Tiwari, Lower Bounds for Integer Greatest Common Divisor Computations, *Proc. 29th IEEE. Symp. on Foundations of Computer Science,* 54–63, IEEE Computer Society Press, 1988.

[MST89] Y. Mansour, B. Scheiber, P. Tiwari, Lower Bounds for Computations with Floor Operation, *Proc. 16th ICALP* (G. Ausiello, M. Dezani-Ciancaglini, S. Ronchi Della Rocca Editors), *Lectures Notes in Computer Science,* **372**, 559–573, Springer, Berlin, 1989.

[Mul] K. Mulmuley, Fast Parallel Algorithm to Compute the Rank of a Matrix, *Combinatorica,* **7, 1**, 101–104, 1987.

[Mul94] K. Mulmuley, Lower Bounds for Parallel Linear Programming and Other Problems, *Proc. 26-th Ann. ACM Symposium on Theory of Computing,* ACM Press, New York, 1994.

[Mus] B.R. Musicus, Levinson and Fast Choleski Algorithms for Toeplitz and Almost Toeplitz Matrices, Internal Report, *Lab. of Electronics, M.I.T.,* Cambridge, Massachusetts, 1981.

[MVV] K. Mulmuley, U. Vazirani, V. Vazirani, Matching Is As Easy As Matrix Inversion, *Combinatorica,* **7, 1**, 105–114, 1987.

[Ne] A.C.R. Newbery, Error Analysis for Polynomial Evaluation, *Math. Comp.,* **28, 127**, 789–793, 1974.

[NSV] H. Narayanan, H.Saran, V.V. Vazirani, Randomized Parallel Algorithms for Matroid Union, Arboresences, and Edge-Disjoint Spanning Trees, *Proc. 3rd Ann. ACM-SIAM Symp. on Discrete Algorithms,* 357–366, ACM Press, New York, SIAM Publications, Philadelphia, Pennsylvania, 1992, and *SIAM J. Comput.,* **23**, **2**, 1994.

[Nus] H. J. Nussbaumer, Fast Polynomial Transform Algorithms for Digital Convolutions, *IEEE Trans. Acoustic, Speech, Signal Proc.,* **ASSP-28**, 205–215, 1980.

[Nus81] H. J. Nussbaumer, *Fast Fourier Transform and Convolution Algorithms,* Springer, Berlin, 1981.

[NWo] Q. L. Nguyen, D. H. Wood, Displacement of Matrix Products, to appear.

[ODR89] S.T. O'Donnell, V. Rokhlin, A Fast Algorithm for Numerical Evaluation of Conformal Mappings, *SIAM J. Sci. Stat. Comput.*, **10**, **3**, 475–487, 1989.

[Of62] Yu. Ofman, On the Algorithmic Complexity of Discrete Functions, *Dokl. Acad. Nauk SSSR,* **145,** **1**, 48–51, 1962. (English translation in *Soviet Physics-Dokl.*, **7**, **7**, 589–591, 1963.)

[Of65] Y.P. Ofman, Universal Automaton (in Russian), *Moscow Math Soc. (Trydy),* **14**, 186–199, 1965.

[Or] J.M. Ortega, *Introduction to Parallel and Vector Solution of Linear Systems,* Plenum Press, New York, 1988.

[OS] A.M. Odlyzko, A. Schönhage, Fast Algorithms for Multiple Evaluations of the Riemann Zeta Function, *Trans. Am. Math. Soc.,* **309**, **2**, 797–809, 1988.

[OV] J.M. Ortega, R.G. Voight, Solution of Partial Differential Equations on Vector and Parallel Computers, *SIAM Rev.,* **27,** **2**, 149–240, 1985.

[P,a] V. Y. Pan, Improved Parallel Solution of Triangular Linear Systems, *Computers & Mathematics (with Applications)*, to appear.

[P,b] V. Y. Pan, Algebraic Improvement of Numerical Algorithms: Interpolation and Economization of Taylor's Series, to appear in *Mathematical and Computer Modelling.*

[P66] V. Y. Pan, On Methods of Computing the Values of Polynomials, *Uspekhi Matematicheskikh Nauk,* **21,** **1(127)**, 103–134, 1966. [Transl. *Russian Mathematical Surveys,* **21,** **1** **(127)**, 105–137, 1966.]

[P78] V. Y. Pan, Computational Complexity of Computing Polynomials over the Fields of Real and Complex Numbers, *Proceedings of the 10th Annual ACM Symposium on Theory of Computing,* 162–172, ACM Press, New York, 1978.

[P80] V. Y. Pan, The Bit-Operation Complexity of the Convolution of Vectors and of the DFT, *Tech. Rep. 80-6, Computer Science Dept., SUNYA,* Albany, New York, 1980 (abstract in *Bulletin of the EATCS,* **14**, p. 95, 1981).

[P84] V. Y. Pan, How Can We Speed Up Matrix Multiplication?, *SIAM Rev.,* **26,** **3**, 393–415, 1984.

[P84a] V. Y. Pan, Trilinear Aggregating and the Recent Progress in the Asymptotic Acceleration of Matrix Operations, *Theoretical Computer Science,* **33**, 117–138, 1984.

[P84b] V. Y. Pan, *How to Multiply Matrices Faster,* Lecture Notes in Computer Science, **179**, Springer, Berlin, 1984.

[P85] V. Y. Pan, Fast and Efficient Parallel Algorithms for the Exact Inversion of Integer Matrices, *Proc. 5-th Ann. Conference on Foundation of Software Technology and Theoretical Computer Science, Lectures Notes in Computer Science,* **206**, 504–521, Springer, Berlin, 1985.

[P87] V. Y. Pan, Complexity of Parallel Matrix Computations, *Theoretical Computer Science,* **54**, 65–85, 1987.

[P87a] V. Y. Pan, Linear Systems of Algebraic Equations, *Encyclopedia of Physical Sciences and Technology (Marvin Yelles Editor),* **7**, 304–329, 1987 (first edition), and **8**, 779–804, 1992 (second edition), Academic Press, San Diego, California.

[P87b] V. Y. Pan, Linear Systems of Equations, in *Encyclopedia of Physical Science and Technology,* (Marvin Yelles Editor), **7**, 304–329, 1987 (first edition) and **8**, 779–804, 1992 (second edition).

[P88] V. Y. Pan, New Methods for Computations with Toeplitz-like Matrices, *Technical Report 88-28, Computer Science Dept., SUNY Albany,* Albany, New York, 1988.

[P89] V. Y. Pan, Fast and Efficient Parallel Evaluation of the Zeros of a Polynomial Having Only Real Zeros, *Computers & Mathematics (with Applications),* **17,** **11**, 1475–1480, 1989.

[P89a] V. Y. Pan, On Computations with Dense Structured Matrices, *Proc. ACM-SIGSAM Intern. Symp. on Symbolic and Alg. Comp.*, 34–42, ACM Press, New

York, 1989, and *Math. Comp.*, **55**, **191**, 179–190, 1990.

[P89b] V. Y. Pan, Parallel Inversion of Toeplitz and Block Toeplitz Matrices, *Operator Theory: Advances and Applications,* **40**, 359–389, Birkhauser, Basel, Switzerland, 1989.

[P89c] V. Y. Pan, Fast Evaluation and Interpolation at the Chebyshev Set of Points, **2**, **3**, 255–258, *Applied Math. Letters*, 1989.

[P89d] V. Y. Pan, Simple Multivariate Polynomial Multiplication, Tech. Report 89-19, *Computer Science Dept., SUNYA,* Albany, New York, 1989, and *J. Symb. Comp.*, to appear.

[P90] V. Y. Pan, Parallel Least-Squares Solution of General and Toeplitz-like Linear Systems, *Proc. 2nd Ann. ACM Symposium on Parallel Algorithms and Architectures,* 244–253, ACM Press, New York, 1990.

[P91] V. Y. Pan, Decreasing the Precision of Matrix Computations, Tech. Report, TR91-11, *Comp. Sci. Dept., SUNYA,* Albany, New York, 1991.

[P92] V. Y. Pan, On Binary Segmentation for Matrix and Vector Operations, *Computers & Mathematics (with Applications)*, **25**, 3, 69–71, 1992.

[P92a] V. Y. Pan, Complexity of Computations with Matrices and Polynomials, *SIAM Rev.,* **34**, **9**, 225–262, 1992.

[P92b] V. Y. Pan, Parametrization of Newton's Iteration for Computations with Structured Matrices and Applications, *Computers & Mathematics (with Applications),* **24**, **3**, 61–75, 1992.

[P92c] V. Y. Pan, Parallel Solution of Toeplitz-like Linear Systems, *J. of Complexity,* **8**, 1–21, 1992.

[P92d] V. Y. Pan, Can We Utilize the Cancellation of the Most Significant Digits?, *Tech. Report 92-061, International Computer Science Institute,* Berkeley, California, 1992.

[P92e] V. Y. Pan, Approximate Evaluation of a Polynomial on a Set of Real Points, Tech. Report TR 92-076, *International Computer Science Institute*, Berkeley, California, 1992.

[P93] V. Y. Pan, Decreasing the Displacement Rank of a Matrix, *SIAM J. Matrix Anal. Appl.,* **14**, **1**, 118–121, 1993.

[P93a] V. Y. Pan, On Randomized Parallel Computations with General and Toeplitz-like Matrices, manuscript, 1993.

[P93b] V. Y. Pan, Parallel Solution of Sparse Linear and Path Systems, in *Synthesis of Parallel Algorithms* (J.H. Reif Editor), 621–678, Morgan Kaufmann publisher, San Mateo, California, 1993.

[P93c] V. Y. Pan, New Progress in Parallel Computations with Matrices and Polynomials, manuscript, 1993.

[P93d] V. Y. Pan, Modular (Residue) Arithmetic for Linear Algebra Computations in the Real Fields, manuscript, 1993.

[P93e] V. Y. Pan, Concurrent Iterative Algorithm for Toeplitz-like Linear Systems, *IEEE Trans. on Parallel and Distributive Systems,* **4**, **5**, 592–600, 1993.

[P94] V. Y. Pan, New Techniques for Approximating Complex Polynomial Zeros, *Proc. 5th Ann. ACM-SIAM Symposium on Discrete Algorithms,* 260–270, ACM Press, New York, and SIAM Publications, Philadelphia, Pennsylvania, 1994.

[P94a] V. Y. Pan, Binary Segmentation for the Acceleration of Parallel Computations over the Rationals, manuscript, 1994.

[Par] B.N. Parlett, *The Symmetric Eigenvalue Problem,* Prentice-Hall, 1980.

[Par92] B. N. Parlett, Reduction to Tridiagonal Form and Minimal Realizations, *SIAM J. Matrix Anal. Appl.*, **13**, **2**, 567–593, 1992.

[Parb] I. Parbery, Parallel Complexity Theory, *Research Notes in Theoretical Computer Science* (R. V. Book Editor), Pitman, London (Wiley, New York), 1987.

[Pease] M. Pease, An Adaptation of the Fast Fourier Transform for Parallel Processing, *J. of ACM,* **15**, 252–264, 1968.

[PFTV] W.H. Press, B.P. Flannery, S.A. Teukolsky, W.T. Vetterling, *Numerical Recipes, The Art of Scientific Computing*, Cambridge University Press, 1986.

[PK] D. Pal, T. Kailath, Fast Triangular Factorization of Hankel and Related Matrices with Arbitrary Rank Profile, Tech. Report, *Information Systems Laboratory,* Stanford University, Stanford, California, 1990.

[PLS] V. Y. Pan, E. Landowne, A. Sadikou, Univariate Polynomial Division with a Remainder by Means of Evaluation and Interpolation, *Information Process. Letters,* **44**, 149–153, 1992, and *Proc. 3rd IEEE Conf. on Parallel and Distributed Proc.*, 212–218, IEEE Computer Society Press, 1991.

[PP] V. Y. Pan, F. P. Preparata, Supereffective Slow-down of Matrix Computations, *Proc. 4th Ann. ACM Symp. on Parallel Algorithms and Architectures*, 353–362, ACM Press, New York, 1992.

[PP,a] V. Y. Pan, F.P. Preparata, Work-Preserving Speed-up of Parallel Matrix Computations, *SIAM J. Comput.* (to appear).

[PR] V. Y. Pan, J. Reif, Fast and Efficient Parallel Solution of Sparse Linear Systems, *SIAM J. Comput.*, **22**, **6**, 1227–1250, 1993.

[Pr] F.P. Preparata, Inverting a Vandermonde Matrix Over a Finite Field in Minimum Parallel Time, *Information Process. Letters,* **38**, **6**, 291–294, 1991.

[PR89] V. Y. Pan, J. Reif, Fast and Efficient Parallel Solution of Dense Linear Systems, *Computers & Mathematics (with Applications),* **17**, **11**, 1481–1491, 1989. (Preliminary version, *Proc. 17th Ann. ACM Symp. on Theory of Computing,* 143–152, ACM Press, New York, 1985.)

[PR90] V. Y. Pan, J. Reif, On the Bit-Complexity of Discrete Solution of PDEs: Compact Multigrid, *Proc. 17th ICALP, Lecture Notes in Computer Science,* **443**, 612–625, Springer, Berlin, 1990, and *Computers & Mathematics (with Applications),* **20**, **2**, 9–16, 1990.

[PR91] V. Y. Pan, J. Reif, The Parallel Computation of the Minimum Cost Paths in Graphs by Stream Contraction, *Information Processing Letters,* **40**, 79–83, 1991.

[PR92] V. Y. Pan, J. Reif, Compact Multigrid, *SIAM J. Sci. Stat. Comput.,* **13**, **1**, 119–127, 1992.

[PR93] V. Y. Pan, J. Reif, Generalized Compact Multigrid, *Computers & Mathematics (with Applications),* **25**,**9**, 3–5, 1993.

[PS] V. Y. Pan, R. Schreiber, An Improved Newton Iteration for the Generalized Inverse of a Matrix, with Applications, *SIAM J. Sci. Stat. Comput.*, **12**, **5**, 1109–1131, 1991.

[PSA] V. Y. Pan, I. Sobze, A. Atinkpahoun, Optimum Parallel Computations with Band Matrices, *Proceedings 5-th Ann. ACM-SIAM Symp. on Discrete Algorithms,* 649–658, ACM Press, New York, and SIAM Publications, Philadelphia, Pennsylvania, 1994.

[PSa] F.P. Preparata, D.V. Sarwate, An Improved Parallel Processor Bound in Fast Matrix Inversion, *Inform. Proc. Lett.,* **7**, **3**, 148–149, 1978.

[PSA,a] V. Y. Pan, I. Sobze, A. Atinkpahoun, Improved Band Matrix Algorithms, Tech. Report 93–060, *Intern. Computer Science Institute,* Berkeley, 1993.

[PSh] F.P. Preparata, M.I. Shamos, Computational Geometry: An Introduction, *Texts and Monographs in Computer Science,* Springer, New York, 1985.

[PSz] G. Polia, G, Szegö, *Problems and Theorems in Analysis,* Springer, Berlin–New York, 1972.

[Pt] V. Ptack, Lyapunov, Bezout and Hankel, *Linear Algebra Appl.,* **58**, 363–390, 1984.

[Q] M.J. Quinn, *Designing Efficient Algorithms for Parallel Computers,* McGraw-Hill, New York, 1987.

[Q94] M.J. Quinn, *Parallel Computing: Theory and Practice,* McGraw-Hill, New York, 1994.

[R88] J. Renegar, A Faster P-Space Algorithm for the Existential Theory of the Reals, *Proc. 29th IEEE Symp. on Foundation of Computer Science,* 291–295, IEEE Computer Society Press, 1988.

[R89] J. Renegar, On the Worst-Case Arithmetic Complexity of Approximating Zeros of Systems of Polynomials, *SIAM J. Comput.,* **18**, 350–370, 1989.

[R92] J. Renegar, On the Computational Complexity and Geometry of the First Order Theory of Reals, *J. Symb. Comp.,* **13, 3**, 255–352, 1992.

[R92a] J. Renegar, On the Computational Complexity of Approximating Solutions for Real Algebraic Formulas, *SIAM J. Comput.,* **21, 6,** 1008–1025, 1992.

[Rab] M. O. Rabin, Probabilistic Algorithm for Testing Primality, *J. Number Theory,* **12**, 128–138, 1980.

[Rab,a] M. Rabin, Probabilistic Algorithms in Finite Fields, *SIAM J. Comput.,* **9, 2,** 273–280, 1980.

[Ran] A.G. Ranade, How to Emulate Shared Memory, *Proc. 28th Ann. IEEE Symp. on Foundation on Computer Science,* 185–194, IEEE Computer Society Press, 1987.

[Raz] A. A. Razborov, Lower Bounds on Monotone Network Complexity of the Logical Permanent, *Math. Zam.*, **37, 6,** 887–900, 1985 (in Russian); English Transl.: *Math. Notes of the Acad. of Sciences of the USSR,* **37**, 485–493, 1985.

[RBFR] P. Rózsa, R. Bevilacqua, P. Favati, F. Romani, On Band Matrices and Their Inverses, *Linear Algebra Appl.*, **150**, 287–296, 1991.

[Rei] R. Reischuck, A Fast Probabilistic Parallel Sorting Algorithm, *Proc. 22nd Ann. IEEE Symp. on Foundation of Computer Science,* 212–219, IEEE Computer Society Press, 1981.

[Reif93] J. H. Reif (Editor), *Synthesis of Parallel Algorithms,* Morgan Kaufmann publishers, San Mateo, California, 1993.

[Reif93a] J. H. Reif, Probabilistic Parallel Prefix Computation, *Computers & Mathematics (with Applications),* **26, 1,** 101–110, 1993.

[Ri] J.R. Rice, *Numerical Methods, Software and Analysis,* McGraw-Hill, New York, 1983.

[Ritt] J.F. Ritt, Prime and Composite Polynomials, *Trans. Amer. Math. Soc.,* **23**, 51–63, 1922.

[Riv] T. J. Rivlin, *The Chebyshev Polynomials*, John Wiley, New York, 1992.

[Riz] S. A. H. Rizvi, Inverses of quasi-tridiagonal matrices, *Linear Algebra Appl.*, **56**, 177–184, 1984.

[RMKW] R.L. Rivest, A.R. Meyer, D.J. Kleitman and K. Winkelman, Coping with Errors in Binary Search Procedures, *J. Comput. Syst. Sci.,* **20**, 396–404, 1980.

[Ro86] F. Romani, On the additive structure of the inverse of band matrices, *Linear Algebra Appl.*, **80**, 131–140, 1986.

[Rok85] F. Rokhlin, Rapid Solution of Integral Equations of Classical Potential Theory, *J. Comput. Physics,* **60**, 187–207, 1985.

[Rok88] V. Rokhlin, A Fast Algorithm for the Discrete Laplace Transformation, *J. of Complexity,* **4**, 12–32, 1988.

[RT] J.H. Reif, S.R. Tate, Optimal Size Division Circuits, *SIAM J. Comput.,* **19, 5**, 912–925, 1990.

[RT94] J.H. Reif, S.R. Tate, Dynamic Algebraic Algorithms, *Proc. 5th Ann. ACM-SIAM Symp. on Discrete Algorithms,* 290–299, ACM Press, New York, and SIAM Publications, Philadalphia, Pennsylvania, 1994.

[RuSu] R. Rubinfeld, M. Sudan, Testing Polynomial Functions Efficiently and over Rational Domains, *Proc. 3rd Ann. ACM-SIAM Symp. on Discrete Algorithms,* 23–32, ACM Press, New York and SIAM Publications, Philadelphia, Pennsylvania, 1992.

[Ruz] W.L. Ruzzo, On Uniform Circuit Complexity, *J. Computer & System Sciences,* **22**, 365–383, 1981.

[S] M. Shub, Some Remarks on Bezout's Theorem and Complexity Theory, in *"From Topology to Computation", Proceedings of the Smalefest,* (M. Hirsch, J. Marsden, M. Shub, Editors), chapter 40, 443–455, 1993.

[Sa] J.E. Savage, *The Complexity of Computing,* Wiley and Sons, New York, 1976.

[SaS] Y. Saad, M. Schultz, GMRES: A Generalized Minimal Residual Algorithm for Solving Nonsymemtric Linear Sytems, *SIAM J. Sci. and Stat. Comput.*, **7**, 856–869, 1986.

[Sc77] A. Schönhage, Schnelle Multiplikation von Polynomen über Korpern der Characteristik 2, *Acta Inf.*, **7**, 395–398, 1977 (in German).

[Sc82] A. Schönhage, The Fundamental Theorem of Algebra in Terms of Computational Complexity, manuscript, *Dept. of Math., University of Tübingen,* Tübingen, Germany, 1982.

[Sc82a] A. Schönhage, Asymptotically Fast Algorithms for the Numerical Multiplication and Division of Polynomials with Complex Coefficients, *Proc. EUROCAM* (J. Calmet Editor), Marseille, 1982, *Lecture Notes in Computer Science,* **144**, 3–15, Springer, Berlin, 1982.

[Sc85] A. Schönhage, Quasi-gcd Computations, *J. of Complexity,* **1**, 118–137, 1985.

[Sc88] A. Schönhage, Probabilistic Computation of Integer Polynomial GCD's, *J. of Algorithms*, **9**, 365–371, 1988.

[Sc93] A. Schönhage, Fast Parallel Computation of Characteristic Polynomials by Leverrier's Power Sum Method Adapted to Fields of Finite Characteristic, *Proc. 20th ICALP, Lecture Notes in Computer Science,* **700**, 410–417, Springer, Berlin, 1993.

[Sch86] A. Schrijver, *Theory of Linear and Integer Programming,* Wiley, New York, 1986.

[Schw] J.T. Schwartz, Fast Probabilistic Algorithms for Verification of Polynomial Identities, *J. of ACM,* **27**, **4**, 701–717, 1980.

[ScSt] A. Schönhage, V. Strassen, Schnelle Multiplikation grosse Zahlen, *Computing,* **7**, 281–292, 1971 (in German).

[Se] S. Serra, On Multi-Iterative Methods, *Computers & Mathematics (with Applications)*, **26**, **4**, 65–87, 1993.

[Se93] S.Serra, Preconditioning Techniques for Ill-Conditioned Block Toeplitz Systems with Nonnegative Generating Functions, Quaderno 40, *Dipartimento di Matematica, Università di Milano*, Milano, Italy, 1993.

[Sei] R. Seidel, On the All-Pairs-Shortest-Path Problem, *Proc. 24th Ann. ACM Symp. on Theory of Computing,* 745–749, ACM Press, New York, 1992.

[Sh] V. Shoup, Fast Construction of Irreducible Polynomials over Finite Fields, *Proc. 4th Ann. ACM-SIAM Symp. on Discrete Algorithms*, 484–492, ACM Press, New York, and SIAM Publications, Philadelphia, Pennsylvania, 1993.

[Sh,a] V. Shoup, Factoring Polynomials Over Finite Fields: Asymptotic Complexity vs. Reality, *Proc. Int. IMACS Symp. on Symbolic Computation, New Trends and Developments,* Lille, France, 124–129, 1993.

[Shp] I. E. Shparlinski, *Computational and Algorithmic Problems in Finite Fields; Series: Mathematic and Its Applications,* **88**, Kluwer Academic Publishers, Norwell, Massachusetts, 1992.

[ShT] M. Shaw, J.F. Traub, On the Number of Multiplications for the Evaluation of a Polynomial and Some of Its Derivatives, *Journal of the ACM*, **21**, 161–167, 1974.

[ShT77] M. Shaw, J.F. Traub, Selection of Good Algorithms from a Family of Algorithms for Polynomial Derivatives, *Information Processing Letters*, **6**, 141–145, 1977.

[SL] J.R. Sendra, J. Llovet, An Extended Polynomial GCD Algorithm using Hankel Matrices, *J. Symbolic Computation* **13**, 25–39, 1992.

[Smo] R. Smolensky, Algebraic Methods in the Theory of Lower Bounds for Boolean Circuit Complexity, *Proc. 19th Ann. ACM Symp. on Theory of Computing,*

77–82, ACM Press, New York, 1987.

[SPL-K] D. Shallcross, V. Y. Pan, Y. Lin-Kriz, The NC Equivalence of Integer Linear Programming and Euclidean GCD, *Proc. 34th Ann. IEEE Symp on Foundation of Computer Science*, 557–564, IEEE Computer Society Press, 1993.

[SS] M. Shub, S. Smale, Complexity of Bezout's Theorem I: Geometric Aspects, *J. of the Amer. Math. Soc.*, **6**, 459–501, 1993.

[SS,a] M. Shub, S. Smale, Complexity of Bezout's Theorem II: Volumes and Probabilities, *Computational Algebraic Geometry* (F. Eyssette, A. Galligo, Editors), *Progress in Mathematics*, **109**, 267–285, Birkhauser, 1993.

[SS,b] M. Shub, S. Smale, Complexity of Bezout's Theorem III: Condition Number and Packing, *J. of Complexity*, **9**, 4–14, 1993.

[SS,c] M. Shub, S. Smale, Complexity of Bezout's Theorem IV: Probability of Success, Extensions, Research Report RC 18921, *IBM T. J. Watson Research Center,* Yorktown Heights, New York, 1993.

[SSmo] V. Shoup, R. Smolensky, Lower Bounds for Polynomial Evaluation and Interpolation Problems, *Proc. 32nd Ann. IEEE Symp. on Foundation of Computer Science*, 378–383, IEEE Computer Society Press, 1991.

[SSt] R. Solovay, V. Strassen, A Fast Monte-Carlo Test for Primality, *SIAM J. Comput.,* **6**, 84–85, 1977.

[St69] V. Strassen, Gaussian Elimination is Not Optimal, *Numer. Math.,* **13**, 354–356, 1969.

[St81] V. Strassen, The Computational Complexity of Continued Fractions, *SIAM J. Comput.,* **12**, 1–27, 1983.

[Ster] P.H. Sterbenz, *Floating-Point Computation,* Prentice Hall, Englewood Cliffs, New Jersey, 1966.

[Str] G. Strang, A Proposal for Toeplitz Matrix Calculations, *Stud. Appl. Math.*, **74**, 171–176, 1986.

[Str80] G. Strang, *Linear Algebra and Its Applications,* Academic Press, New York, 1980.

[Sw] P.N. Swarztrauber, A Parallel Algorithm for Solving General Tridiagonal Equations, *Math. Comp.*, **33**, 185–199, 1979.

[SY] J. Schwartz, C.K. Yap, *Advances in Robotics,* Lawrence Erlbaum Associates, Hillside, New Jersey, 1986.

[T64] W.F. Trench, An Algorithm for Inversion of Finite Toeplitz Matrices, *J. of SIAM,* **12, 3**, 515–522, 1964.

[T65] W.F. Trench, An Algorithm for Inversion of Finite Hankel Matrices, *J. of SIAM,* **13, 4**, 1102–1107, 1965.

[T85] W.F. Trench, Explicit Inversion Formulas for Toeplitz Band Matrices, *SIAM J. Alg. Discr. Meth.,* **6**, **4**, 546–554, 1985.

[T90] W.F. Trench, A Note on a Toeplitz Inversion Formula, *Linear Algebra Appl.,* **29** , 55–61, 1990.

[Ti81] M. Tismenetsky, Bezoutians, Toeplitz and Hankel Matrices in the Spectral Theory of Matrix Polynomials, *Ph.D. Thesis, Technion,* Haifa, Israel, 1981.

[To] A.L. Toom, The Complexity of a Scheme of Functional Elements Realizing the Multiplication of Integers, *Soviet Math. Doklady,* **3**, 714–716, 1963.

[Tur] W. Turin, *Performance Analysis of Digital Transmission Systems,* Computer Society Press, 1990.

[UP] S. Ursic, C. Patarra, Exact Solution of Systems of Linear Equations with Iterative Methods, *SIAM J. Alg. Discr. Meth.,* **4**, 111–115, 1983.

[V] P.M. Vaidya, Solving Linear Equations with Symmetric Diagonally Dominant Matrices by Constructing Good Preconditioners, Tech. Report, *Dept. of Computer Science, University of Illinois,* Urbana, Illinois, 1991.

[Val] L. Valiant, General Purpose Parallel Architectures, *Handbook of Theoretical Computer Science,* (J. van Leeuwen Editor), 943–971, North Holland, Amsterdam, 1990.

[Val90] L. G. Valiant, A Bridging Model for Parallel Computation, *Communication of the ACM*, **33, 8**, 103–111, August 1990.

[Var] R.S. Varga, *Matrix Iterative Analysis,* Prentice-Hall, New Jersey, 1969.

[Vaz] V.V. Vazirani, NC Algorithms for Computing the Number of Perfect Matchings in $K_{3,3}$-free Graphs and Related Problems, *Manuscript,* 1987.

[Vi] G. Villard, Computation of the Smith Normal Form of Polynomial Matrices, *Proc. ACM SIGSAM Int. Symp. on Symb. Alg. Comput. (ISSAC '93),* (M. Bronstein Editor), 209–217, ACM Press, New York, 1993.

[vDR] L. van Dommelen, E. A. Rundensteiner, Fast Adaptive Summation of Point Forces in the 2-dimensional Poisson Equation, *J. of Computational Physics,* **83**, 126–147, 1989.

[vdW] B.L. van der Waerden, *Modern Algebra,* Frederick Ungar Publishing Co., New York, 1953.

[vL] C. van Loan, *Computational Frameworks for the Fast Fourier Transform,* SIAM Publications, Philadelphia, Pennsylvania, 1992.

[VN] M. Vetterli, H.J. Nussbaumer, Simple FFT and DCT Algorithms with Reduced Number of Operations, *Signal Processing,* **6**, 267–278, 1984.

[VSBR] L. Valiant, S. Skyum, S. Berkowitz, C. Rackoff, Fast Parallel Computation of Polynomials Using Few Processors, *SIAM J. Comput.,* **12, 4**, 641–644, 1983.

[VY] V.V. Vazirani, M. Yannakakis, Phaphian Orientations, 0/1 Permanents, and Even Cycles in Directed Graphs, *Manuscript,* 1987.

[W63] J.H. Wilkinson, *Rounding Errors in Algebraic Processes,* Prentice Hall, Englewood Cliffs, New Jersey, 1963.

[W65] J.H. Wilkinson, *The Algebraic Eigenvalue Problem,* Clarendon Press, Oxford, 1965.

[W78] S. Winograd, On Computing the Discrete Fourier Transform, *Math. Comp.,* **32**, 175–199, 1978.

[Wak] A. Waksman, On Winograd's Algorithm for Inner Product, *IEEE Trans. on Computers,* **C-119**, 360–361, 1970.

[Wang] P. Wang, A p-adic Algorithm for Univariate Partial Fractions, *Proc. 1981 ACM Symp. on Symbolic and Algebraic Comp.,* 212–217, ACM Press, New York, 1981.

[WGD] P.S. Wang, M.J.T. Guy, J.H. Davenport, p-adic Reconstruction of Rational Numbers, *SIGSAM Bulletin,* **16**, 2–3, ACM Press, New York, 1982.

[Wi79] S. Winograd, On the Multiplicative Complexity of the Discrete Fourier Transform, *Advances in Math.,* **32**, 83–117, 1979.

[Wied] D.H. Wiedemann, Solving Sparse Linear Equations Over Finite Fields, *IEEE Trans. on Inf. Theory,* **IT-32, 1**, 54–62, 1986.

[Wim] H. K. Wimmer, On the History of the Bezoutian and the Resultant Matrix, *Linear Algebra Appl.,* **128**, 27–34, 1990.

[Wo] D. H. Wood, Extending Four Displacement Principles to Solve Matrix Equations, submitted to *Math. Comp.*

[Wo1] D. H. Wood, Product Rules for the Displacement of Near-Toeplitz Matrices, *Linear Algebra Appl.*, **188, 189**, 641–663, 1993.

[XZ] Xu Zhang, On Moore-Penrose Inverses of Toeplitz Matrices, *Linear Algebra Appl.*, **169**, 9–15, 1992.

[Y] D.Y.Y. Yun, Algebraic Algorithms Using p-adic Constructions, *Proc. 1976 ACM Symp. Symbolic and Algebraic Computation,* 248–259, ACM Press, New York, 1976.

[Yao] A. C-C. Yao, Lower Bounds for Algebraic Computation Trees with Integer Inputs, *Proc. 30th Ann. IEEE Symp. on Foundation of Computer Science,* 308–313, IEEE Computer Society Press, 1989.

[Yao92] A. C.-C. Yao, Algebraic Decision Trees and Euler Characteristics, *Proc. 33rd Ann. IEEE Symp. on Foundation of Computer Science*, 268–277, IEEE Computer Society Press, 1992.

[YG] D.M. Young, R.T. Gregory, *A Survey of Numerical Mathematics, II,* Addison-Wesley, Reading, Massachusetts, 1973.

[You] D.M. Young, *Iterative Solution of Large Linear Systems,* Academic Press, New York, 1971.

[Z] R.E. Zippel, Probabilistic Algorithms for Sparse Polynomials, *Proc. EUROSAM 79, Lecture Notes in Computer Science,* **72**, 216–226, Springer, Berlin, 1979.

[Z90] R.E. Zippel, Interpolating Polynomials from Their Values, *J. Symbolic Comput.,* **9, 3,** 375–403, 1990.

[Z93] R.E. Zippel, *Effective Polynomial Computation,* Kluwer Academic Publishers, Norwell, Massachusetts, 1993.

Index.

Progress in Theoretical Computer Science

Progress in Theoretical Computer Science is a series that focuses on the theoretical aspects of computer science and on the logical and mathematical foundations of computer science, as well as the applications of computer theory. It addresses itself to research workers and graduate students in computer and information science departments and research laboratories, as well as to departments of mathematics and electrical engineering where an interest in computer theory is found.

The series publishes research monographs, graduate texts, and polished lectures from seminars and lecture series. We encourage preparation of manuscripts in some form of TeX for delivery in camera-ready copy, which leads to rapid publication, or in electronic form for interfacing with laser printers or typesetters.

Proposals should be sent directly to the Editor, any member of the Editorial Board, or to: Birkhäuser Boston, 675 Massachusetts Ave., Cambridge, MA 02139. The Series includes:

1. Leo Bachmair, *Canonical Equational Proofs*
2. Howard Karloff, *Linear Programming*
3. Ker-I Ko, *Complexity Theory of Real Functions*
4. Guo-Qiang Zhang, *Logic of Domains*
5. Thomas Streicher, *Semantics of Type Theory: Correctness, Completeness and Independence Results*
6. Julian Charles Bradfield, *Verifying Temporal Properties of Systems*
7. Alistair Sinclair, *Algorithms for Random Generation and Counting*
8. Heinrich Hussmann, *Nondeterminism in Algebraic Specifications and Algebraic Programs*
9. Pierre-Louis Curien, *Categorical Combinators, Sequential Algorithms and Functional Programming*
10. J. Köbler, U. Schöning, and J. Torán, *The Graph Isomorphism Problem: Its Structural Complexity*
11. Howard Straubing, *Finite Automata, Formal Logic, and Circuit Complexity*
12. Dario Bini and Victor Pan, *Polynomial and Matrix Computations, Volume 1 Fundamental Algorithms*
13. James S. Royer and John Case, *Subrecursive Programming Systems: Complexity & Succinctness*
14. Roberto Di Cosmo, *Isomorphisms of Types*